# Feynman Amplitudes, Periods and Motives

CONTEMPORARY MATHEMATICS

648

# Feynman Amplitudes, Periods and Motives

International Research Workshop
Periods and Motives - A Modern Perspective on Renormalization
July 2–6, 2012
Instituto de Ciencias Matemáticas, Madrid, Spain

Luis Álvarez-Cónsul
José Ignacio Burgos-Gil
Kurusch Ebrahimi-Fard
Editors

American Mathematical Society
Providence, Rhode Island

2000 *Mathematics Subject Classification.* Primary 11S40, 11M32, 14C25, 14C15, 14F42, 16T05, 32G20, 81Q30, 81T15, 81T18.

---

**Library of Congress Cataloging-in-Publication Data**

Feynman amplitudes, periods, and motives : international research conference on periods and motives : a modern perspective on renormalization : July 2-6, 2012, Institute de Ciencias Matematicas, Madrid, Spain / Luis Alvarez-Consul, Jose Ignacio Burgos-Gil, Kurusch Ebrahimi-Fard, editors.

pages cm.–(Contemporary mathematics; volume 648)

Includes bibliographical references.

ISBN 978-1-4704-2247-9 (alk. paper)

1. Mathematical physics–Congresses. 2. Perturbation (Quantum dynamics)–Congresses. 3. Perturbation (Mathematics)–Congresses. 4. Quantum field theory–Congresses. I. Álvarez-Cónsul, 1970- editor. II. Burgos-Gil, José I. (José Ignacio), 1962- editor. III. Ebrahimi-Fard, Kurusch, 1973- editor.

QC20.F39 2015

530′.15–dc23    2015010413

Contemporary Mathematics ISSN: 0271-4132 (print); ISSN: 1098-3627 (online)

DOI: http://dx.doi.org/10.1090/conm/648

---



10 9 8 7 6 5 4 3 2 1    20 19 18 17 16 15

# Contents

# Preface

Quantum field theory (QFT) is largely considered as one of the most successful theories in physics. The perturbative approach to QFT, based on power series expansions of probability amplitudes for physical processes, permitted theoretical predictions in breathtaking agreement with experimental data. The coefficients of these power series expansions, known as Feynman amplitudes, are integrals over a large number of variables that can be represented pictorially by the so-called Feynman diagrams. The Feynman rules, named after their inventor, the theoretical physicist and Nobel laureate Richard P. Feynman, provide a systematic way for translating a Feynman diagram into its corresponding amplitude.

The predictive power of perturbative QFT to a great extent hinges on finding efficient ways to calculate Feynman amplitudes. In this context, the past two decades have seen enormous progress on foundational as well as computational aspects of perturbative QFT. Advanced techniques from algebra, geometry and combinatorics have played an increasingly important role, turning the subject into a fruitful playground for mathematicians and theoretical physicists alike. This progress rests primarily on the pioneering work of David Broadhurst and Dirk Kreimer on the occurrence of multiple zeta values in Feynman integral calculations, and on the geometric formulation of Alain Connes and Kreimer of renormalization via a Hopf algebra of Feynman diagrams. This research immediately attracted international attention, leading to the emergence of exciting links between the computation of Feynman amplitudes on the one hand, and the theory of motives of algebraic varieties and their periods on the other hand. The works of Spencer Bloch, Francis Brown, Connes, Hélène Esnault, Alexander Goncharov, Kreimer, Matilde Marcolli, and others, created intensive research collaborations between mathematicians and physicists, with tremendous impact in this area.

The mathematical concepts of periods and motives are pivotal in these developments. The former are numbers that are computed as integrals of algebraic differential forms over topological cycles on algebraic varieties. The term originated from the period of a periodic elliptic function, which can be computed as an elliptic integral. The theory of motives was introduced by Grothendieck as a "universal cohomology theory" to explain the common features of different cohomology theories. One may describe motives as an intermediate step between algebraic varieties and their linear invariants (cohomology). The motives associated to smooth projective varieties are called "pure", while the motives associated to quasi-projective and singular varieties are called "mixed" (they consist of different pure pieces). A complete and satisfactory theory of mixed motives is not yet available, although some pieces of it have already been developed. For instance the category of mixed Tate motives (the simplest mixed motives) is well understood.

In spite of the beautiful links between Feynman amplitudes, motives and periods uncovered in the work of the aforementioned researchers, a precise understanding of the relationship between perturbative QFT and motives remains an open question that has stimulated further research in various directions, encouraging new ways of thinking and leading to unexpected new results. An example demonstrating the fruitfulness of the research collaboration between mathematicians and physicists in this area is the recent work of Goncharov et al. on amplitudes associated to a particular Feynman diagram, whose original expansion spanned an expression covering several pages. Thanks to the motivic interpretation, this expansion could be simplified to just two lines. Such results underline the importance of these developments for progress in physics, and more generally, the remarkable success of this interplay between pure mathematics and modern theoretical physics.

The articles presented in this volume provide an exciting perspective on recent developments in the field. They are based on research talks given at the international workshop "Periods and Motives – A Modern Perspective on Renormalization (PM2012)[1]", which was hosted in Madrid (Spain), from July 2–6, 2012, by the Instituto de Ciencias Matemáticas (ICMAT – Institute of Mathematical Sciences). The organizing committee consisted of David A. Ellwood, Dominique Manchon, Stefan Weinzierl, and the editors. The workshop also included a Young Researchers Session, coordinated by Susama Agarwala and Marina Logares. We would like to thank all the participants as well as the other organizers and coordinators for making this a memorable event.

During the workshop, on July 4, 2012, we received exciting news from the ATLAS and CMS experiments at CERN about the discovery of a particle consistent with the Higgs boson. This caused considerable enthusiasm among the participants. And we are grateful to Claude Duhr, who improvised an inspiring lecture on those LHC experiments, accessible to an audience composed of mathematicians and physicists. We thank ICMAT, Severo Ochoa Excellence Grant SEV-2011-0087 (Spain), and CNRS GDR "Renormalisation" (France) for financial support. In addition, the workshop received generous support on all levels from ICMAT, and we express our deepest gratitude towards ICMAT and its staff for its great professionalism in organizing this meeting. However, saying this, it should be clear that this volume only exists due to the excellent contributions from all authors, and we can not overstate our thankfulness for their patience and dedication that leads to the enjoyment of reading the content of this volume. Thank you also to the referees that helped us uncompromisingly.

Luis Álvarez-Cónsul
José Ignacio Burgos-Gil
Kurusch Ebrahimi-Fard

Madrid, February 20, 2015

[1] `http://www.icmat.es/congresos/periods-and-motives/`

# A note on twistor integrals

Spencer Bloch

ABSTRACT. Generalizing work of Hodges, I show how to express the period, and hence also the motive, associated to the Feynman amplitude of a graph with $n$ loops and $2n+2$ edges in terms of a pfaffian.

## 1. Introduction

This paper is a brief introduction to twistor integrals from a mathematical point of view. It was inspired by a paper of Hodges [**H**] which we studied in a seminar at Cal Tech directed by Matilde Marcolli. The idea is to write the amplitude for a graph with $n$ loops and $2n+2$ propagators using the geometry of pfaffians for sums of rank 2 alternating matrices. (Hodges considers the case of 1 loop and 4 edges). Why is this of interest to a mathematician? The Feynman amplitude is a *period* in the sense of arithmetic algebraic geometry. In parametric form, the amplitude integral associated to a graph $\Gamma$ with $N$ edges and $n$ loops has the form

$$c(N,n)\int_\delta \frac{S_1^{N-2n-2}\Omega}{S_2^{N-2n}}. \tag{1.1}$$

Here $S_1$ and $S_2$ are the first and second Symanzik polynomials [**BK**], [**BEK**], [**IZ**], and $\Omega = \sum \pm A_i dA_1 \wedge \cdots \wedge \widehat{dA_i} \wedge \cdots \wedge dA_N$ is the integration form on $\mathbb{P}^{N-1}$, the projective space with homogeneous coordinates indexed by edges of $\Gamma$. The chain of integration $\delta$ is the locus of points on $\mathbb{P}^{N-1}$ where all the $A_i \geq 0$. Note $\Omega$, $S_1$, $S_2$ are homogeneous of degrees $N$, $n$, $n+1$ in the $A_i$, so the integrand is homogeneous of degree 0 and represents a rational differential form. Finally, $c(N,n)$ is some elementary constant depending only on $N$ and $n$.

Two special cases suggest themselves. In the *log divergent* case when $N=2n$, the integrand is simply $\Omega/S_1^2$. The first Symanzik polynomial depends only on the edge variables $A_i$, so the result in this case is a constant. (If the graph is non-primitive, i.e. has log divergent subgraphs, the integral will diverge. We do not discuss this case.) Inspired by the conjectures of Broadhurst and Kreimer [**BrK**], there has been a great deal of work done on the primitive log divergent amplitudes.

The polynomial $S_1$ itself is the determinant of an $n \times n$-symmetric matrix with entries linear forms in the $A_i$. The linear geometry of this determinant throws an interesting light on the motive of the hypersurface $X(\Gamma): S_1 = 0$. For example, one has a "Riemann-Kempf" style theorem that the dimension of the null space of

2010 *Mathematics Subject Classification.* Primary 14C30, 14M15, 81Q30.

the matrix at a point is equal to the multiplicity of the point on $X(\Gamma)$, **[P]**, **[K]**. Furthermore, the projectivized fibre space $Y(\Gamma)$ of these null lines maps birationally onto $X(\Gamma)$ and in some sense "resolves" the motive. The fibre over a point $x \in X(\Gamma)$ is the projective space associated to the kernel of the corresponding symmetric $n\times n$-matrix. Whereas the motive of $X(\Gamma)$ can be quite subtle, the motive of $Y(\Gamma)$ is quite elementary. In particular, it is mixed Tate **[B]**. (The Riemann-Kempf theorem refers to the map $\pi : Sym^{g-1}C \twoheadrightarrow \Theta \subset J_{g-1}(C)$ where $C$ is a Riemann surface and $\Theta$ is the theta divisor. The dimension of the fibre of $\pi$ at a point of $\Theta$ equals the multiplicity of the divisor $\Theta$ at the point minus one.)

The second case is $N = 2n + 2$, e.g. one loop and 4 edges. The amplitude is $\int_\delta \Omega/S_2^2$ and is a function of external momenta and masses. The second Symanzik has the form

$$S_2 = S_2^0(A,q) - (\sum_{i=1}^{N} m_i^2 A_i)S_1(A) \tag{1.2}$$

Here $q$ denotes the external momenta, and $S_2^0(A,q)$ is homogeneous of degree 2 in $q$ and of degree $n+1$ in the $A$. Moreover, $S_2^0$ is a quaternionic pfaffian associated to a quaternionic hermitian matrix, **[BK]**, so in the case of zero masses there is again the possibility of linking the motive to the geometry of a linear map. In this note we go further and show for the case $N = 2n+2$ that $S_2$ is itself a pfaffian via the calculus of *twistors*.

To avoid issues with convergence for the usual propagator integral, I assume in what follows that the masses are positive and the propagators are euclidean. Note that in (1.4) the pfaffian can vanish where some of the $a_i = 0$. The issues which arise are analogous to issues of divergence already familiar to physicists. They will not be discussed here.

THEOREM 1.1. *Let $\Gamma$ be a graph with $n$ loops and $2n+2$ edges as above. We fix masses $m_i > 0$ and external momenta $q$ and consider the amplitude*

$$\mathcal{A}(\Gamma, q, m) = \int_{\mathbb{R}^{4n}} \frac{d^{4n}x}{\prod_{i=1}^{2n+2} P_i(x,q,m_i)} \tag{1.3}$$

*where the $P_i$ are euclidean. (I.e. $P_i(x,q,m_i) = ||L_i(x,q)||^2 - m_i^2$ where $L_i(x,q)$ is a linear combination of internal and external momenta representing the momentum flow through the i-th edge. Here $||\cdot||$ is the euclidean norm.) Then there exist alternating bilinear forms $Q_i$ on $\mathbb{R}^{2n+2}$ where $Q_i$ depends on $P_i$ and on $m_i$, $1 \le i \le 2n+2$, and a universal constant $C(n)$ depending only on $n$ such that*

$$\mathcal{A}(\Gamma, q, m) = C(n) \int_\delta \frac{\Omega_{2n+1}}{Pfaffian(\sum_{i=1}^{2n+2} a_i Q_i)^2} \tag{1.4}$$

*Here $\Omega_{2n+1} = \sum \pm a_i da_1 \wedge \cdots \widehat{da_i} \cdots da_{2n+2}$ and $\delta$ is the locus on $\mathbb{P}^{2n+1}$ with coordinate functions $a_i$ where all the $a_i \ge 0$.*[1]

By way of analogy, the first Symanzik polynomial is given by

$$S_1(\Gamma)(a_1, \ldots, a_N) = \det(\sum_{e \text{ edge}} a_e M_e) \tag{1.5}$$

[1]The referee asks for the precise value of $C(n)$. It is surely computable, but in addition to the factor $(2n+1)!(2\pi i)^{2n+3}$ from (3.16) and the constant $C'(n)$ in (1.6), the computation will involve the integral (4.10). The author admits to not having worked it out.

where $M_e$ is a rank 1 symmetric $n \times n$-matrix associated to $(e^\vee)^2$, where $e^\vee : H_1(\Gamma, \mathbb{R}) \to \mathbb{R}$ is the functional which associates to a loop the coefficient of $e$ in that loop. Thus, the amplitude in the case of $n$ loops and $2n$ edges is given by

$$\mathcal{A}(\Gamma) = C'(n) \int_\delta \frac{\Omega_{2n-1}}{\det(\sum a_i M_i)^2} \tag{1.6}$$

where $C'(n)$ is another constant depending only on $n$.

This paper may be viewed as a sort of technical "étude". Recent work in physics has been strongly influenced by the geometry of grassmannians, and the author felt the need to better understand. Philosophically, it is nice to have a unified geometric interpretation of the integrand including masses, but the implications of this study, e.g. for the structure of the motives, is unclear.

I want to acknowledge help from S. Agarwala, M. Marcolli, and O. Ceyhan. Much of this work was done during June, 2012 when I was visiting Cal Tech.

## 2. Linear Algebra

Fix $n \geq 1$ and consider a vector space $V = k^{2n+2} = ke_1 \oplus \cdots \oplus ke_{2n+2}$. (Here $k$ is a field of characteristic 0.) We write $O = ke_1 \oplus ke_2$ and $I = ke_3 \oplus \cdots \oplus ke_{2n+2}$, so $V = O \oplus I$. $G(2, V)$ will be the Grassmann of 2-planes in $V$.

We have

$$\mathrm{Hom}_k(O, I) \overset{\iota}{\hookrightarrow} G(2, V) \overset{j}{\hookrightarrow} \mathbb{P}(\bigwedge^2 V). \tag{2.1}$$

Here $\iota(\psi) = k(e_1 + \psi(e_1)) \oplus k(e_2 + \psi(e_2))$ and $j(W) = \bigwedge^2 W \hookrightarrow \bigwedge^2 V$.

Write $V^*$ for the dual vector space with dual basis $e_i^*$. We identify $\bigwedge^2 V^*$ with the dual of $\bigwedge^2 V$ in the evident way, so $\langle e_i^* \wedge e_j^*, e_i \wedge e_j \rangle = 1$. For $\alpha \in \bigwedge^2 V^*$, the assignment

$$\psi \mapsto \langle (e_1 + \psi(e_1)) \wedge (e_2 + \psi(e_2)), a \rangle \tag{2.2}$$

defines a quadratic map $q_\alpha : \mathrm{Hom}(O, I) \to k$.

LEMMA 2.1. *Assume $0 \neq \alpha = v \wedge w$ with $v, w \in V^*$. Then the quadratic map $q_\alpha$ has rank 4.*

PROOF. The assertion of the lemma means that

$$\langle (\sum x_i e_i) \wedge (\sum y_j e_j), v \wedge w \rangle,$$

viewed as a quadric in the $x_i$ and $y_j$ variables, has rank 4. (The rank of a quadric is the rank of the associated symmetric matrix.) By assumption $v, w$ are linearly independent. We can change coordinates so $v = \varepsilon_i^*, w = \varepsilon_j^*$, and $\sum x_i e_i = \sum x'_i \varepsilon_i, \sum y_j e_j = \sum y'_j \varepsilon_j$. The polynomial is then

$$\langle (\sum x'_i \varepsilon_i) \wedge (\sum y'_j \varepsilon_j), \varepsilon_i^* \wedge \varepsilon_j^* \rangle = x'_i y'_j - x'_j y'_i. \tag{2.3}$$

This is a quadratic form of rank 4. □

Returning to the notation in (2.1), we can write $I = \bigoplus_{i=1}^n I_i$ with $I_i = ke_{2i+1} \oplus ke_{2i+2}$. We can think of $\mathrm{Hom}(O, I) = \bigoplus \mathrm{Hom}(O, I_i)$ as the decomposition of momentum space into a direct sum of Minkowski spaces. We identify $\mathrm{Hom}(O, I_i)$ with the space of $2 \times 2$-matrices, and the propagator with the determinant. With these coordinates, an element in $\mathrm{Hom}(O, I)$ can be written as a direct sum $A_1 \oplus \cdots \oplus A_n$ of $2 \times 2$-matrices. The propagators have the form $\det(a_1 A_1 + \cdots a_n A_n)$ with

$a_i \in k$. The map $\psi : O \to I$ given by $\psi(e_1) = x_3e_3 + \cdots + x_{2n+2}e_{2n+2}$ and $\psi(e_2) = y_3e_3 + \cdots + y_{2n+2}e_{2n+2}$ corresponds to the matrices

$$A_i = \begin{pmatrix} x_{2i+1} & x_{2i+2} \\ y_{2i+1} & y_{2i+2} \end{pmatrix}. \tag{2.4}$$

LEMMA 2.2. *Let $A_i$ be as in (2.4). Let*

$$\alpha = (\sum_{i=1}^{n} a_i e_{2i+1}^*) \wedge (\sum_{i=1}^{n} a_i e_{2i+2}^*) \in \bigwedge^2 V^*.$$

*Then the quadratic map $q_\alpha$ in lemma 2.1 is given by*

$$q_\alpha(A_1 \oplus \cdots \oplus A_n) = \det(a_1A_1 + \cdots + a_nA_n). \tag{2.5}$$

PROOF. This amounts to the identity

$$\det \begin{pmatrix} \sum a_i x_{2i+1} & \sum a_i x_{2i+2} \\ \sum a_i y_{2i+1} & \sum a_i y_{2i+2} \end{pmatrix} = \langle (\sum_{i\geq 3} x_i e_i) \wedge (\sum_{i\geq 3} y_i e_i), (\sum_{i=1}^{n} a_i e_{2i+1}^*) \wedge (\sum_{i=1}^{n} a_i e_{2i+2}^*)\rangle. \tag{2.6}$$

For $i = j$ (resp. $i \neq j$) the coefficient of $a_ia_j$ in this expression is

$$x_{2i+1}y_{2i+2} - x_{2i+2}y_{2i+1} \tag{2.7}$$

$$\text{resp. } x_{2i+1}y_{2j+2} - x_{2i+2}y_{2j+1} + x_{2j+1}y_{2j+2} - x_{2j+2}y_{2i+1}. \tag{2.8}$$

□

The full inhomogeneous propagator, which in physics notation would be written $(p_1, \ldots, p_n) \mapsto (\sum a_ip_i + s)^2$ with the $p_i$ and $s$ 4-vectors, becomes in the twistor setup

$$\begin{aligned} &\langle (e_1 + \sum_{i\geq 3} x_i e_i) \wedge (e_2 + \sum_{i\geq 3} y_i e_i), \\ &\quad (c_1e_1^* + c_2e_2^* + \sum_{i\geq 1} a_i e_{2i+1}^*) \wedge (d_1e_1^* + d_2e_2^* + \sum_{i\geq 1} a_i e_{2i+2}^*)\rangle = \\ &\det \begin{pmatrix} c_1 & d_1 \\ c_2 & d_2 \end{pmatrix} + c_1 \sum a_i y_{2i+2} - c_2 \sum a_i x_{2i+2} - d_1 \sum a_i y_{2i+1} + \\ &\quad d_2 \sum a_i x_{2i+1} + \det \begin{pmatrix} \sum a_i x_{2i+1} & \sum a_i x_{2i+2} \\ \sum a_i y_{2i+1} & \sum a_i y_{2i+2} \end{pmatrix} = \\ &\quad \det \begin{pmatrix} \sum a_i x_{2i+1} + c_1 & \sum a_i x_{2i+2} + d_1 \\ \sum a_i y_{2i+1} + c_2 & \sum a_i y_{2i+2} + d_2 \end{pmatrix}. \end{aligned} \tag{2.9}$$

REMARK 2.3. *In (2.9), our $\alpha \in \bigwedge^2 V^*$ is of rank 2, i.e. it is decomposible as a tensor and corresponds to an element in $G(2, V) \subset \mathbb{P}(\bigwedge^2 V^*)$, (2.1). If we want to add mass to our propagator, we simply replace $\alpha$ by $\alpha + m^2 e_1^* \wedge e_2^*$, yielding $(\sum a_ip_i+s)^2+m^2$. The massive $\alpha$ represents a point in $\mathbb{P}(\bigwedge^2 V^*)$ but not necessarily in $G(2, V^*)$.*

## 3. The Twistor Integral

In this section we take $k = \mathbb{C}$. Consider the maps

$$V \times V - S \xrightarrow{\rho} G(2,V) \xrightarrow{j} \mathbb{P}(\bigwedge^2 V). \tag{3.1}$$

Here $S = \{(v,w) \mid v \wedge w = 0\}$ and $\rho(v,w) =$ 2-plane spanned by $v, w$.

LEMMA 3.1. *$V \times V - S/G(2,V)$ is the principal $GL_2(\mathbb{C})$-bundle (frame bundle) associated to the rank 2 vector bundle $\mathcal{W}$ on $G(2,V)$ which associates to $g \in G(2,V)$ the corresponding rank 2 subspace of $V$.*

PROOF. With notation as in (2.1), let $U = \mathrm{Hom}_{\mathbb{C}}(O,I) \subset G(2,V)$. We have

$$\rho^{-1}(U) = \{(z_1,\ldots,z_{2n+2},v_1,\ldots,v_{2n+2}) \mid \det\begin{pmatrix} z_1 & z_2 \\ v_1 & v_2\end{pmatrix} \neq 0\}. \tag{3.2}$$

We can define a section $s_U : U \to \rho^{-1}(U)$ by associating to $a : O \to I$ its graph

$$s_U(a) := (1,0,a_1^1,\ldots,a_{2n}^1;0,1,a_1^2,\ldots,a_{2n}^2). \tag{3.3}$$

Using this section and the evident action of $GL_2(\mathbb{C})$ on the fibres of $\rho$, we can identify $\rho^{-1}(U) = GL_2(\mathbb{C}) \times U$. The fibre $\rho^{-1}(u)$ for $w \in U$ is precisely the set of framings $w = \mathbb{C}z \oplus \mathbb{C}v$ as claimed. □

LEMMA 3.2. *The canonical bundle $\omega_{G(2,V)} = \mathcal{O}(-2n-2)$ where $\mathcal{O}(-1)$ is the pullback $j^*\mathcal{O}_{\mathbb{P}(\bigwedge^2 V)}(-1)$.*

PROOF. The tautological sequence on $G(2,V)$ reads

$$0 \to \mathcal{W} \to V_{G(2,V)} \to V_{G(2,V)}/\mathcal{W} \to 0. \tag{3.4}$$

Here $\mathcal{W}$ is the rank 2 sheaf with fibre over a point of $G(2,V)$ being the corresponding 2-plane in $V$. One has

$$\Omega^1_{G(2,V)} = \underline{Hom}(V_{G(2,V)}/\mathcal{W},\mathcal{W}) = (V_{G(2,V)}/\mathcal{W})^\vee \otimes \mathcal{W}. \tag{3.5}$$

By definition of the Plucker embedding $j$ above we have $\mathcal{O}_G(-1) = \bigwedge^2\mathcal{W}$. The formula for calculating chern classes of a tensor product yields

$$c_1(\Omega^1_G) = c_1((V_{G(2,V)}/\mathcal{W})^\vee)^{\otimes 2} \otimes c_1(\mathcal{W})^{\otimes 2n} = \mathcal{O}_G(-2n-2). \tag{3.6}$$

□

We now fix a point $a \in \mathbb{P}(\bigwedge^2 V^*)$. Upto scale, $a$ determines a non-zero alternating bilinear form on $V$ which we denote by $Q : (x,y) \mapsto \sum_{\nu,\mu} x_\nu Q^{\nu\mu} y_\mu$. By restriction we may view $Q \in \Gamma(G(2,V),\mathcal{O}(1))$. By the lemma $\omega_G \otimes \mathcal{O}(2n+2) \cong \mathcal{O}_G$, so upto scale there is a canonical meromorphic form $\xi$ on $G(2,V)$ of top degree $4n$ with exactly a pole of order $2n+2$ along $Q=0$. We write

$$\xi = \frac{\Xi}{Q^{2n+2}}; \quad 0 \neq \Xi \in \Gamma(G,\omega_G(2n+2)) = \mathbb{C}. \tag{3.7}$$

LEMMA 3.3. *We have*

$$H^i(V \times V - S,\mathbb{Q}) = \begin{cases} \mathbb{Q} & i = 0, 4n+1, 4n+3, 8n+4 \\ (0) & \textit{else} \end{cases}. \tag{3.8}$$

PROOF. We compute the dual groups $H_c^*(V\times V-S,\mathbb{Q})$. Note a complex vector space has compactly supported cohomology only in degree twice the dimension. Also, $H_c^1(V-\{0\})\cong H_c^0(\{0\})=\mathbb{Q}$. Let $p:S\to V$ be projection onto the first factor. The fibre $p^{-1}(v)\cong\mathbb{C}$ for $v\neq 0$ and $p^{-1}(0)=V$. It follows that

$$H_c^i(S-\{0\}\times V)\cong H_c^{i-2}(V-\{0\})=(0);\ i\neq 3,4n+6. \tag{3.9}$$

Now the exact sequence

$$H_c^i(S-\{0\}\times V)\to H_c^i(S,\mathbb{Q})\to H_c^i(V,\mathbb{Q}) \tag{3.10}$$

yields $H_c^i(S)=\mathbb{Q}$, $i=3,4n+4,4n+6$ and vanishes otherwise. Thus, $H_c^j(V\times V-S)=\mathbb{Q}$; $j=4,4n+5,4n+7,8n+8$ and vanishes otherwise. Dualizing, we get the lemma. □

Let $R\subset V\times V$ be the zero locus of the alternating form $Q$ on $V$ defined above. Clearly $S\subset R$.

LEMMA 3.4. *Assume the alternating form $Q$ is non-degenerate. Then we have*

$$H^i(V\times V-R,\mathbb{Q})=\begin{cases}\mathbb{Q} & i=0,1,4n+3,4n+4\\ (0) & \textit{else.}\end{cases} \tag{3.11}$$

PROOF. Again let $p:R\to V$ be projection onto the first factor. We have $p^{-1}(0)=V$ and $p^{-1}(v)\cong\mathbb{C}^{2n+1}$ for $v\neq 0$. It follows that $H_c^i(R-\{0\}\times V)=(0),i\neq 4n+3,8n+6$. As before, this yields $H_c^i(R)=\mathbb{Q}$, $i=4n+3,4n+4,8n+6$ and zero else. Hence $H_c^j(V\times V-R)=\mathbb{Q}$, $j=4n+4,4n+5,8n+7,8n+8$ and the lemma follows by duality. □

Note that in the case $n=0$, $\dim V=2$ we have $S=R$ and the two lemmas give the same information, which also describes the cohomology of the fibres of the map $\rho$. Namely, $H^i(\rho^{-1}(pt))=\mathbb{Q}$, $i=0,1,3,4$ and $H^i=(0)$ otherwise.

The form $Q$ induces a quadratic map on $V\times V$ given by $(v,v')\mapsto vQv'$.

LEMMA 3.5. *Choose a basis for $V$ and write $dv$ for the evident holomorphic form of degree $4n+4$ on $V\times V$. Then $\mu:=dv/Q^{2n+2}$ is homogeneous of degree 0 and represents a non-trivial class in $H_{DR}^{4n+4}(V\times V-R)$.*

PROOF. $V\times V-R$ is affine, so we can calculate de Rham cohomology using algebraic forms. There is an evident $\mathbb{G}_m$-action which is trivial on cohomology. Writing a form $\nu$ as a sum of eigenforms for this action, we can assume the $\mathbb{G}_m$-action is trivial on $\nu$, which therefore is written $\nu=Fdv/Q^{2n+2+N}$ for some $N\geq 0$ and $\deg F=2N$. Since $Q$ is non-degenerate, we can write $F=\sum_i F_i\partial Q/\partial v_i$. Let $(dv)_i$ be the form obtained by contracting $dv$ against $\partial/\partial v_i$. Then

$$\nu+d\Big(\frac{1}{2n+1+N}\sum F_i(dv)_i/Q^{2n+1+N}\Big)=Gdv/Q^{2n+1+N}. \tag{3.12}$$

where $G$ is homogeneous of degree $2(N-1)$. Continuing in this way, we conclude that $\nu$ is cohomologous to a constant times $dv/Q^{2n+2}$. Since by the lemma $H^{4n+4}(V\times V-R)=\mathbb{Q}$, we conclude that $\mu:=dv/Q^{2n+2}$ is not exact. □

If one keeps track of the Hodge structure, lemma 3.4 can be made more precise. One gets e.g. $H^{4n+4}(V\times V-R,\mathbb{Q})\cong\mathbb{Q}(-2n-3)$. For a suitable choice of coordinatizations for the two copies of $V$ and a suitable rational scaling for the chain $\sigma$ representing a class in $H_{4n+4}(V\times V-R,\mathbb{Q})$ we can write the corresponding period as

$$\int_\sigma d^{2n+2}z\wedge d^{2n+2}v/(\sum z_\mu v_\mu)^{2n+2}=(2\pi i)^{2n+3}. \tag{3.13}$$

Now we make the change of coordinates $v_\mu=\sum_p Q^p_\mu w_p$ and deduce

$$\int_\sigma d^{2n+2}z\wedge d^{2n+2}w/(\sum z_\mu Q^{\mu p}w_p)^{2n+2}=\frac{(2\pi i)^{2n+3}}{\det Q}. \tag{3.14}$$

Here $Q$ is alternating in our case, so $\det Q=\text{Pfaffian}(Q)^2$.

The "Feynman trick" in this context is the integral identity

$$\frac{1}{\prod_{i=1}^{2n+2}A_i}=(2n+1)!\int_{0^{2n+2}}^{\infty^{2n+2}}\frac{da_1\cdots da_{2n+2}\delta(1-\sum a_i)}{(\sum a_iA_i)^{2n+2}}. \tag{3.15}$$

We apply the Feynman trick with $A_i=\sum_{\mu,p}z_\mu Q_i^{\mu p}w_p$ and integrate over $\sigma$

$$\int_\sigma\frac{d^{2n+2}z\wedge d^{2n+2}w}{\prod_{i=1}^{2n+2}(\sum_{\mu,p}z_\mu Q_i^{\mu p}w_p)}= \tag{3.16}$$

$$(2n+1)!\int_\sigma d^{2n+2}z\wedge d^{2n+2}w\int_{0^{2n+2}}^{\infty^{2n+2}}\frac{da_1\cdots da_{2n+2}\delta(1-\sum a_i)}{(\sum a_i(\sum_{\mu,p}z_\mu Q_i^{\mu p}w_p))^{2n+2}}\overset{?}{=}$$

$$(2n+1)!\int_{0^{2n+2}}^{\infty^{2n+2}}da_1\cdots da_{2n+2}\delta(1-\sum a_i)\int_\sigma\frac{d^{2n+2}z\wedge d^{2n+2}w}{(\sum_{\mu,p}z_\mu(\sum a_iQ_i^{\mu p})w_p)^{2n+2}}=$$

$$(2n+1)!(2\pi i)^{2n+3}\int_{0^{2n+2}}^{\infty^{2n+2}}\frac{da_1\cdots da_{2n+2}\delta(1-\sum a_i)}{\text{Pfaffian}(\sum a_iQ_i)^2}.$$

The integral on the right in (3.16) can be rewritten as a projective integral as on the right in (1.4):

$$\int_{0^{2n+2}}^{\infty^{2n+2}}\frac{da_1\cdots da_{2n+2}\delta(1-\sum a_i)}{\text{Pfaffian}(\sum a_iQ_i)^2}=\int_\delta\frac{\Omega_{2n+1}}{\text{Pfaffian}(\sum_{i=1}^{2n+2}a_iQ_i)^2}. \tag{3.17}$$

## 4. Proof of theorem 1.1

To finish the proof of theorem 1.1, we need to understand the chain of integration $\sigma$ in (3.16). We also need to choose the alternating forms $Q_i$ on the left side of (3.16) so the resulting integral coincides upto a constant with the Feynman integral in the statement of the theorem (1.3).

Put an hermitian metric $||\cdot||$ on $V$. The induced metric on the bundle of 2-planes defines a submanifold $M\subset V\times V-S$ where $M$ is the set of pairs $(z,v)\in V\times V-S$ such that $||z||=||v||=1$ and $\langle z,v\rangle=0$. $M$ is a $\mathbb{U}_2$-bundle which is a reduction of structure of the $GL_2(\mathbb{C})$ bundle $V\times V-S$. The inclusion $M\subset V\times V-S$ is a homotopy equivalence. In particular, the fibre

$$(R^4\rho_*\mathbb{Z})_w\cong H^4(M_w)=H^4(\mathbb{U}_2)=\mathbb{Z}\cdot[\mathbb{U}_2]. \tag{4.1}$$

($\mathbb{U}_2$ is a compact orientable 4-manifold, so this follows by Poincaré duality.)

For the base, write $G^0:=G(2,V)-\{Q=0\}$ where $Q\in\bigwedge^2V^\vee$ is of rank $2n+2$. $G^0$ is affine (and hence Stein) of dimension $4n$, so $H^i(G^0,\mathbb{Z})=(0)$ for $i>4n$. Let $\rho^0:V\times V-R\to G^0$ be the $GL_2$ principal bundle obtained by restriction from $\rho$. We are interested in the class in $H^{4n+4}(V\times V-R,\mathbb{Q})$ (cf. lemma 3.4) dual to $\sigma$.

The grassmann is simply connected, so by (4.1), necessarily $R^4\rho_*\mathbb{Z} \cong \mathbb{Z}_G$. Since the fibres of $\rho$ have cohomological dimension 4, we have also

$$\mathbb{Q} = H^{4n+4}(V \times V - R, \mathbb{Q}) \cong H^{4n}(G^0, R^4\rho^0_*\mathbb{Q}) \cong H^{4n}(G^0, \mathbb{Q}). \tag{4.2}$$

It is not hard to show in fact that $H^{4n}(G^0, \mathbb{Q}) = \mathbb{Q} \cdot c_2(\mathcal{W})^n$ where $\mathcal{W}$ is the tautological rank 2 bundle on $G(2, V)$ as in (3.4). The interesting question is what if anything this class has to do with the topological closure of real Minkowski space in $G(2, V)$ which is classically the chain of integration for the Feynman integral.

Recall we have $\Gamma$ a graph with no self-loops and no multiple edges. External edges will play no role in our discussion, so assume $\Gamma$ has none. The chain of integration for the Feynman integral is $\mathbb{R}^{4n}$ where $n$ is the loop number of $\Gamma$. This vector space is canonically identified with $H := H_1(\Gamma, \mathbb{R}) \otimes_{\mathbb{R}} \mathbb{R}^4$. In particular, an edge $e \in \mathrm{Edge}(\Gamma)$ yields a functional $e^\vee : H_1(\Gamma, \mathbb{R}) \to \mathbb{R}$ associating to a loop $\ell$ the coefficient of $e$ in $\ell$.

To avoid divergences, the theorem is formulated for euclidean propagators. Let $q : \mathbb{R}^4 \to \mathbb{R}$ be $q(x_1, \ldots, x_4) = x_1^2 + \cdots + x_4^2$. The propagators which appear in the denominator of the integral have the form

$$H = H_1(\Gamma, \mathbb{R}) \otimes_{\mathbb{R}} \mathbb{R}^4 \xrightarrow{e^\vee \otimes id_{\mathbb{R}^4}} \mathbb{R}^4 \xrightarrow{q} \mathbb{R}. \tag{4.3}$$

We take complex coordinates in $\mathbb{C}^4 = \mathbb{R}^4 \otimes \mathbb{C}$ of the form

$$z_1 = x_1 + ix_2,\ z_2 = ix_3 + x_4,\ w_1 = ix_3 - x_4,\ w_2 = x_1 - ix_2; \tag{4.4}$$

$$x_1 = \frac{z_1 + w_2}{2},\ x_2 = \frac{z_1 - w_2}{2i},\ x_3 = \frac{z_2 + w_1}{2i},\ x_4 = \frac{z_2 - w_1}{2}. \tag{4.5}$$

In these coordinates $q = z_1w_2 - z_2w_1$ and the real structure is $\mathbb{R}^4 = \{(z_1, z_2, -\overline{z}_2, \overline{z}_1) \mid z_j \in \mathbb{C}\}$.

Now take real coordinates for $H_1(\Gamma, \mathbb{R})$ and let $(z_1^k, z_2^k, w_1^k, w_2^k),\ k \geq 1$ be the resulting coordinates on $H_{\mathbb{C}}$. It is then the case that for each edge $e$ there are real constants $\alpha_k = \alpha_k(e) \in \mathbb{R}$ not all zero, and the propagator for $e$ is

$$\det\begin{pmatrix} \sum_{k\geq 1} \alpha_k z_1^k & \sum_{k\geq 1} \alpha_k z_2^k \\ -\sum_{k\geq 1} \alpha_k \overline{z}_2^k & \sum_{k\geq 1} \alpha_k \overline{z}_1^k \end{pmatrix} = \Big|\sum_k \alpha_k z_1^k\Big|^2 + \Big|\sum_k \alpha_k z_2^k\Big|^2. \tag{4.6}$$

Since the linear functionals associated to the various edges $e$ span the dual space to $H_1(\Gamma, \mathbb{R})$, we see that a positive linear combination of the propagators is necessarily positive definite on $H_{\mathbb{R}}$ (i.e. $> 0$ except at 0.) Using the cordinates $z_i^k, w_i^k$ we can identify $H_{\mathbb{C}}$ with an open set in $G = G(2, 2n+2)$; namely the point with coordinates $z, w$ is identified with the 2-plane of row vectors

$$\begin{pmatrix} 1 & 0 & z_1^1 & z_2^1 & z_1^2 & z_2^2 & \cdots \\ 0 & 1 & w_1^1 & w_2^1 & w_1^2 & w_2^2 & \cdots \end{pmatrix}. \tag{4.7}$$

We throw in two more coordinates $z_1^0, z_2^0$ (resp. $w_1^0, w_2^0$) and view the $z_j^k$ (resp. $w_j^k$) as coordinates of points in $V_{\mathbb{C}} = \mathbb{C}^{2n+2}$. The fact that the set of non-zero matrices of the form $\begin{pmatrix} z_1 & z_2 \\ -\overline{z}_2 & \overline{z}_1 \end{pmatrix}$ is a group under multiplication means that the set of non-zero $2 \times (2n+2)$-matrices

$$\begin{pmatrix} z_1^0 & z_2^0 & z_1^1 & z_2^1 & \cdots & z_1^n & z_2^n \\ -\overline{z}_2^0 & \overline{z}_1^0 & -\overline{z}_2^1 & \overline{z}_1^1 & \cdots & -\overline{z}_2^n & \overline{z}_1^n \end{pmatrix} \tag{4.8}$$

is closed in $G$. It is clearly the closure in $G$ of the real Minkowski space whose complex points are given in (4.7). It will be convenient to scale the rows by a positive real scalar and assume $\sum_{j,k}|z_j^k|^2 = 1$, so the resulting locus is compact in $V \times V - R$. We also scale the bottom row by a constant $e^{i\theta}$ of norm 1. The resulting locus

$$(4.9)\quad \sigma := \left\{\begin{pmatrix} z_1^0 & z_2^0 & z_1^1 & z_2^1 & \cdots & z_1^n & z_2^n \\ -e^{i\theta}\bar{z}_2^0 & e^{i\theta}\bar{z}_1^0 & -e^{i\theta}\bar{z}_2^1 & e^{i\theta}\bar{z}_1^1 & \cdots & -e^{i\theta}\bar{z}_2^n & e^{i\theta}\bar{z}_1^n \end{pmatrix} \,\middle|\, \sum_{j,k}|z_j^k|^2 = 1\right\} \subset V \times V - R$$

is compact and depends on $4n+4$ real parameters.

Let $Q_e \in \bigwedge^2 V^\vee$ be the form which associates to (4.7) the determinant

$$\det\begin{pmatrix} \sum_{k\geq 1}\alpha_k(e)z_1^k & \sum_{k\geq 1}\alpha_k(e)z_2^k \\ \sum_{k\geq 1}\alpha_k(e)w_1^k & \sum_{k\geq 1}\alpha_k(e)w_2^k \end{pmatrix}.$$

Let $a_e > 0$ be constants, and let $\widetilde{Q} = \sum_e a_e Q_e \in \bigwedge^2 V^\vee$. Finally, let $Q_0 \in \bigwedge^2 V^\vee$ associate to the matrix (4.8) the minor $z_1^0\bar{z}_1^0 + z_2^0\bar{z}_2^0$. It is clear that $Q := Q_0 + \widetilde{Q}$ doesn't vanish on any non-zero matrix of the form (4.8). We conclude:

PROPOSITION 4.1. *Let $G(\mathbb{R}) \subset G$ be the set of points (4.8). Then with $Q$ as above, we have $G(\mathbb{R}) \subset G^0 = G - \{Q = 0\}$.*

The locus $\sigma$, (4.9), projects down to $G(\mathbb{R})$ with fibre the group $\mathbb{U}_2$.

PROPOSITION 4.2. *With this choice of $\sigma$ we have*

$$(4.10)\qquad \int_\sigma \frac{d^{2n+2}z \wedge d^{2n+2}w}{Q^{2n+2}} \neq 0.$$

PROOF. Let $v_j^{k,\vee}$ be the basis of $V^\vee$ which is dual to the coordinate system $z_j^k$ introduced above. Then one checks that $Q$ as described above is associated to an element

$$(4.11)\qquad Q = \sum_{k=0}^{n} b_k v_1^{k,\vee} \wedge v_2^{k,\vee} \in \bigwedge^2 V^\vee; \quad b_k > 0.$$

Applied to the matrix on the right in (4.9),

$$(4.12)\qquad Q(\cdots) = e^{i\theta}\sum_{k=0}^{n} b_k(|z_1^k|^2 + |z_2^k|^2)$$

Computing $d^{2n+2}z \wedge d^{2n+2}w$ on the right hand side of (4.9) yields

$$(4.13)\quad ie^{(2n+2)i\theta}d\theta \wedge \wedge dz_1^0 \wedge \cdots \wedge dz_2^n \wedge \sum_k \Big((\bar{z}_2^k d\bar{z}_1^k - \bar{z}_1^k d\bar{z}_2^k) \wedge \bigwedge_{j\neq k}(d\bar{z}_1^j \wedge d\bar{z}_2^j)\Big).$$

The crucial point is that the $e^{i\theta}$ factor in the integrand (4.10) cancels. Rescaling we can reduce to the case where all the $b_k = 1$. Integrating over $\sigma$ yields a $2\pi i$ from the $id\theta$ and then an integral over the volume form of the $4n+3$ sphere $\sum_{k=0}^n(|z_1^k|^2 + |z_2^k|^2) = 1$. This is non-zero. □

REMARK 4.3. *It also follows from the above discussion that the integral on the left in* (3.16) *is constant along the fibres of* $\sigma$ *over* $G$. *Since the image of* $\sigma$ *in* $G$ *is the compactification of Minkowski space* $\mathbb{R}^{4n}$, *the integral becomes a non-zero constant times the Feynman amplitude* $\mathcal{A}(\Gamma, q, m)$, (1.3).

The proof of theorem 1.1 is now complete. To summarize, given $\Gamma$, one uses the change of coordinates (4.4) in order to rewrite the euclidean propagators $P_i$ as determinants of alternating matrices $Q_i$. One uses the discussion in section 2, particularly formula (2.9) and remark 2.3, to interpret these propagators with external momenta and masses as elements in $\bigwedge^2 V^\vee$, where $V \cong \mathbb{C}^{\mathrm{Edge}(\Gamma)} \cong \mathbb{C}^{2n+2}$. Using (4.6), one sees that a positive linear combination of the $Q_i$ does not vanish on the locus $\sigma$ defined in (4.9). This means that the integrand on the right in (3.16) has poles only on the boundary of the chain of integration where some of the $a_i = 0$. The integral on the left, by remark 4.3, is a constant (depending only on $n$) times the euclidean amplitude integral.

## References


[B] Bloch, S., Unpublished manuscript on the motive of null vectors for the graph matrix, http://math.uchicago.edu/~ bloch/publications.html

[BEK] Spencer Bloch, Hélène Esnault, and Dirk Kreimer, *On motives associated to graph polynomials*, Comm. Math. Phys. **267** (2006), no. 1, 181–225, DOI 10.1007/s00220-006-0040-2. MR2238909 (2008a:81074)

[BK] Spencer Bloch and Dirk Kreimer, *Feynman amplitudes and Landau singularities for one-loop graphs*, Commun. Number Theory Phys. **4** (2010), no. 4, 709–753, DOI 10.4310/CNTP.2010.v4.n4.a4. MR2793425 (2012k:81104)

[BrK] D. J. Broadhurst and D. Kreimer, *Knots and numbers in* $\phi^4$ *theory to* 7 *loops and beyond*, Internat. J. Modern Phys. C **6** (1995), no. 4, 519–524, DOI 10.1142/S012918319500037X. MR1352337 (97a:81143)

[H] Andrew Hodges, *The box integrals in momentum-twistor geometry*, J. High Energy Phys. **8** (2013), 051, front matter+29. MR3106387

[IZ] Itzykson, C., Zuber, J.-B., Quantum Field Theory, Dover Publ. (2005).

[K] George Kempf, *On the geometry of a theorem of Riemann*, Ann. of Math. (2) **98** (1973), 178–185. MR0349687 (50 #2180)

[P] Eric Patterson, *On the singular structure of graph hypersurfaces*, Commun. Number Theory Phys. **4** (2010), no. 4, 659–708, DOI 10.4310/CNTP.2010.v4.n4.a3. MR2793424 (2012f:81093)

[T] Todorov, I., Analytic properties of Feynman diagrams in quantum field theory, monographs in natural philosophy 38, Pergamon Press (1971).


5765 S. BLACKSTONE AVE., CHICAGO, ILLINOIS 60637
*E-mail address*: spencer_bloch@yahoo.com

# Multiple polylogarithms and linearly reducible Feynman graphs

Christian Bogner and Martin Lüders

ABSTRACT. We review an approach for the computation of Feynman integrals by use of multiple polylogarithms, with an emphasis on the related criterion of linear reducibility of the graph. We show that the set of graphs which satisfies the linear reducibility with respect to both Symanzik polynomials is closed under taking minors. As a step towards a classification of Feynman integrals, we discuss the concept of critical minors and exhibit an example at three loops with four on-shell legs.

## 1. Introduction

In recent years we witnessed rapid progress in the developement of techniques for the computation of higher order corrections in perturbative quantum field theory. While other talks at this conference cover progress in the computation of entire amplitudes, our talk refers to the 'classical' approach of computing the amplitude by its Feynman graphs, which is inevitable when meeting the needs of present collider experiments. In this field of research, it has shown to be fruitful to discuss Feynman integrals in their own right, without restrictions to a particular quantum field theory.

Computations of higher order corrections to observables often start from the consideration of hundreds or thousands of Feynman integrals with tensor structure, and proceed via effective standard procedures to reduce the problem, possibly to a relatively small number of scalar integrals. At higher loop-orders, the evaluation of the latter remains to be the hard part of the problem. There is no algorithm which would succeed in the analytical computation of every Feynman integral. However, there is a variety of powerful methods which have been useful for a wide range of relevant cases, such as the Mellin-Barnes approach (see [**10**, **59**, **62**, **63**]), the expansion of hypergeometric functions [**40**, **41**, **49**], differential equation methods [**34**, **44**, **50**, **51**, **55**], difference equations [**45**, **48**, **60**, **61**] or position-space methods [**29**] (also see [**38**]). In this talk we focus on the approach of iteratively integrating out Feynman parameters by use of multiple polylogarithms.

The first author is grateful to Francis Brown for useful communication and hospitality in Paris Jussieu, supported by ERC grant no. 257638. We thank Erik Panzer for useful communication and Dirk Kreimer's group at Humboldt University for hospitality and support. Our graphs were drawn using [**39**].

In order to choose an appropriate strategy for the computation of a given Feynman integral, it would be desirable in general, to know in advance, which are the classes of functions and numbers the integral may evaluate to. As a slightly more refined question of this type we may ask: Which scalar Feynman integrals can be expressed by multiple polylogarithms and multiple zeta values, and for which integrals do we need a wider range of functions and numbers? In the past few years, questions of this type turned out to define a fruitful common field of research for quantum field theorists and algebraic geometers alike. While the physicist's interest in these questions is given by the desire to compute specific integrals or to learn about the 'number content' of a given quantum field theory, the mathematician arrives at the same question from a different direction. In a very general context, Feynman integrals can be viewed as period integrals, and the question of evaluating to multiple zeta values is related to the question whether an underlying motive is mixed Tate over $\mathbb{Z}$ (see [**3–5, 13, 21, 22, 25**]).

A definite classification of Feynman graphs with respect to the above questions is missing. However, for vacuum and two-point graphs important progress was made by considering the first Symanzik polynomial, given by the Feynman parametric representation of a Feynman integral. Even though many vacuum-type Feynman integrals evaluate to multiple zeta values [**18, 19**], this is not the case in general. A first vacuum graph whose period has to belong to a set of numbers beyond multiple zeta values was exhibited in a recent article by Brown and Schnetz [**25**] (also see [**23**]). When allowing the Feynman integrals to depend on kinematical invariants and particle masses, we can ask for graphs where multiple polylogarithms are not sufficient to express the result. Here the first cases show up at much lower loop-order, such as in the case of massive sunrise graphs and related graphs with a cut through three massive edges (see e.g. [**7, 8, 58**]).

In this talk we review a criterion on graphs which is related to the above questions and show that if a graph satisfies the criterion, its minors do so as well. In graph theory such a minor monotony is an important and desireable feature. In section 2 we begin with a brief reminder on scalar Feynman integrals, their two Symanzik polynomials and the approach of integrating out Feynman parameters by use of multiple polylogarithms. In section 3 we briefly review the criterion of linear reducibility of a graph, which is used to decide whether a given integral can be computed by use of the method. If this is the case, the functions and numbers in all intermediate steps and in the result will not exceed combinations of multiple polylogarithms and their values at rational points. In this way the criterion and the corresponding algorithm are useful tools for adressing the above questions. In the case of integrals only involving the first Symanzik polynomial, the criterion was extensively studied in [**21, 22**]. As the iterated integration over Feynman parameters can be expected to be useful in the case of integrals depending on kinematical invariants and particle masses as well, we intend to extend the discussion to the second Symanzik polynomial. In section 4 we consider linear reducibility with respect to both Symanzik polynomials and show that the set of linearly reducible graphs is closed under taking minors. This property is useful for a classification, as it allows us to characterize families of reducible graphs by a small number of graphs not belonging to the family. In a case study we exhibit such a 'forbidden minor' at the level of massless three-loop graphs with four on-shell legs. Section 5 contains our conclusions.

## 2. Multiple Polylogarithms and Feynman Integrals

In this section we recall some general facts about Feynman integrals, Symanzik polynomials and a method to compute period integrals by use of multiple polylogarithms. Let us begin with a generic Feynman graph $G$ with $n$ edges, loop-number (i.e. first Betti number) $L \geq 1$ and with $r$ external half-edges (or 'legs'). We label each edge $e_i$ by an integration variable $\alpha_i$ (Feynman parameter), an integer $\nu_i$ (exponent of the Feynman propagator), a real or complex variable $m_i$ (particle mass). Each leg is labelled by a vector $p_j$ (external momentum).

To this labelled graph $G$ we associate the scalar Feynman integral in Dimensional Regularization:

$$I_G = \frac{\Gamma(\nu - LD/2)}{\prod_{j=1}^{n} \Gamma(\nu_j)} \int_{\alpha_j \geq 0} \delta\left(1 - \sum_{i=1}^{n} \alpha_i\right) \left(\prod_{j=1}^{n} d\alpha_j \alpha_j^{\nu_j - 1}\right) \frac{\mathcal{U}_G^{\nu-(L+1)D/2}}{\mathcal{F}_G^{\nu - LD/2}} \tag{2.1}$$

where $\nu = \sum_{i=1}^{n} \nu_i$. (We omit to write a trivial prefactor by which the integral becomes independent of the physical mass-scale.) The Feynman integral $I_G$ and the function $\mathcal{F}_G$ depend on the particle masses and on certain kinematical invariants, which are quadratic functions of the external momenta. The functions $\mathcal{U}_G$ and $\mathcal{F}_G$ are the first and second Symanzik polynomial of the graph. A definition is given below. Usually a Feynman integral is associated to a Feynman graph by Feynman rules in momentum or position space, and we refer to the literature [**42, 52**] for the standard computation leading from there to the Feynman parametric representation given in eq. 2.1.

Eq. 2.1 defines a very general class of integrals which deserves our attention for several reasons. Firstly, the class contains the Feynman integrals of scalar quantum field theory such as $\phi^3-$ or $\phi^4-$theory. Secondly, any Feynman integral with a tensor-structure, arising from a physical quantum field theory, can in principle be expressed in terms of scalar integrals of the above class [**60, 61**]. Thirdly, as we allow the $\nu_j$ to take arbitrary integer values, there are well-known identities between these scalar integrals which can be used for efficient reduction procedures [**30**]. As a consequence, integrals of the above class appear in a wide range of physical set-ups and their evaluation is the bottleneck of many computational problems in particle physics.

The parameter $D$ can either be fixed to the integer space-time dimension or, as the integral is very often ill-defined in the desired dimension, one may consider $I_G$ in Dimensional Regularisation where $D$ is a complex variable. Then, in order to separate the pole-terms and obtain finite contributions in four-dimenional Minkowski space, one usually attempts to compute the coefficients of a Laurent-expansion

$$I_G = \sum_{j=j_0}^{\infty} c_j \epsilon^j,$$

with $D = 4 - 2\epsilon$, to a desired order. Even though the computation of the functions $c_j$ can be very difficult, we can make a general statement about them. It is shown in [**16**] that if for an arbitrary Feynman graph we evaluate any function $c_j$ at algebraic values of the squared particle masses $m_i^2$ and kinematical invariants $s_i$, where all $m_i^2 \geq 0$ and all $s_i \leq 0$, we obtain a *period* according to the definition of Kontsevich and Zagier [43]. For the special case where the Feynman integral takes the form

$$\mathcal{P}_G = \int_{\alpha_j \geq 0} \delta\left(1 - \sum_{i=1}^{n} \alpha_i\right) \left(\prod_{j=1}^{n} d\alpha_j\right) \frac{1}{\mathcal{U}_G^{D/2}} \tag{2.2}$$

this statement was already proven in [9]. It seems that Feynman integrals in fact evaluate to a restricted subset of periods and it is an important challenge to understand which one this is.

Let us now recall the definition of the Symanzik polynomials of a Feynman graph $G$. The first Symanzik polynomial is defined as

$$\mathcal{U}_G = \sum_T \prod_{e_i \notin T} \alpha_i,$$

where the sum is over all spanning trees of the graph $G$. The second Symanzik polynomial is defined as

$$\mathcal{F}_G = \mathcal{F}_{0,\,G} + \mathcal{U}_G \sum_{i=1}^{n} \alpha_i m_i^2$$

with

$$\mathcal{F}_{0,\,G} = \sum_{(T_1,\,T_2)} \left( \prod_{e_i \notin (T_1,\,T_2)} \alpha_i \right) s_{(T_1,\,T_2)}.$$

Here the sum runs through all spanning two-forests $(T_1,\,T_2)$ of $G$, where $T_1$ and $T_2$ denote the connected components of the forest.

In order to define the kinematical invariants $s_{(T_1,\,T_2)}$, we introduce an arbitrary orientation on $G$. We firstly say that each external momentum $p_j$ is incoming at the vertex at the corresponding leg. We furthermore label each oriented edge by a momentum-vector $q_i$. If the edge $e_i$ is oriented from vertex $v_j$ to $v_k$ then $q_i$ is said to be incoming at $v_k$ and $-q_i$ is incoming at $v_j$. Momentum-conservation on $G$ is reflected in our labels by the condition that the sum of all external momenta $p_j$ is zero, and at each vertex, the sum of all incoming momenta is zero. By these conditions, except for $L$ momenta, each of the $q_i$ can be expressed as a linear combination of external momenta. The kinematical invariants are defined as

$$s_{(T_1,\,T_2)} = \left( \sum_{e_j \notin (T_1,\,T_2)} \pm q_j \right)^2$$

where the sign of $q_j$ is fixed by the condition that we sum over the momenta incoming at the component $T_2$. Note that by momentum conservation, the $s_{(T_1,\,T_2)}$ are functions of the external momenta.

As an alternative to the above construction by spanning trees and forests, there are several ways to obtain both Symanzik polynomials from determinants of certain matrices [14, 17, 22, 54]. To demonstrate such a derivation, let us label each edge $e_i$ by an auxiliary variable $y_i$. Each vertex $v_i$ is labelled by

$$u_i = \begin{cases} z_j \text{ if a leg with incoming momentum } p_j \text{ is attached,} \\ \quad 0 \text{ if no leg is attached.} \end{cases}$$

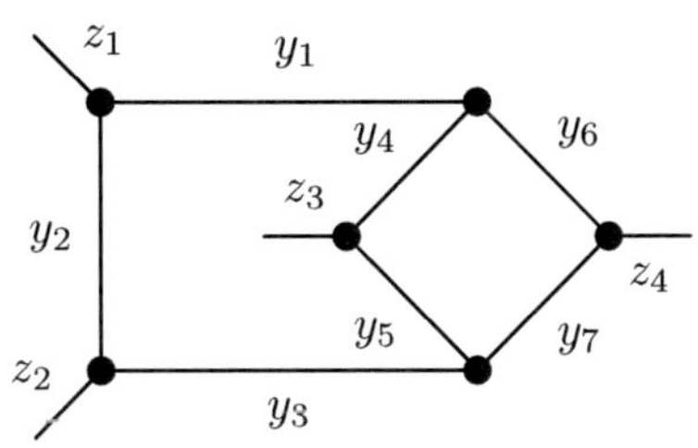


FIGURE 1. The non-planar double-box

For a Feynman graph with vertices $v_1, ..., v_m$ we consider an $m \times m$ matrix $M$ whose entries are:

$$M_{ij} = \begin{cases} u_i + \sum y_k \text{ for } i = j, \, e_k \text{ attached to } v_i \text{ at exactly one end,} \\ -\sum y_k \text{ for } i \neq j, \, e_k \text{ connecting } v_i \text{ and } v_j. \end{cases}$$

We compute the determinant

$$\mathcal{V}(y_1, ..., y_n, z_1, ..., z_r) = \det(M)$$

and consider the function

$$\mathcal{W}(\alpha_1, ..., \alpha_n, z_1, ..., z_r) = \mathcal{V}\left(\alpha_1^{-1}, ..., \alpha_n^{-1}, z_1, ..., z_r\right) \prod_{i=1}^{n} \alpha_i$$

which is a polynomial in the $\alpha-$ and $z-$variables. Note that $M$ depends on a chosen ordering on the vertices but $\mathcal{W}$ does not.

Let us assume that at least two legs are attached to the graph, i.e. $r \geq 2$. We expand $\mathcal{W}$ as $\mathcal{W} = \mathcal{W}^{(1)} + \mathcal{W}^{(2)} + ... + \mathcal{W}^{(r)}$ where $\mathcal{W}^{(k)}$ is homogeneous of degree $k$ in the $z-$variables. We can directly read off the first Symanzik polynomial from the first term in this expansion, as it satisfies

$$\mathcal{W}^{(1)}(\alpha_1, ..., \alpha_n, z_1, ..., z_r) = \mathcal{U}_G(\alpha_1, ..., \alpha_n) \sum_{i=1}^{r} z_i.$$

The massless second Symanzik polynomial $\mathcal{F}_{0,G}$ is directly obtained from $\mathcal{W}^{(2)}$. By construction, $\mathcal{W}^{(2)}$ is homogeneous of degree 2 in the $z-$variables. We replace each product $z_i z_j$ in $\mathcal{W}^{(2)}$ by the scalar-product of the corresponding external momentum vectors $p_i \cdot p_j$. By momentum-conservation, $\sum_{i=1}^{r} p_i = 0$, we express each of the scalar products by the functions $s_{(T_1, T_2)}$. As result we obtain $\mathcal{F}_{0,G}$ [**17**].

As an example let us compute the two Symanzik polynomials of the massless non-planar double-box, shown in figure 1 with auxiliary $y-$ and $z-$variables. For this graph and a chosen ordering on the vertices we have

$$M = \begin{pmatrix} M_{11} & -y_2 & 0 & 0 & -y_1 & 0 \\ -y_2 & M_{22} & -y_3 & 0 & 0 & 0 \\ 0 & -y_3 & M_{33} & -y_7 & 0 & -y_5 \\ 0 & 0 & -y_7 & M_{44} & -y_6 & 0 \\ -y_1 & 0 & 0 & -y_6 & M_{55} & -y_4 \\ 0 & 0 & -y_5 & 0 & -y_4 & M_{66} \end{pmatrix},$$

where $M_{11} = y_1+y_2+z_1$, $M_{22} = y_2+y_3+z_2$, $M_{33} = y_3+y_5+y_7$, $M_{44} = y_6+y_7+z_4$, $M_{55} = y_1+y_4+y_6$, $M_{66} = y_4+y_5+z_3$. Proceeding in the described way we compute

$$\begin{aligned}
\mathcal{U}_G &= (z_1+z_2+z_3+z_4)^{-1}\mathcal{W}^{(1)} \\
&= (\alpha_1+\alpha_2+\alpha_3)(\alpha_4+\alpha_5+\alpha_6+\alpha_7)+(\alpha_4+\alpha_5)(\alpha_6+\alpha_7), \\
\mathcal{F}_{0\,G} &= \mathcal{W}^{(2)}|_{z_iz_j=p_i\cdot p_j,\, \sum_{i=1}^4 p_i=0} \\
&= -p_1^2\alpha_2(\alpha_1(\alpha_4+\alpha_5+\alpha_6+\alpha_7)+\alpha_4\alpha_6) \\
&\quad -p_2^2\alpha_2(\alpha_3(\alpha_4+\alpha_5+\alpha_6+\alpha_7)+\alpha_5\alpha_7) \\
&\quad -p_3^2(\alpha_4\alpha_5(\alpha_1+\alpha_2+\alpha_3+\alpha_6+\alpha_7)+\alpha_3\alpha_4\alpha_7+\alpha_1\alpha_5\alpha_6) \\
&\quad -p_4^2(\alpha_6\alpha_7(\alpha_1+\alpha_2+\alpha_3+\alpha_4+\alpha_5)+\alpha_1\alpha_4\alpha_7+\alpha_3\alpha_5\alpha_6) \\
&\quad -(p_1+p_2)^2(\alpha_1\alpha_3(\alpha_4+\alpha_5+\alpha_6+\alpha_7)+\alpha_1\alpha_5\alpha_7+\alpha_3\alpha_4\alpha_6) \\
&\quad -(p_1+p_3)^2\alpha_2\alpha_5\alpha_6-(p_2+p_3)^2\alpha_2\alpha_4\alpha_7.
\end{aligned}$$

It is often sufficient to consider the Feynman integral after setting some of its legs on-shell, which means that the corresponding external momenta are fixed by setting their square to a squared particle mass. In our example we may assume massless particles and set $p_i^2 = 0$ for all $i = 1, ..., 4$. The corresponding Feynman integral was evaluated in dimensional regularization by classical polylogarithms in reference [**62**]. We will return to Symanzik polynomials of graphs with four on-shell legs in section 4.

Let us now turn to iterated integrals. Let $k$ be the field of either the real or the complex numbers and $M$ a smooth manifold over $k$. We consider a piecewise smooth path on $M$, given by a map $\gamma : [0, 1] \to M$, and some smooth differential 1-forms $\omega_1, ..., \omega_n$ on $M$. The iterated integral of these 1-forms along the path $\gamma$ is defined by

$$\int_\gamma \omega_n...\omega_1 = \int_{0\le t_1\le...\le t_n\le 1} f_n(t_n)dt_n...f_1(t_1)dt_1,$$

where $f_i(t)dt = \gamma^\star(\omega_i)$ is the pull-back of $\omega_i$ to $[0, 1]$. With the term iterated integral we will more generally refer to $k$-linear combinations of such integrals.

We will consider classes of iterated integrals which define the same function for any two homotopic paths. Such integrals are called *homotopy invariant*. They are well-defined functions of variables given by the end-point of $\gamma$. In such iterated integrals the differential forms and the order in which we integrate over them have to satisfy a property known as the *integrability condition*. The condition is best formulated on tensor products of 1-forms over some field $K \subseteq k$, which we denote by $[\omega_1|...|\omega_m]$. Let $D$ denote a $K$-linear map from tensor products of smooth 1-forms on $M$ to tensor products of all forms on $M$, given by

$$D([\omega_1|...|\omega_m]) = \sum_{i=1}^{m}[\omega_1|...|\omega_{i-1}|d\omega_i|\omega_{i+1}|...|\omega_m] + \sum_{i=1}^{m-1}[\omega_1|...|\omega_{i-1}|\omega_i\wedge\omega_{i+1}|...|\omega_m].$$

A $K$-linear combination of tensor products $\xi = \sum_{l=0}^{m} c_{i_1,\dots,i_l}\left[\omega_{i_1}|\dots|\omega_{i_l}\right]$, $c_{i_1,\dots,i_l} \in K$, is called an *integrable word* if it satisfies the equation

$$D\xi = 0.$$

Let $\Omega$ be a finite set of smooth 1-forms and let $B_m(\Omega)$ denote the vector space of integrable words of length $m$ with 1-forms in $\Omega$. Now we return from words to integrals by considering the integration map on integrable words:

$$\sum_{l=0}^{m}\sum_{i_1,\dots,i_l} c_{i_1,\dots,i_l}\left[\omega_{i_1}|\dots|\omega_{i_l}\right] \mapsto \sum_{l=0}^{m}\sum_{i_1,\dots,i_l} c_{i_1,\dots,i_l}\int_\gamma \omega_{i_1}\dots\omega_{i_l}. \tag{2.3}$$

A fundamental theorem of Chen [**28**] states that this map is an isomorphism from $B_m(\Omega)$ to the set of homotopy invariant iterated integrals in 1-forms in $\Omega$ of length less or equal to $m$, if $\Omega$ satisfies further conditions which we do not specify here.

In the following we fix $K = \mathbb{Q}$ and discuss two sets of 1-forms for which the theorem applies. For a coordinate $t_1$ on an open subset of $\mathbb{C}$ we firstly consider the set of closed 1-forms

$$\Omega_n^{\mathrm{Hyp}} = \left\{\frac{dt_1}{t_1}, \frac{dt_1}{t_1-1}, \frac{t_2 dt_1}{t_1t_2-1}, \dots, \frac{\left(\prod_{i=2}^{n} t_i\right)dt_1}{\prod_{i=1}^{n} t_i - 1}\right\}.$$

As a trivial consequence of $dt_1 \wedge dt_1 = 0$, any tensor product of 1-forms in $\Omega_n^{\mathrm{Hyp}}$ is an integrable word. By applying the integration map eq. 2.3 to these words, we obtain the class of hyperlogarithms [**47**]. In particle physics it is very common to use sub-classes of hyperlogarithms. As an example, we may consider $\Omega_2^{\mathrm{Hyp}}$ and fix the constant $t_2 = -1$. To physicists, the iterated integrals obtained from this restriction are well known as harmonic polylogarithms [**56**] and suffice for the evaluation of many Feynman integrals.

We want to focus on a class of functions of severable variables, obtained from another set of 1-forms, where now all the $t_1$, ..., $t_n$ are considered to be coordinates in an open subset of $\mathbb{C}^n$ :

$$\Omega_n^{\mathrm{MPL}} = \left\{\frac{dt_1}{t_1}, \dots, \frac{dt_n}{t_n}, \frac{d\left(\prod_{a\le i\le b} t_i\right)}{\prod_{a\le i\le b} t_i - 1} \text{ where } 1 \le a \le b \le n\right\}.$$

For this set the integrability condition is not trivial and there are words for which it is not satisfied. The homotopy invariant iterated integrals which we obtain via the integration map from the integrable words in $\Omega_n^{\mathrm{MPL}}$ form the vector space $\mathcal{B}(\Omega_n)$ of multiple polylogarithms in $n$ variables. We use the notation $\mathcal{B}_m(\Omega_n)$ for the vector space of such functions obtained from integrable words of length $\le m$. There is an explicit map [**15**] to construct all integrable words in $\Omega_n^{\mathrm{MPL}}$, closely related to the 'symbol' in [**33, 35, 37**].

The functions in $\mathcal{B}(\Omega_n)$ were extensively studied in reference [**20**]. We just want to recall a few statements which are relevant for the following considerations. Firstly, the multiple polylogarithms of Goncharov [**36**], frequently used in the physics literature, are contained in this class. As we want to use the elements of $\mathcal{B}(\Omega_n)$ in an iterative integration procedure, it is important for us to know their primitives and limits. It is proven in [**20**] that $\mathcal{B}(\Omega_n)$ is closed under taking primitives. Furthermore if we take the limits of elements of $\mathcal{B}(\Omega_n)$ at $t_n$ equal to 0 and

1, we obtain $\mathcal{Z}$-linear combinations of elements in $\mathcal{B}(\Omega_{n-1})$, where $\mathcal{Z}$ denotes the $\mathbb{Q}$-vector space of multiple zeta values.

Now let us consider definite integrals of the form

$$I = \int_0^1 dt_n \frac{\beta(\{g_i\})}{f} \tag{2.4}$$

where $f$ is a polynomial and $\beta(\{g_i\}) \in \mathcal{B}(\Omega_n)$ is a multiple polylogarithm whose arguments are some irreducible polynomials $g_i$. Let us call $f$ and the $g_i$ the *critical polynomials* of the integrand. If $f$ and the $g_i$ are linear in $t_n$ we can evaluate the above integral and from the mentioned properties it is clear that the result will be a $\mathcal{Z}$-linear combination of elements in $\mathcal{B}_m(\Omega_{n-1})$. If the result can be again expressed by functions of the form of the above generic integrand and the critical polynomials are linear in $t_{n-1}$ then we can continue and integrate over this variable from 0 to 1, and so on.

Such an iterative procedure can be used to compute Feynman integrals. In reference [**1**] the approach was applied to intricate QCD integrals involving local operators with a dependence on a Mellin variable. A very recent application of the method by use of hyperlogarithms was given in reference [**53**], including the analytical computation of all propagator-type integrals of $\phi^4$-theory to the order $\epsilon^4$ with arbitrary insertions, several four-loop propagators and vacuum integrals up to seven loops. Instead of giving a complete account of the literature on applications, let us furthermore mention references [**6, 27, 31, 32**] where related approaches were applied in various physics contexts.

Aiming at such a computation one has to express the Feynman integral by a finite parametric integral such that the integrand can be written in the above form, where after each integration step, the critical polynomials are linear in at least one of the remaining parameters. The method was introduced systematically in [**21**] and demonstrated for certain Feynman parametric integrals of the type of eq. 2.2, coming from primitive logarithmically divergent vacuum Feynman graphs. However, the approach is not restricted to such graphs. Reference [**24**] presents a method to express Feynman integrals with UV sub-divergences by finite parametric integrals to which the approach may apply. The treatment of graphs with infrared divergences is not excluded in principle, but we are missing a canonical method to express IR-divergent integrals by finite ones. In principle, the method of sector decomposition [**12**] allows us to write down the coefficients of a Laurent expansion for a dimensionally regularized, infrared divergent integral in terms of finite integrals over Feynman parameters, however, the polynomials in these integrals usually become very complicated. In view of the above approach one would ideally wish for a method, where the critical polynomials in the finite integrals could be obtained from the Symanzik polynomials in a rather simple way.

For the following discussion let us assume, that in some way we have already been able to express a given Feynman integral by finite integrals of the type $I$ and that the critical polynomials are the Symanzik polynomials of the graph. We focus on the criterion, that after each integration over a Feynman parameter, the new critical polynomials have to be linear in a next Feynman parameter. The reduction algorithm to be reviewed in section 3 allows us to study this criterion as it computes for each integration step a set in which the critical polynomials are contained.

As a further motivation of the following discussion, let us have a glance at two well-known Feynman graphs in view of the mentioned criterion. For the massless

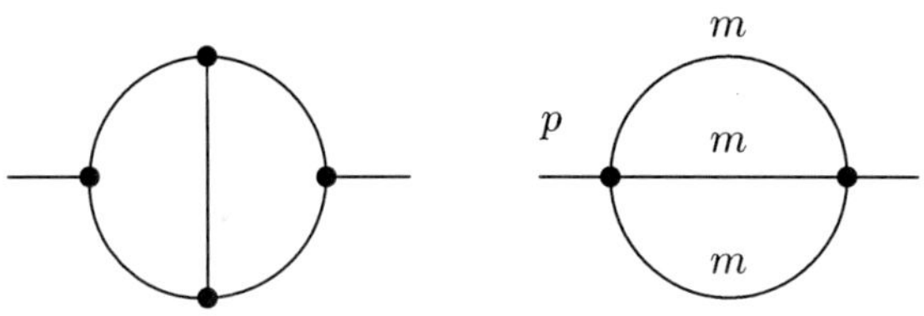


FIGURE 2. (a) Massless two-loop graph, (b) Equal-mass sunrise graph

two-loop graph of figure 2 (a) it was proven by use of the Mellin-Barnes approach and expansions by nested sums that each coefficient of the $\epsilon$-expansion is a combination of multiple zeta values [**11**]. Reference [**21**] confirmed this statement for this two-loop graph and several higher-loop graphs by relating them to integrals of the type of eq. 2.2 whose integrands satisfy the criterion.

The case of the equal-mass two-loop sunrise graph, shown in figure 2 (b), is very different. The desired coefficients in the $\epsilon$-expansion can be derived from the $D = 2$-dimensional version of the Feynman integral,

$$I_{\text{sunrise}} = \int_{\alpha_j \geq 0} d\alpha_1 d\alpha_2 d\alpha_3 \delta\left(1 - \sum_{i=1}^{3} \alpha_i\right) \frac{1}{\mathcal{F}_G},$$

with the second Symanzik polynomial

$$\mathcal{F}_G = -p^2 \alpha_1 \alpha_2 \alpha_3 + m^2 (\alpha_1 \alpha_2 + \alpha_2 \alpha_3 + \alpha_1 \alpha_3)(\alpha_1 + \alpha_2 + \alpha_3),$$

playing the role of the critical polynomial. As $\mathcal{F}_G$ is not linear in any of the Feynman parameters, integral $I_{\text{sunrise}}$ fails the criterion. It is in perfect match with this simple observation, that the known result of the sunrise integral involves elliptic integrals [**46**]. (Also see [**2**] for the case of arbitrary masses.) In general, it is possible that the polynomials of an integrand fail the criterion and still the integral can be expressed by multiple polylogarithms. However, the criterion may provide a useful first classification and give a hint where to look for integrals, which exceed the class of multiple polylogarithms. We also want to mention reference [**26**], where different criteria are used to search for such integrals.

## 3. Linear Reducibility

Let us briefly review the polynomial reduction algorithm of [**21**]. Let $S = \{f_1, ..., f_N\}$ be a set of polynomials in the variables $\alpha_1, ..., \alpha_n$ with rational coefficients.

(1) If there is an index $1 \leq r_1 \leq n$ such that all polynomials in $S$ are linear in $\alpha_{r_1}$ we can write

$$f_i = g_i \alpha_{r_1} + h_i \text{ for all } 1 \leq i \leq N,$$

where $g_i = \frac{\partial f_i}{\partial \alpha_{r_1}}$ and $h_i = f_i|_{\alpha_{r_1}=0}$. We define

$$S'_{(r_1)} = \left\{ (g_i)_{1 \leq i \leq N}, (h_i)_{1 \leq i \leq N}, (h_i g_j - g_i h_j)_{1 \leq i < j \leq N} \right\} \tag{3.1}$$

and furthermore we define $S_{(r_1)}$ to be the set of irreducible polynomials in $S'_{(r_1)}$. In $S_{(r_1)}$ we neglect all constants and monomials.

(2) If there is a $1 \leq r_2 \leq n$ such that all polynomials in $S_{(r_1)}$ are linear in $\alpha_{r_2}$, we repeat the above step, now with $S_{(r_1)}$ and $\alpha_{r_2}$ in the roles of $S$ and $\alpha_{r_1}$, and obtain a new set of polynomials which we call $S_{[r_1](r_2)}$. Then, assuming that starting from $S$ the above steps can be done first for $\alpha_{r_2}$ and then for $\alpha_{r_1}$, we compute $S_{[r_2](r_1)}$ and take the intersection of both sets:

$$S_{[r_1,r_2]} = S_{[r_1](r_2)} \cap S_{[r_2](r_1)}.$$

Whenever we speak of intersections here and in the following, we mean the common zero loci, such that if a polynomial appears in two sets with a different constant prefactor, it nevertheless belongs to the intersection. Then we choose a next variable in which all polynomials of $S_{[r_1,r_2]}$ are linear and continue in the same way. At each iteration we apply step (1) and take the intersection

$$S_{[r_1,r_2,\ldots,r_k]} = \cap_{1\leq i\leq k} S_{[r_1,\ldots,\hat{r_i},\ldots,r_k](r_i)}.$$

If a set $S_{[r_1,\ldots,\hat{r_i},\ldots,r_k]}$ contains a polynomial which is non-linear in $\alpha_{r_i}$, the set $S_{[r_1,\ldots,\hat{r_i},\ldots,r_k](r_i)}$ is undefined and omitted in the intersection. If this happens for all $1 \leq i \leq k$ the set $S_{[r_1,r_2,\ldots,r_k]}$ is undefined and the algorithm stops. Unless this situation occurs, we obtain for each sequence of variables $(\alpha_{r_1}, \ldots, \alpha_{r_k})$, $k \leq n$, a sequence of sets $S_{(r_1)}, S_{[r_1,r_2]}, \ldots, S_{[r_1,r_2,\ldots,r_k]}$.

DEFINITION 3.1. We say that $S$ is *Fubini reducible* (or *linearly reducible*) if there is an ordering of all $n$ variables $(\alpha_{r_1}, \ldots, \alpha_{r_n})$ such that every polynomial in $S_{[r_1,r_2,\ldots,r_k]}$ is linear in $\alpha_{r_{k+1}}$ for all $1 \leq k < n$.

The linear reducibility of the set of critical polynomials of an integrand as in eq. 2.4 is a criterion for the integral to be computable by the above approach. The criterion is sufficient but not neccessary. The sets $S_{[r_1,\ldots,r_k]}$ contain the critical polynomials of the integrand after the first $k$ integrations, but might as well contain spurious polynomials which drop out in the integration procedure. A more refined reduction algorithm presented in reference [**22**] omits such cases, but not necessarily all of them. Furthermore the occurrence of a quadratic polynomial does not always forbid us to continue with the computation.

By applying the above algorithm to first Symanzik polynomials, it was shown in [**21**] that several vacuum Feynman integrals can be computed and evaluate to combinations of multiple zeta values. Moreover the same is true for coefficients of a dimensional series expansion of Feynman integrals with two legs, obtained from cutting one edge in one of these vacuum graphs, as in the case of figure 2 (a). The linear reducibility of first Symanzik polynomials is extensively studied in terms of Dodgson polynomials in [**22**]. It is shown that the first five iterations of the reduction succeed for any first Symanzik polynomial. Moreover, a first Symanzik polynomial is reducible, if its graph has vertex width less or equal three. (This is the class of graphs which decompose into two connected components after removing three vertices or less.) These results explain, why one has to go up to complicated graphs at high loop orders to find first examples for vacuum-type Feynman integrals which exceed the set of multiple zeta values [**23, 25**].

In the following, we want to consider the above algorithm applied to both Symanzik polynomials. In order to include $\mathcal{F}$ which may depend on particle masses and external momenta, we slightly extend the above formulation of step (1), allowing $f_i$ to be polynomials whose coefficients are rational numbers or algebraic functions of additional parameters $s_1, \ldots, s_m$. The rest of the algorithm is not affected by this change.

It will be useful to consider polynomial reduction in coordinates for products of $\mathbb{P}^1$. Let $P(\alpha_1, \ldots, \alpha_n)$ be a polynomial in $n$ Feynman parameters and consider an associated function of new variables $x_1, \ldots, x_n, y_1, \ldots, y_n$ defined by

$$\bar{P}(x_1, \ldots, x_n, y_1, \ldots, y_n) = \left(\prod_{i-1}^{n} y_i^{d_i(P)}\right) P\left(\frac{x_1}{y_1}, \ldots, \frac{x_n}{y_n}\right),$$

where $d_i(P)$ denotes the maximal degree of $P$ in $\alpha_i$ for all $1 \leq i \leq n$. We obtain our original polynomial back by setting

$$\bar{P}(\alpha_1, \ldots, \alpha_n, 1, \ldots, 1) = P(\alpha_1, \ldots, \alpha_n). \tag{3.2}$$

The function $\bar{P}$ is by definition a polynomial in $x_1, \ldots, x_n, y_1, \ldots, y_n$ and it is linear in $x_k$ if and only if $P$ is linear in $\alpha_k$.

Let us consider a set of polynomials $S = \{P_1, \ldots, P_N\}$ which are all linear in $\alpha_k$. We can write

$$P_i = \alpha_k \frac{\partial}{\partial \alpha_k} P_i + P_i|_{\alpha_k = 0}.$$

Consider the corresponding set of polynomials $\bar{S} = \{\bar{P}_1, \ldots, \bar{P}_N\}$, obtained from $S$ by changing to the projective coordinates. Each of these polynomials is linear in $x_k$ and we can write

$$\bar{P}_i = x_k \bar{P}_i|_{x_k=1,\, y_k=0} + y_k \bar{P}_i|_{x_k=0,\, y_k=1}.$$

Let us use the convention that in a Fubini reduction with respect to the $x$-variables, we neglect the above prefactor $y_k$ in the sense that in eq. 3.1 the terms are given by $g_i = \bar{P}_i|_{x_k=1,\, y_k=0}$ and $h_i = \bar{P}_i|_{x_k=0,\, y_k=1}$, which does not affect the linear reducibility. Step (1) of the reduction algorithm applied to $\bar{S}$ with respect to $x_k$ then gives the set $\bar{S}_{(k)}$ consisting of the irreducible factors of

$$\begin{aligned} \bar{S}'_{(k)} = & \Big\{ \left(\bar{P}_i|_{x_k=1,\, y_k=0}\right)_{1\leq i\leq N}, \left(\bar{P}_i|_{x_k=0,\, y_k=1}\right)_{1\leq i\leq N}, \\ & \left(\bar{P}_i|_{x_k=0,\, y_k=1} \cdot \bar{P}_j|_{x_k=1,\, y_k=0} - \bar{P}_i|_{x_1=1,\, y_k=0} \cdot \bar{P}_j|_{x_k=0,\, y_k=1}\right)_{1\leq i<j\leq N} \Big\}. \end{aligned} \tag{3.3}$$

Considering a Fubini reduction in $x$-variables instead of $\alpha$-variables, we convince ourselves that the factorizations into irreducible polynomials are not affected by the change of variables. Indeed if $P$ factorizes as

$$P = f_1 \cdot f_2$$

into two polynomials $f_1$, $f_2$ then $\bar{P}$ factorizes as

$$\bar{f}_1 \cdot \bar{f}_2 = \left(\prod_{i=1}^{n} y_i^{d_i(f_1)}\right) f_1\left(\frac{x_1}{y_1}, \ldots, \frac{x_n}{y_n}\right) \cdot \left(\prod_{i=1}^{n} y_i^{d_i(f_2)}\right) f_2\left(\frac{x_1}{y_1}, \ldots, \frac{x_n}{y_n}\right) = \bar{P},$$

because of $d_i(f_1) + d_i(f_2) = d_i(P)$ for all $1 \leq i \leq n$. As a consequence we obtain:

LEMMA 3.2. *$P$ is linearly reducible with respect to the variables $\alpha_1, \ldots, \alpha_n$ if and only if $\bar{P}$ is linearly reducible with respect to the variables $x_1, \ldots, x_n$.*

Now let us choose some coordinate $x_l$, $1 \leq l \leq n$, and define two new sets of polynomials by restrictions $x_l = 0$, $y_l = 1$ and $x_l = 1$, $y_l = 0$ respectively:

$$\begin{aligned}\bar{S}^{(l,\,0,\,1)} &= \bar{S}|_{x_l=0,\, y_l=1},\\ \bar{S}^{(l,\,1,\,0)} &= \bar{S}|_{x_l=1,\, y_l=0}.\end{aligned}$$

To these sets we apply one reduction step with respect to $x_k$. At first let us assume $l \neq k$. Step (1) of the algorithm gives

$$\begin{aligned}\bar{S}^{(l,\,0,\,1)}_{(k)} &= \text{irreducible factors of } \bar{S}'_{(k)}|_{x_l=0,\, y_l=1},\\ \bar{S}^{(l,\,1,\,0)}_{(k)} &= \text{irreducible factors of } \bar{S}'_{(k)}|_{x_l=1,\, y_l=0}.\end{aligned}$$

If a polynomial $\bar{P}$ factorizes as $\bar{P} = f_1 \cdot f_2$ then furthermore

$$\begin{aligned}\bar{P}|_{x_l=0,\, y_l=1} &= f_1|_{x_l=0,\, y_l=1} \cdot f_2|_{x_l=0,\, y_l=1},\\ \bar{P}|_{x_l=1,\, y_l=0} &= f_1|_{x_l=1,\, y_l=0} \cdot f_2|_{x_l=1,\, y_l=0},\end{aligned}$$

and the maximal degrees with respect to another variable $x_i$ satisfy $d_i(f_j) \geq d_i\left(f_j|_{x_l=0,\, y_l=1}\right)$ and $d_i(f_j) \geq d_i\left(f_j|_{x_l=1,\, y_l=0}\right)$ for $j = 1, 2$ and for all $i = 1, \ldots, n$. This means that for each polynomial $f \in \bar{S}_{(k)}$ there is at most one polynomial $f' \in \bar{S}^{(l,\,0,\,1)}_{(k)}$, and $d_i(f) \geq d_i(f')$ for all $i = 1, \ldots, n$. The same is true for $\bar{S}^{(l,\,1,\,0)}_{(k)}$.

Lastly we observe that in the case of $l = k$, the irreducible factors of $\bar{S}^{(l,\,0,\,1)}$ and $\bar{S}^{(l,\,1,\,0)}$ are already contained in $\bar{S}_{(k)}$. This proves the following lemma:

LEMMA 3.3. If $\bar{S}$ is linearly reducible with the ordering $(x_{r_1}, \ldots, x_{r_n})$ then for any $1 \leq l \leq n$ the sets $\bar{S}|_{x_l=0,\, y_l=1}$ and $\bar{S}|_{x_l=1,\, y_l=0}$ are linearly reducible with $(x_{r_1}, \ldots, \hat{x}_l, \ldots, x_{r_n})$.

In combination with the previous lemma we obtain:

LEMMA 3.4. *Let* $S = \{P_1, \ldots, P_N\}$ *be a set of polynomials which is linearly reducible with* $(\alpha_{r_1}, \ldots, \alpha_{r_n})$ *and whose members are linear in* $\alpha_l$. *Then the sets* $S^l = \left\{\frac{\partial P_1}{\partial \alpha_l}, \ldots, \frac{\partial P_N}{\partial \alpha_l}\right\}$ *and* $S_l = \{P_1|_{\alpha_l=0}, \ldots, P_N|_{\alpha_l=0}\}$ *are linearly reducible with* $(\alpha_{r_1}, \ldots, \hat{\alpha}_l, \ldots, \alpha_{r_n})$.

## 4. Towards a Classification by Critical Minors

Let $G$ be a graph and $E_G$ its set of edges. For $e \in G$ we denote by $G \backslash e$ the graph obtained from $G$ by *deletion* of $e$. Furthermore we write $G//e$ for the graph obtained from $G$ by *contraction* of $e$. This is the graph where the end-points of $e$ are identified and then $e$ is removed. For any distinct edges $e_1, e_2 \in E_G$ the operations of deleting (or contracting) $e_1$ and deleting (or contracting) $e_2$ commute. Therefore we can more generally write $\gamma = G \backslash D//C$ with distinct $D, C \subset E_G$, for the unique graph obtained from $G$ by deleting all edges in $D$ and contracting all edges in $C$. Any such $\gamma$ is called a *minor* of $G$.

If $G$ is connected and there is an edge $e$ such that $G \backslash e$ is disconnected then $e$ is called a *bridge*. When speaking of Feynman graphs, we may ignore disconnected graphs, so we introduce the convention that the corresponding Symanzik polynomial are zero:

$$\mathcal{U}_{G\backslash e} = 0,\ \mathcal{F}_{0,\, G\backslash e} = 0 \text{ if } e \text{ is a bridge.}$$

Let us furthermore call $e_t \in E_G$ a *tadpole* if it is attached to the same vertex at both ends. In this case we have a factorization in the corresponding Feynman parameter:

$$\begin{aligned} \mathcal{U}_G &= \mathcal{U}_{G\backslash e_t}\alpha_t, \\ \mathcal{F}_{0,\,G} &= \mathcal{F}_{0,\,G\backslash e_t}\alpha_t. \end{aligned}$$

For any $e$ which is not a tadpole we have the well-known deletion/contraction identities:

$$\begin{aligned} \mathcal{U}_G &= \mathcal{U}_{G\backslash e}\alpha_e + \mathcal{U}_{G//e}, \\ \mathcal{F}_{0,\,G} &= \mathcal{F}_{0,\,G\backslash e}\alpha_e + \mathcal{F}_{0,\,G//e}. \end{aligned}$$

Now we consider a set of arbitrary particle masses $m_1$, ..., $m_n$ distributed over the edges of $G$, with the restriction that at least one edge $e_k$ is massless, $m_k = 0$. Then the second Symanzik polynomial

$$\mathcal{F}_G = \mathcal{F}_{0,\,G} + \mathcal{U}_G \cdot \sum_{i\neq k} \alpha_i m_i^2$$

is linear in $\alpha_k$ and the above relations are extended by

$$\begin{aligned} \mathcal{F}_G &= \mathcal{F}_{G\backslash e_k}\alpha_k \text{ if } e_k \text{ is a tadpole,} \\ \mathcal{F}_G &= \mathcal{F}_{G\backslash e_k}\alpha_k + \mathcal{F}_{G//e_k} \text{ if } e_k \text{ is not a tadpole.} \end{aligned}$$

Equivalently, if $e_k$ is a massless edge of $G$ and not a tadpole then

(4.1) $$\frac{\partial}{\partial\alpha_k}\mathcal{U}_G = \mathcal{U}_{G\backslash e_k},\ \mathcal{U}_G|_{\alpha_k=0} = \mathcal{U}_{G//e_k},$$

(4.2) $$\frac{\partial}{\partial\alpha_k}\mathcal{F}_G = \mathcal{F}_{G\backslash e_k},\ \mathcal{F}_G|_{\alpha_k=0} = \mathcal{F}_{G//e_k}.$$

In the following let us call a Feynman graph $G$ linearly reducible, if $\{\mathcal{U}_G, \mathcal{F}_G\}$ is linearly reducible. We arrive at the main statement of these notes:

THEOREM 4.1. *If $G$ is a linearly reducible Feynman graph then any minor of $G$ is linearly reducible as well.*

PROOF. Let $G$ be an arbitrary Feynman graph, $e_j$ any of its edges and $\tilde{G} = G|_{m_j=0}$ the graph obtained from $G$ by setting the mass $m_j$ associated to $e_j$ equal to zero. Assume that $G$ is linearly reducible. Then $G$ is linearly reducible for any value of $m_j$ and therefore $\tilde{G}$ is linearly reducible as well. As $e_j$ is massless in $\tilde{G}$ we can apply equations 4.1 and 4.2 and obtain, if $e_j$ is not a tadpole,

$$\begin{aligned} S_{\tilde{G}\backslash e_j} &= \left\{\mathcal{U}_{\tilde{G}\backslash e_j}, \mathcal{F}_{\tilde{G}\backslash e_j}\right\} = \left\{\frac{\partial}{\partial\alpha_j}\mathcal{U}_{\tilde{G}}, \frac{\partial}{\partial\alpha_j}\mathcal{F}_{\tilde{G}}\right\}, \\ S_{\tilde{G}//e_j} &= \left\{\mathcal{U}_{G//e_j}, \mathcal{F}_{G//e_j}\right\} = \left\{\mathcal{U}_{\tilde{G}}|_{\alpha_j=0}, \mathcal{F}_{\tilde{G}}|_{\alpha_j=0}\right\}. \end{aligned}$$

If $e_j$ is a tadpole we have

$$S_{\tilde{G}//e_j} = S_{\tilde{G}\backslash e_j} = \left\{\frac{\partial}{\partial\alpha_j}\mathcal{U}_{\tilde{G}}, \frac{\partial}{\partial\alpha_j}\mathcal{F}_{\tilde{G}}\right\}.$$

By lemma 3.4 it follows from the linear reducibility of $\tilde{G}$ that $\tilde{G}\backslash e_j$ and $\tilde{G}//e_j$ are linearly reducible. $G$ and $\tilde{G}$ have the same minors with respect to deleting or contracting $e_j$ :

$$G\backslash e_j = \tilde{G}\backslash e_j,\ G//e_j = \tilde{G}//e_j$$

and therefore $S_{G\setminus e_j} = S_{\tilde{G}\setminus e_j}$ and $S_{G//e_j} = S_{\tilde{G}//e_j}$. By induction, this proves the theorem. $\square$

Let us remark again that our discussion strictly restricts to the question of the reducibility of polynomials and leaves the problem of possible divergences of corresponding Feynman integrals untouched. For example, when we set a particle mass equal to zero, as in the proof above, we do not change the reducibility behaviour of a polynomial for the worse, but on the level of integrals, such an operation could introduce infrared divergences.

A set of graphs $\mathcal{G}$ is called *minor closed* if for all $G \in \mathcal{G}$ every minor of $G$ belongs to $\mathcal{G}$ as well. Let $\mathcal{H}$ be any set of graphs and let $\mathcal{G}_{\mathcal{H}}$ be the set of all graphs which do not have a minor in the set $\mathcal{H}$. Then the set $\mathcal{G}_{\mathcal{H}}$ is minor closed, and the graphs in $\mathcal{H}$ are called *forbidden minors* of $\mathcal{G}_{\mathcal{H}}$. A theorem of Robertson and Seymour [**57**] states that any minor closed set of graphs can be defined in such a way by a finite set of forbidden minors.

A graph $H$ is called a *critical minor* of a set of graphs $\mathcal{G}$ if the minors of $H$ belong to $\mathcal{G}$ but $H$ does not. By removing a graph from $\mathcal{H}$ which is a minor of another graph in $\mathcal{H}$, we do not change the set $\mathcal{G}_{\mathcal{H}}$. Therefore, to define a minor closed set $\mathcal{G}_{\mathcal{H}}$, it is sufficient to let $\mathcal{H}$ consist only of critical minors. A well-known example for a characterization by critical minors is given by the set of planar graphs. Due to a theorem of Wagner [**64**], the set of planar graphs is the set with forbidden minors $\{K_{3,3}, K_5\}$.

In general we have to be careful when adapting notions from pure graph theory to the study of Feynman graphs, which are equipped with labels and special properties. However, theorem 4.1 suggests to attempt an analogous characterization of the set of linearly reducible Feynman graphs by critical minors, i.e. by a set of Feynman graphs which are not linearly reducible but have linearly reducible minors. We conclude these notes with a first step towards such a characterization for Feynman graphs with a dependence on external momenta. Let $\Lambda$ be the set of massless Feynman graphs with four on-shell legs, attached to four distinct vertices. The on-shell condition restricts the four external momenta to satisfy $p_i^2 = 0$ such that the dependence on the external momenta can be expressed by two Mandelstam variables. By use of a Maple-implementation of the Fubini algorithm, we find linear reductions for all two-loop graphs of this class. At three loops there are several graphs in this class for which our program fails to find a linear reduction. Most of these graphs have the graph of figure 3 as a minor. This is the complete four-vertex graph $K_4$ where we attached an on-shell leg at each vertex. From direct observation of the two Symanzik polynomials one can see that the graph is not linearly reducible. As on the other hand all its minors are linearly reducible, the graph plays the role of a critical minor for the set of all linearly reducible graphs. Our case study suggests that only a few further critical minors are needed to distinguish all linearly reducible graphs of $\Lambda$ at three loops. In this way, a large class of graphs can be separated into linearly reducible and irreducible members by knowing only a small number of critical minors.

## 5. Conclusions

In this talk we reviewed the criterion of linear reducibility of Symanzik polynomials, which can be used to decide whether a corresponding Feynman integral can be computed by iteratively introducing hyperlogarithms or multiple polylogarithms.

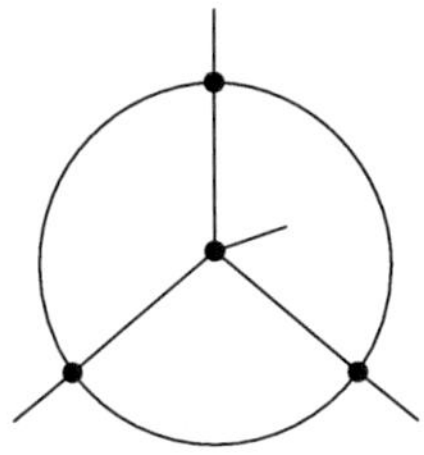

FIGURE 3. $K_4$ with four legs

Brown showed in [22] that with respect to the first Symanzik polynomial, the set of linearly reducible graphs is closed under taking minors and that therefore a characterization of this set by forbidden, critical minors is possible. We extended this line of argument to the case of graphs with masses and kinematical invariants, by showing, that minor closedness is true for both Symanzik polynomials. We exhibit a first critical minor with a non-trivial dependence on kinematical invariants.

We expect a classification with respect to the criterion of linear reducibility to be useful for the more difficult question of which Feynman integrals evaluate to multiple polylogarithms. Due to the simplicity of its derivation and the property of minor closedness, such a systematic classification of a large class of Feynman graphs appears feasible.

## References


[1] Jakob Ablinger, Johannes Blümlein, Alexander Hasselhuhn, Sebastian Klein, Carsten Schneider, and Fabian Wißbrock, *Massive 3-loop ladder diagrams for quarkonic local operator matrix elements*, Nuclear Phys. B **864** (2012), no. 1, 52–84, DOI 10.1016/j.nuclphysb.2012.06.007. MR2950979

[2] Luise Adams, Christian Bogner, and Stefan Weinzierl, *The two-loop sunrise graph with arbitrary masses*, J. Math. Phys. **54** (2013), no. 5, 052303, 18, DOI 10.1063/1.4804996. MR3098926

[3] Paolo Aluffi and Matilde Marcolli, *Feynman motives of banana graphs*, Commun. Number Theory Phys. **3** (2009), no. 1, 1–57, DOI 10.4310/CNTP.2009.v3.n1.a1. MR2504753 (2010d:81109)

[4] Paolo Aluffi and Matilde Marcolli, *Feynman motives and deletion-contraction relations*, Topology of algebraic varieties and singularities, Contemp. Math., vol. 538, Amer. Math. Soc., Providence, RI, 2011, pp. 21–64, DOI 10.1090/conm/538/10594. MR2777815 (2012a:81100)

[5] Paolo Aluffi and Matilde Marcolli, *Parametric Feynman integrals and determinant hypersurfaces*, Adv. Theor. Math. Phys. **14** (2010), no. 3, 911–963. MR2801414 (2012e:81075)

[6] C. Anastasiou, C. Duhr, F. Dulat and B. Mistlberger, *Soft triple-real radiation for Higgs production at N3LO*, [hep-ph/1302.4379].

[7] S. Bauberger, F.A. Berends, M. Böhm, M. Buza and G. Weiglein, *Calculation of two-loop self-energies in theelectroweak Standard Model*, Nucl. Phys. B (Proc. Suppl.) 37B (1994) 95-114.

[8] S. Bauberger, F.A. Berends, M. Böhm and M. Buza, *Analytical and numerical methods for massive two-loop self-energy diagrams*, Nucl. Phys. B 434 (1995) 383-407.

[9] Prakash Belkale and Patrick Brosnan, *Periods and Igusa local zeta functions*, Int. Math. Res. Not. **49** (2003), 2655–2670, DOI 10.1155/S107379280313142X. MR2012522 (2004h:11097)

[10] M. C. Bergère and Yuk-Ming P. Lam, *Asymptotic expansion of Feynman amplitudes. I. The convergent case*, Comm. Math. Phys. **39** (1974), 1–32. MR0366271 (51 #2519)

[11] I. Bierenbaum and S. Weinzierl, *The massless two-loop two-point function*, Eur.Phys.J. C32 (2003) 67-78, hep-ph/0308311.

[12] T. Binoth and G. Heinrich, *An automatized algorithm to compute infrared divergent multi-loop integrals*, Nuclear Phys. B **585** (2000), no. 3, 741–759, DOI 10.1016/S0550-3213(00)00429-6. MR1784111 (2001i:81180)

[13] Spencer Bloch, Hélène Esnault, and Dirk Kreimer, *On motives associated to graph polynomials*, Comm. Math. Phys. **267** (2006), no. 1, 181–225, DOI 10.1007/s00220-006-0040-2. MR2238909 (2008a:81074)

[14] Spencer Bloch and Dirk Kreimer, *Feynman amplitudes and Landau singularities for one-loop graphs*, Commun. Number Theory Phys. **4** (2010), no. 4, 709–753, DOI 10.4310/CNTP.2010.v4.n4.a4. MR2793425 (2012k:81104)

[15] C. Bogner and F. Brown, *Symbolic integration and multiple polylogarithms*, PoS(LL2012)053, arXiv:1209.6524 [hep-ph].

[16] Christian Bogner and Stefan Weinzierl, *Periods and Feynman integrals*, J. Math. Phys. **50** (2009), no. 4, 042302, 16, DOI 10.1063/1.3106041. MR2513973 (2010g:81110)

[17] Christian Bogner and Stefan Weinzierl, *Feynman graph polynomials*, Internat. J. Modern Phys. A **25** (2010), no. 13, 2585–2618, DOI 10.1142/S0217751X10049438. MR2651267 (2012b:81071)

[18] D. J. Broadhurst and D. Kreimer, *Knots and numbers in $\phi^4$ theory to 7 loops and beyond*, Internat. J. Modern Phys. C **6** (1995), no. 4, 519–524, DOI 10.1142/S012918319500037X. MR1352337 (97a:81143)

[19] D. J. Broadhurst and D. Kreimer, *Association of multiple zeta values with positive knots via Feynman diagrams up to 9 loops*, Phys. Lett. B **393** (1997), no. 3-4, 403–412, DOI 10.1016/S0370-2693(96)01623-1. MR1435933 (98g:11101)

[20] Francis C. S. Brown, *Multiple zeta values and periods of moduli spaces $\overline{\mathfrak{M}}_{0,n}$* (English, with English and French summaries), Ann. Sci. Éc. Norm. Supér. (4) **42** (2009), no. 3, 371–489. MR2543329 (2010f:32013)

[21] Francis Brown, *The massless higher-loop two-point function*, Comm. Math. Phys. **287** (2009), no. 3, 925–958, DOI 10.1007/s00220-009-0740-5. MR2486668 (2009k:81094)

[22] F. Brown, *On the periods of some Feynman integrals*, math.AG/0910.0114.

[23] F. Brown and D. Doryn, *Framings for graph hypersurfaces*, math.AG/1301.3056.

[24] F. Brown and D. Kreimer, *Angles, scales and parametric renormalization*, hep-th/1112.1180.

[25] Francis Brown and Oliver Schnetz, *A K3 in $\phi^4$*, Duke Math. J. **161** (2012), no. 10, 1817–1862, DOI 10.1215/00127094-1644201. MR2954618

[26] S. Caron-Huot and K.J. Larsen, *Uniqueness of two-loop master contours*, JHEP 1210 (2012) 026, hep-ph/1205.0801.

[27] Federico Chavez and Claude Duhr, *Three-mass triangle integrals and single-valued polylogarithms*, J. High Energy Phys. **11** (2012), 114, front matter + 31. MR3036464

[28] Kuo Tsai Chen, *Iterated path integrals*, Bull. Amer. Math. Soc. **83** (1977), no. 5, 831–879. MR0454968 (56 #13210)

[29] K. G. Chetyrkin, A. L. Kataev, and F. V. Tkachov, *New approach to evaluation of multiloop Feynman integrals: the Gegenbauer polynomial $x$-space technique*, Nuclear Phys. B **174** (1980), no. 2-3, 345–377, DOI 10.1016/0550-3213(80)90289-8. MR591619 (83g:81031)

[30] K. G. Chetyrkin and F. T. Tkatchov, *Integration by parts: The algorithm to calculate $\beta$-functions in 4 loops*, Nucl. Phys. B192 (1981), 159.

[31] Vittorio Del Duca, Claude Duhr, E. W. Nigel Glover, and Vladimir A. Smirnov, *The one-loop pentagon to higher orders in $\epsilon$*, J. High Energy Phys. **1** (2010), 042, 61. MR2660822 (2011g:81157)

[32] Vittorio Del Duca, Claude Duhr, and Vladimir A. Smirnov, *The two-loop hexagon Wilson loop in $\mathcal{N} = 4$ SYM*, J. High Energy Phys. **5** (2010), 084, i, 119. MR2721123 (2011j:81235)

[33] C. Duhr, H. Gangl and J. Rhodes, *From polygons and symbols to polylogarithmic functions*, math-ph/1110.0458.

[34] T. Gehrmann and E. Remiddi, *Differential Equations for Two-Loop Four-Point Functions*,

[35] A. B. Goncharov, *A simple construction of Grassmannian polylogarithms*, math.AG/0908.2238v2.

[36] A. B. Goncharov, *Multiple polylogarithms and mixed Tate motives*, math.AG/0103059v4.

[37] A. B. Goncharov, M. Spradlin, C. Vergu, and A. Volovich, *Classical polylogarithms for amplitudes and Wilson loops*, Phys. Rev. Lett. **105** (2010), no. 15, 151605, 4, DOI 10.1103/PhysRevLett.105.151605. MR2734443 (2011k:81206)

[38] J. Gracey, *Conformal methods for massless Feynman integrals and large $N_f$ methods*, [hep/ph:1301.7583].

[39] T. Hahn and P. Lang, *FeynEdit - a tool for drawing Feynman diagrams*, Comput.Phys.Commun.179:931-935,2008, hep-ph/0711.1345.

[40] T. Huber and D. Maitre, *HypExp: A Mathematica package for expanding hypergeometric functions around integer-valued parameters*, Comput.Phys.Commun. 175 (2006) 122-144, hep-ph/0507094.

[41] Tobias Huber and Daniel Maître, HypExp 2, *expanding hypergeometric functions about half-integer parameters*, Comput. Phys. Comm. **178** (2008), no. 10, 755–776, DOI 10.1016/j.cpc.2007.12.008. MR2672095

[42] C. Itzykson and J.-B. Zuber, *Quantum field theory*, Dover Publications, 2006.

[43] Maxim Kontsevich and Don Zagier, *Periods*, Mathematics unlimited—2001 and beyond, Springer, Berlin, 2001, pp. 771–808. MR1852188 (2002i:11002)

[44] A. V. Kotikov, *Differential equations method. New technique for massive Feynman diagram calculation*, Phys. Lett. B **254** (1991), no. 1-2, 158–164, DOI 10.1016/0370-2693(91)90413-K. MR1092911 (92k:81124)

[45] S. Laporta, *High-precision calculation of multiloop Feynman integrals by difference equations*, Internat. J. Modern Phys. A **15** (2000), no. 32, 5087–5159, DOI 10.1142/S0217751X00002157. MR1811366 (2001m:81194)

[46] S. Laporta and E. Remiddi, *Analytic treatment of the two loop equal mass sunrise graph*, Nuclear Phys. B **704** (2005), no. 1-2, 349–386, DOI 10.1016/j.nuclphysb.2004.10.044. MR2122179 (2005h:81238)

[47] I. Lappo-Danilevskii, *Mémoires sur la théorie des systèmes des équations différentielles linéaires*, Chelsea, New York (1953).

[48] R. N. Lee, *Space-time dimensionality $\mathcal{D}$ as complex variable: calculating loop integrals using dimensional recurrence relation and analytical properties with respect to $\mathcal{D}$*, Nuclear Phys. B **830** (2010), no. 3, 474–492, DOI 10.1016/j.nuclphysb.2009.12.025. MR2594391 (2011d:81116)

[49] Sven Moch, Peter Uwer, and Stefan Weinzierl, *Nested sums, expansion of transcendental functions, and multiscale multiloop integrals*, J. Math. Phys. **43** (2002), no. 6, 3363–3386, DOI 10.1063/1.1471366. MR1902486 (2004c:33005)

[50] Stefan Müller-Stach, Stefan Weinzierl, and Raphael Zayadeh, *A second-order differential equation for the two-loop sunrise graph with arbitrary masses*, Commun. Number Theory Phys. **6** (2012), no. 1, 203–222, DOI 10.4310/CNTP.2012.v6.n1.a5. MR2955935

[51] S. Müller-Stach, S. Weinzierl and R. Zayadeh, *Picard-Fuchs equations for Feynman integrals*, arXiv:1212.4389, (2012).

[52] N. Nakanishi, *Graph Theory and Feynman Integrals*, Gordon and Breach, 1971.

[53] Erik Panzer, *On the analytic computation of massless propagators in dimensional regularization*, Nuclear Phys. B **874** (2013), no. 2, 567–593, DOI 10.1016/j.nuclphysb.2013.05.025. MR3079414

[54] Eric Patterson, *On the singular structure of graph hypersurfaces*, Commun. Number Theory Phys. **4** (2010), no. 4, 659–708, DOI 10.4310/CNTP.2010.v4.n4.a3. MR2793424 (2012f:81093)

[55] Ettore Remiddi, *Differential equations for Feynman graph amplitudes*, Geometry Seminars, 1996–1997 (Italian) (Bologna), Univ. Stud. Bologna, Bologna, 1998, pp. 187–200. MR1635646 (99e:81163)

[56] E. Remiddi and J. A. M. Vermaseren, *Harmonic polylogarithms*, Internat. J. Modern Phys. A **15** (2000), no. 5, 725–754, DOI 10.1142/S0217751X00000367. MR1746652 (2001i:33003)

[57] N. Robertson and P. Seymour, in the series *Graph Minors*, IV-VII, IX-XII and XIV-XX, Journal of Combinatorial Theory, Series B, since 1983.

[58] R. Scharf, *Zur Berechnung von skalaren Ein- und Zwei-Schleifen-Integralen*, Universität Würzburg, diploma thesis, 1991.

[59] V.A. Smirnov, *Analytical result for dimensionally regularized massless on-shell double box*, Phys. Lett. B460 (1999), 397–404, hep-ph/9905323.

[60] O. V. Tarasov, *Connection between Feynman integrals having different values of the space-time dimension*, Phys. Rev. D (3) **54** (1996), no. 10, 6479–6490, DOI 10.1103/PhysRevD.54.6479. MR1423586 (97m:81030)

[61] O.V. Tarasov, *Generalized Recurrence Relations for Two-loop Propagator Integrals with Arbitrary Masses*, Nucl. Phys. B502 (1997), 455, hep-ph/9703319.

[62] J. B. Tausk, *Non-planar massless two-loop Feynman diagrams with four on-shell legs*, Phys. Lett. B **469** (1999), no. 1-4, 225–234, DOI 10.1016/S0370-2693(99)01277-0. MR1733866 (2000h:81073)

[63] N.I. Ussyukina, *On a Representation for Three Point Function*, Teor.Mat.Fiz. 22 (1975) 300-306.

[64] K. Wagner, *Über eine Eigenschaft der ebenen Komplexe* (German), Math. Ann. **114** (1937), no. 1, 570–590, DOI 10.1007/BF01594196. MR1513158

Institute of Physics, Humboldt Universität zu Berlin, Unter den Linden 6, 10099 Berlin, Germany

*E-mail address*: `bogner@math.hu-berlin.de`

Institute of Physics, Humboldt Universität zu Berlin, Unter den Linden 6, 10099 Berlin, Germany

*E-mail address*: `lueders@physik.hu-berlin.de`

# Comparison of motivic and simplicial operations in mod-$l$-motivic and étale cohomology

Patrick Brosnan and Roy Joshua

ABSTRACT. In this paper we explore the relationships between the motivic and simplicial cohomology operations defined on mod-$l$ motivic cohomology. We also explore similar relationships in étale cohomology and conclude by considering certain operations that commute with proper push-forwards.

## 1. Introduction

Throughout the paper $k$ will denote a fixed perfect field of characteristic $p \geq 0$. We will restrict to the category, $\mathrm{Sm}/k$, of smooth separated schemes of finite type over $k$. If $X$ denotes such a scheme, $H^n_{\mathcal{M}}(X, \mathbb{Z}(r))$ denotes the motivic cohomology with *degree* $n$ and *weight* $r$ and $H^n_{\mathcal{M}}(X, \mathbb{Z}/l(r))$ denotes the corresponding mod-$l$-variant. In [**Voev1**], Voevodsky defined a sequence of operations on motivic cohomology analogous to the operations defined by Steenrod on the singular cohomology of topological spaces. We call Voevodsky's operations *motivic operations*. They have the form

$$
\begin{aligned}
(1.0.1) \qquad & P^r_{\mathcal{M}} : H^i_{\mathcal{M}}(X, \mathbb{Z}/l(j)) \rightarrow H^{i+2r(l-1)}_{\mathcal{M}}(X, \mathbb{Z}/l(j + r(l-1))) \text{ and} \\
& \beta P^r_{\mathcal{M}} : H^i_{\mathcal{M}}(X, \mathbb{Z}/l(j)) \rightarrow H^{i+2r(l-1)+1}_{\mathcal{M}}(X, \mathbb{Z}/l(j + r(l-1))),
\end{aligned}
$$

where $l$ is a prime not equal to $p$.

When $i = 2j$, the above motivic cohomology groups identify with the usual mod-$l$ Chow groups. In [**Bros**], there is a construction of Steenrod operations on mod-$l$ Chow groups, which is independent of Voevodsky's (and somewhat simpler).

Voevodsky's operations are defined by making use of a geometric model for the classifying spaces of finite groups. By using a simplicial model for these classifying spaces, one obtains another sequence of operations in mod-$l$ motivic cohomology, which we call *simplicial* and which are defined even if $l = p$. These operations, which were originally introduced by I. Kriz and J.P. May in [**Kr-May**], have the

2010 *Mathematics Subject Classification.* Primary 14F42, 14F20.

The first author was supported by a grant from the NSF.

The second author was supported by the IHES and grants from the NSA and NSF at different stages of this work.

form:

$$(1.0.2)\qquad P^r_s : H^i_{\mathcal{M}}(X,\mathbb{Z}/l(j)) \to H^{i+2r(l-1)}_{\mathcal{M}}(X,\mathbb{Z}/l(jl)) \text{ and}$$
$$\beta P^r_s : H^i_{\mathcal{M}}(X,\mathbb{Z}/l(j)) \to H^{i+2r(l-1)+1}_{\mathcal{M}}(X,\mathbb{Z}/l(jl)).$$

The main goal of this paper is to explore the relationship between these two types of operations in mod-$l$ motivic cohomology. In doing so, we elaborate on the results in an earlier preprint by the authors studying these relations after inverting the Bott-element.

The comparison of the total power operations, yields a straightforward comparison between the motivic and simplicial operations for classes with degree equal to twice the weight. By observing that both the simplical and motivic operations are stable with respect to suspension in the degree, we can then deduce the comparison between classes with degree $\leq$ twice the weight. The case when the degree is strictly greater than twice the weight is more involved and makes use of the Cartan formula. The main comparison results may be summarized as follows.

THEOREM 1.1. *(See section 6 for more details.) Assume the base field has a primitive $l$-th root of unity and let $B \in H^0_{\mathcal{M}}(\mathrm{Spec}\, k, \mathbb{Z}/l(1))$ denote the motivic Bott element. Let $F$ denote a pointed simplicial sheaf on* $\mathrm{Sm}/k_{\mathrm{Nis}}$ .

(i) *Let $\alpha \in \tilde{H}^i_{\mathcal{M}}(F,\mathbb{Z}/(q)$ for any $i \leq 2q$. Then $P^r_s(\alpha) = B^{(q-r)(l-1)}P^r_{\mathcal{M}}(\alpha)$, $\beta P^r_s(\alpha) = B^{(q-r)(l-1)}\beta P^r_{\mathcal{M}}(\alpha)$.*

(ii) *Let $\alpha \in \tilde{H}^{2q+t}_{\mathcal{M}}(F,\mathbb{Z}/l(q))$, with $t > 0$.*

*If $t = 2t'$ for some integer $t'$, then*

$$B^{t'l}P^r_s(x) = B^{(q+t'-r)(l-1)}.B^{t'}P^r_{\mathcal{M}}(x) \quad \text{and}$$
$$B^{t'l}\beta P^r_s(x) = B^{(q+t'-r)(l-1)}.B^{t'}\beta P^r_{\mathcal{M}}(x), \quad 0 \leq r \leq q+t'.$$

*If $t = 2t' + 1$ for some integer $t'$, then*

$$B^{(t'+1)l}P^r_s(x) = B^{(q+t'+1-r)(l-1)}.B^{t'+1}P^r_{\mathcal{M}}(x) \quad \text{and}$$
$$B^{(t'+1)l}\beta P^r_s(x) = B^{(q+t'+1-r)(l-1)}.B^{t'+1}\beta P^r_{\mathcal{M}}(x), \quad 0 \leq r \leq q+t'+1.$$

(iii) *Both the motivic and simplicial operations extend to operations on étale cohomology with respect to the sheaf $\mu_l$. If $F$ denotes a pointed simplicial sheaf on $\mathrm{Sm}/k_{\acute{e}t}$ and $\alpha \in \tilde{H}^i_{\acute{e}t}(F,\mu_l(q))$, for any $i \geq 0$, then*

$$P^r_s(\alpha) = B^{(q-r)(l-1)}P^r_{\mathcal{M}}(\alpha), \quad \beta P^r_s(\alpha) = B^{(q-r)(l-1)}\beta P^r_{\mathcal{M}}(\alpha).$$

B. Guillou and C. Weibel have obtained a similar comparison of motivic and simplicial operations in section 5 of [**GW**] using results of Voevodsky on the motivic cohomology of Eilenberg-MacLane spaces: the results in [**GW**, sections 5, 6 and 7] seem to complement some of the results in this paper. Loosely speaking, our proof of Theorem 1.1 works by comparing the motivic classifying spaces of symmetric and cyclic groups with the simplicial ones. On the other hand, Guillou and Weibel use theorems of Voevodsky to show that, in appropriate bi-degrees, the group of operations on motivic cohomology injects into the group of operations on étale cohomology. (See [**GW**, Proposition 5.8].)

One may consult Examples 6.5 for various examples of the relations in Theorem 1.1.

**Outline.** We begin section 2 by reviewing quickly the cohomology of the classifying spaces of finite groups, using both the geometric and simplicial models for the classifying spaces. We discuss the total power operations in detail in the next two sections. First we recall the total power operations defined by Voevodsky for defining the motivic operations. We show that this may be modified to define total power operations for the simplicial operations, at least for classes whose degree is less than or equal to twice their weight. Both of these are first defined for algebraic cycles whose degree is twice their weight. Since the motivic operations are stable with respect to suspension in both the degree and the weight this suffices to define the motivic operations for all classes. However, since the simplicial operations are stable with respect to suspension in only the degree, the above total power operations do not define simplicial operations except for classes with degree less than or equal to twice their weight. Therefore we define total power operations in a different manner to be able to define simplicial operations without the above restriction and then show that these new total power operations agree with the ones defined above for classes with degree equal to twice their weight. This is carried out in detail in section 4.

At this point, the usual relations among the simplicial operations, like the Cartan formulae and Adem relations are not obvious. The quickest approach to establishing these for the simplicial operations is to show that the simplicial operations defined here identify with the operations defined operadically as in [**J1**, section 5] (making use of [**May**]), where such relations are known to hold. We prove this in section 5. The next section contains the comparison theorem relating the motivic operations with the simplicial ones.

We explore some applications of the above results in the last section. Motivic operations that commute with proper push-forward have played a major role in various degree formulae. Therefore, with a view towards such applications, we construct simplicial cohomology operations that commute with proper push-forwards and work out several examples of such push-forward formulae. The existence of mod$-p$ cohomology operations in characteristic $p$ is not known, however, the simplicial operations exist and have expected properties even in this case. Therefore, as an application, we consider simplicial operations in mod$-p$ motivic cohomology in characteristic $p$, which identify with the cohomology with respect to the logarithmic de-Rham Witt complex.

Another role of the simplicial operations is the following phenomenon, which the first author first learned about in conversations with C. Weibel. Though the simplicial operations may be obtained from the motivic ones when considering the motivic cohomology of smooth schemes, this is no longer true for the motivic cohomology of general simplicial schemes. (See Examples 6.5.) Therefore, it is likely that the simplicial operations will play a non-trivial role in understanding the motivic cohomology of simplicial schemes and simplicial presheaves.

**Conventions.** We restrict to smooth separated schemes of finite type over a field $k$ and $l$ will be a fixed prime. Usually this will be assumed to be different from the characteristic of $k$, and $k$ will be assumed to be provided with a primitive $l$-th root of unity, though such hypotheses are not required to define the simplicial operations. We will also consider simplicial schemes $X_\bullet$ over such a base field, where each $X_n$ will be assumed to be a smooth scheme of finite type over $k$. The symbol $\mathrm{Sm/k}_{\mathrm{Nis}}$ (resp. $\mathrm{Sm/k}_{\acute{e}t}$) will denote the category Sm/k provided with the big Nisnevich (resp. étale) topology. If C denotes any one of these sites, SSh(C) will denote the category of all pointed simplicial sheaves on C. The symbol $\mathcal{H}$C will denote the corresponding homotopy category obtained by inverting all $\mathbb{A}^1$-weak-equivalences. Any pointed simplicial scheme $X_\bullet$, with each $X_n \in (\mathrm{Sm}/\mathrm{k})$, as well as any pointed simplicial set will be viewed as an object in each of the categories $\mathrm{SSh}(\mathrm{Sm}/\mathrm{k}_{\mathrm{Nis}})$ and $\mathrm{SSh}(\mathrm{Sm}/\mathrm{k}_{\acute{e}t})$ in the obvious manner.

The mod-$l$ motivic complex of weight $i$ will be denoted $\mathbb{Z}/l(i)$. This should be distinguished from the integers mod-$l$, which will be denoted $\mathbb{F}_l$. The symbol $H^*_{\mathcal{M}}$ will denote cohomology with respect to the motivic or mod-$l$ motivic complex computed on the Nisnevich site; $H^*_{\acute{e}t}$ will denote cohomology computed on the étale site. Often, when certain computations hold in any of these cases, we will simply use $H^*$ to denote cohomology computed on any of these sites.

## 2. Cohomology of the classifying space for a finite group

**2.1. Geometric classifying spaces.** We begin by recalling briefly the construction of the *geometric classifying space of a linear algebraic group*, which is originally due to Totaro [**Tot**]. Let G be a linear algebraic group over $S = \mathrm{Spec}\, k$, i.e., a closed subgroup-scheme in $\mathbf{GL}_n$ over $S$ for some n. For a faithful representation $i : G \to \mathbf{GL}_n$, the *geometric classifying space* $B_{gm}(G;i)$ of $G$ with respect to $i$ is defined as follows. For $m \geq 1$ let $U_m$ be the open sub-scheme of $\mathbb{A}^{nm}$ where the diagonal action of $G$ determined by $i$ is free. Let $V_m = U_m/G$ be the quotient $S$-algebraic space of the action of $G$ on $U_m$ induced by the (diagonal) action of $G$ on $\mathbb{A}^{nm}$. Since the action of $G$ on $U_m$ is free, the quotient $V_m$ is smooth. We have closed embeddings $U_m \to U_{m+1}$ and $V_m \to V_{m+1}$ corresponding to the embeddings $(Id \times \{0\})^n : \mathbb{A}^{mn} \to (\mathbb{A}^{m+1})^n$ and we set $EG^{gm} = \lim_{m\to\infty} U_m$ and $BG^{gm} = \lim_{m\to\infty} V_m$ where the colimit is taken in the category of sheaves on $(\mathrm{Sm}/\mathrm{k})_{\mathrm{Nis}}$ or on $(\mathrm{Sm}/\mathrm{k})_{\acute{e}t}$. Observe that if $G = \Sigma_n$ (or a subgroup of it) acting on $\mathbb{A}^n$ by permuting the $n$-coordinates and acting on $\mathbb{A}^{nm}$ diagonally, we may take $U_m = \{(u_1, \cdots, u_n) | u_i \in \mathbb{A}^m, u_i \neq u_j, i \neq j\}$.

**2.2. Affine replacement via the the Jouanolou trick.** In general, $U_m$ is only quasi-projective, but by the Jouanolou trick, one may replace $U_m$ by an affine scheme with a free action by $G$, which maps $G$-equivariantly onto $U_m$ and which is an affine-space bundle over $U_m$. (See, for example, [**Voev1**, p. 15].) Therefore, we will henceforth denote the above affine replacement of $U_m$ by $U_m$ itself. Then [**Nag**, Theorem 2, p.7] shows that $V_m$ (which is the geometric quotient of $U_m$ by $G$) is also affine. (These observations will be rather important for the construction of the total power operations constructed below.)

**2.3.** The equivariant motivic (étale cohomology) of a scheme $X$ with an action by $\Sigma_n$ will be defined to be $H^*_{\mathcal{M}}(E\Sigma_n^{gm} \underset{\Sigma_n}{\times} X, \mathbb{Z}/l(\star))$ $(H^*_{\acute{e}t}(E\Sigma_n^{gm} \underset{\Sigma_n}{\times} X, \mathbb{Z}/l(\star))$, respectively). The results in [**MV**, section 4] show that one may also define equivariant étale cohomology using the simplicial construction for $E\Sigma_n$.

**2.4. Motivic cohomology of the geometric classifying space.** We recall the computation of the reduced equivariant motivic cohomology of $F$, where $F$ is any pointed simplicial sheaf on $\mathrm{Sm}/k)_{\mathrm{Nis}}$ and $l$ is a prime different from $char(k) = p$. (For example, $F = X_+$ where $X$ is a given scheme with trivial action by $\Sigma_l$). (See [**Voev1**, Section 6].) The following result is Voevodsky's [**Voev1**, Proposition 6.16].

THEOREM 2.5 (Voevodsky). *Let $F$ be a pointed simplicial sheaf. Then $\tilde{H}^*_{\mathcal{M}}(F \wedge (B\Sigma_l^{gm})_+; \mathbb{Z}/l(\star))$ is a free module over $\tilde{H}^*_{\mathcal{M}}(F; \mathbb{Z}/l(\star))$with a basis $\{c\bar{d}^i, d^i | i \geq 0\}$ where*

$$\bar{d} \text{ is a class in } \tilde{H}^{2l-2}_{\mathcal{M}}((B\Sigma_l^{gm})_+; \mathbb{Z}/l(l-1))$$

*which is the mod-$l$ reduction of a class $d \in \tilde{H}^{2l-2}_{\mathcal{M}}((B\Sigma_l^{gm})_+; \mathbb{Z}(l-1))$ and*

$$c \text{ is a class in } \tilde{H}^{2l-3}_{\mathcal{M}}((B\Sigma_l^{gm})_+; \mathbb{Z}/l(l-1)) \text{ so that } \delta(c) = \bar{d}.$$

Let *cycl* denote the cycle map from motivic cohomology to étale cohomology. (By identifying the motivic cohomology with the higher Chow groups, these cycle maps identify with those defined in [**Bl**].) Now one may observe that the same computation as above holds in étale cohomology with the classes $c$ and $d$ replaced by their images under the above cycle map.

**2.6. Motivic cohomology of the simplicial classifying space.** For any discrete group $G$, we let $EG$ and $BG$ denote the standard simplicial sets (with $EG_n = G^{n+1}$ and $BG_n = G^n$). These (and other) simplicial sets give rise to simplicial sheaves in the obvious way, and we use the same notation for the simplicial set as for the corresponding simplicial sheaf.

THEOREM 2.7. *Suppose $F$ is a pointed simplicial sheaf on $(\mathrm{Sm}/k)_{\mathrm{Nis}}$. Then $\tilde{H}^*_{\mathcal{M}}(F \wedge B\Sigma_{l,+}; \mathbb{Z}/l(\star))$ is a free module over $\tilde{H}^*_{\mathcal{M}}(F; \mathbb{Z}/l(\star))$ with a basis $\{x\bar{y}^i, \bar{y}^i | i \geq 0\}$ where $\bar{y}$ is a class in $\tilde{H}^{2l-2}_{\mathcal{M}}(B\Sigma_{l,+}; \mathbb{Z}/l(0))$ which is the mod-$l$ reduction of a class $y \in \tilde{H}^{2l-2}_{\mathcal{M}}(B\Sigma_{l+}; \mathbb{Z}(0))$ and $x$ is a class in $\tilde{H}^{2l-3}_{\mathcal{M}}(B\Sigma_{l+}; \mathbb{Z}/l(0))$ so that $\delta(x) = \bar{y}$.*

PROOF. This follows from the Künneth formula of Dugger and Isaksen [**DI**, Theorem 8.6 and Remark 8.7] and the (easy) fact that

$$H^*_{\mathcal{M}}(B\Sigma_l; \mathbb{Z}/l(\star)) = H^*(\Sigma_l, \mathbb{Z}/l) \otimes H^*_{\mathcal{M}}(k, \mathbb{Z}/l(\star)),$$

where $H^*(\Sigma_l, \mathbb{Z}/l)$ denotes the group cohomology of $\Sigma_l$. □

REMARK 2.8. One may observe that the main difference between the computations in 2.5 and in 2.7 is that the classes $x$, $y$ and $\bar{y}$ have weight 0. Also it is important to observe the computations in 2.7 hold even if $l = char(k) = p$.

One may replace $F$ above with a pointed simplicial sheaf on $(\mathrm{Sm}/k)_{\acute{e}t}$ and the above cohomology with $H^*_{\acute{e}t}$ to obtain:

**2.9. Étale cohomology.** $\tilde{H}^*_{\acute{e}t}(F \wedge B\Sigma_{l,+}; \mathbb{Z}/l(\star))$ is a free module over $\tilde{H}^*_{\acute{e}t}(F; \mathbb{Z}/l(\star))$ with a basis $\{x\bar{y}^i, \bar{y}^i | i \geq 0\}$ where $\bar{y}$ is a class in $\tilde{H}^{2l-2}_{\acute{e}t}(B\Sigma_{l+}; \mathbb{Z}/l(0))$ and $x$ is a class in $\tilde{H}^{2l-3}_{\acute{e}t}(B\Sigma_{l+}; \mathbb{Z}/l(0))$ so that $\delta(x) = \bar{y}$.

REMARKS 2.10. We add some remarks here concerning the computation of motivic cohomology in Voevodsky's Theorem 2.5. First we consider the question of stability, i.e. the effect on the above computations when $B\Sigma_l^{gm}$ $(B\Sigma_l)$ is replaced by $U_N/\Sigma_l$ $(E\Sigma_l \underset{\Sigma_l}{\times} U_N)$ for $N$ large. Key results here are [**Voev1**, Lemma 3.5 and Proposition 6.1]. These first of all show the following: for any smooth scheme $X$ of finite type over $k$, the maps

(2.10.1) $\mathbb{H}_{Nis}(X_+ \wedge (U_{N+1}/\Sigma_l)_+, \mathbb{Z}/l(q)) \to \mathbb{H}_{Nis}(X_+ \wedge (U_N/\Sigma_l)_+, \mathbb{Z}/l(q))$ and

$$\mathbb{H}_{Nis}(X_+ \wedge (E\Sigma_l \underset{\Sigma_l}{\times} U_{N+1})_{+,n}, \mathbb{Z}/l(q)) \to \mathbb{H}_{Nis}(X_+ \wedge (E\Sigma_l \underset{\Sigma_l}{\times} U_N)_{+,n}, \mathbb{Z}/l(q))$$

are quasi-isomorphisms for all $n$, provided $N > q$. $\mathbb{H}_{Nis}$ denotes a hypercohomology complex, i.e. $RHom(X_+ \wedge (U_{N+1}/\Sigma_l)_+, \mathbb{Z}/l(q))$ in the first case and $RHom(X_+ \wedge (E\Sigma_l \underset{\Sigma_l}{\times} U_{N+1})_n, \mathbb{Z}/l(q))$ in the second case both computed on the Nisnevich site. If $F$ is a pointed simplicial sheaf, and $X_{\bullet,+} \to F$ is a resolution, then the same quasi-isomorphisms hold on taking the homotopy inverse limit over the $X_{m,+}$ in the first line and on taking the homotopy inverse limit over the $X_{m,+}$ and the $n$ in the second line. (If one prefers, one may replace the motivic complex $\mathbb{Z}/l(q)$ by a corresponding sheaf of abelian group spectra so that one may work in the category of spectra, where homotopy inverse limits are more familiar.) Therefore, one obtains the isomorphisms (observe the $\lim^1$-terms vanish):

$$(2.10.2)\qquad \begin{aligned} \tilde{H}^*_{\mathcal{M}}(F \wedge B\Sigma^{gm}_{l,+}, \mathbb{Z}/l(\star)) &\cong \lim_{\infty \leftarrow N} \tilde{H}^*_{\mathcal{M}}(F \wedge (U_N/\Sigma_l)_+, \mathbb{Z}/l(\star)) \text{ and} \\ \tilde{H}^*_{\mathcal{M}}(F \wedge B\Sigma_{l,+}, \mathbb{Z}/l(\star)) &\cong \lim_{\infty \leftarrow N} \tilde{H}^*_{\mathcal{M}}(F \wedge (E\Sigma_l \underset{\Sigma_l}{\times} U_N/\Sigma_l)_+, \mathbb{Z}/l(\star)). \end{aligned}$$

Now the computation in 2.5 follows by first doing a similar computation for $\Sigma_l$ replaced by $\mu_l$ and adopting a transfer argument as shown in [**Voev1**, Theorems 6.10 and 6.16].

## 3. The total power operations: I

A key step in the comparison between the motivic and simplicial cohomology operations is a thorough understanding of the total power operation. We proceed to discuss this in detail. *We will assume throughout this section that $l$ is a prime different from $char(k) = p$.*

If one only considers the case $i = 2j$, then $\tilde{H}^{2j}(X_+, \mathbb{Z}/l(j))$, for a smooth scheme $X$, identifies with the usual Chow groups of $X$ reduced mod-$l$. Then, as a first approximation, one could define the total power operation by simply sending a class

$$\alpha \in \tilde{H}^{2j}(X_+, \mathbb{Z}/l(j)) \mapsto \alpha^l$$

which defines a class in $\tilde{H}^{2jl}(B\Sigma^{gm}_{l,+} \wedge X_+, \mathbb{Z}/l(jl))$.

In order to be able to extend this total power operation as a natural transformation

$$\tilde{P} : \tilde{H}^{2j}(\quad ; \mathbb{Z}/l(j)) \to \tilde{H}^{2jl}(\quad \wedge B\Sigma^{gm}_{l+}, \mathbb{Z}/l(jl))$$

defined on all pointed simplicial sheaves on the big Zariski, Nisnevich or étale site over $k$, one needs to adopt the construction in [**Voev1**, section 5]. We will adopt this suitably modified to also define a version of total power operations when $B\Sigma^{gm}_l$ is replaced by the simplicial model $B\Sigma_l$.

Next recall the following. An augmented simplicial object $X_\bullet$ in a category C consists of a simplicial object $Y_\bullet$ in C with $Y_i = X_i, i \geq 0$ together with an object $X_{-1} \in$ C and an augmentation $\epsilon : Y_0 \to X_{-1}$ so that $d_0 \circ \epsilon = d_1 \circ \epsilon$, $i = 0, 1$.

Let $X_\bullet$ denote an augmented simplicial scheme with each $X_i$, $i \geq -1$, an scheme. Let $k[X_\bullet] = \{k[X_n]|n\}$ denote the corresponding co-ordinate ring. A *finitely generated $k[X_\bullet]$-module* is given by a collection $\{M_n|n\}$ where each $M_n$ is a finitely generated $k[X_n]$-module and provided with a compatible collection of maps $\{\phi_\alpha : \alpha^*(M_n) \to M_m\}$ for each structure map $\alpha : X_m \to X_n$ of $X_\bullet$. $M_\bullet$ will be called *a finitely generated projective module (or a vector bundle on $X_\bullet$) (a finitely generated free module (or a trivial vector bundle))* if each $M_m$ is a finitely generated projective (free, respectively) $k[X_m]$-module and each of the structure maps $\phi_\alpha$ is an isomorphism of $k[X_m]$-modules.

PROPOSITION 3.1. *Let $X_\bullet$ denote an augmented simplicial scheme so that each $X_i$, $i \geq -1$, is affine. If $M_\bullet = \{M_m | m\}$ is a finitely generated module on $X_\bullet$ which is the pull-back of a finitely generated $k[X_{-1}]$-module, then there exists a finitely generated free module $F_\bullet$ on $X_\bullet$ and a map $\phi : F_\bullet \to M_\bullet$ which is an epimorphism in each degree. In case $M_\bullet$ is the pull-back of a finitely generated projective $k[X_{-1}]$-module, one may also find a finitely generated projective $k[X_\bullet]$-module $N_\bullet$ so that $M_m \oplus N_n \cong F_n$ for all $n$ and where the last isomorphism is compatible with the structure maps of the augmented simplicial scheme.*

PROOF. Since $M_\bullet$ is the pull-back of a finitely generated $k[X_{-1}]$-module, it suffices to prove the first statement when the augmented simplicial scheme $X_\bullet$ has been replaced by the affine scheme $X_{-1}$. This is then clear since $X_{-1}$ is affine. If $N_{-1}$ is the kernel of the surjection, then $M_{-1} \oplus N_{-1} \cong F_{-1}$. This isomorphism pulls-back to a similar isomorphism $M_n \oplus N_n \cong F_n$ for each $n$ and compatible with the structure maps of the augmented simplicial scheme $X_\bullet$. □

The following results relate the geometric classifying space $B\Sigma_l^{gm}$ with the simplicial classifying space $B\Sigma_l$.

PROPOSITION 3.2. *Let $U_N$ denote the open subscheme*

$$\{(u_1, \cdots, u_l) | u_i \in \mathbb{A}^N, u_j \neq u_k, \quad j \neq k\} \text{ of } \mathbb{A}^{Nl}.$$

*For any fixed integer $N > 0$, let $gs_N : E\Sigma_l \underset{\Sigma_l}{\times} U_N \to E\Sigma_l \underset{\Sigma_l}{\times} \mathrm{Spec}\, k = B\Sigma_l$ denote the obvious map of simplicial schemes. This map induces a weak-equivalence in $\mathcal{H}\mathrm{Sm/k}_{\mathrm{Nis}}$ (and also in $\mathcal{H}\mathrm{Sm/k}_{\acute{e}t}$) on taking the colimit as $N \to \infty$. (The above homotopy categories are the homotopy categories of simplicial sheaves on the appropriate sites with $\mathbb{A}^1$-inverted.)*

PROOF. By Proposition 2.3 on page 134 of [**MV**], the colimit $U_\infty := \mathrm{colim}_N\, U_N$ in the category of Nisnevich (or étale) sheaves is $\mathbb{A}^1$-homotopy equivalent to $\mathrm{Spec}\, k$. Therefore $\mathrm{colim}_N\, E\Sigma_l \underset{\Sigma_l}{\times} U_N$ is $\mathbb{A}^1$-homotopy equivalent to $E\Sigma_l \underset{\Sigma_l}{\times} \mathrm{Spec}\, k = B\Sigma_l$. □

In view of the above proposition, we may approximate $B\Sigma_l$ by $E\Sigma_l \underset{\Sigma_l}{\times} U_N$ by taking $N$ high enough. We will let $U$ denote $U_N$ and $V$ denote $V_N$ for a large $N$.

Observe that one has an obvious augmentation

$$E\Sigma_l \underset{\Sigma_l}{\times} U_N \to U_N/\Sigma_l \tag{3.2.1}$$

One may view this diagram as an augmented simplicial scheme. Observe that the schemes $U_N$ and $V_N = U_N/\Sigma_l$ may be replaced equivariantly by affine schemes as shown in 2.2, so that Proposition 3.1 applies.

3.2.2. *A key construction used in the (motivic) total power construction.* Recall from [**Voev1**, Theorem 2.1] that the functor $X \to H^{2n}(X, \mathbb{Z}/l(n))$ is represented by the sheaf (in the $\mathbb{A}^1$-localized homotopy category of complexes of sheaves on the Nisnevich topology): $U \mapsto \mathbb{F}_{ltr}(\mathbb{A}^n)(U)/\mathbb{F}_{ltr}(\mathbb{A}^n - \{0\})(U), U \in (\mathrm{Sm/k})_{\mathrm{Nis}}$. We will denote the *presheaf* $U \mapsto \mathbb{F}_{ltr}(\mathbb{A}^n)(U)/\mathbb{F}_{ltr}(\mathbb{A}^n - \{0\})(U), U \in (\mathrm{Sm/k})_{\mathrm{Nis}}$ by $\mathrm{K}(n)^{pre}$ and the corresponding sheaf by $\mathrm{K}(n)$.

Let $X$ denote a scheme in (Sm/k) and $E, L$ vector bundles on $X$ provided with an isomorphism $\phi : E \times_X L \to \mathbb{A}^N_X$ which is the $N$-dimensional trivial bundle on $X$. Given a cycle $Z$ on $E$ with coefficients in $\mathbb{F}_l$ and finite over $X$, we consider the cycle on $L \times_X E \times_X L$ whose fiber over a point $(x, l)$ of $L$ is $(l, Z_x, l)$, where $Z_x$ denotes the fiber of $Z$ over $x \in X$. It is observed in [**Voev1**, Construction 5.1] that this is a cycle finite over $L$ and that by identifying $E \times_X L$ with $\mathbb{A}^N_X$ using the isomorphism $\phi$, one obtains a map of pointed sheaves $Th(L) \to \mathbb{F}_{l\,tr}(\mathbb{A}^N)/\mathbb{F}_{l\,tr}(\mathbb{A}^N - \{0\})$ (where $Th(L)$ denotes the Thom-space of $L$), i.e. a class in $\tilde{H}^{2N}(Th(L), \mathbb{Z}/l(N))$.

REMARK 3.3. The main point of this construction is that the vector bundle $E$ is usually a *non-trivial* vector bundle, and $X$ will be an affine scheme so that one can find a complementary vector bundle $L$ so that $E \oplus L$ is a trivial vector bundle. See 3.3.4 below where $\bar{E}$ plays the role of $E$ above with the scheme $X$ being $V = U/\Sigma_l$. In case the vector bundle $E$ was trivial, one could take $L$ to be zero-dimensional, i.e. $Th(L) = X_+$.

3.3.1. We will denote the cycle constructed above in 3.2.2 $\Delta_*(q^*(Z))$, where $q : E \underset{X}{\times} L \to E$ is the obvious projection and $\Delta : E \underset{X}{\times} L \to L \underset{X}{\times} E \underset{X}{\times} L$ is the diagonal.

Moreover, making use of the Thom-isomorphism,

$$\tilde{H}^{2rk(E)}(X_+, \mathbb{Z}/l(rk(E))) \cong \tilde{H}^{2N}(Th(L), \mathbb{Z}/l(N)),$$

one observes that this defines a class in $\tilde{H}^{2rk(E)}(X_+, \mathbb{Z}/l(dim(E)))$ which is denoted $a(Z)$ and shown to be independent of the choice of the isomorphism $\phi$ and the vector bundle $L$: see [**Voev1**, Construction 5.1]. (Here $rk(E)$ denotes the rank of $E$ as a vector bundle.)

3.3.2. *An avatar of the motivic Thom-class for trivial bundles.* For later use, we make the following observation. The Thom class of the trivial bundle $\mathbb{A}^c$ on $\operatorname{Spec} k$ corresponds to the class of the identity in $Hom(\mathbb{F}_{l\,tr}(\mathbb{A}^c)/\mathbb{F}_{l\,tr}(\mathbb{A}^c - 0), \mathbb{F}_{l\,tr}(\mathbb{A}^c)/\mathbb{F}_{l\,tr}(\mathbb{A}^c - 0))$. In fact, this corresponds to the class of the diagonal $\Delta \subseteq \mathbb{A}^c \times \mathbb{A}^c$, which is a correspondence that is finite for the projection to the first factor. We will denote this class by $\Delta_*(1)$. Next let $E$ denote a trivial vector bundle of rank $c$ on the smooth scheme $X$. Then $X \times \Delta_{\mathbb{A}^c}$ defines a correspondence on $X \times \mathbb{A}^c \times \mathbb{A}^c = E \times \mathbb{A}^c$ whose projection to $E = X \times \mathbb{A}^c$ is finite. This defines the Thom-class, $t_E$, for $E$ in $H^{2c}(Th(E), \mathbb{F}_{l\,tr}(\mathbb{A}^c)/\mathbb{F}_{l\,tr}(\mathbb{A}^c - 0))$ and will be denoted $\Delta_*(q^*(1))$ where $q : E = X \times \mathbb{A}^c \to X$ is the obvious projection.

Next let $L$ denote a vector bundle on the smooth scheme $X$. Then correspondences $C$ on $L \times \mathbb{A}^n$ whose projection to $L$ is finite and so that $C$ restricted to $(L - X) \times \mathbb{A}^n$ in fact is contained in $(L - X) \times (\mathbb{A}^n - 0)$ define classes in $Hom(Th(L), \mathbb{F}_{l\,tr}(\mathbb{A}^n)/\mathbb{F}_{l\,tr}(\mathbb{A}^n - 0))$ and hence in $H^{2n}(Th(L), \mathbb{Z}/l(n))$. Then cup-product with the Thom-class $t_E \cup [C]$ may be identified with the class $\Delta_*(q_E^*(C)) = \{(e, C, e) | e \in \mathbb{A}^c\}$ where $q_E : (L \oplus E) \times \mathbb{A}^n \to L \times \mathbb{A}^n$ is the obvious projection and $\Delta$ is the diagonal $L \times \mathbb{A}^c \times \mathbb{A}^n = (L \oplus E) \times \mathbb{A}^n \to (L \oplus E \oplus E) \times \mathbb{A}^n = L \times \mathbb{A}^c \times \mathbb{A}^c \times \mathbb{A}^n$.

Observe from 3.2.2, that one obtains the isomorphism:

$$\tilde{H}^{2i}(X_+, \mathbb{Z}/l(i)) \cong \tilde{H}^0(X_+, C_*(\mathrm{K}(i)))$$

3.3.3. *Construction of the total power operation.* We will start with a section of the presheaf $\mathrm{K}^{pre}(i)$, i.e. a cycle $Z$ on $X \times \mathbb{A}^i$ finite over $X$. Let $Z^l$ denote the $l$-th external power of $Z$: this is now a cycle on $(X \times \mathbb{A}^i)^l$. We will let $p^*(Z^l)$ denote its pull-back to $(X \times \mathbb{A}^i)^l \times U$, where $U = U_N$ for some suitably large $N$. Since $\Sigma_l$

acts freely on $U$, one may observe that the cycle $p^*(Z^l)$ descends to a unique cycle $Z'$ on $((X \times \mathbb{A}^i)^l \times U)/\Sigma_l$ equi-dimensional and finite over $(X^l \times U)/\Sigma_l$. On pulling back by the diagonal, $\Delta : X \to X^l$, one obtains the cycle $Z''$ on $X \times (\mathbb{A}^{il} \times U)/\Sigma_l$.

3.3.4. Now $\bar{E} = (\mathbb{A}^l \times U)/\Sigma_l$ is a vector bundle on $V = U/\Sigma_l$ (where $\Sigma_l$ acts diagonally) and $V$ is affine: see 2.2. Moreover, recall that we have the augmented simplicial schemes: $E\Sigma_l \underset{\Sigma_l}{\times} (\mathbb{A}^l \times U) \to (\mathbb{A}^l \times U)/\Sigma_l$ and $E\Sigma_l \underset{\Sigma_l}{\times} U \to U/\Sigma_l$ .

The pull-back of the vector bundle $\bar{E}$ to the simplicial scheme $E\Sigma_l \underset{\Sigma_l}{\times} U$ defines a vector bundle we will denote by $\hat{E}$. Since the pull-back of $\bar{E}$ to $U$ is a trivial vector bundle, it follows that $\hat{E}$ is also a trivial bundle, i.e. trivial on restriction to each $(E\Sigma_l \underset{\Sigma_l}{\times} U)_n$. (We define a vector bundle on a simplicial scheme $X_\bullet$ to be trivial, if its restriction to each $X_n$ is a trivial vector bundle.) By invoking Proposition 3.1, one may find a vector bundle $\bar{L}$ on $V = U/\Sigma_l$ so that $\bar{E} \times_V \bar{L} \cong \mathbb{A}^N_V$ for some $N$. Therefore, the pull-back $\hat{L}$ of $\bar{L}$ to $E\Sigma_l \underset{\Sigma_l}{\times} U$ also has the property that $\hat{E} \times_{E\Sigma_l \underset{\Sigma_l}{\times} U} \hat{L}$ is a trivial bundle of rank $N$.

Next we let $\tilde{E} = X \times \bar{E}$, $\tilde{L} = X \times \bar{L}$ denote the pull-backs of $\bar{E}$ and $\bar{L}$ to $X \times V$. We also let $E = X \times \hat{E}$, $L = X \times \hat{L}$ denote the corresponding vector bundles on the simplicial scheme $X \times (E\Sigma_l \underset{\Sigma_l}{\times} U)$. Then $E \times_{X \times (E\Sigma_l \underset{\Sigma_l}{\times} U)} L$ is a trivial bundle of rank $N$ on the simplicial scheme $X \times (E\Sigma_l \underset{\Sigma_l}{\times} U)$ and $\tilde{E} \times_X \tilde{L}$ is a trivial bundle of rank $N$ on $X \times V$. Moreover, $E^{\oplus i}$ ($L^{\oplus i}$) will denote the corresponding $i$-fold sums. Observe that there is a natural map $Th(L^{\oplus i}) = X_+ \wedge Th(\hat{L}^{\oplus i}) \to Th(\tilde{L}^{\oplus i}) = X_+ \wedge Th(\bar{L}^{\oplus i})$.

In this context, the same arguments as above show that a cycle $Z$ on $X \times \mathbb{A}^i$ finite over $X$ defines (pointed) maps

(3.3.5)

$$\mathcal{P}_{\mathcal{M}}(Z) : X_+ \wedge Th(\bar{L}^{\oplus i}) = Th(\tilde{L}^{\oplus i}) \to \mathbb{F}_{l_{tr}}(\mathbb{A}^{iN})/Z_{tr}(\mathbb{A}^{iN} - 0) = \mathrm{K}(iN) \quad \textit{and}$$
$$\mathcal{P}_s(Z) : X_+ \wedge Th(\hat{L}^{\oplus i}) = Th(L^{\oplus i}) \to \mathbb{F}_{l_{tr}}(\mathbb{A}^{iN})/Z_{tr}(\mathbb{A}^{iN} - \{0\}) = \mathrm{K}(iN)$$

with the latter being obtained by pre-composing the first map with the obvious map $Th(\hat{L}^{\oplus i}) \to Th(\bar{L}^{\oplus i})$.

The contravariant functoriality of the above constructions in $X$ shows first of all that the above arguments suffice to define $\mathcal{P}_{\mathcal{M}}(Z)$ and $\mathcal{P}_s(Z)$ associated to section of the sheaf $\mathrm{K}(i)$). (To see this, one recalls the sheafification process starts with a presheaf $P$ and first takes $P^+$ whose sections over a $U$ are $\underset{\to}{\lim} ker(\Gamma(U_i, P) \overset{pr_1^* - pr_2^*}{\to} \Gamma(U_i \underset{U}{\times} U_j, P))$, where $\{U_i\}$ is an open cover of $U$ and the colimit is over a cofinal system of open covers of $U$. To obtain the associated sheaf from $P$, one takes $(P^+)^+$.) Next consider the simplicial resolution of the sheaf $\mathrm{K}(i)$ by pointed smooth simplicial schemes defined as follows. (See [**Voev1**, p. 9] for details.) In degree $n$, it is given by the pointed smooth scheme

$$G_n\mathrm{K}(i) = (\sqcup_{X_0 \to X \to \cdots X_n, Z \in \Gamma(X_n, \mathrm{K}(i)_0} X_0)_+$$

Making use of the above resolution, the same contravariant functoriality of the above constructions shows that one may replace $X$ in (3.3.5) by the sheaf $\mathrm{K}(i)$) to obtain maps of (pointed) sheaves in $(\mathrm{Sm/k})_{\mathrm{Nis}}$ (and also in $SSh\mathrm{Sm/k}_{Zar}$ and

$SSh\mathrm{Sm}/\mathrm{k}_{\acute{e}t}$):

$$\begin{aligned}(3.3.6)\qquad &\mathcal{P}_{\mathcal{M}} : \mathrm{K}(i) \wedge Th(\bar{L}^{\oplus i}) \to \mathrm{K}(iN) \quad and \\ &\mathcal{P}_s : \mathrm{K}(i) \wedge Th(\hat{L}^{\oplus i}) \to \mathrm{K}(iN))\end{aligned}$$

with the latter being obtained by composing the first map with the map induced by the map $Th(\hat{L}^{\oplus i}) \to Th(\bar{L}^{\oplus i})$.

The above natural transformations of simplicial sheaves induce the natural transformations (defined on the categories $\mathcal{H}\mathrm{Sm}/\mathrm{k}_{\mathrm{Nis}}$ and $\mathcal{H}\mathrm{Sm}/\mathrm{k}_{\acute{e}t}$):

$$P'_{\mathcal{M}} : \tilde{H}^{2i}(\quad, \mathbb{Z}/l(i)) \to \tilde{H}^{2iN}(\quad \wedge Th(\bar{L}^{\oplus i}), \mathbb{Z}/l(iN)) \text{ and}$$

$$P'_s : \tilde{H}^{2i}(\quad, \mathbb{Z}/l(i)) \to \tilde{H}^{2iN}(\quad \wedge Th(\bar{L}^{\oplus i}), \mathbb{Z}/l(iN))$$

Making use of Thom-isomorphisms and observing also that $rk(E) = l$, these then correspond to

$$P'_{\mathcal{M}} : \tilde{H}^{2i}(\quad, \mathbb{Z}/l(i)) \to \tilde{H}^{2il}(\quad \wedge (U_N/\Sigma_l)_+, \mathbb{Z}/l(il)) \text{ and}$$

$$P'_s : \tilde{H}^{2i}(\quad, \mathbb{Z}/l(i)) \to \tilde{H}^{2il}(\quad \wedge (E\Sigma_l \underset{\Sigma_l}{\times} U_N)_+, \mathbb{Z}/l(il))$$

so that the latter is obtained from the former by composing with the map induced by the augmentation $E\Sigma_l \underset{\Sigma_l}{\times} U_N \to U_N/\Sigma_l$. Moreover, the isomorphisms in (2.10.2) along with [**Voev1**, Lemma 5.7] show that these are compatible as $N \to \infty$ which provides the following result.

PROPOSITION 3.4. *We obtain natural transformations:*

$$\begin{aligned}P_{\mathcal{M}} : \tilde{H}^{2i}(\quad, \mathbb{Z}/l(i)) &\to \underset{\infty \leftarrow N}{lim}\, \tilde{H}^{2il}(\quad \wedge (U_N/\Sigma_l)_+, \mathbb{Z}/l(il)) \\ &\cong \tilde{H}^{2il}(\quad \wedge (B\Sigma_l)_+^{gm}, \mathbb{Z}/l(il)) \text{ and} \\ P_s : \tilde{H}^{2i}(\quad, \mathbb{Z}/l(i)) &\to \underset{\infty \leftarrow N}{lim}\, \tilde{H}^{2il}(\quad \wedge (E\Sigma_l \underset{\Sigma_l}{\times} U_{N+}), \mathbb{Z}/l(i)) \\ &\cong \tilde{H}^{2il}(\quad \wedge B\Sigma_{l+}, \mathbb{Z}/l(il))\end{aligned}$$

*on $\mathcal{H}(\mathrm{Sm}/\mathrm{k}_{\mathrm{Nis}})_+$ (and also on $\mathcal{H}(\mathrm{Sm}/\mathrm{k}_{\acute{e}t})_+$) so that the latter is obtained from the former by composing with the map induced by the augmentation $E\Sigma_l \underset{\Sigma_l}{\times} U_N \to U_N/\Sigma_l$.*

DEFINITION 3.5. The natural transformation $P_{\mathcal{M}}$ ($P_s$) will be called the geometric total power operation (the simplicial total power operation, respectively).

**3.6. Motivic operations.** Next we recall the definition of the cohomology operations of Voevodsky. Let $F$ denote a pointed simplicial sheaf on $(\mathrm{Sm}/\mathrm{k})_{\mathrm{Nis}}$ (or on $(\mathrm{Sm}/\mathrm{k})_{Zar}$).

One starts with the total power operation :

$$(3.6.1)\qquad P_{\mathcal{M}} : \tilde{H}^{2i}_{\mathcal{M}}(F, \mathbb{Z}/l(i)) \to \tilde{H}^{2il}_{\mathcal{M}}(F \wedge (U/\Sigma_l)_+, \mathbb{Z}/l(il))$$

By the results in 2.5, $\underset{i,j}{\oplus}\tilde{H}^{il}_{\mathcal{M}}(F \wedge (U/\Sigma_l)_+, \mathbb{Z}/l(jl))$ is a free module over $\tilde{H}^{*}_{\mathcal{M}}(F, \mathbb{Z}/(\star))$ with basis given by the elements $\bar{d}^r$ and $c\bar{d}^r$, $r \geq 0$. The operations $P^r_{\mathcal{M}}$ and $\beta P^r_{\mathcal{M}}$ are defined by the formula:

$$(3.6.2) \qquad P_{\mathcal{M}}(w) = \Sigma_{r\geq 0} P^r_{\mathcal{M}}(w)\bar{d}^{i-r} + \beta P^r_{\mathcal{M}}(w) c\bar{d}^{i-r-1}, \quad w \in \tilde{H}^{2i}(F, \mathbb{Z}/l(i))$$

Observe that, so defined $P^r_{\mathcal{M}} : \tilde{H}^{2i}_{\mathcal{M}}(F, \mathbb{Z}/l(i)) \to \tilde{H}^{2i+2r(l-1)}_{\mathcal{M}}(F, \mathbb{Z}/l(i+r(l-1)))$ and
$\beta P^r_{\mathcal{M}} : \tilde{H}^{2i}_{\mathcal{M}}(F, \mathbb{Z}/l(j)) \to \tilde{H}^{2i+2r(l-1)+1}_{\mathcal{M}}(X, \mathbb{Z}/l(j + r(l-1)))$.

**Behavior under suspension**: A key observation is that, since the motivic cohomology operations are stable with respect to shifting degrees by 1, and also both degrees and weights by 1, this defines the operations $P^r_{\mathcal{M}}$ and $\beta P^r_{\mathcal{M}}$ on all $\tilde{H}^{i}_{\mathcal{M}}(F, \mathbb{Z}/l(j))$.

The simplicial operations are *not* stable with respect to suspension of weights, and therefore, one cannot define simplicial operations in general using the total power operations considered above. For this, we define a new total power operation when the simplicial model is used for the classifying spaces of finite groups. We also show that, when applied to classes with degree = twice their weight, with $l \neq char(k) = p$, these total power operations identify with the ones considered above. All of these are discussed in detail in the next section.

## 4. The total power operations: II

We proceed to define total power operations in a somewhat different manner so as to be able to define the simplicial operations on all classes. Let $\Sigma_l$ denote the symmetric group on $l$-letters and let $\pi$ denote a subgroup of $\Sigma_l$. Let $B\pi$ denote the simplicial classifying space of $\pi$ with $E\pi \to B\pi$ denoting the associated principal $\pi$-fibration. We let $\mathbb{F}_l(E\pi)$ denote the chain complex obtained by taking the free $\mathbb{F}_l$-vector space in each simplicial degree and viewing that as a chain-complex in the usual manner using the alternating sums of the face maps as the differential. We let $\mathbb{F}_l(E\pi)^\vee = Hom(\mathbb{F}_l(E\pi), \mathbb{F}_l)$ which now forms a co-chain complex (i.e. with differentials of degree +1) trivial in negative degrees.

Let $K$ denote a possibly unbounded co-chain complex. Now $K^{\otimes^l}$ is the $l$-fold tensor product of $K$:the symmetric group $\Sigma_l$ acts in the obvious manner on $K^{\otimes^l}$. Therefore, one may now form the co-chain complex:

$$\mathbb{F}_l(E\pi)^\vee \underset{\mathbb{F}_l[\pi]}{\otimes} K^{\otimes^l}$$

where the differentials of the tensor-product are induced by the differentials of the two factors in the usual manner. (Since $K$ is allowed to be an unbounded complex, one needs to exercise care in taking the above complex: strictly speaking one needs to take the homotopy inverse limit of the double co-chain complexes obtained this way: see [**J1**]. However, one may identify this with a suitable total chain-complex as is shown in [**Brow**, Appendix].) In particular, the differential, $((\mathbb{F}_l(E\pi))^\vee)^0 \underset{\mathbb{F}_l[\pi]}{\otimes} K^{\otimes^l})^n \to (\mathbb{F}_l(E\pi)^\vee \underset{\mathbb{F}_l(\pi)}{\otimes} (K^{\otimes^l})^{n+1}$ is such that if $z \in K^n$ is a cycle, then its $l$-th power $z^{\otimes^l}$ defines a cycle of degree $nl$ in the above total complex we denote by

$$\tilde{Q}(z) \in (\mathbb{F}_l(E\Sigma_l)^\vee \underset{\mathbb{F}_l(\Sigma_l)}{\otimes} K^{\otimes^l})^{nl} \tag{4.0.3}$$

We will choose the complex $K$ as follows. First we allow three distinct contexts:

(i) We work throughout on the site $(\mathrm{Sm}/k)_{Zar}$ with $H^*$ denoting cohomology on the Zariski site.

(ii) We work throughout on the site $(\mathrm{Sm}/k)_{\mathrm{Nis}}$ with $H^*$ denoting cohomology on the Nisnevich site.

(iii) We work throughout on the site $(\mathrm{Sm}/k)_{ét}$ with $H^*$ denoting cohomology on the étale site.

Next observe that the category of (possibly unbounded) co-chain complexes of abelian sheaves on any of the above two sites is a quasi-simplicial model category in the sense of [**Fausk**] and therefore it is closed under homotopy inverse limits. Let $\mathcal{H}om$ denote the internal hom in this category. Then, given co-chain complexes of abelian sheaves $M$, $N$, we let $\mathcal{RH}om(M, N) = \mathcal{H}om(M, GN)$ with $G$ denoting the homotopy inverse limit of the cosimplicial object defined by the Godement resolution computed on the appropriate site. $RHom$ will denote the corresponding external hom, i.e. where $\mathcal{H}om$ in the above definition of $\mathcal{RH}om$ has been replaced by the external hom, $Hom$.

Let $X \in (\mathrm{Sm/k})$. We let $K = \Gamma(X, \mathcal{RH}om(M, \{\mathbb{F}_l(i)))$ where $M$ is any chain complex of abelian sheaves trivial in negative degrees. Moreover, now $K = RHom(M \otimes Z(X), \{\mathbb{F}_l(i))$, where $Z(X)$ denotes the co-chain complex associated to the simplicial abelian presheaf defined by $\Gamma(U, Z(X)) = Z(\Gamma(U, X))$.

In fact we may start with a pointed simplicial sheaf $F$ in $(\mathrm{Sm/k})_{\mathrm{Nis}}$ and let $M$ denote the normalized co-chain complex obtained by taking the associated free simplicial sheaf $\mathbb{F}_l(F)$ of $\mathbb{F}_l$-vector spaces (with the base point of $F$ identified with 0) and re-indexing so that we obtain a co-chain complex. Then we define $\mathcal{RH}om(F, \{\mathbb{F}_l(i)) = \mathcal{RH}om(M, \{\mathbb{F}_l(i)) = \mathcal{H}om(M, G\{\mathbb{F}_l(i))$. The above definition makes implicit use of the adjunction between the free $\mathbb{F}_l$-vector space functor and the underlying functor sending a $\mathbb{F}_l$-vector space to the underlying set. A useful observation in this context is that the natural map $\mathbb{F}_l(S)\underset{\mathbb{F}_l}{\otimes}\mathbb{F}_l(T) \to \mathbb{F}_l(S \bigwedge T)$ is a weak-equivalence for any pointed simplicial presheaves $S$ and $T$. (One may prove this as follows. First this is clear if $S$ is a presheaf of pointed sets, i.e. it is true if $S$ is replaced by its 0-th skeleton. One may prove using ascending induction on $n$, that the above map is a weak-equivalence when $S$ is replaced by its $n$-skeleton. Finally take the colimit as $n \to \infty$ over the $n$-skeleta of $S$.) This will enable one to obtain the weak-equivalence $\mathcal{RH}om(M' \otimes M'', \{\mathbb{F}_l(i)) \simeq \mathcal{RH}om(F' \bigwedge F'', \{\mathbb{F}_l(i))$, when $M' = \mathbb{F}_l(F')$ and $M'' = \mathbb{F}_l(F'')$.

Then $K^{\otimes^l} = \Gamma(X, \mathcal{RH}om(M, \{\mathbb{F}_l(i)))^{\otimes^l} = \Gamma(X^l, \mathcal{RH}om(M, \{\mathbb{F}_l(i))^{\boxtimes^l})$ maps to $\Gamma(X, \mathcal{RH}om(M, \{\mathbb{F}_l(il)))$ by pull-back by the diagonal $\Delta : X \to X^l$. (In fact this makes use of the diagonal map $\mathbb{F}_l(F) \to \mathbb{F}_l(F)^{\otimes^l}$ and the pairing $\{\mathbb{F}_l{}^{\otimes^l} \to \{\mathbb{F}_l$.) We proceed to show this pairing is compatible with the obvious action of $\Sigma_l$. First observe that $M$ being the normalized chain complex obtained from the simplicial abelian sheaf $\mathbb{F}_l(F)$ (re-indexed so as to become a co-chain complex), has the structure of a co-algebra over the Barratt-Eccles operad as shown in [**B-F**, 2.1.1 Theorem]. $\{\mathbb{F}_l$ has the structure of an algebra over the same operad as shown in [**J1**, Theorem 1.1]. Therefore, one may readily show that these structures provide

$\mathcal{RH}om(M, \{\mathbb{F}_l)$ the structure of an algebra over the tensor product of the Barratt-Eccles operad and the Eilenberg-Zilber operad: see [**J1**, Proposition 6.4]. Therefore, the above pairing is compatible with the obvious action of $\Sigma_l$ and one obtains the obvious map

$$\mathbb{F}_l(E\Sigma_l)^\vee \underset{\mathbb{F}_l(\Sigma_l)}{\otimes} \Gamma(X, \mathcal{RH}om(\mathbb{F}_l(F), \{\mathbb{F}_l(i)))^{\otimes l}$$
$$\to \mathbb{F}_l(E\Sigma_l)^\vee \underset{\mathbb{F}_l(\Sigma_l)}{\otimes} \Gamma(X, \mathcal{RH}om(\mathbb{F}_l(F), \{\mathbb{F}_l(il))).$$

(See for example (5.1.3), which explains such pairings in more detail.) One may identify the last term with

$$\Gamma(X, \mathcal{RH}om(\mathbb{F}_l(F) \underset{\mathbb{F}_l(\Sigma_l)}{\otimes} \mathbb{F}_l(E\Sigma_l), \{\mathbb{F}_l(il)) = \Gamma(X, \mathcal{RH}om(\mathbb{F}_l(F) \otimes \mathbb{F}_l(B\Sigma_l), \{\mathbb{F}_l(il)).$$

We denote the above composition

$$\text{(4.0.4)} \qquad \begin{aligned} &\mathbb{F}_l(E\Sigma_l)^\vee \underset{\mathbb{F}_l(\Sigma_l)}{\otimes} \Gamma(X, \mathcal{RH}om(\mathbb{F}_l(F), \{\mathbb{F}_l(i)))^{\otimes l} \\ &\quad \to \Gamma(X, \mathcal{RH}om(\mathbb{F}_l(F) \otimes \mathbb{F}_l(B\Sigma_l), \{\mathbb{F}_l(il))) \end{aligned}$$

by $\bar{Q}_s$. As observed above any cycle $z \in \Gamma(X, RHom(M, \{\mathbb{F}_l(i)))$ in degree $n$ provides a cycle in degree $nl$ in the source of the last map: this cycle was denoted $\tilde{\mathcal{Q}}(z)$. Therefore, $\bar{Q}_s(\tilde{\mathcal{Q}}(z))$ defines a cycle in the target of the last map in degree $nl$. Moreover, one may show readily that if two cycles $z$ and $z' \in \Gamma(X, RHom(M, \{\mathbb{F}_l(i)))$ are such that their difference is a co-boundary, then the same holds for the cycles $\tilde{\mathcal{Q}}(z)$ and $\tilde{\mathcal{Q}}(z')$ as well as $\bar{Q}_s(\tilde{\mathcal{Q}}(z)$ and $\bar{Q}_s(\tilde{\mathcal{Q}}(z'))$: this may be proven as in [**St-Ep**, Chapter VII, Lemma 2.2]. Therefore, the above discussion provides the natural transformation

$$\text{(4.0.5)} \qquad Q_s : \tilde{H}^j(\quad, \mathbb{Z}/l(i)) \to \tilde{H}^{jl}(\quad \wedge B\Sigma_{l+}, \mathbb{Z}/l(il))$$

for all $j$ and all $i \geq 0$ on the category $\mathcal{H}\mathrm{Sm/k}_{\mathrm{Nis}}$ and $\mathcal{H}\mathrm{Sm/k}_{\acute{e}t}$. (We call this the (second) *simplicial total power operation.*)

**4.1. The simplicial operations.** The book [**Kr-May**] first introduced operations of the form (1.0.2). A detailed construction along the lines of *loc. cit.* is given in [**J1**]. However, for the comparison with the motivic cohomology operations, it is more convenient for us to define simplicial operations using the total operation $Q_s$ defined above in (4.0.5). We then compare the operations to those of [**Kr-May**] and [**J1**] in Section 5.

Let $F$ denote a pointed simplicial sheaf. The computation of $\tilde{H}^*_{\mathcal{M}}(F \wedge B\Sigma_{l+}; \mathbb{Z}/l)$ in 2.7 shows that it is a free module over $\tilde{H}^*_{\mathcal{M}}(F; \mathbb{Z}/l)$ with basis $\{x\bar{y}^i, \bar{y}^i | i \geq 0\}$. The operation $P^r_s$ $(\beta P^r_s)$ is defined by the formula:

(4.1.1)

$$Q_s(w) = \Sigma_{r \geq 0} P^r_s(w) \bar{y}^{j/2-r} + \beta P^r_s(w) x \bar{y}^{j/2-r-1}, \quad w \in \tilde{H}^j(F, \mathbb{Z}/l(i)), \text{ for all j.}$$

Observe that, so defined,

$$P^r_s : \tilde{H}^j(F, \mathbb{Z}/l(i)) \to \tilde{H}^{j+2r(l-1)}(F, \mathbb{Z}/l(il)) \text{ and}$$

$$\beta P^r_s : \tilde{H}^j(F, \mathbb{Z}/l(i)) \to \tilde{H}^{j+2r(l-1)+1}(X, \mathbb{Z}/l(il)).$$

**Behavior under suspension**. In contrast to the motivic operations, these operations are compatible with shifting the degree alone by 1. This will follow from the comparison theorem in the next section.

Next we proceed to show that, for $l \neq char(k) = p$ and for classes with degree $=$ twice the weight, the total power operation $Q_s$ identifies with the total power operation $P_s$ defined above in Proposition 3.4.

PROPOSITION 4.2. *Let $l \neq char(k) = p$. Let $\alpha \in \tilde{H}^{2i}_{\mathcal{M}}(F, \mathbb{Z}/l(i))$ denote a class. Then $Q_s(\alpha) = P_s(\alpha)$.*

PROOF. First we observe from [**Voev1**, Theorem 2.1] that since we are only considering cycles whose degree equals twice their weight, one may replace the motivic complex $\mathbb{Z}/l(m)[2m]$ by the complex of sheaves $C_*(\mathrm{K}(m))$ defined earlier.

The next key step to is to invoke the following result proved in Proposition 3.2: the map $gs_N : E\Sigma_l \underset{\Sigma_l}{\times} U_N \to E\Sigma_l \underset{\Sigma_l}{\times} \mathrm{Spec}\, k = B\Sigma_l$ of simplicial schemes induces a weak-equivalence in $\mathcal{H}\mathrm{Sm/k}_{\mathrm{Nis}}$ (and also in $\mathcal{H}\mathrm{Sm/k}_{Zar}$ and $\mathcal{H}\mathrm{Sm/k}_{\acute{e}t}$) on taking the colimit as $N \to \infty$. Therefore, and in view of Proposition 3.4, one may replace $B\Sigma_l$ in the above definition of the simplicial operations by $E\Sigma_l \underset{\Sigma_l}{\times} U$, where $U = U_N$, $N >> 0$. i.e. First the map in (4.0.4) may be replaced by the map

$$(4.2.1)\quad \mathbb{F}_l(E\Sigma_l)^\vee \underset{\mathbb{F}_l(\Sigma_l)}{\otimes} \Gamma(X, \mathcal{RH}om(\mathbb{F}_l(F) \otimes \mathbb{F}_l(U), C_*(\mathrm{K}(i))))^{\otimes l}$$
$$\to \Gamma(X, \mathcal{RH}om(\mathbb{F}_l(F) \otimes \mathbb{F}_l(E\Sigma_l) \underset{\mathbb{F}_l(\Sigma_l)}{\otimes} \mathbb{F}_l(U), C_*(\mathrm{K}(il))))$$

Therefore, the total power operation $Q_s$ may be defined as a map

$$(4.2.2)\qquad Q_s : \tilde{H}^j(\quad, C_*(\mathrm{K}(i))) \to \tilde{H}^{jl}(\quad \wedge (E\Sigma_l \underset{\Sigma_l}{\times} U)_+, C_*(\mathrm{K}(il))).$$

Next will consider the case when $F = X$ which is a smooth scheme and with a class in $\tilde{H}^{2i}(F, \{\mathbb{F}_l(i))$ coming from a section in $\Gamma(X, C_*(\mathrm{K}(i)^{pre}))$. Since $\Gamma(X, C_n(\mathrm{K}(i)^{pre}) \subseteq \Gamma(X \times \Delta[n], \mathrm{K}(i)^{pre})$, we may assume without loss of generality that $n = 0$. Therefore, such a class is represented by a cycle $Z$ on $X \times \mathbb{A}^i$ equi-dimensional and finite over $X$. One first pulls-back the cycle $Z$ to $p^*(Z^l)$ on $X \times \mathbb{A}^{il} \times U$. This cycle is invariant under the obvious action of the symmetric group $\Sigma_l$ on $\mathbb{A}^{il} \times U$ and therefore defines a cycle in

$$\mathbb{F}_l(E\Sigma_l)^\vee \underset{\mathbb{F}_l(\Sigma_l)}{\otimes} \Gamma(X, \mathcal{H}om(\mathbb{F}_l(U), \mathrm{K}(il))) = \Gamma(X, \mathcal{H}om(\mathbb{F}_l(E\Sigma_l) \underset{\mathbb{F}_l(\Sigma_l)}{\otimes} \mathbb{F}_l(U), \mathrm{K}(il))).$$

Observe that this is the total complex of the double complex defined by the cosimplicial co-chain complex: $\{\Gamma(X, \mathcal{H}om(\mathbb{F}_l(\Sigma_l^{\times^n}) \otimes \mathbb{F}_l(U), \mathrm{K}(il)))|n\}$.

This cycle will be denoted $\hat{Q}_s(Z)$. Since the cycle $Z^l$ is clearly stable under the permutation action of the symmetric group $\Sigma_l$, $\hat{Q}_s(Z)$ corresponds to a class in $\Gamma(X, \mathcal{H}om(\mathbb{F}_l(U), \mathrm{K}(il)))$ so that its pull-back to classes in $\Gamma(X, \mathcal{H}om(\mathbb{F}_l(\Sigma_l) \otimes \mathbb{F}_l(U), \mathrm{K}(il)))$ by the group-action $\mu : \Sigma_l \times U \to U$ and the projection $pr_2 : \Sigma_l \times U \to U$ are the same, i.e. it is an invariant cycle.

4.2.3. A *key observation* is that the assignment $Z \mapsto \hat{\mathcal{Q}}_s(Z)$ is contravariantly functorial in $X$ and that $\hat{\mathcal{Q}}_s(Z)$ represents the class $Q_s(Z)$.

Next let $\bar{E}$ and $\bar{L}$ denote the vector bundles on $V = U/\Sigma_l$ defined as in 3.3.4. Recall $rk(\bar{E}) = l$, $l + rk(L) = N$, $\hat{E}$, $\hat{L}$ $(E, L)$ are the corresponding pull-backs to $E\Sigma_l \underset{\Sigma_l}{\times} U$ $(X \times E\Sigma_l \underset{\Sigma_l}{\times} U$, respectively). Therefore, the same cycle as above defines the cycle

$$\begin{aligned}\Delta_*(q^*(\hat{\mathcal{Q}}_s(Z))) &\in \Gamma(X, \mathcal{H}om(\mathbb{F}_l(Th(\hat{L}_0^{\oplus i})), \mathrm{K}(iN)) \\ &= Hom(Z(X) \otimes (\mathbb{F}_l(Th(\hat{L}^{\oplus i})), \mathrm{K}(iN)).\end{aligned}$$

where $q : \hat{E}^{\oplus i} \oplus \hat{L}^{\oplus i} \to \hat{E}^{\oplus i}$ is the obvious projection and $\Delta : \hat{E}^{\oplus i} \oplus \hat{L}^{\oplus i} \to \hat{L}^{\oplus i} \oplus \hat{E}^{\oplus i} \oplus \hat{L}^{\oplus i}$ is the diagonal. Observe that the vector bundle $\bar{L}$ on $V$ corresponds to a $\Sigma_l$-equivariant vector bundle on $U$, which is none other than the degree-0-term $\hat{L}_0$ of $\hat{L}$. Now the diagonal of $\hat{L}_0$ is clearly stable by the $\Sigma_l$-action. Therefore, the class denoted $\Delta_*(q^*(\hat{\mathcal{Q}}_s(Z)))$ corresponds to a class in

$$Hom(Z(X) \otimes (\mathbb{F}_l(Th(\hat{L}_0^{\oplus i})), \mathrm{K}(iN))$$

so that its pull-back by $d_0$ and $d_1$ to classes in $\Gamma(X, \mathcal{RH}om(\mathbb{F}_l(Th(\hat{L}_1^{\oplus i})), \mathrm{K}(iN)))$ identify, i.e. it again defines a $\Sigma_l$-invariant cycle in

$$Hom(Z(X) \otimes (\mathbb{F}_l(Th(\hat{L}^{\oplus i})), \mathrm{K}(iN)).$$

Moreover, the definition of the class $\mathcal{P}_s(Z)$ as in (3.3.5) shows that, $\mathcal{P}_s(Z)$ is the class in

$$\tilde{H}^{2il}(\quad \wedge (E\Sigma_l \underset{\Sigma_l}{\times} U)_+, C_*(\mathrm{K}(il)))$$

that maps to the class $\Delta_*(q^*(\hat{\mathcal{Q}}_s(Z)))$, under the Thom-isomorphism:

$$\tilde{H}^{2il}(\quad \wedge (E\Sigma_l \underset{\Sigma_l}{\times} U)_+, C_*(\mathrm{K}(il))) \overset{Thom-isom}{\to} \tilde{H}^{2iN}(\quad \wedge Th(\hat{L}^{\oplus i}), C_*(\mathrm{K}(iN))).$$

4.2.4. Observe that it suffices to show that the class $\Delta_*(q^*(\hat{\mathcal{Q}}_s(Z)))$ is also the image under Thom-isomorphism of the class $\hat{\mathcal{Q}}_s(Z)$ in

$$H^{2il}(X_+ \wedge (E\Sigma_l \underset{\Sigma_l}{\times} U)_+, \mathbb{Z}/l(il)).$$

For this, we begin by observing that we have the following commutative diagram (where $Hom$ denotes Hom in the category of simplicial sheaves)

$$\begin{array}{ccc} Hom(X_+ \wedge Th(\hat{L}^{\oplus i}), C_*(\mathrm{K}(iN))) & \xrightarrow{\Delta_* q_E^*(\ )} & Hom(\Sigma^{2iN,iN} X_+ \wedge (E\Sigma_l \underset{\Sigma_l}{\times} U)_+, C_*(\mathrm{K}(i(N+l)))) \\ \uparrow {\scriptstyle \Delta_*(q_L^*(\ ))} & \nearrow {\scriptstyle \Delta_* q_{\mathbb{A}^{iN}}^*(\ )} & \\ Hom(X_+ \wedge (E\Sigma_l \underset{\Sigma_l}{\times} U)_+, C_*(\mathrm{K}(il))) & & \end{array}$$

where the top horizontal map denoted $\Delta_* q_E^*$ (the left-vertical map denoted $\Delta_* q_L^*$) is the map defined by the construction in 3.3.1 where $q_L : E^{\oplus i} \oplus L^{\oplus i} \to E^{\oplus i}$ $(q_E : (L^{\oplus i} \oplus E^{\oplus i}) \times \mathbb{A}^{iN} \to L^{\oplus i} \times \mathbb{A}^{iN})$ is the obvious projection. One also obtains a similar commutative diagram using the simplicial mapping space functor $Map$ instead of $Hom$.

In fact one may begin with a similar diagram involving the corresponding simplical presheaves $\mathrm{K}(s)^{pre}$ and make use of the contravariant functoriality of the construction in 3.3.1 to obtain the above diagram of simplicial sheaves. (To see this, one may again recall the sheafification process starts with a presheaf $P$ and first takes $P^+$ whose sections over a $U$ are $\lim_{\to} ker(\Gamma(U_i, P) \overset{pr_1^* - pr_2^*}{\to} \Gamma(U_i \underset{U}{\times} U_j, P))$, where $\{U_i\}$ is an open cover of $U$ and the colimit is over a cofinal system of open covers of $U$. To obtain the associated sheaf from $P$, one takes $(P^+)^+$.)

Observe that the vector bundles $\hat{E}$ and $\hat{E} \oplus \hat{L}$ and hence $\hat{E}^{\oplus i}$ and $\hat{E}^{\oplus i} \oplus \hat{L}^{\oplus i}$ are all trivial, i.e. $\hat{E}_0$ ($\hat{E}_0 \oplus \hat{L}_0$) corresponds to $U \times \mathbb{A}^l$ ( $U \times \mathbb{A}^N$, respectively) where $\Sigma_l$ acts in the obvious way by permutation of the coordinates on $\mathbb{A}^l$ and the action of $\Sigma_l$ on $\mathbb{A}^N$ is induced by its actions on $\hat{E}_0$ and $\hat{L}_0$ (which is some $\Sigma_l$-equivariant vector bundle on $U$ obtained by pull-back from the vector bundle $\bar{L}$ on $V$.) Since $E$ ($L$) is the pull-back of $\hat{E}$ ($\hat{L}$) to vector bundles over $X \times E\Sigma_l \underset{\Sigma_l}{\times} U$ it follows that $E$, $E \oplus L$, $E^{\oplus i}$ and $E^{\oplus i} \oplus L^{\oplus i}$ are all trivial vector bundles.

The map $q_{\mathbb{A}^{iN}}$ appearing in the slant map above is the projection map $q_{\mathbb{A}^{iN}}$ : $\mathbb{A}^{iN} \times E^{\oplus i} = L^{\oplus i} \oplus E^{\oplus i} \oplus E^{\oplus i} \to E^{\oplus i}$ and $\Delta$ there is the diagonal $\Delta : \mathbb{A}^{iN} \times E^{\oplus i} \to \mathbb{A}^{iN} \times \mathbb{A}^{iN} \times E^{\oplus i}$. Therefore, as observed in 3.3.2, the top horizontal map and the slant map are infact concrete realizations of taking cup-product with the corresponding Thom-classes. Finally, the commutativity of the above triangle follows from the naturality of the construction 3.2.2. It may also be seen more explicitly as follows. Let $\mathcal{Y}$ denote a section of the sheaf $C_*(\mathrm{K}(il))$ over $X_+ \wedge (E\Sigma_l \underset{\Sigma_l}{\times} U)_+$. Then we already observed that the class $\Delta_*(q_L^*(\mathcal{Y}))$ is its image in the top left corner. Therefore, on applying the map in the top horizontal row to the class $\Delta_*(q_L^*(\mathcal{Y}))$, one obtains $\Delta_*(q_E^*(\Delta_*(q_L^*(\mathcal{Y}))))$. Let the fiber over a point $(x, u) \in X \times U$ of the cycle $\mathcal{Y}$ be denoted $\mathcal{Y}_{x,u}$. Then the corresponding fiber of $\Delta_* q_L^*(\mathcal{Y})$, will be $(l_u, \mathcal{Y}_{x,u}, l_u)$ where $l_u$ is a point in the fiber of $L^{\oplus i}$ over $u$. Now the corresponding fiber of $\Delta_* q_E^*(\Delta_*(q_L^*(\mathcal{Y})))$ over $(x, u)$ will be $(l_u, e_u, \mathcal{Y}_{x,u}, l_u, e_u)$ where $e_u$ is a point of $E^{\oplus i}$ in the fiber over $u$.

Since the construction in 3.2.2 is contravariantly functorial in $X$, one may now let $X$ be replaced by any smooth scheme so that we obtain the commutative diagram of simplicial presheaves:

$$\begin{array}{ccc}
Map((\ \ )_+ \wedge Th(\hat{L}^{\oplus i}), C_*(\mathrm{K}(iN))) & \xrightarrow{\Delta_* q_E^*(\ \ )} & Map(\Sigma^{2iN, iN}(\ \ )_+ \wedge (E\Sigma_l \underset{\Sigma_l}{\times} U)_+, C_*(\mathrm{K}(i(N+l)))) \\
\uparrow {\scriptstyle \Delta_*(q_L^*(\ \ ))} & \nearrow {\scriptstyle \Delta_* q_{\mathbb{A}^{iN}}^*(\ \ )} & \\
Map((\ \ )_+ \wedge (E\Sigma_l \underset{\Sigma_l}{\times} U)_+, C_*(\mathrm{K}(il))) & &
\end{array}$$

Since the above diagram of simplicial presheaves strictly commutes, one will obtain a similar commutative triangle, when the simplicial presheaves above have been replaced by *fibrant* simplicial presheaves. (Here one may assume an injective model structure for simplcial presheaves, where every simplcial presheaf is cofibrant and the fibrations are global fibrations.) One may now observe that this corresponds upto weak-equivalence to replacing the simplicial sheaves $C_*(\mathrm{K}(s))$ appearing above, for varying $s$, with globally fibrant simplicial presheaves upto weak-equivalence. These observations result in a similar commutative triangle when the simplicial sheaves $C_*(\mathrm{K}(s))$ all have been replaced by globally fibrant simplicial sheaves upto weak-equivalence and the argument (  ) can be any simplicial scheme which is smooth in all degrees. Since any pointed simplicial sheaf has a simplicial resolution by pointed smooth schemes, it follows one may put any pointed simplicial sheaf in the argument ( ). (Observe that the vector bundles $\bar{E}$ and $\bar{L}$ as in 3.2.2 on $V$ as well as the associated vector bundles $\hat{E}$ and $\hat{L}$ on $E\Sigma_l \underset{\Sigma_l}{\times} U$ are defined independently of $X$, so that one may pull back these bundles to $(X_\bullet \times E\Sigma_l) \underset{\Sigma_l}{\times} U$. The pull-back of $\hat{E}$ ($\hat{L}$) will be denoted $E$ ($L$, respectively): clearly $L \oplus E$ will be trivial.) These result in the commutative triangle

(4.2.5)

$$\begin{array}{ccc}
H^*((\ \ )_+ \wedge Th(\hat{L}^{\oplus i}), C_*(\mathrm{K}(iN))) & \xrightarrow{\Delta_* q_E^*(\ \ )} & H^*(\Sigma^{2iN,iN}(\ \ )_+ \wedge (E\Sigma_l \underset{\Sigma_l}{\times} U)_+, C_*(\mathrm{K}(i(N+l)))) \\
\uparrow{\scriptstyle \Delta_*(q_L^*(\ \ ))} & \nearrow{\scriptstyle \Delta_* q^*_{\mathbb{A}iN}(\ \ )} & \\
H^*((\ \ )_+ \wedge (E\Sigma_l \underset{\Sigma_l}{\times} U)_+, C_*(\mathrm{K}(il))) & &
\end{array}$$

On the other hand, the naturality of Thom-isomorphisms now provides us with the following commutative triangle:

$$\begin{array}{ccc}
\tilde{H}^{2iN}((\ \ )_+ \wedge Th(\hat{L}^{\oplus i}), \mathbb{Z}/l(iN)) & \longrightarrow & \tilde{H}^{2i(N+l)}(\Sigma^{2iN,iN}(\ \ )_+ \wedge (E\Sigma_l \underset{\Sigma_l}{\times} U)_+, \mathbb{Z}/l(i(N+l))) \\
\uparrow & \nearrow{\scriptstyle \Sigma^{2i(N+l),N+l}} & \\
\tilde{H}^{2il}((\ \ )_+ \wedge (E\Sigma_l \underset{\Sigma_l}{\times} U)_+, \mathbb{Z}/l(il)) & &
\end{array}$$

The left-vertical map is Thom-isomorphism for the vector bundle $L^{\oplus i}$, the top horizontal map is Thom-isomorphism for the vector bundle $E^{\oplus i}$ and the slant-map corresponds to Thom-isomorphism with respect to the trivial bundle $L^{\oplus i} \oplus E^{\oplus i}$. As observed above, making use of 3.3.2, the top horizontal map and the slant map in the diagram (4.2.5) are in fact Thom-isomorphism. Comparing the two commutative triangles above, it follows therefore, that the left-vertical map in (4.2.5) also identifies with the corresponding Thom-isomorphism, i.e. taking cup-product with the corresponding Thom-class. Therefore, we obtain:

$$\mathcal{P}_s(Z) = \hat{\mathcal{Q}}_s(Z)).$$

Since $\hat{\mathcal{Q}}_s(Z)$ represents the class $Q_s(Z)$, this completes the proof of the proposition first when the simplicial sheaf $F$ is a smooth scheme and $\alpha \in \tilde{H}^{2i}_{\mathcal{M}}(F, \mathbb{Z}/l(i))$ . In the general case where $F$ is any simplicial sheaf, the construction in 3.2.2 extended to smooth simplicial schemes defines a class $\Delta_*(q^*(\alpha)) \in H^{2iN}(F \wedge Th(\hat{L}^{\oplus i}), \mathbb{Z}/l(iN))$. The class $P_s(\alpha)$ is the class in $H^{2il}(F \wedge E\Sigma_l \underset{\Sigma_l}{\times} U_+, \mathbb{Z}/l(il))$ that corresponds to the class $\Delta_*(q^*(\alpha))$ under Thom-isomorphism. Comparison of the above two commutative triangles will show again that the map $\alpha \mapsto \Delta_* q^*(\alpha)$ identifies with the Thom-isomorphism. This completes the proof of Proposition 4.2. □

## 5. Comparison with the operadic definition of simplicial cohomology operations: properties of simplicial operations

An $E^\infty$-structure on the motivic complex $\mathbb{A} = \oplus_i \mathbb{Z}/l(i))$ is shown to lead to a somewhat different definition of the *simplicial* cohomology operations on mod$-l$

motivic cohomology as discussed in [**J1**, Section 5] and based on the earlier work [**May**]. We will presently show that these operations are in fact identical to the simplicial operations defined above. Since the simplicial operations defined operadically readily inherit several well-known properties, we are thereby able to carry over such properties to the simplicial cohomological operations defined above. Some of these properties of the simplicial cohomology operations, for example, the Cartan formulae are used in an essential manner in the comparison results in the next section.

The only other way to establish such properties for the simplicial cohomology operations would be by a tedious step-by-step verification of these properties following the approach in [**St-Ep**]. Therefore we prefer the approach adopted here, which is far simpler.

PROPOSITION 5.1. *The cohomology operations defined above coincide with the simplicial cohomology operations defined on mod-l motivic cohomology in* [**J1**, Section 5].

PROOF. Throughout the proof we will denote the pairing between a vector-space over $\mathbb{F}_l$ and its dual by $< \;,\; >$. Recall the simplicial Barratt-Eccles operad is the operad $\{NZ(E\Sigma_n)|n\}$ where $E\Sigma_n$ denotes the simplicial bar-resolution of the finite group $\Sigma_n$ and $NZ(E\Sigma_n)$ denotes the normalized chain complex associated to the simplicial abelian group $Z(E\Sigma_n)$. The operad structure obtained this way is discussed in [**J1**]. We will assume that it is an action by the simplicial Barratt-Eccles operad on the motivic complex that provides its $E_\infty$-structure. The above action of the operad $\{NZ(E\Sigma_n)|n\}$ on the complex $\mathcal{A} = \oplus_{n\geq 0}\mathbb{Z}/l(n)$ provides us maps

$$\theta_l : NZ(E\Sigma_l) \otimes \mathcal{A}^{\otimes^l} \to \mathcal{A} \tag{5.1.1}$$

Recall that $K^\vee$ denotes $\mathcal{H}om(K, \mathbb{F}_l)$, if $K$ is any complex of $\mathbb{F}_l$-vector spaces. From the above pairing we obtain

$$\theta_l^* : NZ(E\Sigma_l) \otimes \mathcal{A}^\vee \to (\mathcal{A}^\vee)^{\otimes^l}$$

where we define $\theta_l^*(h, a^\vee)(a_1 \otimes \cdots \otimes a_l) =< \theta_l(h \otimes a_1 \otimes \cdots \otimes a_l), a^\vee >$, $a_i \in \mathcal{A}$, $a^\vee \in \mathcal{A}^\vee$ and $h \in NZ(E\Sigma_l)$. In fact these pairings provide the dual $\mathcal{A}^\vee$ with the structure of a co-algebra over the operad $\{NZ(E\Sigma_l|l\}$. It is a standard result in this situation (i.e. for co-algebras over acyclic operads) that the map $\theta_l^*$ is a chain map and is an *approximation to the diagonal map* (i.e. homotopic to the diagonal map) $\Delta : \mathcal{A}^\vee \to (\mathcal{A}^\vee)^{\otimes^l}$. (Here, as well as elsewhere in this section, we use the observation that for any vector space $V$ over $\mathbb{F}_l$, a vector $v \in V$ ( a vector $v^\vee \in V^\vee$) is determined by its pairing $< v, w >$ with all vectors $w \in V^\vee$ (its pairing $< u, v^\vee >$ with all vectors $u \in V$, respectively.).)

We next take the dual of the pairing $\theta_l^*$ to define a chain-map:

$$((\mathcal{A}^\vee)^\vee)^{\otimes^l} \to NZ(E\Sigma_l)^\vee \otimes (\mathcal{A}^\vee)^\vee.$$

The formula defining the chain map $\theta_l^*$ shows that this map sends $\mathbb{A}^{\otimes^l} \subseteq ((\mathcal{A}^\vee)^\vee)^{\otimes^l}$ to $NZ(E\Sigma_l)^\vee \otimes \mathcal{A}$. Clearly there is a pairing $NZ(E\Sigma_l)^\vee \otimes NZ(E\Sigma_l)^\vee \to NZ(E\Sigma_l)^\vee$ induced by the diagonal $\Delta : E\Sigma_l \to E\Sigma_l \times E\Sigma_l$. Tensoring the last map with $NZ(E\Sigma_l)^\vee$ and making use of this pairing provides us with the map:

(5.1.2) $$d : (NZ(E\Sigma_l))^\vee \otimes \mathcal{A}^{\otimes^l} \to (NZ(E\Sigma_l))^\vee \otimes \mathcal{A}$$

One may recall that the action of $\sigma \in \Sigma_l$ on $NZ(E\Sigma_l)$ and of $\sigma^{-1}$ on $\mathcal{A}^{\otimes^l}$ cancel out. Tracing through these actions of $\Sigma_l$ on the maps in the above steps, one concludes that the map $d$ induces a map on the quotients:

(5.1.3) $$\bar{d} : (NZ(E\Sigma_l))^\vee \underset{Z\Sigma_l}{\otimes} \mathcal{A}^{\otimes^l} \to (NZ(E\Sigma_l))^\vee \underset{Z\Sigma_l}{\otimes} \mathcal{A}$$

Now the cohomology of the complex $(NZ(E\Sigma_l))^\vee \underset{N(Z(\Sigma_l))}{\otimes} \mathcal{A}$ identifies with $H^*(B\Sigma_l; \mathbb{F}_l) \otimes H^*(\mathcal{A})$ whereas the cohomology of the complex $(NZ(E\Sigma_l))^\vee \underset{Z\Sigma_l}{\otimes} \mathcal{A}^{\otimes^l}$ identifies with the equivariant cohomology: $H^*(\mathcal{A}^{\otimes^l}, \Sigma_l; \mathbb{F}_l)$. Therefore, the map $\bar{d}$ defines a map

(5.1.4) $$\bar{d}^* : H^*(\mathcal{A}^{\otimes^l}, \Sigma_l; \mathbb{F}_l) \to H^*(B\Sigma_l; \mathbb{F}_l) \otimes H^*(\mathcal{A})$$

The formula defining $d$ also shows that the map $\bar{d}^*$ is a map of $H^*(B\Sigma_l, \mathbb{F}_l)$-modules. One may also observe readily that the $l$-th power map defines a map $H^*(\mathcal{A}) \to H^*(\mathcal{A}^{\otimes^l}, \Sigma_l; \mathbb{F}_l)$, $a \mapsto a^l$. Let $\{e_i, fe_i | i \geq 0\}$ denote a basis of the $\mathbb{F}_l$-vector space $H_*(B\Sigma_l; \mathbb{F}_l)$ dual to the basis $\{y^i, xy^i | i \geq 0\}$ for $H^*(B\Sigma_l; \mathbb{F}_l)$, i.e. $< e_i, y^j >= 0$, if $i \neq j$ and $= 1$ if $i = j$. Also $< fe_i, y^j >= 0$ for all $i$ , $j$, $< fe_i, xy^j >= 0$ for $i \neq j$ and $= 1$ for $i = j$. Observe that now we have the following computation for a class $\alpha \in H^q(\mathcal{A})$:

$$< \bar{d}^*(\alpha^l), e_i \otimes (-)^\vee >=< \bar{\theta}_l^*(e_i, (-)^\vee), \alpha^l >=< (-)^\vee, \bar{\theta}_l(e_i, \alpha^l) > \text{ and}$$

$$< \bar{d}^*(\alpha^l), fe_i \otimes (-)^\vee >=< \bar{\theta}_l^*(fe_i, (-)^\vee), \alpha^l >=< (-)^\vee, \bar{\theta}_l(fe_i, \alpha^l) >$$

where $(-)^\vee \in H^*(\mathcal{A})^\vee$ and $\bar{\theta}_l^*$ is the map induced by $\theta_l^*$ on taking homology of the corresponding complexes. (One may prove the above equalities, by observing that the map $\bar{d}^*$ is essentially the dual of $\theta_l^*$.) Since the map $\theta_l^*$ was observed to be chain homotopic to the diagonal, it follows that $\bar{d}^* = \Delta^*$ where $\Delta$ is the obvious diagonal. Therefore, the coefficient of $y^i$ $(xy^i)$ in the expansion of $\bar{d}^*(\alpha^l) \in H^*(B\Sigma_l; \mathbb{Z}/l) \otimes H^*(\mathcal{A})$ identifies with $\bar{\theta}_i(e_i, \alpha^l)$ $(\bar{\theta}_i(fe_i, \alpha^l)$, respectively). This completes the proof of the proposition □

The main point of the above comparison is to provide the following corollary where the corresponding results are shown to hold for the simplicial operations defined operadically in [**J1**, Theorem 5.3] invoking the results of [**May**].

THEOREM 5.2. *Let $F$ denote a pointed simplicial sheaf on $(\mathrm{Sm}/k)_{\mathrm{Nis}}$ in which case $H^*$ will denote cohomology computed on the Nisnevich site or on $(\mathrm{Sm}/k)_{\text{ét}}$ in which case $H^*$ will denote cohomology computed on the étale site.*

*The simplicial cohomology operations*

$Q^s : \tilde{H}^q(F, \mathbb{Z}/l(t)) \to \tilde{H}^{q+2s(l-1)}(F, \mathbb{Z}/l(l.t))$ *and*

$\beta Q^s : \tilde{H}^q(F, \mathbb{Z}/l(t)) \to \tilde{H}^{q+2s(l-1)+1}(F, \mathbb{Z}/l(l.t))$.

*satisfy the following properties:*

*(i) Contravariant functoriality: if $f : F' \to F$ is a map between simplicial sheaves, $f^* \circ Q^s = Q^s \circ f^*$*

*(ii) Let $x \in \tilde{H}^q(F, \mathbb{Z}/l(t))$. $Q^s(x) = 0$ if $2s > q$, $\beta Q^s(x) = 0$ if $2s \geq q$ and if $(q = 2s)$, then $Q^s(x) = x^l$.*

*(iii) If $\beta$ is the Bockstein, $\beta \circ Q^s = \beta Q^s$.*

*(iv) Cartan formulae: For all primes $l$,*

$$Q^s(x \otimes y) = \sum_{i+j=s} Q^i(x) \otimes Q^j(y) \text{ and}$$

$$\beta Q^s(x \otimes y) = \sum_{i+j=s} \beta Q^i(x) \otimes Q^j(y) + Q^i(x) \otimes \beta Q^j(y).$$

*(v) Adem relations For each pair of integers $i \geq 0, j \geq 0$, we let $(i,j) = \frac{(i+j)!}{i!j!}$ with the convention that $0! = 1$. We will also let $(i,j) = 0$ if $i < 0$ or $j < 0$. (See* [**May**, p. 183].*) With this terminology we obtain:*

*If $(l > 2,\ a < lb$, and $\epsilon = 0, 1)$ or if $(l = 2,\ a < lb$ and $\epsilon = 0)$ one has*

$$\beta^\epsilon Q^a Q^b = \Sigma_i (-1)^{a+i}(a - li, (l-1)b - a + i - 1)\beta^\epsilon Q^{a+b-i} Q^i \tag{5.2.1}$$

*where $\beta^0 Q^s = Q^s$ while $\beta^1 Q^s = \beta Q^s$. If $l > 2$, $a \leq lb$ and $\epsilon = 0, 1$, one also has*

$$\begin{aligned}\beta^\epsilon Q^a \beta Q^b &= (1-\epsilon)\Sigma_i (-1)^{a+i}(a - li, (l-1)b - a + i - 1)\beta Q^{a+b-i} Q^i \\ &\quad - \Sigma_i (-1)^{a+i}(a - li - 1, (l-1)b - a + i)\beta^\epsilon Q^{a+b-i} \beta Q^i\end{aligned} \tag{5.2.2}$$

*(vi) The operations $Q^s$ commute with the simplicial suspension isomorphism in $H^n(F; \mathbb{Z}/l(r)) \cong H^{n+1}(S^1_s F; \mathbb{Z}/l(r))$.*

*(vii) The operation $Q^s$ commutes with change of base fields and also with the higher cycle map into mod–$l$ étale cohomology.*

REMARK 5.3. It is important to observe that $Q^0$ is not the identity. The property (ii) above shows that in general $Q^0(x) = x^l$, if $x \in H^0(X, \mathbb{Z}/l(t)) = \tilde{H}^0(X_+, \mathbb{Z}/l(t))$ for any smooth scheme $X$ and any $t \geq 0$. This will play a major role in the comparison results in the next section.

## 6. Comparison between the motivic and simplicial operations

In view of the results established in the earlier sections we are able to provide a nearly complete comparison of the motivic and simplicial operations.

6.0.1. **The Motivic Bott element.** Throughout the rest of this section, we will assume that the field $k$ has a primitive $l$-th root of unity. Recall that we have:

$$\begin{aligned} H^p_{\mathcal{M}}(\mathrm{Spec}, \mathbb{Z}(1)) &= 0, p \neq 1 \\ &= k^*, p = 1 \end{aligned} \tag{6.0.2}$$

Now the universal coefficient sequence associated to the short exact sequence $0 \to \mathbb{Z}(1) \stackrel{\times l}{\to} \mathbb{Z}(1) \to \mathbb{Z}/l(1) \to 0$ of motivic complexes, provides the isomorphism

$$H^0_{\mathcal{M}}(\mathrm{Spec}\, k, \mathbb{Z}/l(1)) \cong \mu_l(k) \tag{6.0.3}$$

The *Motivic Bott element* is the class in $H^0_{\mathcal{M}}(\operatorname{Spec} k, \mathbb{Z}/l(1))$ corresponding under the above isomorphism to the primitive $l$-th root of unity $\zeta$. We will denote this element by $B$. Since $cycl(B) = \zeta$ in $H^*_{\acute{e}t}(\ \ , \mu_l(*))$, multiplication by the class $cycl(B)$ induces an isomorphism: $H^*_{\acute{e}t}(\ \ , \mu_l(r)) \to H^*_{\acute{e}t}(\ \ , \mu_l(r+1))$. It follows that the cycle map *cycl* induces a map of cohomology functors:

$$cycl(B^{-1}) : H^*_{\mathcal{M}}(\ \ , \mathbb{Z}/l(\star))[B^{-1}] \to H^*_{\acute{e}t}(\ \ , \mu_l(\star)). \tag{6.0.4}$$

It is shown in [**Lev**] that this map is an isomorphism on smooth schemes.

As observed above, the cohomology $H^*(B\Sigma_l^{gm}; \mathbb{Z}/l)$ maps naturally to $H^*(B\Sigma_l; \mathbb{Z}/l)$ under which the total power operation $\mathcal{P}_{\mathcal{M}}$ maps to the total power operation $\mathcal{P}_s$. Therefore, a simple comparison of the degrees and weights of the classes involved provides the following proposition.

PROPOSITION 6.1. *Assume that the base field $k$ has a primitive $l$-th root of unity. Let $\alpha \in H^{2q}_{\mathcal{M}}(X, \mathbb{Z}/l(q)$ for some $q \geq 0$ with $X \in (\mathrm{Sm}/\mathrm{k})$. Then*

$$P^r_s(\alpha) = B^{(q-r).(l-1)}.P^r_{\mathcal{M}}(\alpha), \quad \beta P^r_s(\alpha) = B^{(q-r).(l-1)}.\beta P^r_{\mathcal{M}}(\alpha)$$

*for $r \leq q$. For $r > q$, $P^r_s(\alpha) = 0 = P^r_{\mathcal{M}}(\alpha)$.*

COROLLARY 6.2. *The same relation holds for any class $\alpha \in \tilde{H}^i_{\mathcal{M}}(F, \mathbb{Z}/l(q))$ when $F$ is any pointed simplicial sheaf on $(\mathrm{Sm}/\mathrm{k})_{\mathrm{Nis}}$ provided $i \leq 2q$.*

PROOF. We will first observe that the relations hold when $i = 2q$ and $F$ is any pointed simplicial sheaf on $(\mathrm{Sm}/\mathrm{k})_{\mathrm{Nis}}$. This follows readily in view of the observation that the two total power operations $\mathcal{P}_{\mathcal{M}}$ and $\mathcal{P}_s$ are compatible as natural transformations defined on the category of all pointed simplicial sheaves on $(\mathrm{Sm}/\mathrm{k})$. Next we consider the statement when $i < 2q$. For example, if $i = 2q-1$, $\tilde{H}^{2q-1}_{\mathcal{M}}(F, \mathbb{Z}/l(q)) \cong \tilde{H}^{2q}_{\mathcal{M}}(\Sigma^1_s \wedge F, \mathbb{Z}/l(q))$. Now using the observation that both the motivic and simplicial operations are stable with respect to the suspension $\Sigma^1_s \wedge \ \ $, such a degree-suspension reduces this to the case when $i = 2q$, which has been proved already. Observe also that when $i \leq 2q$, one knows that $P^r_s(\alpha) = 0 = P^r_{\mathcal{M}}(\alpha)$ for $r > q$, (see [**Voev1**, Lemma 9.9] for a proof of the last equality) so that for the classes $\alpha$ for which $P^r_s$ is non-zero, the exponent $(q-r)(l-1)$ of $B$ is $\geq 0$. (If $i > 2q$ this may no longer be true apriori.)

In case $F$ is in fact a scheme $X \in (\mathrm{Sm}/\mathrm{k})$, the identification $H^i_{\mathcal{M}}(X, \mathbb{Z}/(q)) \cong CH^j(X, 2q-i; \mathbb{F}_l)$ shows that these groups are trivial if $i > 2q$. Therefore, it suffices to consider the case when $i \leq 2q$ in case $F$ is in fact a scheme $X \in (\mathrm{Sm}/\mathrm{k})$. □

Next we will consider what may be said about the case $i > 2q$. First observe that the Bott element $B$ defines a class in $H^0_{\mathcal{M}}(X, \mathbb{Z}/l(1))$ for any smooth scheme $X$ by pull-back. Next consider a a pointed simplicial sheaf $F$. Then one finds a resolution of $F$ by pointed simplicial schemes $X_{\bullet,+}$: see [**Voev1**, section 3]. The structure map $X_1 \to \operatorname{Spec} k$ factors through the structure map $X_0 \to \operatorname{Spec} k$, so that $B$ pulls-back to define a class (still denoted) $B \in \tilde{H}^0_{\mathcal{M}}(F, \mathbb{Z}/l(1))$.

LEMMA 6.3. *Let $F$ denote a pointed simplicial sheaf on $(\mathrm{Smt}/k)_{\mathrm{Nis}}$. Then*

(i) $Q^0(B) = B^l$.

(ii) *if $x \in \tilde{H}^q_{\mathcal{M}}(F, \mathbb{Z}/l(t))$, then $P^r_s(B.x) = B^l P^r_s(x)$ and $\beta P^r_s(B.x) = B^l \beta P^r_s(x)$ for all $x \in \tilde{H}^q_{\mathcal{M}}(F, \mathbb{Z}/l(t))$.*

PROOF. (i) Take $x = B$ in Theorem 5.2(ii). Then $q = 0 = s$ there so that $P_s^r(B) = 0$ for $r > 0$ and $Q^0(B) = B^l$. This proves (i). (ii) now follows from (i) making use of the Cartan formula in Theorem 5.2(iv). □

Our basic technique to handle the case where the degree > twice the weight (i.e. $i > 2q$) is to apply suitable weight and degree suspensions so as to reduce to the case where the degree = twice the weight. Then we handle this case by the comparison above. Both the motivic and simplicial operations commute with degree suspension, and the motivic operations commute with weight suspensions as well. The simplicial operations do not, however, commute with weight suspensions. But weight suspensions may effected by multiplying with the class $B$ and the behavior of the simplicial operations with respect to tensoring with $B$ is explained by the results above. Therefore, we obtain the extension of our comparison to classes of all degree and weight as explained below.

PROPOSITION 6.4. *Suppose* $x \in \tilde{H}_{\mathcal{M}}^{2q+t}(F, \mathbb{Z}/l(q))$, *with* $t > 0$.
*(i) If* $t = 2t'$ *for some integer* $t'$, *then*

$$B^{t'l} P_s^r(x) = B^{(q+t'-r)(l-1)}.B^{t'} P_{\mathcal{M}}^r(x) \ \ and$$
$$B^{t'l} \beta P_s^r(x) = B^{(q+t'-r)(l-1)}.B^{t'} \beta P_{\mathcal{M}}^r(x), \quad 0 \leq r \leq q + t'.$$

*(ii) If* $t = 2t' + 1$,

$$B^{(t'+1)l} P_s^r(x) = B^{(q+t'+1-r)(l-1)}.B^{t'+1} P_{\mathcal{M}}^r(x) \ \ and$$
$$B^{(t'+1)l} \beta P_s^r(x) = B^{(q+t'+1-r)(l-1)}.B^{t'+1} \beta P_{\mathcal{M}}^r(x), \quad 0 \leq r \leq q + t' + 1.$$

PROOF. To obtain (i), one first applies an iterated weight suspension $t'$-times: this is effected by multiplying $x$ by $B^{t'}$. Now the class $B^{t'} x \in \tilde{H}_{\mathcal{M}}^{2q+2t'}(F, \mathbb{Z}/l(q+t'))$, so that one may apply the comparison in Proposition 6.1 to it and obtain:

$$P_s^r(B^{t'} x) = B^{(q+t'-r)(l-1)} P_{\mathcal{M}}^r(B^{t'} x) \text{ and } \beta P_s^r(B^{t'} x) = B^{(q+t'-r)(l-1)} \beta P_{\mathcal{M}}^r(B^{t'} x).$$

Making use of Lemma 6.3, we see that $P_s^r(B^{t'} x)$ simplifies to $B^{t'l} P_s^r(x)$ while $\beta P_s^r(B^{t'} x)$ simplifies to $B^{t'l} \beta P_s^r(x)$. $P_{\mathcal{M}}^r(B^{t'} x) = B^{t'} P_{\mathcal{M}}^r(X)$ and $\beta P_{\mathcal{M}}^r(B^{t'} x) = B^{t'} \beta P_{\mathcal{M}}^r(x)$. These prove (i). To obtain (ii), one needs to apply an iterated weight suspension $t' + 1$-times followed by a degree suspension once. This produces the class $\Sigma_s^1 B^{t'+1} x \in \tilde{H}_{\mathcal{M}}^{2q+2t'+2}(\Sigma_s^1 F, \mathbb{Z}/l(q + t' + 1))$. Now one applies the comparison in Proposition 6.1 to it. Then one makes use of Lemma 6.3 to pull-out the $B$ from the left-hand-side. □

EXAMPLES 6.5. (i) Take $t = 1$. In this case one obtains

$$B^l P_s^r(x) = B^{(q+1-r)(l-1)} B P_{\mathcal{M}}^r(x) \text{ and}$$
$$B^l \beta P_s^r(x) = B^{(q+1-r)(l-1)} B \beta P_{\mathcal{M}}^r(x).$$

One may now also take $r = q$ to obtain, $B^l Q^q(x) = B^l P^q(x)$ and $B^l \beta Q^q(x) = B^l \beta P^q(x)$. Since $B$ is not invertible, multiplication by $B$ need not be injective and therefore, one cannot conclude that therefore $Q^q(x) = P^q(X)$ or that $\beta Q^q(x) = \beta P^q(x)$.

(ii) Take $t = 2$. In this case one obtains $B^l P_s^r(x) = B^{(q+1-r)(l-1)} B P_{\mathcal{M}}^r(x)$ and $B^l \beta P_s^r(x) = B^{(q+1-r)(l-1)} B \beta P_{\mathcal{M}}^r(x)$.

(iii) Take $t = 3$. In this case one obtains $B^{2l}P^r_s(x) = B^{(q+2-r)(l-1)}B^2P^r_{\mathcal{M}}(x)$ and $B^{2l}\beta P^r_s(x) = B^{(q+2-r)(l-1)}B^2\beta P^r_{\mathcal{M}}(x)$. If, in addition, $r = q+1$, then this becomes $B^{2l}P^r_s(x) = B^{l+1}P^r_{\mathcal{M}}(x)$ and $B^{2l}\beta P^r_s(x) = B^{l+1}\beta P^r_{\mathcal{M}}(x)$. Once again, since $B$ in not invertible, one cannot conclude that therefore $B^{l-1}P^r_s(x) = P^r_{\mathcal{M}}(x)$ or that $B^{l-1}\beta P^r_s(x) = \beta P^r_{\mathcal{M}}(x)$.

Observe that by the multiplicative properties of the operations and the observation that $P^r_{\mathcal{M}}(B) = 0$ if $r \geq 1$ ([**Voev1**, Lemma 9.8]):

(6.5.1)
$$P^r_{\mathcal{M}}(B^j\alpha) = B^jP^r_{\mathcal{M}}(\alpha),$$
$$\beta P^r_{\mathcal{M}}(B^j\alpha) - B^j\beta P^r_{\mathcal{M}}(\alpha).$$

The above relations show that the motivic cohomology operations above induce operations on $H^*(\quad, \mathbb{Z}/l(\star))[B^{-1}]$ in the obvious manner: we define $P^r_{\mathcal{M}}(\alpha.B^{-1}) = P^r_{\mathcal{M}}(\alpha).B^{-1}$ and $\beta P^r_{\mathcal{M}}(\alpha.B^{-1}) = \beta P^r_{\mathcal{M}}(\alpha).B^{-1}$. Next we proceed to compare these induced motivic and simplicial operations on mod$-l$ étale cohomology,

Proposition 6.6. *(Comparison of operations in mod$-l$ étale cohomology.) Assume that the base field $k$ has a primitive $l$-th root of unity. Let $F$ denote a pointed simplicial sheaf on $(\mathrm{Sm}/k)_{\acute{e}t}$. Let $\alpha \in H^i_{\acute{e}t}(F, \mu_l(q))$ for some $q \geq 0$. Then*

$$P^r_s(\alpha) = B^{(q-r).(l-1)}.P^r_{\mathcal{M}}(\alpha), \quad \beta P^r_s(\alpha) = B^{(q-r).(l-1)}.\beta P^r_{\mathcal{M}}(\alpha)$$

*for $r \leq i/2$ and all $i \geq 0$. $P^r_s(\alpha) = 0$ and $P^r_{\mathcal{M}}(\alpha) = 0$ for $r > i/2$.*

Proof. For the case $r \leq q$ this follows from Proposition 6.1. For the other cases it follows by expanding the exponents of $B$ on both sides of the formulae in Proposition 6.4 and canceling out all the powers of $B$ on the left-hand-side. □

## 7. Cohomological operations that commute with proper push-forwards and Examples

The operations considered so far commute with pull-backs only and do not commute with push-forwards by proper maps. In this section we modify the above operations to obtain operations that commute with proper-push-forwards. The goal of this discussion is to consider the analogues of degree formulae in mod-$p$ motivic cohomology: such degree formulae have played a major role in some of the applications of motivic cohomology operations. The key to this is the following formula, which follows by a deformation to the normal cone argument as shown in [**FL**, Chapter VI]. We state this for the convenience of the reader. Recall that motivic cohomology is a contravariant functor on smooth schemes. By identifying motivic cohomology with higher Chow groups, one may show the former is also covariant for proper maps.

Proposition 7.1. *Let*

$$\begin{array}{ccc} X & \xrightarrow{i} & W \\ \downarrow{\scriptstyle f} & & \downarrow{\scriptstyle g} \\ X' & \xrightarrow{i'} & W' \end{array}$$

*denote a cartesian square with all schemes smooth and with the vertical maps either regular closed immersions or projections from a projective space. Let the normal bundle associated to i (i) be N (N′, respectively). Then the square commutes:*

$$\begin{array}{ccc} H^*(X', \mathbb{Z}/l(\bullet)) & \xrightarrow{i'_*} & H^*(W', \mathbb{Z}/l(\bullet)) \\ \downarrow{\scriptstyle e(N)f^*} & & \downarrow{\scriptstyle g^*} \\ H^*(X, \mathbb{Z}/l(\bullet)) & \xrightarrow{i_*} & H^*(W, \mathbb{Z}/l(\bullet)) \end{array}$$

*where $N = f^*(N')/N$ is the excess normal bundle and $e(N)$ denotes the Euler-class of $N$. In case $g$ and hence $f$ are also closed immersions with normal bundles $N_g$ and $N_f$, respectively, then $N \cong N_{g|X}/N_f$. Moreover if a finite constant group scheme $G$ acts on the above schemes, the corresponding assertions holds in the $G$-equivariant motivic cohomology defined below.*

DEFINITION 7.2. Let $G$ denote a finite group acting on a scheme $X$. Then we let $\mathbb{H}_G(X, \mathbb{Z}/(r)) = \underset{\Delta}{\text{holim}}\, R\Gamma(EG \underset{G}{\times} X, \mathbb{Z}/l(r))$ following the terminology in [**J2**, Section 6]. We let $H^n_G(X, \mathbb{Z}/l(r)) = \pi_{-n}(\mathbb{H}_G(X, \mathbb{Z}/l(r))$ .

REMARKS 7.3. 1. One may now verify that if $G = Z/l$, for a fixed prime $l$, then

$$H^*_G(\mathrm{Spec}\, k, \mathbb{Z}/l(\bullet)) \cong H^*(\mathrm{Spec}\, k, \mathbb{Z}/l(\bullet)) \otimes H^*_{sing}(BG, Z/l)$$

where $H^*(\mathrm{Spec}\, k, \mathbb{Z}/l(\bullet))$ denotes the motivic cohomology of Spec $k$ and $H^*_{sing}(BG, Z/l)$ denotes the singular cohomology of the space $BG$ with $\mathbb{F}_l$-coefficients. Recall that if $l = 2$, $H^*_{sing}(BG, Z/l)$ is a polynomial ring in one variable and when $l > 2$, $H^*_{sing}(BG, Z/l) = Z/l[t] \otimes \Lambda[\nu]$ where $\beta t = \nu$ and $\Lambda[\nu]$ denotes an exterior algebra in one generator $\nu$.

2. The situation where will apply the above proposition will be the following: $X$ will denote a given smooth scheme and $X'$ will denote $X^{\times^l}$. $W$ will denote another smooth scheme provided with a closed immersion $X \to W$ and $W'$ will denote $W^{\times^l}$. In this case the normal bundle associated to the diagonal imbedding of $X$ in $X^{\times^l}$ is $T_{X^{\times^{l-1}}}$ (the normal bundle associated to the diagonal imbedding of $W$ in $W^{\times^l}$ is $T_{W^{\times^{l-1}}}$,respectively). As equivariant vector bundles for the obvious permutation action of $\mathbb{Z}/l$ on $X^{\times^l}$ and $W^{\times^l}$ these identify with $R \otimes_k T_X$ and $R \otimes_k T_W$ where $R$ is the representation of $\mathbb{Z}/l$ given by $(k[x]/(x^l - 1))/k$. For a line bundle $\mathcal{E}$, let $w(\mathcal{E}, t) = 1 + c_1(L)^{l-1}t$. One extends the definition of $w(\mathcal{E}, t)$ to all vector bundles $\mathcal{E}$ by making this class take short exact sequences to products. Then the Euler-class $e(R \otimes_k T_X) = t^{dim(X)} w(T_X, 1/t)$ and $e(R \otimes_k T_W) = t^{dim(W)} w(T_W, 1/t)$.

At this point, we may adopt the arguments as in [**Bros**] to define cohomology operations that are compatible with push-forwards by proper maps between quasi-projective schemes. i.e. Let $Q^\bullet : H^*(X, \mathbb{Z}/l(\bullet)) \to H^*(X, \mathbb{Z}/l(\bullet))$ denote the total operation defined by $Q^\bullet = \Sigma_s Q^s$. Now we define the *covariantly functorial* operations $Q_s$ by letting

$$Q_\bullet = \Sigma_s Q_s = Q^\bullet \cap w(T_X)^{-1} \tag{7.3.1}$$

(Recall that the class $w(T_X)$ is invertible.) If we re-index motivic cohomology homologically, (i.e. if $X$ is proper and of pure dimension $d$, we let $H_n(X, \mathbb{Z}/l(r)) =$

$H^{2d-n}(X, \mathbb{Z}/(d-r)))$ the operations $Q_s$ map $H_n(X, \mathbb{Z}/l(t))$ to $H_{n-2s(l-1)}(X, \mathbb{Z}/l(tl-d(l-1)))$.

PROPOSITION 7.4. *Let $f : X \to Y$ denote a proper map between quasi-projective schemes over* Spec $k$. *Then $Q_\bullet \circ f_* = f_* \circ Q_\bullet$.*

PROOF. Since $X$ and $Y$ are quasi-projective, $f$ may be factored as a closed immersion $i : X \to Y \times \mathbb{P}^n$ for some projective space $\mathbb{P}^n$ and the obvious projection $\pi : Y \times \mathbb{P}^n \to Y$. Therefore, it suffices to prove the assertion separately for $f = i$ and for $f = \pi$. The case $f = i$ is clear from the statements above. Next observe that $\mathbb{P}^n$ is a linear scheme and therefore the motivic cohomology of $X \times \mathbb{P}^n$ is given by an obvious Kunneth formula: see [**AJ**, Appendix] for example. Therefore the Cartan formula immediately implies the required assertion for the case $f = \pi$. □

We proceed to consider various examples.

**7.5. Examples.** The first example we consider is an operation

$$Q_s : H_q(X, \mathbb{Z}/l(t)) \to H_{q-2s(l-1)}(X, \mathbb{Z}/l(tl-d(l-1)))$$

on a projective smooth scheme $X$ of dimension $d$ so that the composition with the proper map $\pi_* : H_{q-2s(l-1)}(X, \mathbb{Z}/l(tl-d(l-1))) \to H_{q-2s(l-1)}(\mathrm{Spec}\, k, \mathbb{Z}/l(tl-d(l-1)))$ is in fact zero.

For example, one may take $dim(X) = 3$, $q = 2$, $t = 1$, $s = 1$ and $l = 2$. Now we have the operation

$$Q_1 : H_2(X, \mathbb{Z}/2(1)) \to H_0(X, \mathbb{Z}/2(-1)).$$

In cohomology notation this identifies with an operation $Q_1 : H^4(X, \mathbb{Z}/2(2)) \to H^6(X, \mathbb{Z}/2(4))$. The projection to Spec $k$ sends the source to the group $H_2(\mathrm{Spec}\, k, \mathbb{Z}/2(1)) \cong H^{-2}(\mathrm{Spec}\, k, \mathbb{Z}/2(-1)) \cong CH^{-1}(Speck\, k, \mathbb{Z}/2) = 0$. It follows that $\pi_* \circ Q_1 = 0$. Recall that $H^4(X, \mathbb{Z}/2(2))$ identifies with $CH^2(X, \mathbb{Z}/2)$. Therefore any closed integral sub-scheme of $X$ of codimension 2 defines a class in this group. If $\alpha$ is such a class, our conclusion is that $\pi_*(Q_1(\alpha)) = 0$.

So far we did not put any restriction on the prime $l$. Next we assume $l = p$. Let $\nu(r)$ be the sheaf that is kernel of $W^* - C : Z\Omega^r_{X/S} \to \Omega^r_{X^{(p)}/S}$. Here $X^{(p)}$ is the scheme obtained as the pull-back of $X \underset{S}{\times} S$ where the map $S \to S$ is the absolute Frobenius and $S = \mathrm{Spec}\, k$ is the base field. Moreover $W^*$ is defined as the adjoint to the obvious map $\Omega^r_{X/S} \to W_* \Omega^r_{X^{(p)}/S}$ and $Z\Omega^r_{X/S}$ denotes the kernel of the differential $d : \Omega^r_{X/S} \to \Omega^{r+1}_{X/S}$. (See [**Ill**, 2.4] for more details.) It is known that $\nu(0) =$ the constant sheaf $Z/p$, $\nu(1) = dlog(\mathcal{O}^*_X)$ and that $\nu(r)$, viewed as a sheaf on $X_{\acute{e}t}$ is generated locally by $dlog(x_1) \cdots dlog(x_r)$, $x_i \in \mathcal{O}^*_X$.

It is shown in [**GL**, Theorem 8.4] that if $X$ is a smooth integral scheme over $k$ and $k$ is perfect, then one has the natural isomorphism (induced by a quasi-isomorphism $\nu(r)[-r] \simeq \mathbb{Z}/p(r)$) $H^s(X, \nu(r)) \cong H^{s+r}(X, \mathbb{Z}/p(r))$, where cohomology denotes cohomology computed either on the Zariski or étale sites.

Therefore, if we require $l = p$ and the field $k$ is perfect, the last operation takes on the form

$$Q_1 : H^2(X, \nu(2)) \to H^2(X, \nu(4))$$

where cohomology denotes cohomology computed either on the Zariski or étale sites.

As another example, we may assume $dim(X) = 4$, $q = 3$, $t = 1$, $s = 1$ and $l = 3$. Now we obtain the operation $Q_1 : CH^3(X, \mathbb{Z}/3, 1) \cong H^5(X, \mathbb{Z}/3(3)) \rightarrow H^9(X, \mathbb{Z}/3(9))$. Re-indexing homologically this identifies with

$$Q_1 : H_3(X, \mathbb{Z}/3(1)) \rightarrow H_{-1}(X, \mathbb{Z}/3(-5)).$$

Now $\pi_* \circ Q_1 = Q_1 \circ \pi_*$ and $\pi_*$ maps the group $H_3(X, \mathbb{Z}/3(1))$ to $H_3(\operatorname{Spec} k, \mathbb{Z}/3(1)) \cong H^{-3}(\operatorname{Spec} k, \mathbb{Z}/3(-1)) \cong CH^{-1}(\operatorname{Spec} k, \mathbb{Z}/3, 1) = 0$ since the higher Chow groups indexed by the codimension are trivial for negative codimension. Therefore, the composition $\pi_* \circ Q_1 = 0$. In case $l = p$, this operation now takes on the form

$$Q_1 : H^2(X, \nu(3)) \rightarrow H^0(X, \nu(9)).$$

As yet another example, we will presently show that the only simplicial operations that send the usual mod-$l$ Chow groups to the usual mod-$l$ Chow groups are the power operations. Recall that the usual mod$-l$ Chow groups are given by the mod$-l$ motivic cohomology groups $H^{2n}(X, \mathbb{Z}/l(n))$. Now let $Q^s : H^{2t}(X, \mathbb{Z}/l(t)) \rightarrow H^{2t+2s(l-1)}(X, \mathbb{Z}/l(lt))$ be given so that the $2t + 2s(l-1) = 2lt$. Then $2s(l-1) = 2t(l-1)$ so that $s = t$. Therefore we see from Theorem 5.2(ii) that the given operation is none other than the $l$-th power operation.

## References


[AJ] Reza Akhtar and Roy Joshua, *Künneth decompositions for quotient varieties*, Indag. Math. (N.S.) **17** (2006), no. 3, 319–344, DOI 10.1016/S0019-3577(06)80035-2. MR2321103 (2008i:14008)

[B-F] C. Berger and B. Fresse, *Combinatorial operad actions on co-chains*, preprint, math.AT/0109158.

[Bl] Spencer Bloch, *Algebraic cycles and the Beĭlinson conjectures*, The Lefschetz centennial conference, Part I (Mexico City, 1984), Contemp. Math., vol. 58, Amer. Math. Soc., Providence, RI, 1986, pp. 65–79, DOI 10.1090/conm/058.1/860404. MR860404 (88e:14006)

[Bros] Patrick Brosnan, *Steenrod operations in Chow theory*, Trans. Amer. Math. Soc. **355** (2003), no. 5, 1869–1903 (electronic), DOI 10.1090/S0002-9947-03-03224-0. MR1953530 (2004f:14008)

[Brow] Kenneth S. Brown, *Abstract homotopy theory and generalized sheaf cohomology*, Trans. Amer. Math. Soc. **186** (1973), 419–458. MR0341469 (49 #6220)

[DI] Daniel Dugger and Daniel C. Isaksen, *Motivic cell structures*, Algebr. Geom. Topol. **5** (2005), 615–652, DOI 10.2140/agt.2005.5.615. MR2153114 (2007c:55015)

[Ep] D. B. A. Epstein, *Steenrod operations in homological algebra*, Invent. Math. **1** (1966), 152–208. MR0199240 (33 #7389)

[Fausk] H. Fausk, *T-model structures on chain complexes of presheaves*, preprint, math.AG/0612414v1 (14 Dec 2006).

[FL] William Fulton and Serge Lang, *Riemann-Roch algebra*, Grundlehren der Mathematischen Wissenschaften [Fundamental Principles of Mathematical Sciences], vol. 277, Springer-Verlag, New York, 1985. MR801033 (88h:14011)

[GL] Thomas Geisser and Marc Levine, *The K-theory of fields in characteristic p*, Invent. Math. **139** (2000), no. 3, 459–493, DOI 10.1007/s002220050014. MR1738056 (2001f:19002)

[GW] Christian Hasemeyer and Jens Hornbostel, *Motives and etale motives with finite coefficients*, K-Theory **34** (2005), no. 3, 195–207, DOI 10.1007/s10977-005-1563-6. MR2182375 (2007g:19003)

[Ill] Luc Illusie, *Complexe de de Rham-Witt et cohomologie cristalline* (French), Ann. Sci. École Norm. Sup. (4) **12** (1979), no. 4, 501–661. MR565469 (82d:14013)

[J1] Roy Joshua, *K-theory and G-theory of DG-stacks*, Regulators, Contemp. Math., vol. 571, Amer. Math. Soc., Providence, RI, 2012, pp. 175–217, DOI 10.1090/conm/571/11328. MR2953415

[J2] Roy Joshua, *Higher intersection theory on algebraic stacks. I*, $K$-Theory **27** (2002), no. 2, 133–195, DOI 10.1023/A:1021116524762. MR1942183 (2003k:14019a)

[Kr-May] Igor Kříž and J. P. May, *Operads, algebras, modules and motives* (English, with English and French summaries), Astérisque **233** (1995), iv+145pp. MR1361938 (96j:18006)

[Lev] Marc Levine, *Inverting the motivic Bott element*, $K$-Theory **19** (2000), no. 1, 1–28, DOI 10.1023/A:1007874218371. MR1740880 (2001k:14048)

[Mac] Saunders Mac Lane, *Categories for the working mathematician*, 2nd ed., Graduate Texts in Mathematics, vol. 5, Springer-Verlag, New York, 1998. MR1712872 (2001j:18001)

[May] J. Peter May, *A general algebraic approach to Steenrod operations*, The Steenrod Algebra and its Applications (Proc. Conf. to Celebrate N. E. Steenrod's Sixtieth Birthday, Battelle Memorial Inst., Columbus, Ohio, 1970), Lecture Notes in Mathematics, Vol. 168, Springer, Berlin, 1970, pp. 153–231. MR0281196 (43 #6915)

[Mi] James S. Milne, *Étale cohomology*, Princeton Mathematical Series, vol. 33, Princeton University Press, Princeton, N.J., 1980. MR559531 (81j:14002)

[MFK] D. Mumford, J. Fogarty, and F. Kirwan, *Geometric invariant theory*, 3rd ed., Ergebnisse der Mathematik und ihrer Grenzgebiete (2) [Results in Mathematics and Related Areas (2)], vol. 34, Springer-Verlag, Berlin, 1994. MR1304906 (95m:14012)

[MV] Fabien Morel and Vladimir Voevodsky, $\mathbf{A}^1$*-homotopy theory of schemes*, Inst. Hautes Études Sci. Publ. Math. **90** (1999), 45–143 (2001). MR1813224 (2002f:14029)

[Nag] M. Nagata, *Lectures on Hilbert's fourteenth problem*, TIFR, Bombay, (1965).

[New] P. E. Newstead, *Introduction to moduli problems and orbit spaces*, Tata Institute of Fundamental Research Lectures on Mathematics and Physics, vol. 51, Tata Institute of Fundamental Research, Bombay; by the Narosa Publishing House, New Delhi, 1978. MR546290 (81k:14002)

[MVW] Carlo Mazza, Vladimir Voevodsky, and Charles Weibel, *Lecture notes on motivic cohomology*, Clay Mathematics Monographs, vol. 2, American Mathematical Society, Providence, RI; Clay Mathematics Institute, Cambridge, MA, 2006. MR2242284 (2007e:14035)

[St-Ep] N. E. Steenrod, *Cohomology operations*, Lectures by N. E. STeenrod written and revised by D. B. A. Epstein. Annals of Mathematics Studies, No. 50, Princeton University Press, Princeton, N.J., 1962. MR0145525 (26 #3056)

[Tot] Burt Totaro, *The Chow ring of a classifying space*, Algebraic $K$-theory (Seattle, WA, 1997), Proc. Sympos. Pure Math., vol. 67, Amer. Math. Soc., Providence, RI, 1999, pp. 249–281, DOI 10.1090/pspum/067/1743244. MR1743244 (2001f:14011)

[Voev1] Vladimir Voevodsky, *Reduced power operations in motivic cohomology*, Publ. Math. Inst. Hautes Études Sci. **98** (2003), 1–57, DOI 10.1007/s10240-003-0009-z. MR2031198 (2005b:14038a)

[Voev2] Vladimir Voevodsky, *Motivic cohomology with* $\mathbf{Z}/2$*-coefficients*, Publ. Math. Inst. Hautes Études Sci. **98** (2003), 59–104, DOI 10.1007/s10240-003-0010-6. MR2031199 (2005b:14038b)

Department of Mathematics, University of Maryland, College Park, Maryland 20740

*E-mail address*: pbrosnan@math.umd.edu

Department of Mathematics, Ohio State University, Columbus, Ohio 43210

*E-mail address*: joshua@math.ohio-state.edu

# On the Broadhurst-Kreimer generating series for multiple zeta values

Sarah Carr, Herbert Gangl, and Leila Schneps

ABSTRACT. Let $\mathcal{F}$ denote the free polynomial algebra $\mathcal{F} = \mathbb{Q}\langle s_3, s_5, s_7, \ldots\rangle$ on non-commutative variables $s_i$ for odd $i \geq 3$. The algebra $\mathcal{F}$ is weight-graded by letting $s_n$ be of weight $n$; we write $\mathcal{F}_n$ for the weight $n$ part. In this paper we put a "special" decreasing depth filtration

$$\mathcal{F} = \mathcal{F}^1 \supset \mathcal{F}^2 \supset \cdots \supset \mathcal{F}^d \supset \mathcal{F}^{d+1} \ldots$$

on $\mathcal{F}$, based on the period polynomials associated to cusp forms on $\mathrm{SL}_2(\mathbb{Z})$. We define a lattice $L$ of particular combinatorially defined subspaces of $\mathcal{F}$, and conjecture that this lattice is distributive. Assuming this conjecture, we show that the dimensions of the weight $n$ filtered quotients $\mathcal{F}_n^d/\mathcal{F}_n^{d+1}$ are given by the coefficients of the well-known Broadhurst-Kreimer generating series, defined by them to predict dimensions for the algebra of multiple zeta values. We end by explaining the expected relationship between $\mathcal{F}$ equipped with the special depth filtration and the algebras of formal and motivic multiple zeta values.

## 1. Introduction

Let $\mathcal{F} = \mathbb{Q}\langle s_3, s_5, s_7, \ldots\rangle$ be the free polynomial algebra generated by elements $s_i$, $i \geq 3$ odd. The algebra $\mathcal{F}$ is equipped with a weight grading $\mathcal{F} = \oplus_{n\geq 3}\mathcal{F}_n$ given by associating the weight $i$ with each $s_i$ and extending the weight additively to products. Also, $\mathcal{F}$ is equipped with a standard depth grading for which a monomial is of depth $d$ if it is a product $s_{i_1}\cdots s_{i_d}$. In this text we will equip $\mathcal{F}$ with another structure, called the *special depth filtration*, which is a decreasing filtration $\mathcal{F}^1 \supset \mathcal{F}^2 \supset \cdots$ on $\mathcal{F}$, which thus induces a filtration $\mathcal{F}_n^1 \supset \mathcal{F}_n^2 \supset \cdots$ also on each graded part of $\mathcal{F}$.

The main goal of this paper is to compare the dimensions of the associated graded of $\mathcal{F}$ for the special depth filtration with the coefficients of the famous Broadhurst-Kreimer generating series for multiple zeta values. Broadhurst and Kreimer defined the following 2-variable rational function:

$$BK(X,Y) = \frac{1}{1 - O(X)Y + S(X)Y^2 - S(X)Y^4}, \tag{1.1}$$

where

$$O = O(X) = \frac{X^3}{1-X^2} = \sum_{i=1}^{\infty} X^{2i+1}$$

and

$$S = S(X) = \frac{X^{12}}{(1-X^4)(1-X^6)} = \sum_{\substack{n=12 \\ n \text{ even}}}^{\infty} sp(n)X^n,$$

where $sp(n) = \lfloor \frac{n-4}{4} \rfloor - \lfloor \frac{n-2}{6} \rfloor$ is the dimension of the space $S_n(\mathrm{SL}_2(\mathbb{Z}))$ of weight $n$ cusp forms on the full modular group. Their conjecture states that the coefficient of $X^nY^d$ in the Taylor expansion of this function around $(0,0)$ is equal to the number of linearly independent multiple zeta values of weight $n$ and depth $d$ (see §7). In this paper, we state a conjecture on a certain lattice of combinatorially defined subspaces of the free algebra $\mathcal{F}$, and then, assuming this conjecture, prove that the coefficients of the Broadhurst-Kreimer series do actually give the dimensions of the graded pieces of the associated graded of $\mathcal{F}$ with respect to the special depth filtration.

In §2, we recall the definition of the period polynomials associated to cusp forms on $\mathrm{SL}_2(\mathbb{Z})$, and use them to define the special depth filtration on the free algebra $\mathcal{F}$. In §3 we give a preliminary version of the distributivity conjecture and state our main results on the Broadhurst-Kreimer generating series and the dimensions of the associated graded pieces of $\mathcal{F}$ for the special depth filtration. In §4 we define the lattice of subspaces we need to consider in $\mathcal{F}$, and give the precise statement of our main conjecture. Sections §§5-6 are devoted to the proofs. Finally, in §7, we discuss the relation of this theorem to formal and motivic multiple zeta values.

ACKNOWLEDGEMENTS. The first two authors wish to thank the MPI in Bonn for its hospitality during the earliest stages of the work in this paper, in 2009. The second two authors thank the Isaac Newton Institute in Cambridge for its hospitality during the preparation of the final version. Both institutes provided ideal and stimulating working conditions. We are also grateful to the referees for their helpful remarks, especially concerning an error in the first version of this paper, to F. Brown for clarifying comments on the period map and to P. Lochak and B. Enriquez for their explanations concerning Koszulity.

## 2. Period polynomials and the special depth filtration

DEFINITION 2.1. A polynomial $P(X) \in \mathbb{Q}[X]$ is said to be a *period polynomial of weight $n$ with respect to* $\mathrm{SL}_2(\mathbb{Z})$ if it is a polynomial of degree $\leq n-2$ satisfying the following two equalities:

$$P(X) + X^{n-2}P(\frac{-1}{X}) = 0,$$

$$P(X) + X^{n-2}P(1-\frac{1}{X}) + (X-1)^{n-2}P(\frac{1}{1-X}) = 0.$$

We say that $P(X)$ is an *even* period polynomial if $P(X) = P(-X)$, and a *reduced even* period polynomial if $P(X)$ is even, $\deg P \leq n-4$, and $P(0) = 0$. The reduced even period polynomials of weight $n$ form a vector space over $\mathbb{Q}$ which we denote by $\mathcal{P}_n$.

REMARK 2.2. Consider the integral

$$r_f(X) = \int_0^{i\infty} f(z)(z-X)^{n-2}dz$$

attached to a weight $n$ modular form $f$ on $\mathrm{SL}_2(\mathbb{Z})$. Set

$$r_f^+(X) = \frac{1}{2}\big(r_f(X) + r_f(-X)\big);$$

then $r_f^+(X)$ is an even polynomial. Let $R_f^+(X)$ denote the *reduced* even period polynomial obtained from the even period polynomial $r_f^+(X)$ by removing the term of degree $n-2$ and the constant term (if any). Up to scalar multiple, the even period polynomial associated to the weight $n$ Eisenstein series $G_n$ is $X^{n-2}-1$, so that in fact the reduced even period polynomial $R_f^+(X)$ is always the even period polynomial $r_g^+(X)$ of a unique linear combination $g$ of $f$ and $G_n$. The Eichler-Shimura correspondence shows that $f \mapsto r_f^+(X)$ yields an isomorphism between the vector space $M_n\big(\mathrm{SL}_2(\mathbb{Z})\big)$ of modular forms of weight $n$ on $\mathrm{SL}_2(\mathbb{Z})$ and the vector space of even period polynomials, which then induces an isomorphism $f \mapsto R_f^+(X)$ between the space $S_n\big(\mathrm{SL}_2(\mathbb{Z})\big)$ of cusp forms of weight $n$ on $\mathrm{SL}_2(\mathbb{Z})$ and the space $\mathcal{P}_n$ of reduced even period polynomials. In particular, it follows that

$$\dim \mathcal{P}_n = \dim S_n\big(\mathrm{SL}_2(\mathbb{Z})\big) = sp(n) = \lfloor (n-4)/4 \rfloor - \lfloor (n-2)/6 \rfloor.$$

In particular, the first values of $n$ for which $\dim \mathcal{P}_n \neq 0$ are $n = 12, 16, 18, \ldots$ The identity of the period polynomials (resp. the even, odd, reduced even ones) defined this way and those of Definition 2.1 can be found for example in [**20**].

EXAMPLE 2.3. In weight 12, there is a degree 8 period polynomial given by

$$P(X) = X^8 - 3X^6 + 3X^4 - X^2 \tag{2.1}$$

which is the reduced period polynomial associated to the Ramanujan $\Delta$ cusp form.

Using the period polynomials, we can now give the definition of the special depth filtration on the free algebra $\mathcal{F}$. We first need to define the subspaces of *special elements* of $\mathcal{F}$.

DEFINITION 2.4. Let $\mathcal{S}_n$ be the image of $\mathcal{P}_n$ under the map $\rho_n : \mathcal{P}_n \to \mathcal{F}$ given by

$$P(X) = \sum_i a_i(X^{2i} - X^{n-2-2i}) \mapsto \sum_i a_i\big(s_{2i+1}s_{n-2-2i+1} - s_{n-2-2i+1}s_{2i+1}\big).$$

By definition, $\mathcal{S}_n$ is a vector subspace of the graded weight $n$ part $\mathcal{F}_n$ of $\mathcal{F}$. Since the maps $\rho_n$ are obviously injective, we have $\dim(\mathcal{S}_n) = \mathrm{sp}(n) = \lfloor (n-4)/4 \rfloor - \lfloor (n-2)/6 \rfloor$. For convenience, we write $\mathcal{S}$ for the union of the underlying sets of the spaces $\mathcal{S}_n$, $n \geq 12$. Elements of $\mathcal{S}$ are *special elements* of $\mathcal{F}$.

Let us also introduce a notation for the standard depth filtration: for $j \geq 1$, we write $D^j\mathcal{F}$ for the $j$-th part of the ordinary depth grading on $\mathcal{F}$, so $D^j\mathcal{F}$ is the vector subspace of $\mathcal{F}$ generated by all monomials $s_{i_1}\cdots s_{i_j}$.

DEFINITION 2.5. The special depth filtration on $\mathcal{F}$ is defined as follows. For $j = 1, 2, 3$, let $V^j\mathcal{F} = D^j\mathcal{F}$. Let $V^4\mathcal{F}$ be the vector subspace generated by $D^4\mathcal{F}$ and also by the elements of all the subspaces $\mathcal{S}_n \subset V^2\mathcal{F} \subset \mathcal{F}$. For $j \geq 5$, let $V^j\mathcal{F}$ be generated by products $ST$ where $S \in V^k\mathcal{F}$ and $T \in V^l\mathcal{F}$ with $k + l = j$, with

$k, l \geq 1$.[1] We define the *special depth filtration* on $\mathcal{F}$ by

$$\mathcal{F}^d = \langle \bigcup_{i \geq d} V^i \mathcal{F} \rangle.$$

## 3. Distributivity conjecture and Broadhurst-Kreimer dimensions

In this section, we state our main results on the relation between the coefficients of the expansion around $(0,0)$ of the Broadhurst-Kreimer generating function

$$BK(X,Y) = \sum_{n \geq 0, d \geq 0} c_n^d X^n Y^d \tag{4.1}$$

and the dimensions of the weight-graded parts of the associated graded of $\mathcal{F}$ for the special depth filtration. Our results depend on conjectural properties of certain subspaces $\mathcal{F}_n^K \subset \mathcal{F}$, the *shuffle subspaces*, whose precise definition can be found in §4. The statement of our conjecture is as follows. Let $\mathcal{L}$ denote the lattice (in the sense of order theory) of subspaces of $\mathcal{F}$ generated by the shuffle subspaces, i.e. $\mathcal{L}$ contains the shuffle subspaces and is closed under intersections and sums.

CONJECTURE 3.1. *The lattice $\mathcal{L}$ is distributive, i.e. for any $U, V, W \in \mathcal{L}$, we have*

$$U \cap (V + W) = U \cap V + U \cap W.$$

REMARK 3.2. This conjecture is equivalent to the statement that the algebra $\mathcal{F}/I(\mathcal{S})$ is Koszul, where $I(\mathcal{S})$ is the ideal of $\mathcal{F}$ generated by $\mathcal{S}$. For a discussion of this and other equivalent formulations related to the Broadhurst-Kreimer conjecture, see [**9**].

THEOREM 3.3. *Assume Conjecture 3.1. Then the coefficient $c_n^d$ of the Taylor expansion of $BK(X,Y)$ around $(0,0)$ is equal to the dimension of the quotient of successive filtered pieces of $\mathcal{F}_n$ by the special depth filtration:*

$$c_n^d = \dim \mathcal{F}_n^d / \mathcal{F}_n^{d+1}.$$

*In particular, if $n \not\equiv d$ mod 2, then*

$$c_n^d = \dim \mathcal{F}_n^d / \mathcal{F}_n^{d+1} = 0,$$

*i.e. $\mathcal{F}_n^d = \mathcal{F}_n^{d+1}$ if $n \not\equiv d$ mod 2.*

Theorem 3.3 will follow as an easy corollary of the more detailed Theorem 3.4, which makes use of the standard depth grading on $\mathcal{F}$. Let $\mathcal{F}_{n,k}$ denote the subspace of $\mathcal{F}$ generated by monomials of standard depth $k$ (i.e. products of $k$ $s_i$'s) of weight $n$, so $\mathcal{F}_n = \oplus_{k \geq 1} \mathcal{F}_{n,k}$. In fact, the standard depths that can occur for weight $n$ are limited. In particular, no elements of standard depth $k$ can occur in weight $n$ if $n \not\equiv k$ mod 2. Moreover, we have

$$\mathcal{F}_n = \bigoplus_{k=1}^{\lfloor n/3 \rfloor} \mathcal{F}_{n,k}, \tag{3.1}$$

as the maximal possible standard depth in weight $n$ occurs when $n \equiv 0$ mod 3 for the monomial $s_3^{n/3}$.

[1] Note in particular that $\mathcal{S}_n = V^2 \mathcal{F}_n \cap V^4 \mathcal{F}_n$ for each $n$, and in general, $V^j \mathcal{F} \cap V^{j+2} \mathcal{F} \neq 0$.

The special depth filtration induces a filtration on each $\mathcal{F}_n$ and on each $\mathcal{F}_{n,k}$.

For the next theorem, we introduce a refined version of the Broadhurst-Kreimer function, with three variables, as follows:

$$\widetilde{BK}(X,Y,Z) = \frac{1}{1 - O(X)YZ + S(X)Y^2Z^2(1-Z^2)}.$$

Clearly $\widetilde{BK}(X,1,Y) = BK(X,Y)$. Furthermore, $\widetilde{BK}(X,Y,1)$ is a generating series for the filtered quotients of $\mathcal{F}$ equipped with the standard filtration.

THEOREM 3.4. (i) *Unless* $n \equiv k \equiv d$ *mod 2, we have* $\mathcal{F}^d_{n,k}/\mathcal{F}^{d+1}_{n,k} = 0$.

(ii) *Assume Conjecture* 3.1, *and let* $e^d_{n,k}$ *be the coefficient of* $X^nY^kZ^d$ *in the Taylor series of* $\widetilde{BK}(X,Y,Z)$. *Then*

$$e^d_{n,k} = \dim \mathcal{F}^d_{n,k}/\mathcal{F}^{d+1}_{n,k}. \tag{3.2}$$

PROOF OF THEOREM 3.3, using Theorem 3.4. By (3.1), we have

$$\mathcal{F}^d_n = \bigoplus_{k=1}^{\lfloor n/3 \rfloor} \mathcal{F}^d_{n,k}. \tag{3.3}$$

We first observe that if $k > d$, then $\mathcal{F}^d_{n,k} = \mathcal{F}^{d+1}_{n,k}$. Indeed, the special depth of an element is always greater than or equal to its standard depth; in particular the special depth of an element of standard depth $k$ is at least $k$, so if $k > d$, then the special depth is also greater than $d$. Thus if $k > d$, any element of $\mathcal{F}^d_{n,k}$ also lies in $\mathcal{F}^{d+1}_{n,k}$, so these spaces are equal.

Thus we find that quotienting $\mathcal{F}^d_n$ by $\mathcal{F}^{d+1}_n$ gives the direct sum of quotients

$$\mathcal{F}^d_n/\mathcal{F}^{d+1}_n = \bigoplus_{k=1}^{\lfloor n/3 \rfloor} \mathcal{F}^d_{n,k}/\mathcal{F}^{d+1}_{n,k} = \bigoplus_{k=1}^{d} \mathcal{F}^d_{n,k}/\mathcal{F}^{d+1}_{n,k}, \tag{3.4}$$

using (3.3) for the first equality and the above observation to reduce the upper limit of the sum from $\lfloor n/3 \rfloor$ to $d$. But it follows from (3.2) of Theorem 3.4 that the dimension of the right-hand term of (4.5) is $\sum_{k=1}^d e^d_{n,k}$, and by the equality $\widetilde{BK}(X,1,Y) = BK(X,Y)$, we have

$$\sum_{k=1}^{d} e^d_{n,k} = c^d_n. \tag{3.5}$$

This proves the first statement of Theorem 3.3.

The second statement of Theorem 3.3 follows immediately from Theorem 3.4 (i) by summing over $k$. □

## 4. Shuffle subspaces of $\mathcal{F}$

In this section we make the statement of Conjecture 3.1 precise by defining the *shuffle subspaces* of $\mathcal{F}$, and prove some properties that will be necessary for the proof of Theorem 3.4.

We start by considering the subspaces $\mathcal{F}^d_{n,k}$ with $d \equiv k$ mod 2 (indeed, it will be shown in the proof of part (i) of Theorem 3.4 below that these spaces are zero when $d \not\equiv k$ mod 2). The main remark is that $\mathcal{F}^d_{n,k}$ is generated by the set of possible products, in any order, of $a = (d-k)/2$ special elements and $d - 4a$ single

$s_i$'s, of total weight $n$. The definition of $a$ ensures that the standard depth of such a product is given by $d-2a=k$, and the special depth is equal to $d$. We say that such a product is of type $(a, d-4a)$, so these products generate $\mathcal{F}^d_{n,k} = \mathcal{F}^d_{n,d-2a}$. Let us illustrate this with an example.

Note that given a monomial in elements $s_i$ and special elements $S \in \mathcal{S}$, its degree $k$ as a polynomial in the $s_i$ is given by the number of $s_i$ plus twice the number $a$ of special elements $S$ in the monomial. We write $d = 2a+k$; this $d$ measures the special depth of the monomial, i.e. its position in the special depth filtration. Each $s_i$ in the monomial adds 1 to the count of $d$, and by the definition of the special depth filtration, each $S$ adds 4.

EXAMPLE 4.1. Let $n=24$. In the case where $d=6$ and $a=1$ (so $k=4$), the space $\mathcal{F}^6_{24,4}$ is generated by "monomials" that are products of two single $s_i$ and one special element in $\mathcal{S}$; the possibilities are

$$s_3s_9S_{12},\ s_5s_7S_{12},\ s_7s_5S_{12},\ s_9s_3S_{12},\ s_3S_{12}s_9,\ s_5S_{12}s_7,\ s_7S_{12}s_5,\ s_9S_{12}s_3,\ S_{12}s_3s_9,$$

$$S_{12}s_5s_7,\ S_{12}s_7s_5, S_{12}s_9s_3,\ s_3s_5S_{16},\ s_5s_3S_{16},\ s_3S_{16}s_5,\ s_5S_{16}s_3,\ S_{16}s_3s_5,\ S_{16}s_5s_3,$$

$$s_3^2S_{18},\ s_3S_{18}s_3,\ S_{18}s_3^2,$$

where $S_{12} = [s_3, s_9] - 3[s_5, s_7]$ generates $\mathcal{S}_{12}$, $S_{16} = 2[s_3, s_{13}] - 7[s_5, s_{11}] + 11[s_7, s_9]$ generates $\mathcal{S}_{16}$, and $S_{18} = 8[s_3, s_{15}] - 25[s_5, s_{13}] + 26[s_7, s_{11}]$ generates $\mathcal{S}_{18}$.

If we now consider the case where $d=8$ and $a=2$ (so again $k=4$), the space $\mathcal{F}^8_{24,4}$ is one-dimensional, generated by the single element $S_{12}^2$.

This example shows that each of the vector spaces $\mathcal{F}^d_{n,d-2a}$ possesses natural subspaces determined by the order of the terms, more precisely by the positions or slots occupied by the special elements $S$ between the single $s_i$. Decomposition into these subspaces will be essential in computing Broadhurst-Kreimer dimensions.

Let us introduce a notation to describe the subspace of $\mathcal{F}^d_{n,d-2a}$ spanned by products of $a$ special elements and $d-4a$ single $s_i$ in which the special elements occur in prescribed places.

DEFINITION 4.2. For each choice of $d, a \geq 0$ with $d \geq 4a$, let $sh(a, d-4a)$ denote the set of shuffles of $a$ $S$'s and $d-4a$ $s$'s, giving words of total length $d-2a$. For example, $sh(2,3)$ is given by

$$\{SSsss, SsSss, SssSs, SsssS, sSSss, sSsSs, sSssS, ssSSs, ssSsS, sssSS\}$$

is the set of shuffles of two $S$'s and three $s$'s, corresponding to the values $a=2$, $d=11$. Another way of viewing these shuffles is as the set of ordered partitions of a sequence of length $d-2a$ into blocks of length 1 (the $s$'s) and 2 (the $S$'s).

For each $K \in sh(a, d-4a)$, let $\mathcal{F}^K_n \subset \mathcal{F}^d_{n,d-2a}$ be the *shuffle subspace* spanned by monomials of total weight $n$ and of the form of $K$, where each $S$ represents an element of $\mathcal{S}$ and each $s$ a single $s_i$. We write $\mathcal{L}$ for the lattice of subspaces of $\mathcal{F}$ generated by the $\mathcal{F}^K_n$; in other words, $\mathcal{L}$ is the set of subspaces $\mathcal{F}^K_n$ and all subspaces obtained from them by sums and intersections.

EXAMPLE 4.3. In the case $n=24$, $a=1$ and $d=6$ as in the example above, we have the decomposition according to the shuffle set $sh(a, d-4a) = sh(1,2) = \{Sss, sSs, ssS\}$, so

$$\mathcal{F}^6_{24,4} = \mathcal{F}^{Sss}_{24} + \mathcal{F}^{sSs}_{24} + \mathcal{F}^{ssS}_{24},$$

where

$$\begin{cases} \mathcal{F}_{24}^{ssS} = \langle s_3s_9S_{12}, s_5s_7S_{12}, s_7s_5S_{12}, s_9s_3S_{12}, s_3s_5S_{16}, s_5s_3S_{16}, s_3s_3S_{18}\rangle, \\ \mathcal{F}_{24}^{sSs} = \langle s_3S_{12}s_9, s_5S_{12}s_7, s_7S_{12}s_5, s_9S_{12}s_3, s_3S_{16}s_5, s_5S_{16}s_3, s_3S_{18}s_3\rangle, \\ \mathcal{F}_{24}^{Sss} = \langle S_{12}s_3s_9, S_{12}s_5s_7, S_{12}s_7s_5, S_{12}s_9s_3, S_{16}s_3s_5, S_{16}s_5s_3, S_{18}s_3s_3\rangle. \end{cases}$$

In general, since $\mathcal{F}^d_{n,d-2a}$ is precisely spanned by the shuffles of $d-4a$ $s$'s and $a$ $S$'s, we obtain a decomposition into subspaces

$$\mathcal{F}^d_{n,d-2a} = \sum_{K\in sh(a,d-4a)} \mathcal{F}_n^K. \tag{4.1}$$

The sum (4.1) is not always a direct sum; indeed the shuffle spaces $\mathcal{F}_{24}^{Sss}$ and $\mathcal{F}_{24}^{ssS}$ intersect in the one-dimensional subspace $\mathcal{F}_{24}^{SS}$ generated by $S_{12}^2$ (see Example 4.1). The last result of this section, Proposition 4.6, determines these intersections precisely. To state it, we first need one more definition.

DEFINITION 4.4. Let $K_i \in sh(a_i, d_i-4a_i)$, $i=1,\dots,r$, be shuffles of $a_i$ $S$'s with $d_i-4a_i$ $s$'s, and assume that the numbers $d_i-2a_i$ are all equal to some fixed number $k>0$. The family $K_1,\dots,K_r$ is said to be *compatible* if there exists a shuffle $\mathbf{K}$, called the *intersection* $\mathbf{K}=\cap K_i$ of the $K_i$, such that each $K_i$ can be obtained from $\mathbf{K}$ by repeating the operation of replacing an $S$ by $ss$, and $\mathbf{K}$ is the unique shuffle having this property and containing a maximal number of $S$'s.

An equivalent way to define the notions of compatibility and intersection (which justifies the use of the term "intersection") is the following. For $i=1,\dots,r$, let $K_i'$ be the word with $k$ letters obtained from $K_i$ by substituting a double symbol, say $TT$, for each $S$. Let $B$ be the set of pairs $(b,b+1)$ of indices, with $1\le b\le d-2a-1$, that index pairs of $TT$ coming from a single $S$ in any of the $K_i$. Then the family is compatible if the set $B$ consists of disjoint pairs. If so, let $\mathbf{K}'$ be the word in $k$ letters such that all pairs of letters indexed by pairs in $B$ are $T$'s and the rest are $s$'s. By definition, the $T$'s appear in strings of even length in $\mathbf{K}'$, and the intersection $\mathbf{K}$ is obtained from $\mathbf{K}'$ by replacing adjacent pairs of $TT$ by $S$, starting from the left. If the family is not compatible, we set the intersection $\mathbf{K}=\emptyset$.

EXAMPLE 4.5. The shuffles $ssSssSs, SssssSs$ and $SSsssss \in sh(2,5)$ form a compatible family, since all can be obtained from $\mathbf{K}=SSssSs$ by replacing one $S$ by $ss$; the associated words in $s,T$ are $ssTTssTTs$, $TTssssTTs$ and $TTTTsssss$, so $B=\{(3,4),(7,8),(1,2)\}$, $\mathbf{K}'=TTTTssTT$ and $\mathbf{K}=SSssSs$. The shuffles $Sss$ and $sSs$ are not compatible (associated $B=\{(1,2),(2,3)\}$ with common member 2), nor are $sSsssS$ and $sSSSs$ ($B=\{(2,3),(7,8),(4,5),(6,7)\}$ with common member 7). In terms of the ordered partitions into blocks of length 1 or 2 discussed earlier, a compatible family can be viewed as having a common refinement partition of that type.

PROPOSITION 4.6. *Fix $d,a\ge 0$ with $d\ge 4a$. Let $K_1,\dots,K_r$ be distinct shuffles in the shuffle set $sh(a,d-4a)$, and let $\mathbf{K}$ denote their intersection. We write $\mathcal{F}_n^{\mathbf{K}}=\{0\}$ if $\mathbf{K}=\emptyset$. Then*

$$\mathcal{F}_n^{K_1}\cap\cdots\cap\mathcal{F}_n^{K_r} = \mathcal{F}_n^{\mathbf{K}}. \tag{4.2}$$

PROOF. We use induction on $r$. If $r=1$ the spaces are identical. Let $\mathbf{H}$ be the intersection of $K_2,\dots,K_r$, and assume that $\cap_{i=2}^r\mathcal{F}_n^{K_i}=\mathcal{F}_n^{\mathbf{H}}$. The two shuffles

$K_1$ and $\mathbf{H}$ form a two-member family whose intersection is still $\mathbf{K}$. By an abuse of notation, let us write $K_2 = \mathbf{H}$ and $\mathbf{H} \in sh(a_2, d_2 - 2a_2)$ (even if these are not the same $a_2, d_2$ as before), so that $\mathbf{K}$ is the intersection of $K_1$ and $K_2$. Thus we only have to prove that $\mathcal{F}_n^{K_1} \cap \mathcal{F}_n^{K_2} = \mathcal{F}_n^{\mathbf{K}}$.

Let $V$ denote the vector space generated by the $s_i$, and $R$ the vector space generated by the $S \in \mathcal{S}$, so $R \subset V \otimes V$. Then the shuffle spaces $\mathcal{F}^{K_i}$ are equal to the corresponding tensor products of $V$ and $R$ with $s$ replaced by $V$ and $S$ by $R$. For example, if $K_1 = ssSsss$ and $K_2 = SsssS$, we have

$$\begin{cases} \mathcal{F}^{ssSsss} = V \otimes V \otimes R \otimes V \otimes V \otimes V \\ \mathcal{F}^{SsssS} = R \otimes V \otimes V \otimes V \otimes R. \end{cases}$$

In the case where $K_1$ and $K_2$ are compatible families, this precisely means that the two tensor products can be "lined up" in such a way that each $R$ in either $\mathcal{F}^{K_1}$ or $\mathcal{F}^{K_2}$ lines up over either an $R$ or a $V \otimes V$ in the other space. In the example above, which is compatible, we illustrate it by bracketing factors $V \otimes V$ in each tensor product that line up over an $R$ in the other one.

$$\begin{cases} \mathcal{F}^{ssSsss} = \big(V \otimes V\big) \otimes R \otimes V \otimes \big(V \otimes V\big) \\ \mathcal{F}^{SsssS} = R \otimes \big(V \otimes V\big) \otimes V \otimes R. \end{cases}$$

Then since $R \subset V \otimes V$, the intersection of the two spaces is obtained from either one of the two tensor products by replacing $V \otimes V$ by $R$ for each factor of $V \otimes V$ that lines up with an $R$ in the other one. Thus, in the example above, the intersection is $R \otimes R \otimes V \otimes R$. But this is nothing other than the tensor product of $V$'s and $R$'s corresponding to $\mathcal{F}^{\mathbf{K}}$ with $\mathbf{K} = K_1 \cap K_2$ (indeed, in the example, $\mathbf{K} = SSsS$).

In the case where $K_1$ and $K_2$ are non-compatible, this means that it is impossible to line up each $R$ in one factor with a $V \otimes V$ in the other; at some point from left to right, this line-up must fail. This means that moving from left to right, at some point in the line-up we must have a factor of $V \otimes R$ in one of the tensor products lying above a factor of $R \otimes V$ in the other. Thus, to prove that $\mathcal{F}^{K_1} \cap \mathcal{F}^{K_2} = \{0\}$, it is enough to show that $V \otimes R \cap R \otimes V = \{0\}$. This is not always true for general vector spaces with $R \subset V \otimes V$, and is quite difficult to show in the present case, but it was proven by Goncharov ([**14**], Theorem 1.5), see also [**9**], Prop. 5.4. □

## 5. Proof of Theorem 3.4.

PROOF OF THEOREM 3.4. (i) To show that $\mathcal{F}_{n,k}^{d} = \mathcal{F}_{n,k}^{d+1}$ if $k \not\equiv d$ mod 2, it suffices to consider what elements of $\mathcal{F}$ can exist which are of standard depth $k$ and special depth $d$. The space $\mathcal{F}_{n,k} = \mathcal{F}_{n,k}^{k}$ is generated by products $s_{i_1} \cdots s_{i_k}$ with $i_1 + \cdots + i_k = n$. By definition, the only elements in $\mathcal{F}_{n,k}^{k}$ which have special depth $> k$ are those in the intersection of $\mathcal{F}_{n,k}^{k}$ with the ideal of $\mathcal{F}$ generated by the "special elements" (elements of $\mathcal{S}$); thus they are linear combinations of products of $a$ special elements and $c$ single $s_i$'s. Such a product is of standard depth $k = 2a + c$ and special depth $d = 4a + c$, so $d = 2a + k$, so $d \equiv k$ mod 2. This proves (i).

(ii) The subspace decomposition (4.1) together with Conjecture 3.1 positing the distributivity property of the lattice $\mathcal{L}$ ensures that we may compute the dimension of $\mathcal{F}^d_{n,d-2a}$ via the standard inclusion-exclusion formula:

$$\dim(\mathcal{F}^d_{n,d-2a}) = \dim\Big(\sum_{K\in sh(a,d-4a)} \mathcal{F}_n^K\Big) \tag{5.1}$$

$$= \sum_K \dim(\mathcal{F}_n^K) - \frac{1}{2!}\sum_{K_1,K_2} \dim(\mathcal{F}_n^{K_1}\cap\mathcal{F}_n^{K_2}) + \frac{1}{3!}\sum_{K_1,K_2,K_3} \dim(\mathcal{F}_n^{K_1}\cap\mathcal{F}_n^{K_2}\cap\mathcal{F}_n^{K_3}) - \cdots$$

Indeed, the distributivity of the lattice $\mathcal{L}$ is equivalent to the existence of a basis $\mathcal{B}$ for $\mathcal{F}$ such that $\mathcal{B}\cap\mathcal{F}_n^K$ forms a basis of $\mathcal{F}_n^K$.

We do the computation using the following three lemmas (the proofs of the second and third, somewhat technical, are relegated to §6).

LEMMA 5.1. *Fix $d \geq 4a$. Then the dimension of $\mathcal{F}_n^K$ is independent of the shuffle $K \in sh(a, d-4a)$, and is given by the coefficient*

$$\dim(\mathcal{F}_n^K) = (S^a O^{d-4a} | X^n).$$

PROOF. The rational function $S(X)$ is the generating series whose coefficients determine the dimensions of the spaces of special elements $\mathcal{S}_m$, and the rational function $O(X)$ is the generating series whose coefficients determine the dimensions of the spaces generated by monomials in the $s_i$. In other words, $(S(X)^a|X^n)$ determines the dimension of the space of products of $a$ special elements of total weight $n$, and $(O(X)^b|X^n)$ determines the dimension of the space of products of $b$ $s_i$ of total weight $n$. Clearly the dimension of $\mathcal{F}_n^K$ is independent of the actual choice of $K \in sh(a, d-4a)$; for each $K$, the number of monomials is identical, corresponding simply to inserting the $a$ special elements in different positions among the $d-2a$ single $s_i$. Thus to determine the dimension of $\mathcal{F}_n^K$, we may assume that

$$K = \underbrace{S\cdots S}_{a}\underbrace{s\cdots s}_{d-4a},$$

so that $\mathcal{F}_n^K$ is spanned by monomials $S^1\cdots S^a s^1 \cdots s^{d-2a}$. The rational function $S(X)^a O(X)^{d-4a}$ describes exactly the number of these monomials in each weight $n$. □

LEMMA 5.2. *Fix $a$ and $d$ with $d \geq 4a$, and fix $b \geq 1$. Then the number of compatible ordered families $K_1, \ldots, K_r$ of pairwise distinct shuffles in $sh(a, d-4a)$ such that the intersection $\mathbf{K}$ lies in $sh(a', d'-4a')$ with $a' = a+b$ and $d' = d+2b$ is given by*

$$S_{a,b}(r) = \binom{d-3a-b}{a+b} r! \Big(\sum_{j=0}^{b} (-1)^j \binom{a+b}{j}\binom{\binom{a+b-j}{a}}{r}\Big).$$

LEMMA 5.3. *The expressions $S_{a,b}(r)$ from Lemma 5.2 satisfy:*

$$\sum_{r=1}^{\binom{a+b}{a}} (-1)^{r-1}\frac{S_{a,b}(r)}{r!} = (-1)^b \binom{d-3a-b}{a+b}\binom{a+b-1}{b}. \tag{5.2}$$

Let us now use these to complete the proof of Theorem 3.4 by calculating the dimensions from (5.1). Note that the result of Lemma 5.2 shows in particular that

$$b > [(d-4a)/2] \Rightarrow S_{a,b}(r) = 0, \tag{5.3}$$

since the binomial coefficient $\binom{d-3a-b}{a+b}$ is zero for such $b$, and similarly, that

$$r > \binom{a+b}{a} \Rightarrow S_{a,b}(r) = 0. \tag{5.4}$$

As above, if $K_1, \ldots, K_r$ form a compatible family, we write $\mathbf{K}$ for the intersection, but if they do not form a compatible family, we set $\mathbf{K} = \emptyset$; then we write $\mathcal{F}^{\emptyset} = \{0\}$. If the family is compatible, then there exists some $b \geq a$ such that $\mathbf{K} \in sh(a', d'-4a')$ with $a' = a+b$, $d' = d+2b$; in this case, by Lemma 5.1 above, we have

$$\dim(\mathcal{F}_n^{\mathbf{K}}) = (S^{a'}O^{d'-4a'}|X^n) = (S^{a+b}O^{d-4a-2b}|X^n). \tag{5.5}$$

We now fix $d, a \geq 0$ with $d \geq 4a$, and a weight $n$ such that $n \geq 3d$.

The preceding results give us a way to compute the dimensions of the key intersections that we need to calculate $\dim(\mathcal{F}^d_{n,d-2a})$ using (5.1).

We rewrite (5.1) in terms of shuffles $K \in sh(a, d-4a)$ as:

$$\begin{aligned}
\dim(\mathcal{F}^d_{n,d-2a}) &= \dim\Big(\sum_K \mathcal{F}_n^K\Big) \\
&= \sum_{r\geq 1} (-1)^{r-1}\frac{1}{r!} \sum_{K_1,\ldots,K_r} \dim\big(\mathcal{F}_n^{K_1}\cap\cdots\cap\mathcal{F}_n^{K_r}\big) \\
&= \sum_{r\geq 1} (-1)^{r-1}\frac{1}{r!} \sum_{b=1}^{[(d-4a)/2]} \sum_{\substack{K_1,\ldots,K_r \ s.t.\\ \mathbf{K}\in sh(a+b,d-4a-2b)}} \underbrace{\dim\big(\mathcal{F}_n^{K_1}\cap\cdots\cap\mathcal{F}_n^{K_r}\big)}_{} \quad \text{by (5.3)} \\
&\qquad\qquad\qquad\qquad = \underbrace{\dim(\mathcal{F}_n^{\mathbf{K}})}_{} \quad \text{by (4.2)} \\
&\qquad\qquad\qquad\qquad = (S^{a+b}O^{d-4a-2b}|X^n) \quad \text{by (5.5)} \\
&= \sum_{r\geq 1} (-1)^{r-1}\frac{1}{r!} \sum_{b=1}^{[(d-4a)/2]} S_{a,b}(r)(S^{a+b}O^{d-4a-2b}|X^n) \quad \text{by Lemma 5.8} \\
&= \sum_{b=1}^{[(d-4a)/2]} \Bigg(\sum_{r=1}^{\binom{a+b}{a}} (-1)^{r-1}\frac{1}{r!}S_{a,b}(r)\Bigg)(S^{a+b}O^{d-4a-2b}|X^n) \quad \text{by (5.4)} \\
&= \sum_{b=1}^{[(d-4a)/2]} (-1)^b\binom{d-3a-b}{a+b}\binom{a+b-1}{b}(S^{a+b}O^{d-4a-2b}|X^n) \quad \text{by (5.2).}
\end{aligned}$$

Setting $c = d-4a-2b$, we rewrite this formula as

$$\dim(\mathcal{F}^d_{n,d-2a}) = \sum_{\substack{2b+c=d-4a\\ b,c\geq 0}} (-1)^b\binom{a+b+c}{c}\binom{a+b-1}{b}(S^{a+b}O^c|X^n). \tag{5.6}$$

Applying (5.6) to the situation $(d, a)$ and also $(d+2, a+1)$ for the same $n$ and taking the difference, and using the standard binomial identity $\binom{k}{j} = \binom{k-1}{j} + \binom{k-1}{j-1}$, we immediately obtain

$$\dim(\mathcal{F}^d_{n,d-2a}/\mathcal{F}^{d+2}_{n,d-2a}) = \sum_{\substack{2b+c=d-4a\\ b,c\geq 0}} (-1)^b \binom{a+b+c}{c}\binom{a+b}{b}(S^{a+b}O^c|X^n). \tag{5.7}$$

To conclude the proof of Theorem 3.4, we show that

$$\sum_{\substack{2b+c=d-4a\\ b,c\geq 0}} (-1)^b \binom{a+b+c}{c}\binom{a+b}{b}(S^{a+b}O^c|X^n) = e^d_{n,d-2a},$$

the coefficient of the monomial $X^nY^{d-2a}Z^d$ in the three-variable Broadhurst-Kreimer function. The Taylor expansion of this function in $Y$ and $Z$ is given by

$$\sum_{k\geq 0}(SY^2Z^4 - SY^2Z^2 + OYZ)^k. \tag{5.8}$$

Using the standard formula for expanding trinomials

$$(x+y+z)^k = \sum_{a+b+c=k} \binom{a+b+c}{a,b} x^a y^b z^c,$$

where

$$\binom{a+b+c}{a,b} = \frac{(a+b+c)!}{a!\,b!\,c!},$$

we find that the monomial $Y^{d-2a}Z^d$ appears in (5.8) as

$$\sum_{\substack{4A+2B+C=d\\ C+2B+2A=d-2a}} \binom{A+B+C}{A,B}(SY^2Z^4)^A(-SY^2Z^2)^B(OYZ)^C =$$

$$\sum_{\substack{4A+2B+C=d\\ C+2B+2A=d-2a}} (-1)^B\binom{A+B+C}{A,B} S^{A+B}O^CY^{C+2B+2A}Z^{C+2B+4A}$$

$$\sum_{2B+C=d-4a} (-1)^B\binom{a+B+C}{a,B} O^CS^{a+B}Y^{d-2a}Z^d.$$

Replacing $B, C$ by $b, c$ and using the identity of binomial and trinomial coefficients

$$\binom{a+b+c}{c}\binom{a+b}{b} = \binom{a+b+c}{a,b}$$

completes the proof of Theorem 3.4. □

## 6. Proofs of Lemmas 5.2 and 5.3.

PROOF OF LEMMA 5.2. Though a little complicated, this lemma is proved by the same basic inclusion-exclusion technique as in formula (5.1). Consider the set of all compatible families of $r$ shuffles $K_1, \ldots, K_r \in sh(a, d-4a)$ whose intersection $\mathbf{K}$ lies in $sh(a', d'-4a')$ with $a' = a+b$, $d' = d+2b$. Since $\mathbf{K}$ is a shuffle of $a' = a+b$ $S$'s with $d' - 4a' = d - 4a - 2b$ $s$'s, there are exactly $\binom{d-3a-b}{a+b}$ possibilities for $\mathbf{K}$.

The statement of the lemma is now reduced to the claim that the number of compatible $r$-families $K_1, \ldots, K_r \in sh(a, d-4a)$ whose intersection $\mathbf{K}$ is a *fixed* element of $sh(a+b, d-4a-2b)$ is equal to $r!$ (corresponding to all possible orders of a given unordered family $\{K_1, \ldots, K_r\}$) times

$$\sum_{j=0}^{b}(-1)^j\binom{a+b}{j}\binom{\binom{a+b-j}{a}}{r}. \tag{6.1}$$

In the word $\mathbf{K}$ in letters $s$ and $S$, whose total number of letters is $d-3a-b$, let $p_1, \ldots, p_{a+b}$ denote the positions of the $a+b$ letters $S$. Since $\mathbf{K}$ is the intersection of the $K_i$, each $K_i$ can be obtained from $\mathbf{K}$ by replacing exactly $b$ $S$'s in $\mathbf{K}$ by $ss$, leaving $a$ $S$'s intact. Let $\{p_1, \ldots, p_{a+b}\} \subset \{1, \ldots, d-3a-b\}$ denote the indices of the letters $S$ in the word $\mathbf{K}$. Thus, the number of families $K_1, \ldots, K_r$ with intersection $\mathbf{K}$ is the same as the number of families of $r$ subsets of order $a$ of $\{p_1, \ldots, p_{a+b}\}$ whose union covers $\{p_1, \ldots, p_{a+b}\}$; this is the same as asking for families of $r$ subsets of order $a$ of $\{1, \ldots, a+b\}$ whose union covers $\{1, \ldots, a+b\}$. Thus, we may identify each $K_i$ with a subset of order $a$ of $\{1, \ldots, a+b\}$.

The total number of families of $r$ subsets of order $a$ of $\{1, \ldots, a+b\}$ is given by

$$\binom{\binom{a+b}{a}}{r}. \tag{6.2}$$

From this we have to subtract off the number of "bad" families, those whose union does not cover the whole set $\{1, \ldots, a+b\}$.

For any family $K_1, \ldots, K_r$ that does not cover, its intersection $\mathbf{K}$ must necessarily be contained in one of the $a+b$ subsets of $\{1, \ldots, a+b\}$ of order $a+b-1$. Let $R_1, \ldots, R_{a+b}$ denote these subsets, and for $1 \le i \le a+b$, let $\mathcal{R}_i$ denote the collection of "bad" families of order $r$ whose union lives inside $R_i$. The bad families are thus those in the union $\mathcal{R}_1 \cup \cdots \cup \mathcal{R}_{a+b}$. To compute the order of this union, we use the inclusion-exclusion formula

$$|\mathcal{R}_1 \cup \cdots \cup \mathcal{R}_{a+b}| = \sum_{j=1}^{a+b} (-1)^{j-1} \sum_{1 \le i_1 < \cdots < i_j \le a+b} |\mathcal{R}_{i_1} \cap \cdots \cap \mathcal{R}_{i_j}|.$$

But the intersection $\mathcal{R}_{i_1} \cap \cdots \cap \mathcal{R}_{i_j}$ is just the set of families which live in the subset $R_{i_1} \cap \cdots \cap R_{i_j}$, which is of order $a+b-j$, so

$$|\mathcal{R}_{i_1} \cap \cdots \cap \mathcal{R}_{i_j}| = \binom{\binom{a+b-j}{a}}{r},$$

which vanishes when $j > b$, so the number of bad families is equal to

$$\begin{aligned} |\mathcal{R}_1 \cup \cdots \cup \mathcal{R}_{a+b}| &= \sum_{j=1}^{a+b} (-1)^{j-1} \sum_{1 \le i_1 < \cdots < i_j \le a+b} \binom{\binom{a+b-j}{a}}{r} \\ &= \sum_{j=1}^{b}(-1)^{j-1}\binom{a+b}{j}\binom{\binom{a+b-j}{a}}{r}. \end{aligned} \tag{6.3}$$

Thus, the number of good families is given by the difference between (6.2) and (6.3), which is precisely equal to (6.1). This proves the lemma. □

PROOF OF LEMMA 5.3. The statement comes down to proving that

$$\sum_{r=1}^{\binom{a+b}{a}} (-1)^{r-1} \sum_{j=0}^{b} (-1)^j \binom{a+b}{j} \left( \binom{\binom{a+b-j}{a}}{r} \right) = (-1)^b \binom{a+b-1}{b}.$$

By switching the sums and removing the zero terms, we write this as

$$\sum_{j=0}^{b} (-1)^j \binom{a+b}{j} \sum_{r=1}^{\binom{a+b-j}{a}} (-1)^{r-1} \left( \binom{\binom{a+b-j}{a}}{r} \right) = (-1)^b \binom{a+b-1}{b}.$$

Written this way, it is clear that the inner sum over $r$ is equal to 1, so the equality becomes

$$\sum_{j=0}^{b} (-1)^j \binom{a+b}{j} = (-1)^b \binom{a+b-1}{b}.$$

This identity follows easily from the standard identities

$$\binom{m}{n} = \sum_{k=0}^{n} \binom{m-n-1+k}{k},$$

together with

$$\binom{-a}{b} = (-1)^b \binom{a+b-1}{b}.$$

□

## 7. Multiple zeta values and their duals

In this final section, we explain the relation between our main theorem and the theory of multiple zeta values, whether formal, motivic or real. We keep the exposition as brief as possible, however we do need to introduce a certain number of well-known facts and properties concerning multiple zeta values. The goal is to explain what we see as the meaning of the Broadhurst-Kreimer dimension conjecture, and relate it to the results of Theorem 3.3 on the dimensions of $\mathcal{F}_n^d$.

### 7.1. Real and formal multiple zeta values.

DEFINITION 7.1. The *multiple zeta algebra* $\mathcal{Z}$ is generated over $\mathbb{Q}$ by the real numbers known as *multiple zeta values*, given by

$$\zeta(k_1, \ldots, k_d) = \sum_{n_1 > \cdots > n_d > 0} \frac{1}{n_1^{k_1} \cdots n_d^{k_d}}, \qquad (7.1)$$

where the $k_i$ are strictly positive integers, with $k_1 > 1$ in order to ensure convergence of the series, and $k_1 + \cdots + k_d \geq 2$. The number $d$ is the *y-depth* of the multiple zeta value, and its weight $n$ is given by $k_1 + \cdots + k_d = n$. We let $\overline{\mathcal{Z}}$ denote the quotient of $\mathcal{Z}$ modulo the ideal generated by $\zeta(2)$; this is the *reduced* multiple zeta algebra.

Let $\mathcal{Z}_n^d$ denote the vector space over $\mathbb{Q}$ spanned by multiple zeta values of weight $n$ and $y$-depth $d$. The Broadhurst-Kreimer function was first identified by Broadhurst and Kreimer [**3**] when they used a computer to calculate the dimensions of the subspace $\overline{\mathcal{Z}}_n^d$ modulo $\overline{\mathcal{Z}}_n^{d-1}$ for quite high values of $n$ and $d$ (it has successively been corroborated numerically considerably further by Blümlein, Broadhurst and

Vermaseren [**2**]); they realized that these dimensions formed the coefficients of the Taylor series of $BK(X,Y)$ (cf. (1.1)), and conjectured that $BK(X,Y)$ yields the correct dimensions for all $n, d$, i.e. that

$$c_n^d = \overline{\mathcal{Z}}_n^d / \overline{\mathcal{Z}}_n^{d-1} \tag{7.2}$$

for $n \geq 3$, $d \geq 1$, where $c_n^d$ is the coefficient of $X^n Y^d$ in $BK(X,Y)$.

In particular, the function $BK(X,1)$ then conjecturally determines the dimensions of the weight $n$ subspaces $\overline{\mathcal{Z}}_n$. It is obviously not known whether the weight forms a grading of $\mathcal{Z}$ or $\overline{\mathcal{Z}}$ – indeed, not much is known even about the transcendence of the single zeta values $\zeta(n)$ for odd $n \geq 3$ – but the conjecture implies that the weight is indeed a grading, and that the dimensions of $\overline{\mathcal{Z}}_n$ are given by the generating series

$$BK(X,1) = (1 - X^2)/(1 - X^2 - X^3) \tag{7.1}$$

first observed by Zagier [**26**], [**27**].

We now need to introduce some basic notions. Define the *shuffle product* of words in non-commutative variables $x, y$ recursively by the formula

$$sh(u,1) = sh(1,u) = u,$$

$$sh(Xu, Yv) = X \cdot sh(u, Yv) + Y \cdot sh(Xu, v),$$

where $X, Y \in \{x, y\}$. We write $\mathrm{Lie}_n[x,y]$ for the vector space of weight $n$ (i.e. degree $n$) Lie polynomials. For any polynomial $f \in \mathbb{Q}\langle x,y\rangle$, let $(f|w)$ denote the coefficient of a word $w$ in the polynomial $f$, and extend it linearly, i.e. $(f|au+bv) = a(f|u) + b(f|v)$. It is a well-known fact that for a polynomial $f \in \mathbb{Q}\langle x,y\rangle$, we have

$$f \in \mathrm{Lie}[x,y] \Leftrightarrow \big(f|sh(u,v)\big) = 0 \text{ for all words } u, v \text{ in } x, y. \tag{7.2}$$

We define the *stuffle product* of words in $x, y$ ending in $y$ as follows. Set $y_i = x^{i-1}y$ for $i \geq 1$; all words ending in $y$ can be written $y_{i_1} \cdots y_{i_r}$. We define the stuffle product of two such words by

$$st(v,1) = st(1,v) = v,$$

$$st(y_{i_1}y_{i_2}\cdots y_{i_r}, y_{j_1}y_{j_2}\cdots y_{j_s}) = y_{i_1} \cdot st(y_{i_2}\cdots y_{i_r}, y_{j_1}y_{j_2}\cdots y_{j_s})$$
$$+ y_{j_1} \cdot st(y_{i_1}y_{i_2}\cdots y_{i_r}, y_{j_2}\cdots y_{j_s}) + y_{i_1+j_1} \cdot st(y_{i_2}\cdots y_{i_r}, y_{j_2}\cdots y_{j_s}).$$

We say that a word is convergent if it starts in $x$ and ends in $y$. Any convergent word can be written $w = y_{k_1}\cdots y_{k_d}$ with $k_1 > 1$; we set $\zeta(w) = \zeta(k_1, \ldots, k_d)$. Following [**10**], we extend the definition of $\zeta(w)$ from convergent words to all words in $x, y$ by the following formula. If $w$ is not convergent, write $w = y^a u x^b$ where $u$ is convergent; then set

$$\zeta(w) = \sum_{r=0}^{a}\sum_{s=0}^{b}(-1)^{a+b}\zeta\Big(\pi\big(sh(y^r, y^{a-r}ux^{b-s}, x^s)\big)\Big), \tag{7.3}$$

where $\zeta$ is considered to be linear on words, and $\pi$ denotes the projector sending a polynomial onto only its part consisting of convergent words (and $sh(w_1, w_2, w_3)$ is defined as $sh\big(w_1, sh(w_2, w_3)\big)$).

We also define a different extension of $\zeta(w)$, this time from convergent words only to words starting and ending in $y$. We begin by defining values $\zeta_*(\underbrace{1,\ldots,1}_{n})$ by the equality

$$1+\sum_{n\geq 1}\zeta_*(\underbrace{1,\ldots,1}_{n})y^n=\exp\Big(\sum_{n\geq 2}\frac{(-1)^{n-1}}{n}\zeta(x^{n-1}y)y^n\Big).$$

In particular, we have $\zeta_*(1)=\zeta(1)=0$, $\zeta_*(1,1)=-\frac{1}{2}\zeta(2)$, $\zeta_*(1,1,1)=\frac{1}{3}\zeta(3)$ and $\zeta_*(1,1,1,1,1)=\frac{1}{5}\zeta(5)-\frac{1}{6}\zeta(3)\zeta(2)$. It is clear from the definition that $\zeta_*(\underbrace{1,\ldots,1}_{n})\in\mathcal{Z}$ is always an algebraic expression in the $\zeta(x^{n-1}y)$ for $n\geq 1$.

Now, for any word $w$ starting and ending in $y$, we write $w=y^k u$ with $u$ convergent, and set

$$\zeta_*(w)=\sum_{i=0}^{k}\zeta_*(\underbrace{1,\ldots,1}_{i})\zeta(y^{k-i}u).$$

Again, the $\zeta_*(w)$ are all algebraic expressions in the $\zeta(w)$.

It is a well-known and fundamental result of the theory of multiple zeta values that the $\zeta$ and $\zeta_*$ satisfy the following two families of algebraic relations:

**(Regularized) shuffle relation:** For any two words $u$ and $v$ in $x,y$,

$$\zeta\big(sh(u,v)\big)=\zeta(u)\zeta(v). \tag{7.4}$$

**(Regularized) stuffle relation:** Let $W_y$ denote the set of words in $x,y$ ending in $y$. Then for all $u,v\in W_y$,

$$\zeta_*\big(st(u,v)\big)=\zeta_*(u)\zeta_*(v). \tag{7.5}$$

DEFINITION 7.2. Let $\mathcal{FZ}$ denote the $\mathbb{Q}$-algebra generated by formal symbols $Z(w)$ for all words $w$ in $x,y$, subject to the relations above. For any convergent word $w$, we again write $w=y_{k_1}\cdots y_{k_d}$ and $Z(k_1,\ldots,k_d)=Z(w)$.

This $\mathcal{FZ}$ is called the *formal multiple zeta algebra*, and it is one of the major conjectures in the theory of multiple zeta values that $\mathcal{FZ}\simeq\mathcal{Z}$. Obviously $\mathcal{FZ}$ surjects onto $\mathcal{Z}$ since the real multizeta values are known to satisfy (7.4) and (7.5) but the injectivity seems out of reach for the present. To give an idea of the difficulty of this conjecture, it would imply in particular that the weight $n$ gives a grading on $\mathcal{Z}$, which in turn would imply that all multiple zeta values are transcendental (indeed, if a non-zero multiple zeta value is algebraic, then expanding out its minimal polynomial according to (7.4) would give a linear combination of multiple zetas in different weights equal to zero, contradicting the weight grading).

Let $\overline{\mathcal{FZ}}$ be the quotient of $\mathcal{FZ}$ by the ideal generated by $Z(xy)=Z(2)$. It is known that $\overline{\mathcal{FZ}}$ is a Hopf algebra, because it arises as the dual of the enveloping algebra of a Lie algebra (see the next section). It is weight-graded since the relations take place within weight-graded parts. The weight 0 part is given by $\overline{\mathcal{FZ}}_0=\mathbb{Q}$, then $\overline{\mathcal{FZ}}_1=\overline{\mathcal{FZ}}_2=\{0\}$, and $\overline{\mathcal{FZ}}_3=\{Z(3)\}$.

Following Furusho **[10]**, we set $\mathfrak{nfz}=\overline{\mathcal{FZ}}/\mathcal{I}$ where $\mathcal{I}$ is the ideal generated by constants and products, i.e. by $\overline{\mathcal{FZ}}_0$ and $(\overline{\mathcal{FZ}}_{\geq 3})^2$. A priori, $\mathfrak{nfz}$ is simply a vector space, but because it is known that its graded dual is a Lie algebra (again, see next section), it follows that $\mathfrak{nfz}$ has the structure of a weight-graded Lie coalgebra, and that the dimensions of its graded pieces are equal to those of the dual, so that the dimensions of the graded pieces of $\overline{\mathcal{FZ}}$ are equal to those of the universal enveloping algebra of the dual to the Lie coalgebra. We write $z(w)$ for the image in $\mathfrak{nfz}$ of $Z(w) \in \mathcal{FZ}$. It follows from the definitions that for every word $w$ ending in $y$ which is not a power of $y$, the image of $Z(w)$ is equal to the image of $Z^*(w)$ in $\mathfrak{nfz}$. These $z(w)$ satisfy linearized versions of the double shuffle relations (7.4) and (7.5), namely

$$z\big(sh(u,v)\big) = 0 \tag{7.6}$$

for all pairs of words $u, v$ in $x, y$, and

$$z\big(st(u,v)\big) = 0 \tag{7.7}$$

for all pairs of words $u, v$ not both powers of $y$.

**7.2. The double shuffle Lie algebra.** In this subsection, we give an explicit definition of the *double shuffle* Lie algebra $\mathfrak{ds}$, which is the graded dual of the Lie coalgebra $\mathfrak{nfz}$ of formal multiple zetas defined in the preceding section. The advantage of studying the dual Lie algebra is that its structure can be given more economically in terms of Lie algebra generators, while the dimensions of its graded parts are of course the same as those of $\mathfrak{nfz}$. The definition of $\mathfrak{ds}$ is given by formally dualizing (7.6) and (7.7).

Let $W_y$ denote the set of words ending in $y$ as above, and let $W'_y$ denote the set of pairs $(u, v)$ of words in $W_y$ that are not both powers of $y$.

DEFINITION 7.3. Let $\mathfrak{ds}$ denote the *double shuffle* Lie algebra defined as follows

$$\mathfrak{ds} = \Big\{ f \in \mathbb{Q}_{\geq 3}\langle x, y\rangle \;\Big|\; \big(f|sh(u,v)\big) = 0 \;\; \forall\, u, v, \;\; \big(f|st(u,v)\big) = 0 \;\; \forall\, (u,v) \in W'_y \Big\}, \tag{7.8}$$

where $\mathbb{Q}_{\geq 3}\langle x, y\rangle$ is the set of polynomials in $x, y$ of degree $\geq 3$.

The terminology *double shuffle* comes from the fact that $f \in \mathfrak{ds}$ satisfies the shuffle equations (7.4) and the stuffle equations (7.5). Note that, by (7.2), elements of $\mathfrak{ds}$ lie in $\mathrm{Lie}[x, y]$.

The proof that $\mathfrak{ds}$ is a Lie algebra was discovered by Racinet in his thesis **[22]**. He worked with the Poisson bracket defined as follows: to any $f \in \mathrm{Lie}[x, y]$, associate a derivation $D_f$ of $\mathrm{Lie}[x, y]$ defined by $D_f(x) = 0$, $D_f(y) = [y, f]$. Then $[D_f, D_g] = D_{\{f,g\}}$ where $\{f, g\} = [f, g] + D_f(g) - D_g(f)$. Racinet showed that $\mathfrak{ds}$ is stable under the Poisson Lie bracket defined by $\{\cdot, \cdot\}$ (also called the Ihara bracket). This implies that $\mathcal{U}\,\mathfrak{ds}$ is a Hopf algebra, so its dual $\mathcal{FZ}$ is a Hopf algebra and the quotient by products $\mathfrak{nfz}$ is a Lie coalgebra; in other words, $\mathfrak{nfz}^* \simeq \mathfrak{ds}$.

The Lie algebra $\mathfrak{ds}$ is graded by the weight (i.e. degree) of the Lie polynomials. The first interesting elements, written in the Lyndon-Lie basis, are

$$f_3 = [x, [x, y]] + [y, [x, y]],$$

$$f_5 = [x, [x, [x, [x, y]]]] + 2[x, [x, [[x, y], y]]] - \frac{3}{2}[[x, [x, y]], [x, y]]$$
$$+2[x, [[[x, y], y], y]] + \frac{1}{2}[[x, y], [x, [x, y]]] + [[[[x, y], y], y], y].$$

The dimensions of $\mathfrak{ds}_n$ are equal to 0 for $n = 4, 6$ and to 1 for $n = 3, 5, 7, 8, 9$. Thus there is a canonical generator $f_n$ (up to scalar multiple) for $n = 3, 5, 7, 9$. The Lie algebra $\mathfrak{ds}$ is also equipped with a decreasing *y-depth filtration*, defined by letting the $y$-depth of a polynomial be equal to the smallest number of $y$'s in any of its monomials. We write $\mathfrak{ds}^1 \supset \mathfrak{ds}^2 \supset \cdots$ for this filtration.

The elements $f_n$ for $n = 3, 5, 7, 9$ are all of $y$-depth 1. Indeed, it is known (cf. [R]) that for each odd weight $n \geq 3$, there exists an element of $y$-depth 1 in $\mathfrak{ds}$; $y$-depth 1 means that the Lie polynomial contains at least one of the Lie words $\mathrm{ad}(x)^i(y)$ with non-zero coefficient. A well-known conjecture states that any choice $f_3, f_5, f_7, \ldots$ of such elements provides a free generating set for $\mathfrak{ds}$.

We relate our work in the previous sections to the theory of multiple zeta values via the following conjecture. Let $\mathrm{Lie}\,\mathcal{F}$ denote the free algebra $\mathrm{Lie}[s_3, s_5, \ldots] \subset \mathcal{F}$. The special depth filtration on $\mathcal{F}$ restricts naturally to one on $\mathrm{Lie}\,\mathcal{F}$.

CONJECTURE 7.4. *Let $f_3, f_5, \ldots$ be a chosen set of $y$-depth 1 elements of $\mathfrak{ds}$ in each odd weight $\geq 3$. Then the Lie algebra homomorphism*

$$\rho : \mathrm{Lie}\,\mathcal{F} \to \mathfrak{ds} \tag{7.9}$$

*mapping $s_i \mapsto f_i$ is a bijection, and furthermore it carries the special depth filtration on $\mathrm{Lie}\,\mathcal{F}$ to the $y$-depth filtration on $\mathcal{U}\,\mathfrak{ds}$.*

Thus the conjecture implies that the universal enveloping algebra of $\mathfrak{ds}$, which is dual to the Hopf algebra of formal multiple zeta values $\mathcal{FZ}$, is isomorphic to $\mathcal{F}$, and so our results on the dimensions of the associated graded pieces of $\mathcal{F}$ would lead to

$$\dim \mathcal{U}\,\mathfrak{ds}_n^d / \mathcal{U}\,\mathfrak{ds}_n^{d+1} = c_n^d,$$

where $c_n^d$ is the coefficient of the monomial $X^n Y^d$ in the Broadhurst-Kreimer function $BK(X, Y)$. By duality, this would mean that

$$\dim \overline{\mathcal{FZ}}_n^d / \overline{\mathcal{FZ}}_n^{d-1} = c_n^d,$$

which corresponds exactly to Broadhurst-Kreimer's original motivation given in (7.2), except that it would hold for formal multizeta values whereas their original conjecture concerns the real ones.

**7.3. Evidence for the conjecture.** The Lie algebra $\mathfrak{ds}$ has been calculated explicitly up to about weight 20, and the resulting dimensions bear out the correctness of the Broadhurst-Kreimer dimensions for $\mathfrak{ds}_n^d/\mathfrak{ds}_n^{d+1}$. However, there is also some more theoretical evidence in favor of Conjecture 7.4.

7.3.1. *Injectivity of the homomorphism $\rho : \mathrm{Lie}\,\mathcal{F} \to \mathfrak{ds}$.* Although we do not know much about the Lie homomorphism (7.9) for a general choice of $f_3, f_5, \ldots$ in $\mathfrak{ds}$, we can show that there exist choices of $y$-depth 1 elements such that $\rho$ is injective. Let us sketch how this is known.

In a series of articles (see e.g. [**18**], but also [**12**], [**16**], [**13**], [**8**]), Goncharov (and co-authors) defined *motivic multiple zeta values*, which are (framed) mixed Tate motives with the motivic version $\zeta^{\mathfrak{m}}(2)$ of $\zeta(2)$ being zero, and which form a Hopf algebra denoted $\mathcal{MZ}$ under the tensor product, with a coproduct defined explicitly by Goncharov ([**13**], foreshadowed by [**12**]). It is known that the motivic multiple zeta values satisfy relations (7.4)–(7.5) (see [**13**], for (7.5) see also [**24**]). Furthermore, F. Brown defined [**4**] a graded algebra comodule of motivic multiple zeta values $\mathcal{H}$ in which $\zeta^{\mathfrak{m}}(2)$ is non-zero, and showed that $\mathcal{H}$ is non-canonically isomorphic to $\mathcal{MZ} \otimes \mathbb{Q}[\zeta^{\mathfrak{m}}(2)]$ and that Goncharov's period map (with values in a quotient of $\mathbb{R}$ only) can be lifted to a surjection $\mathcal{H} \twoheadrightarrow \mathcal{Z}$. Choosing an isomorphism identifies $\mathcal{MZ}$ as a subspace of $\mathcal{H}$, and the image of $\mathcal{MZ}$ under the period map is then a subspace of $\mathcal{Z}$ isomorphic to $\overline{\mathcal{Z}}$. Thus there are surjections

$$\overline{\mathcal{FZ}} \to \mathcal{MZ} \to \overline{\mathcal{Z}},$$

where the second surjection comes from taking the period of a framed mixed Tate motive in the above sense.

In a remarkable paper, Brown [**5**], with help on a key point from D. Zagier [**28**], recently succeeded in proving that the dimensions of the weight-graded parts of $\mathcal{MZ}$ are given by the coefficients of the generating series $BK(X,1)$ (cf. (7.1)). Thus, as was pointed out earlier by Goncharov ([**17**], or see 25.7.3.1 of [**1**] for more detail, as well as [**25**]), these dimensions form an upper bound for the dimensions of the weight $n$ parts $\overline{\mathcal{Z}}_n$ of $\overline{\mathcal{Z}}$, and a lower bound for the dimensions of $\overline{\mathcal{FZ}}_n$. Writing $\mathfrak{nmz}$ for the quotient of $\mathcal{MZ}$ by the degree 0 part and products, the surjection $\overline{\mathcal{FZ}} \to\to \mathcal{MZ}$ induces a surjection of quotients $\mathfrak{nfz} \twoheadrightarrow \mathfrak{nmz}$. Thanks to [**22**], Goncharov's coproduct passes to a cobracket on $\mathfrak{nmz}$, which is in fact a Lie coalgebra, so by duality this yields an injection of Lie algebras

$$\mathfrak{nmz}^* \hookrightarrow \mathfrak{nfz}^* = \mathfrak{ds}.$$

It follows directly from Brown's dimension result that the Lie algebra $\mathfrak{nmz}^*$ is free on one $y$-depth 1 generator for each odd $n \geq 3$.

These observations thus show that there is at least one choice of a set of $y$-depth 1 elements $f_n$ for odd $n \geq 3$ in $\mathfrak{ds}$ (namely the images of a set of Lie generators for $\mathfrak{nmz}^*$) which generate a free Lie subalgebra inside $\mathfrak{ds}$. Thus, for this choice of $f_i$, the map (7.9) is injective.

7.3.2. *Compatibility of $\rho : \mathrm{Lie}\,\mathcal{F} \to \mathfrak{ds}$ with the depth filtrations.* There is also some evidence to support the part of Conjecture 7.4 concerning the depth filtrations. Assume that generators $f_i$ are chosen as above, so that $\rho$ is injective.

By the definition of the $y$-depth filtration, a Poisson bracket of $d$ $f_i$'s lies in $\mathfrak{ds}^d$, so we have maps

$$\rho : \mathrm{Lie}\,\mathcal{F}^i \to \mathfrak{ds}^i$$

for $i = 1, 2, 3$.

Racinet [**22**] proved that $\mathfrak{ds}^1/\mathfrak{ds}^2$ is generated by one element in each odd rank $\geq 3$, so the induced map

$$\overline{\rho} : \mathrm{Lie}\,\mathcal{F}^1/\mathrm{Lie}\,\mathcal{F}^2 \to \mathfrak{ds}^1/\mathfrak{ds}^2$$

is an isomorphism.

In $y$-depth 2, it was shown in [**23**] that the only relations in $\mathfrak{ds}^2/\mathfrak{ds}^3$ between (images of) double Poisson brackets $\{f_i, f_j\}$ are of the form

$$\sum_{i,j} a_{ij}\{f_i, f_j\} \equiv 0 \quad \text{modulo } \mathfrak{ds}^3, \tag{7.10}$$

where $\sum_{i,j} a_{ij}(X^{i-1}Y^{j-1} - X^{j-1}Y^{i-1})$ is a reduced even period polynomial. Thus the definition of the special depth filtration shows that

$$\overline{\rho} : \mathrm{Lie}\,\mathcal{F}^2/\mathrm{Lie}\,\mathcal{F}^3 \to \mathfrak{ds}^2/\mathfrak{ds}^3$$

is an isomorphism.

In $y$-depth 3, a combination of impressive dimension results due to Goncharov [**15**] with Brown's theorem mentioned above shows that $\mathfrak{ds}^3$ is generated by triple

Poisson brackets $\{f_i, \{f_j, f_k\}\}$ and that the induced map

$$\overline{\rho} : \mathrm{Lie}\,\mathcal{F}^3/\mathrm{Lie}\,\mathcal{F}^4 \to \mathfrak{ds}^3/\mathfrak{ds}^4$$

is also an isomorphism.

The interesting phenomenon starts in $y$-depth 4. Naturally, the relations (7.10) give rise to many relations of higher $y$-depth in $\mathfrak{ds}$ by bracketing; for example the period polynomial (2.1) corresponds to $\{f_3, f_9\} - 3\{f_5, f_7\}$, and we also have the relation

$$\{f_3, \{f_3, f_9\}\} - 3\{f_3, \{f_5, f_7\}\} \equiv 0 \text{ modulo } \mathfrak{ds}^4.$$

In fact, it is known ([**19**]) that the "period-polynomial elements" (7.10) lie not just in $\mathfrak{ds}^3$ but in $\mathfrak{ds}^4$. Conjecture 7.4 is tantamount to assuming that (i) these period-polynomial elements lie in $\mathfrak{ds}^4$ and are non-zero modulo $\mathfrak{ds}^5$, and (ii) all relations in the associated graded of $\mathfrak{ds}$ for the $y$-depth filtration come from the Lie ideal of $\mathfrak{ds}$ generated by the relations (7.10). In [**6**], Brown considers similar questions relating this type of assertion to the Broadhurst-Kreimer function, but rather than with $\mathrm{Lie}\,\mathcal{F}$ and $\mathfrak{ds}$, he works with the Lie algebra $\mathfrak{nmz}$ and another Lie algebra, called "linearized double shuffle", which is known to contain the associated graded of $\mathfrak{ds}$. He displays an explicit and elegant family of Lie elements depth 4 which are conjecturally candidates for the images of period polynomial elements in $\mathfrak{ds}^4/\mathfrak{ds}^5$ (in fact in the motivic Lie subalgebra $\mathfrak{nmz}^*$) corresponding to some choice of generators $f_i$.

Because the period polynomial elements lie in $\mathfrak{ds}^4$, the map $\rho$ sends $\mathrm{Lie}\,\mathcal{F}^4 \to \mathfrak{ds}^4$. In order for the induced map $\overline{\rho} : \mathrm{Lie}\,\mathcal{F}^4/\mathrm{Lie}\,\mathcal{F}^5 \to \mathfrak{ds}^4/\mathfrak{ds}^5$ to be an isomorphism, we would need to know (i) and (ii) above in $y$-depth 4, i.e. that the period polynomial elements are not in the kernel, and that there are no relations in $\mathfrak{ds}^4/\mathfrak{ds}^5$ other than those coming from the Lie ideal generated by the period polynomial elements.

Although we cannot prove any of these results at present, it is at least easy to show that for each period polynomial, there does exist a choice of free $y$-depth 1 generators $f_3, f_5, \ldots$ of the motivic Lie subalgebra $\mathfrak{nmz}^*$ in $\mathfrak{ds}$ such that the period-polynomial element in those generators is non-zero in $\mathfrak{ds}^4/\mathfrak{ds}^5$. Indeed, if we had

$$\sum_{\substack{i<j\\ i+j=n}} a_{ij}\{f_i, f_j\} \in \mathfrak{ds}_n^5 \tag{7.11}$$

for a given choice of $f_k$'s, it would be sufficient to pick the smallest $i$ such that $a_{ij}$ is non-zero, and change $f_j = f_{n-i}$ to $f_j + \{f_a, \{f_b, f_c\}\}$ with $a+b+c = j$, $b < c$. Such triple brackets exist as soon as $j \geq 11$. If $j < 11$, there is no such triple bracket, but we can deal with that situation case by case. Indeed, since $i < j$ in all the pairs $\{f_i, f_j\}$ appearing in (7.11), we must have $i < j < 11$, so the only possibilities are $(i, j) = (3, 5), (3, 7), (3, 9), (5, 7), (5, 9), (7, 9)$ in weights 8,10,12,14,16. There are no period polynomials in weights 8,10 and 14, and only one up to scalar multiple in each of weights 12 and 16; the corresponding period polynomial elements in $\mathfrak{ds}$ can be calculated explicitly, and it turns out that they do not vanish in $\mathfrak{ds}^4/\mathfrak{ds}^5$. Thus, this method yields at least the partial result that for each period polynomial, there exists a choice of motivic generators $f_3, f_5, \ldots$, such that the corresponding period polynomial element does not vanish in $\mathfrak{ds}^4/\mathfrak{ds}^5$. It would be very helpful to prove that this result holds independently of the choice of motivic generators $f_n$.

## References


[1] Yves André, *Une introduction aux motifs (motifs purs, motifs mixtes, périodes)* (French, with English and French summaries), Panoramas et Synthèses [Panoramas and Syntheses], vol. 17, Société Mathématique de France, Paris, 2004. MR2115000 (2005k:14041)

[2] J. Blümlein, D. J. Broadhurst, and J. A. M. Vermaseren, *The multiple zeta value data mine*, Comput. Phys. Comm. **181** (2010), no. 3, 582–625, DOI 10.1016/j.cpc.2009.11.007. MR2578167 (2011a:11163)

[3] D. J. Broadhurst and D. Kreimer, *Association of multiple zeta values with positive knots via Feynman diagrams up to 9 loops*, Phys. Lett. B **393** (1997), no. 3-4, 403–412, DOI 10.1016/S0370-2693(96)01623-1. MR1435933 (98g:11101)

[4] Francis C. S. Brown, *On the decomposition of motivic multiple zeta values*, Galois-Teichmüller theory and arithmetic geometry, Adv. Stud. Pure Math., vol. 63, Math. Soc. Japan, Tokyo, 2012, pp. 31–58. MR3051238

[5] Francis Brown, *Mixed Tate motives over* $\mathbb{Z}$, Ann. of Math. (2) **175** (2012), no. 2, 949–976, DOI 10.4007/annals.2012.175.2.10. MR2993755

[6] Brown, F., *Depth-graded motivic multiple zeta values*, arXiv:1301.3053.

[7] Carr, S., *Multizeta values: Lie algebras and periods on* $M_{0,5}$, Ph.D. thesis, arXiv:0911.2643.

[8] Pierre Deligne and Alexander B. Goncharov, *Groupes fondamentaux motiviques de Tate mixte* (French, with English and French summaries), Ann. Sci. École Norm. Sup. (4) **38** (2005), no. 1, 1–56, DOI 10.1016/j.ansens.2004.11.001. MR2136480 (2006b:11066)

[9] Enriquez, B., Lochak, P., *On the Broadhurst-Kreimer conjectures*, preprint 2014.

[10] Hidekazu Furusho, *The multiple zeta value algebra and the stable derivation algebra*, Publ. Res. Inst. Math. Sci. **39** (2003), no. 4, 695–720. MR2025460 (2005h:11196)

[11] Herbert Gangl, Masanobu Kaneko, and Don Zagier, *Double zeta values and modular forms*, Automorphic forms and zeta functions, World Sci. Publ., Hackensack, NJ, 2006, pp. 71–106, DOI 10.1142/9789812774415_0004. MR2208210 (2006m:11138)

[12] Goncharov, A.B., *Hyperlogarithms, mixed Tate motives and multiple* $\zeta$*-numbers*, MIT preprint 1993.

[13] A. B. Goncharov, *Galois symmetries of fundamental groupoids and noncommutative geometry*, Duke Math. J. **128** (2005), no. 2, 209–284, DOI 10.1215/S0012-7094-04-12822-2. MR2140264 (2007b:11094)

[14] A. B. Goncharov, *Multiple polylogarithms, cyclotomy and modular complexes*, Math. Res. Lett. **5** (1998), no. 4, 497–516, DOI 10.4310/MRL.1998.v5.n4.a7. MR1653320 (2000c:11108)

[15] A. B. Goncharov, *The dihedral Lie algebras and Galois symmetries of* $\pi_1^{(l)}(\mathbb{P}^1-(\{0,\infty\}\cup\mu_N))$, Duke Math. J. **110** (2001), no. 3, 397–487, DOI 10.1215/S0012-7094-01-11031-4. MR1869113 (2002m:14020)

[16] Alexander B. Goncharov, *Multiple* $\zeta$*-values, Galois groups, and geometry of modular varieties*, European Congress of Mathematics, Vol. I (Barcelona, 2000), Progr. Math., vol. 201, Birkhäuser, Basel, 2001, pp. 361–392. MR1905330 (2003g:11073)

[17] Goncharov, A.B., *Multiple polylogarithms and mixed Tate motives,* arXiv:math/0103059.

[18] A. B. Goncharov and Yu. I. Manin, *Multiple* $\zeta$*-motives and moduli spaces* $\overline{\mathcal{M}}_{0,n}$, Compos. Math. **140** (2004), no. 1, 1–14, DOI 10.1112/S0010437X03000125. MR2004120 (2005c:11090)

[19] Kentaro Ihara, Masanobu Kaneko, and Don Zagier, *Derivation and double shuffle relations for multiple zeta values*, Compos. Math. **142** (2006), no. 2, 307–338, DOI 10.1112/S0010437X0500182X. MR2218898 (2007e:11110)

[20] W. Kohnen and D. Zagier, *Modular forms with rational periods*, Modular forms (Durham, 1983), Ellis Horwood Ser. Math. Appl.: Statist. Oper. Res., Horwood, Chichester, 1984, pp. 197–249. MR803368 (87h:11043)

[21] Ju. I. Manin, *Parabolic points and zeta functions of modular curves* (Russian), Izv. Akad. Nauk SSSR Ser. Mat. **36** (1972), 19–66. MR0314846 (47 #3396)

[22] Georges Racinet, *Doubles mélanges des polylogarithmes multiples aux racines de l'unité* (French, with English and French summaries), Publ. Math. Inst. Hautes Études Sci. **95** (2002), 185–231, DOI 10.1007/s102400200004. MR1953193 (2004c:11117)

[23] Leila Schneps, *On the Poisson bracket on the free Lie algebra in two generators*, J. Lie Theory **16** (2006), no. 1, 19–37. MR2196410 (2007a:17034)

[24] Ismael Soudères, *Motivic double shuffle*, Int. J. Number Theory **6** (2010), no. 2, 339–370, DOI 10.1142/S1793042110002995. MR2646761 (2012c:11139)

[25] Tomohide Terasoma, *Mixed Tate motives and multiple zeta values*, Invent. Math. **149** (2002), no. 2, 339–369, DOI 10.1007/s002220200218. MR1918675 (2003h:11073)

[26] Don Zagier, *Values of zeta functions and their applications*, First European Congress of Mathematics, Vol. II (Paris, 1992), Progr. Math., vol. 120, Birkhäuser, Basel, 1994, pp. 497–512. MR1341859 (96k:11110)

[27] Don Zagier, *Periods of modular forms, traces of Hecke operators, and multiple zeta values*, Sūrikaisekikenkyūsho Kōkyūroku **843** (1993), 162–170. Research into automorphic forms and $L$ functions (Japanese) (Kyoto, 1992). MR1296720 (95i:11048)

[28] D. Zagier, *Evaluation of the multiple zeta values* $\zeta(2, \ldots, 2, 3, 2, \ldots, 2)$, Ann. of Math. (2) **175** (2012), no. 2, 977–1000, DOI 10.4007/annals.2012.175.2.11. MR2993756

Mathematisches Institut der Universität München, Theresienstr. 39, D-80333 München, Germany

*E-mail address*: Sarah.Carr@mathematik.uni-muenchen.de

Department of Mathematical Sciences, South Road, Durham DH1 3LE, United Kingdom

*E-mail address*: herbert.gangl@durham.ac.uk

Institut de Mathématiques de Jussieu, 4 place Jussieu, Case 247, 75252 Paris Cedex, France

*E-mail address*: leila@math.jussieu.fr

# Dyson–Schwinger equations in the theory of computation

Colleen Delaney and Matilde Marcolli

ABSTRACT. Following Manin's approach to renormalization in the theory of computation, we investigate Dyson–Schwinger equations on Hopf algebras, operads and properads of flow charts, as a way of encoding self-similarity structures in the theory of algorithms computing primitive and partial recursive functions and in the halting problem.

## 1. Introduction

There are many deep connections between theoretical physics and information theory, and in particular the theory of computation, see for instance the account given in [**1**]. In the recent papers [**24**] and [**25**], Manin developed a new approach to the theory of computation and the halting problem, based on importing ideas and techniques from the Hopf-algebraic formulation of renormalization in perturbative quantum field theory, [**3**], [**9**], [**17**]. The purpose of this paper is to show that the Hopf algebra of flow charts introduced by Manin in [**24**] exhibits self-similarity structures given by solutions of combinatorial Dyson–Schwinger equations, defined as in perturbative renormalization, [**2**], [**10**], [**11**], [**12**], [**21**], [**34**]. These can be thought of as a notion of "equations of motion" in the theory of computation.

Dyson–Schwinger equations [**7**], [**29**] are a formulation of equations of motion in perturbative quantum field theory, expressed in the form of relations between Green functions. In recent years, an algebraic formulation of the combinatorial structure of perturbative renormalization for scalar field theories was developed, starting with the work of Kreimer on the Hopf-algebraic structure of renormalization [**17**], followed by the Connes–Kreimer formulation [**3**] of the BPHZ renormalization procedure, and the formulation in terms of a Riemann–Hilbert correspondence for categories of differential systems in Connes–Marcolli [**5**], [**6**]. The Connes–Kreimer formulation of renormalization was given a very general algebraic formulation in terms of Rota–Baxter algebras in the work of Ebrahimi-Fard, Guo and Kreimer [**9**]. Correspondingly, the Dyson–Schwinger equations were also given a combinatorial form, reflecting the Hopf-algebraic structure of renormalization, in the work of Bergbauer, Kreimer, and Yeats, [**2**], [**21**], [**34**] and more recently with an extensive study by Foissy, [**10**], [**11**], [**12**].

In this paper we investigate the formulation of Dyson–Schwinger equations in the context of Manin's approach to renormalization and computation. Our main application will be an alternative formulation of a possible "Galois theory of algorithms", similar in spirit to the approach proposed by Yanofsky in [**33**], but more directly related to the Hopf algebra structure of algorithms.

In §2 we review some variants of the construction of a Hopf algebra of *flow charts.* These are diagrams, consisting of decorated planar rooted trees, that compute primitive recursive functions. In §3, we discuss how different version of the Hopf algebra, of flow charts or of "templates", give rise to slightly different notions of *algorithms*, meant in the sense of [**32**] as intermediate levels between programs and functions. In §4 we discuss grafting operators and Dyson–Schwinger equations for flow charts, as a way of identifying certain self-similarity structures in the computation of primitive recursive functions. We review some known results about existence and uniqueness of solutions and on conditions under which the coordinates of the solution generate a Hopf subalgebra. We show that one can also consider ideals generated by the coordinates of solutions. Under the condition that the grafting operator is a cocycle, these can give rise to Hopf ideals. The quotient Hopf algebra determines a subgroup of the affine group scheme dual to the (commutative) Hopf algebra of flow charts. We propose this subgroup as a suitable notion of Galois group, following a point of view closely related to Yanofsky's Galois theory of algorithm [**33**], although the groups we consider are of a different nature than those considered in [**33**]. The group we consider consists of those symmetries that are compatible with the self-similarity structure imposed by the Dyson–Schwinger equation. We then show an explicit example of a Dyson–Schwinger equation based on computation by binary trees that does not satisfy the Hopf algebra condition. In order to accommodate this type of example, that is relevant to the setting of computation, we show that it is convenient to take an operadic viewpoint on Dyson–Schwinger equations, which was already suggested in the work of Bergbauger–Kreimer [**2**]. In §5 we show that Dyson–Schwinger equations can be formulated in the operad setting, using a family of grafting operations, which is not required to satisfy a cocycle condition. We consider Dyson–Schwinger equations on an *operad of flow charts.* We prove the existence and uniqueness of solutions for these operadic Dyson–Schwinger equations and also for a *properad* version. The extension from operads to properads is motivated by the more recent formulation, given by Manin in [**27**], of a *properad of flow charts* based on directed (acyclic) graphs instead of rooted trees. Finally, §6 is more speculative in nature and it focuses on an algebraic Feynman rule associated to the halting problem proposed in [**25**]. We also formulate some general questions about the meaning of the BPHZ renormalization procedure and the role of Dyson–Schwinger equations in the halting problem.

## 2. Primitive recursive functions and the Hopf algebra of flow charts

We recall here the setup of Manin [**24**] on the Hopf algebra of flow charts. First recall from [**23**] the following facts about primitive recursive functions.

**2.1. Primitive recursive functions.** As in §V.2 of [**23**], we consider the class of *primitive recursive functions* as generated by the *basic functions*

- Successor $s : \mathbb{N} \to \mathbb{N}$, $s(x) = x + 1$;
- Constant $c^n : \mathbb{N}^n \to \mathbb{N}$, $c^n(x) = 1$ (for $n \geq 0$);
- Projection $\pi_i^n : \mathbb{N}^n \to \mathbb{N}$, $\pi_i^n(x) = x_i$ (for $n \geq 1$);

With the *elementary operations*

- Composition (substitution) $\mathfrak{c}_{(m,n,p)}$: for $f : \mathbb{N}^m \to \mathbb{N}^n$, $g : \mathbb{N}^n \to \mathbb{N}^p$,
$$g \circ f : \mathbb{N}^m \to \mathbb{N}^p, \quad \mathcal{D}(g \circ f) = f^{-1}(\mathcal{D}(g));$$
- Bracketing (juxtaposition) $\mathfrak{b}_{(k,m,n_i)}$: for $f_i : \mathbb{N}^m \to \mathbb{N}^{n_i}$, $i = 1, \dots, k$,
$$f = (f_1, \dots, f_k) : \mathbb{N}^m \to \mathbb{N}^{n_1+\cdots+n_k}, \quad \mathcal{D}(f) = \mathcal{D}(f_1) \cap \cdots \cap \mathcal{D}(f_k);$$
- Recursion $\mathfrak{r}_n$: for $f : \mathbb{N}^n \to \mathbb{N}$ and $g : \mathbb{N}^{n+2} \to \mathbb{N}$,
$$h(x_1, \dots, x_n, 1) := f(x_1, \dots, x_n),$$
$$h(x_1, \dots, x_n, k+1) := g(x_1, \dots, x_n, k, h(x_1, \dots, x_n, k)), \quad k \geq 1,$$
where recursively $(x_1, \dots, x_n, 1) \in \mathcal{D}(h)$ iff $(x_1, \dots, x_n) \in \mathcal{D}(f)$ and $(x_1, \dots, x_n, k+1) \in \mathcal{D}(h)$ iff $(x_1, \dots, x_n, k, h(x_1, \dots, x_n, k)) \in \mathcal{D}(g)$.

**2.2. Hopf algebras of decorated rooted trees.** We recall here the construction of Hopf algebras of decorated rooted trees and of planar decorated rooted trees and their main properties, see **[11]** and references therein. Throughout this paper vector spaces and algebras are over a field $\mathbb{K}$ of characteristic zero.

2.2.1. *Rooted trees.* A rooted tree is a finite graph $\tau$, whose geometric realization $|\tau|$ is simply connected, defined by combinatorial data $(F_\tau, V_\tau, v_\tau, \delta_\tau, j_\tau)$, with $F_\tau$ a set of half-edges (also called flags or tails), $V_\tau$ the set of vertices with a distinguished element $v_\tau \in V_\tau$ (the root), a boundary map $\partial_\tau : F_\tau \to V_\tau$ that associates to half-edge its boundary vertex and an involution $j_\tau : F_\tau \to F_\tau$, $j_\tau^2 = 1$ that performs the matching of half-edges that determines the edges of $\tau$. The resulting graph has edges (or internal edges) given by pairs of half-edges matched by the involution, and tails (or external edges) given by half-edges that are fixed by the involution. We denote by $E_\tau$ the set of internal edges and by $E_\tau^{ext}$ the set of external edges.

2.2.2. *Orientations.* We consider rooted trees as endowed with the orientation that the root vertex as the *output*, namely where all edges are oriented in the direction of the unique path to the root.

2.2.3. *External edges.* We also assume that the root vertex has an outgoing half-edge and a number of incoming edges, while all other vertices have one outgoing edge and a number of incoming edges and possibly a number of incoming tails. Each leaf of the tree has an incoming tail.

2.2.4. *Planarity.* A *planar* rooted tree is a rooted tree $\tau$ together with a fixed embedding $\iota : |\tau| \hookrightarrow \mathbb{R}^2$ of its geometric realization in the plane.

2.2.5. *Decorations.* A (planar) decorated rooted tree is rooted tree and with a map $\phi_V : V_\tau \to \mathcal{D}_V$ of the set of vertices to a set $\mathcal{D}_V$ of vertex-labels and a map $\phi_F : F_\tau \to \mathcal{D}_F$ to a set $\mathcal{D}_F$ of labels of flags. The assignment of a decoration $(\phi_V, \phi_F)$ to a rooted tree $\tau$ is in general subject to constraints. For example, if $f$ and $f'$ in $F_\tau$ are matched by the involution $j_\tau$ to form an edge $e \in E_\tau$, then they should carry the same decoration $\phi_F(f) = \phi_F(f')$. This defines a labeling of edges. We will see other constraints below, in the specific case of the Hopf algebra of flow charts. We also consider the forgetful map that forgets the flag labels $\mathcal{D}_F$ and retains the vertex labels $\mathcal{D}_V$ and a further forgetful map to unlabeled trees.

2.2.6. *Noncommutative Hopf algebra of planar rooted trees.* Given a (planar) rooted tree $\tau$, an *admissible cut* $C$ of $\tau$ is a modification of the involution $j_\tau$ that *cuts* a subset of internal edges $e_i \in E_\tau$ into two flags $f_i$, $f_i'$. Namely, instead of having $j_\tau(f_i) = f_i'$, one modifies the involution so that the new $\hat{\jmath}(f_i) = f_i$ and $\hat{\jmath}(f_i) = f_i'$, where the set of edges $e_i$ is chosen in such a way that every oriented path in $\tau$ from a leaf vertex to the root contains at most one $e_i$. The new graph obtained using the involution $\hat{\jmath}$ is a forest

$$C(\tau) = \rho_C(\tau) \amalg \pi_C(\tau)$$

consisting of one (planar) rooted tree $\rho_C(\tau)$ containing the root vertex of $\tau$ and a finite disjoint union of other oriented planar trees $\pi_C(\tau) = \amalg_i \pi_{C,i}(\tau)$, where each tree $\pi_{C,i}(\tau)$ has a single output external edge, to which we assign as root the boundary vertex of this output.

The *noncommutative* Hopf algebra of planar rooted trees $\mathcal{H}^{nc}$ is defined, as an algebra, as the free algebra generated by the planar rooted trees, with the coalgebra structure given by the *admissible-cuts* coproduct

$$\Delta(\tau) = \tau \otimes 1 + 1 \otimes \tau + \sum_C \pi_C(\tau) \otimes \rho_C(\tau). \tag{2.1}$$

The coproduct is coassociative and the Hopf algebra is graded by the number of vertices of rooted trees, hence the antipode is defined inductively by the formula

$$S(x) = -x - \sum S(x')x'', \quad \text{for } \Delta(x) = x \otimes 1 + 1 \otimes x + \sum x' \otimes x''$$

with $x'$, $x''$ terms of lower degree. See **[13]**, **[15]** for more details on this Hopf algebra. An element $x$ in a Hopf algebra is primitive if $\Delta(x) = 1 \otimes x + x \otimes 1$.

2.2.7. *Commutative Hopf algebra of rooted trees.* In the Hopf algebraic approach to perturbative renormalization, as formulated by Kreimer **[17]** and Connes–Kreimer **[3]**, one considers a *commutative* Hopf algebra of rooted trees decorated by Feynman graphs, with the admissible-cuts coproduct (2.1), or else a commutative Hopf algebra of Feynman graphs, with a coproduct that corresponds to inclusions of subgraphs (subdivergences) and quotient graphs.

In the case of planar rooted trees, one can also consider a commutative Hopf algebra $\mathcal{H}$, which is a quotient of $\mathcal{H}^{nc}$ by imposing the commutativity of multiplication. The generators of $\mathcal{H}$ are still planar rooted trees, but monomials in these generators no longer identify with *embedded* forests. The grading and antipode are as above, induced by those of $\mathcal{H}^{nc}$.

**2.3. Manin's Hopf algebra of flow charts.** We consider, as in **[24]**, the set of planar labelled rooted trees, where the label set $\mathcal{D}_V$ of vertices is given by the set of elementary operations

$$\{\mathfrak{c}_{(m,n,p)}, \mathfrak{b}_{(k,m,n_i)}, \mathfrak{r}_n\}$$

(composition, bracketing, recursion) and the label set $\mathcal{D}_F$ of the flags is the set of primitive recursive functions. Notice that, because the inputs of the composition, bracketing and recursion operations are ordered, we need to work with *planar* rooted trees.

The data $(\phi_V(v), \phi_F(f_i))$ of a labelings $\phi_V(v) \in \mathcal{D}_V$ and $\phi_F(f_i)$ of a vertex $v \in V_\tau$ and of all the half-edges $f_i \in F_\tau$ with $\partial(f_i) = v$ are *admissible* if they satisfy the following conditions:

- If $\phi_V(v) = \mathfrak{c}_{(m,n,p)}$, then $v$ must have valence three; the labels $h_1 = \phi_F(f_1)$ and $h_2 = \phi_F(f_2)$ of the two incoming flags must have domains and ranges $h_1 : \mathbb{N}^m \to \mathbb{N}^n$ and $h_2 : \mathbb{N}^n \to \mathbb{N}^p$ and the outgoing flag must be labeled by the composition $h_2 \circ h_1 = \mathfrak{c}_{(m,n,p)}(h_1, h_2)$.
- If $\phi_V(v) = \mathfrak{r}_n$, then $v$ must have valence three; the labels $h_1 = \phi_F(f_1)$ and $h_2 = \phi_F(f_2)$ of the two incoming flags must have domains and ranges $h_1 : \mathbb{N}^n \to \mathbb{N}$ and $h_2 : \mathbb{N}^{n+2} \to \mathbb{N}$ and the outgoing flag must be labeled by the recursion $h = \mathfrak{r}_n(h_1, h_2)$.
- If $\phi_V(v) = \mathfrak{b}_{(k,m,n_i)}$ then $v$ must have valence $k+1$ and all the functions $h_i = \phi_F(f_i)$ associated to the incoming flags must have domain in the same $\mathbb{N}^m$. The outgoing flag must be labeled by the bracketing $f = (f_1, \ldots, f_k) = \mathfrak{b}_{(k,m,n_i)}(f_1, \ldots, f_k)$.

DEFINITION 2.1. *An* admissible labeling *of a planar rooted tree $\tau$ by primitive recursive functions is a collection of labelings*

$$\{(\phi_V(v), \phi_F(f_i)) \,|\, v \in V_\tau \text{ and } f_i \in F_\tau \text{ with } \partial(f_i) = v\}$$

*such that the admissibility condition above is satisfied at all vertices and such that $\phi_F(f) = \phi_F(f')$ whenever $j_\tau(f) = f'$.*
*A* vertex labeling *of a planar rooted tree $\tau$ by elementary operations is a collection of labeling $\{\phi_V(v) \in \{\mathfrak{c}, \mathfrak{r}, \mathfrak{b}_k\} \,|\, v \in V_\tau\}$ with the only constraint that labels of type $\mathfrak{c}$ and $\mathfrak{r}$ can only be assigned to vertices with two incoming flags and one outgoing flag and labels of type $\mathfrak{b}_k$ can be assigned to vertices with $k$ incoming flags and one outgoing flag.*

Via the forgetful map that keeps the vertex labels and forgets the flag labels, a planar rooted tree with an admissible labeling by primitive recursive functions determines one with vertex labeling by elementary operations, while not all vertex labeling by elementary operations on a given planar rooted tree will admit a compatible flag labeling. Notice that, in the case of vertex labelings, we no longer distinguish between labels $\mathfrak{c}_{(m,n,p)}$, with different choices of $m$, $n$ and $p$, since we do not fix the flag labels: all these vertex labels in $\mathcal{D}_V$ correspond to the same label $\mathfrak{c}$. Similarly, all the labels $\mathfrak{r}_n$ of admissible labeling correspond to the same label $\mathfrak{r}$ of vertex labeling, and all the $\mathfrak{b}_{(k,m,n_i)}$ correspond to the same $\mathfrak{b}_k$. We retain the different $\mathfrak{b}_k$ as vertex labelings as those correspond to different valences of the vertex.

Thus, a planar rooted tree with an admissible labeling by primitive recursive functions can be interpreted as a *flow chart* that computes the output function starting from the functions associated to the incoming external edges according to the operations performed at the vertices.

A planar rooted tree $\tau$ with a vertex labeling by elementary operations, on the other hand, should be thought of as a template of a possible computational scheme: when applied to a tuple of primitive recursive functions $(h_1, \ldots, h_k)$, with $k$ the number of incoming flags of $\tau$, it either computes an output functions $h$, if the inputs $(h_1, \ldots, h_k)$ together with the assigned vertex labels determine an admissible labeling of $\tau$ by partial recursive function, or else we set its output to be the empty function.

We then define two slightly different versions of the Hopf algebra of flow charts, depending on the type of labeling that we want to use on trees.

DEFINITION 2.2. *The noncommutative Hopf algebra* $\mathcal{H}^{nc}_{\text{flow},\mathcal{P}}$ *of flow charts is the free algebra generated by the planar rooted trees with admissible labelings by primitive recursive functions, with the coproduct* (2.1). *We also denote by* $\mathcal{H}_{\text{flow},\mathcal{P}}$ *the commutative quotient of* $\mathcal{H}^{nc}_{\text{flow},\mathcal{P}}$. *Similarly, the noncommutative Hopf algebra* $\mathcal{H}^{nc}_{\text{flow},\mathcal{V}}$ *is the free algebra generated by planar rooted trees with vertex labeling by elementary operations, again with the admissible-cuts coproduct. We denote by* $\mathcal{H}_{\text{flow},\mathcal{V}}$ *its commutative quotient.*

Notice that if $\tau$ has an admissible labeling, then the rooted trees $\rho_C(\tau)$ and $\pi_{C,i}(\tau)$ also inherit induced admissible labelings, so the Hopf algebras are well defined. There is a Hopf algebra homomorphism $\mathcal{H}^{nc}_{\text{flow},\mathcal{P}} \to \mathcal{H}_{\text{flow},\mathcal{V}}$, determined by forgetting the flag labels, but as observed above not all generators of $\mathcal{H}_{\text{flow},\mathcal{V}}$ are in the image of this map. Since the rooted trees in $\mathcal{H}_{\text{flow},\mathcal{V}}$ have no labels attached to flags, we can represent them as rooted trees without external edges. This will be implicitly done in the description of the grafting operators in Lemma 4.1 below.

In the following, whenever a statement applies to both types of tree decorations on the Hopf algebras of flow charts, we will drop the $\mathcal{P}$ and $\mathcal{V}$ subscripts.

2.3.1. *Binary versus k-ary operations.* It is possible to reformulate the Hopf algebra of flow charts by using only *binary* trees. In fact, it is shown in §2 of **[32]** that the description of primitive recursive functions in terms of basic functions (successor, constants and projections) and elementary operations (composition, bracketing and recursion), as recalled above, can be reformulated as repeated applications of *binary* operations of these same forms. In fact, the composition and recursion operations already are binary (they label vertices with two input flags and one output) so one only needs to verify that the bracketing operation $\mathfrak{b}_{(k,m,n_i)}$ assigning to $k$ functions $f_i : \mathbb{N}^m \to \mathbb{N}^{n_i}$, $i = 1, \dots, k$ their juxtaposition $f = (f_1, \dots, f_k) : \mathbb{N}^m \to \mathbb{N}^{n_1+\cdots+n_k}$ is a composition

$$\mathfrak{b}_{(k,m,n_i)} = \mathfrak{b}_{(2,m,n_1,n_2+\cdots+n_k)} \circ \cdots \circ \mathfrak{b}_{(2,m,n_{k-1},n_k)}.$$

This replaces a single vertex with $k$ inputs and one output with $k$ vertices with two inputs and one output. Thus, without loss of generality in the algorithmic representation of primitive recursive functions, one can consider a sub-Hopf algebra $\mathcal{H}^{nc,0}_{\text{flow}} \subset \mathcal{H}^{nc}_{\text{flow}}$ generated by *binary* planar rooted tree with an admissible labeling by primitive recursive functions, and the corresponding commutative quotient $\mathcal{H}^{0}_{\text{flow}} \subset \mathcal{H}_{\text{flow}}$.

One the other hand, one can also extend the two binary operations of composition and recursion to $k$-ary operations for arbitrary $k \geq 2$. In the case of composition, we can introduce a new set of vertex labels $\mathfrak{c}_{(k,n_i)}$ that correspond to the $k$-ary compositions $\mathfrak{c}_{(k,m,n_i)}(h_i) = h_k \circ \cdots \circ h_1$ of functions $h_i : \mathbb{N}^{n_{i-1}} \to \mathbb{N}^{n_i}$, for $i = 1, \dots, k$, with $n_0 = m$. In the case of recursion, for $k \geq 1$, one can consider $(k+1)$-ary recursions where the recursion depends on $k$ initial conditions. Namely, we define $\mathfrak{r}_{k+1,n}$ as the operation that computes the function $h = \mathfrak{r}_{k,n}(h_1, \dots, h_{k+1}) : \mathbb{N}^{n+1} \to \mathbb{N}$, from the input $h_i : \mathbb{N}^n \to \mathbb{N}$ for $i = 1, \dots, k$

and $h_{k+1} : \mathbb{N}^{2n+1} \to \mathbb{N}$, as
(2.2)

$$\begin{aligned} &h(x_1, \dots, x_n, 1) = h_1(x_1, \dots, x_n), \\ &\vdots \\ &h(x_1, \dots, x_n, k) = h_k(x_1, \dots, x_n), \\ &h(x_1, \dots, x_n, k+\ell) = h_{k+1}(x_1, \dots, x_n, h_1(x_1, \dots, x_n), \dots, h_k(x_1, \dots, x_n), k+\ell-1), \\ &\text{for } \ell \geq 1 \end{aligned}$$

These $k$-ary operations are clearly reducible to compositions of binary operations, hence any primitive recursive function that can be computed using these operations can also be computed by the presentation in terms of binary compositions and recursions. However, as we shall discuss more in detail below, it is convenient to include these explicitly in the structure of the Hopf algebra of flow charts in order to have well defined grafting operators that give rise to Hochschild cocycles and to Dyson–Schwinger equations.

Thus, we propose another minor modification of Manin's Hopf algebra of flow charts, as follows.

DEFINITION 2.3. *We denote by $\mathcal{H}^{nc}_{\text{flow},\mathcal{P}'}$ and $\mathcal{H}^{nc}_{\text{flow},\mathcal{V}'}$ the noncommutative Hopf algebras of flow charts where the vertex label sets is enlarged to contain all the $k$-ary compositions and recursions $\mathfrak{c}_{(k,n_i)}$ and $\mathfrak{r}_{k,n}$, and the admissibility condition is restated accordingly. We denote by $\mathcal{H}_{\text{flow},\mathcal{P}'}$ and $\mathcal{H}_{\text{flow},\mathcal{V}'}$ their commutative quotients.*

## 3. Flow charts, templates, and algorithms

In the above, we have introduced different versions of rooted trees with vertex and edge decorations. We discuss here briefly the meaning of the objects considered above from the point of view of computation.

**3.1. Flow charts versus templates.** A rooted tree with vertices decorated by elementary operations and edges compatibly decorated by recursive functions is a *flow chart*. It describes a concrete realization of a primitive recursive function (the output) in terms of a computation that starts with other primitive recursive functions, possibly basic (the inputs of the tree) with the operations performed at the vertices.

By extension, we also refer to arbitrary elements of the Hopf algebra $\mathcal{H}^{nc}_{\text{flow},\mathcal{P}}$ as "flow charts". These are formal sums of forests, with a planar embedding, decorated as above by primitive recursive functions at edges and elementary operations at vertices. In this extended sense a flow charts computes a formal sum of ordered sequences of primitive recursive functions, given by the collection of the outputs of each of the trees.

We also considered a different type of objects, given by rooted trees where only the vertices are decorated by elementary operations, while the edges are undecorated. These are computational architectures where the inputs are open: once a set of primitive recursive functions is assigned as input of such a tree, one obtains a primitive recursive function as output, computed according to the operations specified at the vertices. We refer to these trees with vertex decorations as *templates*.

By extension we also refer to an arbitrary element of the Hopf algebra $\mathcal{H}^{nc}_{\text{flow},\mathcal{V}}$ as a *template*. These are formal sums of forests with vertex decorations by elementary operations.

The difference between working with flow charts (vertex and edge decorations) or with what we call here templates (vertex decorations only) will become more transparent when one considers implementing certain "reasonable equivalence relations", which describe, in the general philosophy of Yanofsky [**32**], an intermediate level of *algorithms*, in between programs (flow charts) and primitive recursive functions.

**3.2. The notion of algorithm.** Yanofsky's work [**32**] provides an in depth discussion of the notion of *algorithm*. To summarize quickly his general approach, one considers equivalence classes consisting of different programs that implement "the same algorithm". In this view, algorithms describe any intermediate level between the programs (the flow charts in our language above) and the primitive recursive functions they compute. Rather than fixing precisely what the intermediate level should be, Yanofsky allows for some freedom, by considering all the possible quotients of the set of programs by "reasonable equivalence relations" and organizing all these intermediate levels into a hierarchy governed by a suitable kind of Galois correspondence, [**32**], [**33**].

In Section 3 of [**32**], Yanofsky gives a list of examples of "reasonable equivalence relations" one can impose on programs. These include: associativity of composition, projections acting as identity, distributivity of composition over bracketing, associativity of bracketing, almost-commutativity of bracketing, functoriality of bracketing, idempotency of twists, Reidemeister moves, relation between recursion and bracketing, relation between recursion and composition, relation between recursion and the successor function.

Our viewpoint here follows closely this idea of Yanofsky, but with one important difference. We consider the existence of a Hopf algebra structure on flow charts as an essential part of the data and we therefore require that a notion of "reasonable equivalence relation" should include the compatibility with the Hopf algebra structure. This means requiring that equivalence relations among flow charts should define Hopf ideals.

Notice how some of the relations listed above from [**32**] reflect properties (associativity, distributivity, almost-commutativity) of the elementary operations, which are independent of the input functions, while others involve relations between the elementary operations and certain specific input functions (projections, successor). The different nature of these two types of relations will be reflected, in our viewpoint, in the fact that the first type of relations can be implemented both at the level of the Hopf algebra of flow charts $\mathcal{H}^{nc}_{\text{flow},\mathcal{P}}$ and at the level of the Hopf algebra of templates $\mathcal{H}^{nc}_{\text{flow},\mathcal{V}}$, while the second type of relation can only be implemented in $\mathcal{H}^{nc}_{\text{flow},\mathcal{P}}$.

To see a concrete example of both cases, let us first consider the relation that corresponds to associativity of composition, as stated in §3.1.1 of [**32**]. This implies identifying trees of the form

(3.1)
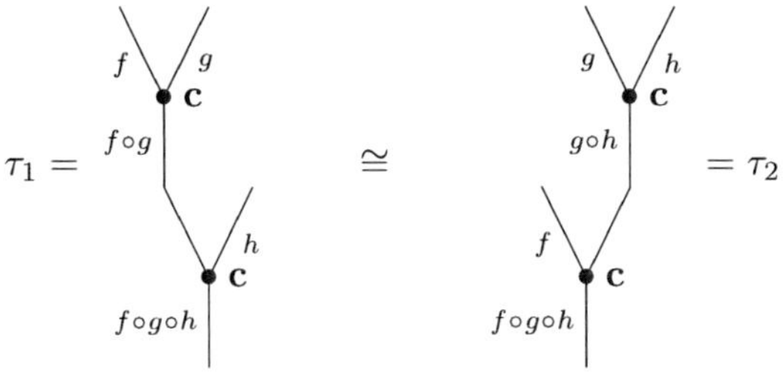


In order to implement this relation at the level of the Hopf algebra, we need to construct a Hopf ideal. The ideal $\mathcal{I} = (\tau_1 - \tau_2)$ in the underlying algebra would not, by itself, be a Hopf ideal, because the non-primitive part of the coproduct $\Delta(\tau_1 - \tau_2)$ would not be in $\mathcal{H} \otimes \mathcal{I} \oplus \mathcal{I} \otimes \mathcal{H}$, but if we also consider the relation identifying the trees

(3.2)
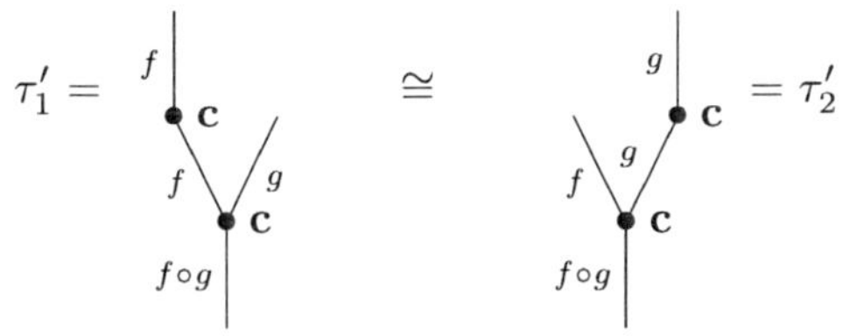


and the ideal $\mathcal{I} = (\tau_1 - \tau_2, \tau_1' - \tau_2')$, then the coproduct satisfies $\Delta(\mathcal{I}) \subset \mathcal{H} \otimes \mathcal{I} \oplus \mathcal{I} \otimes \mathcal{H}$. The resulting Hopf ideal is the appropriate way, in our setting, to implement the "associativity of composition" relation.

Notice that the relations defining $\mathcal{I}$ do not depend on particular input functions being assigned to the incoming edges, hence they not only define a Hopf ideal in the Hopf algebra $\mathcal{H}^{nc}_{\mathrm{flow},\mathcal{P}}$ of flow charts, but they also define a Hopf ideal in the Hopf algebra $\mathcal{H}^{nc}_{\mathrm{flow},\mathcal{V}}$ of templates.

We then consider, as a second example, the relation defined in §3.1.2 of [**32**], which represents the fact that projection $\pi^n = (\pi_i^n) : \mathbb{N}^n \to \mathbb{N}^n$ is the identity. This relation is implemented by identifying, for any primitive recursive function $f : \mathbb{N}^n \to \mathbb{N}^m$, the decorated rooted trees

(3.3)
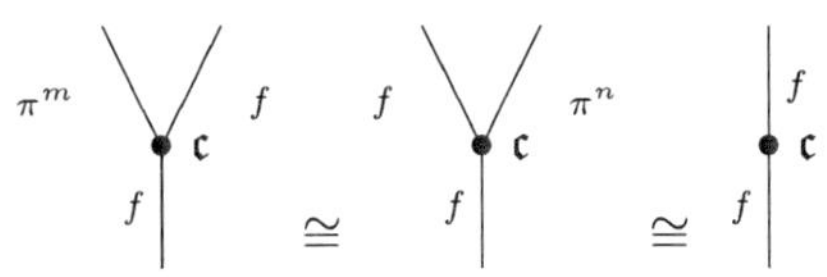


As in the previous case, one can associate to this relation a Hopf ideal in $\mathcal{H}^{nc}_{\mathrm{flow},\mathcal{P}}$. However, in this case, the relation involves not only a property of the elementary operations that label the vertices but also of a specific choice of input functions, hence it does not define a Hopf ideal in $\mathcal{H}^{nc}_{\mathrm{flow},\mathcal{V}}$.

Thus, from the point of view of computation and algorithms, the difference between working with flow charts or with templates corresponds to two different notions of algorithms. In the first case algorithms are equivalence classes of flow charts under relations that can involve both properties of the elementary operations that hold for arbitrary input functions and properties that hold for specific input functions, while in the case of templates, one only allows the first type of relations in the definition of algorithms.

**3.3. Symmetries: a physics motivated perspective.** We will be discussing, especially in relation to Dyson–Schwinger equations, notions of *symmetry* in the context of Hopf algebras. The main motivating example for the notion of symmetry we adopt is again coming from the setting of quantum field theory, where gauge symmetries are implemented at the quantum level in the form of Ward (or Slavonov–Taylor) identities. It was shown by van Suijlekom in **[30]** that the Slavonov–Taylor identities define a Hopf ideal in the Connes–Kreimer Hopf algebra of Feynman graphs. It is therefore a natural approach from the physics perspective to require that a good notion of symmetry in a setting where the main objects of interest (Feynman graphs, flow charts, etc.) are described in terms of a Hopf algebra should be expressed in terms of Hopf ideals. We adopt this viewpoint in the following, in §4.5, where we propose a different approach to Yanofsky's Galois theory of algorithms, using Hopf ideals in the Hopf algebra of flow charts as a good notion of equivalence relations defining intermediate levels (algorithms) between flow charts and primitive recursive functions, and in §4.4.1, where we consider Hopf ideals associated to solutions of Dyson–Schwinger equations.

## 4. Dyson–Schwinger equations in the Hopf algebra of flow charts

Recall here the formulation of combinatorial Dyson–Schwinger equations in Hopf algebras of decorated rooted trees, following **[2]**, **[34]** and especially **[10]**, **[11]**, **[12]**. We then focus on the specific case of the Hopf algebras of flow charts described above.

**4.1. Insertion operators for decorated planar rooted trees.** We work here with the Hopf algebra of flow charts $\mathcal{H}^{nc}_{\text{flow},\mathcal{V}'}$. By analogy with the setting of perturbative renormalization **[2]**, **[34]**, we define, for each type of elementary operation $\delta \in \{\mathfrak{b}, \mathfrak{c}, \mathfrak{r}\}$, a grafting operator $B^+_\delta$ defined in the following way. Given a monomial $T$ in $\mathcal{H}^{nc}_{\text{flow},\mathcal{V}'}$, which consists of a forest with vertex labeling by elementary operations, we define $B^+_\delta(T)$ as the sum of planar graphs obtained by adding a new root vertex with a number of incoming flags equal to the number of trees in $T$ and a single output flag, with the vertex decorated by $\delta$.

4.1.1. *Hochschild 1-cocyles.* In the case of perturbative renormalization, it is well known **[2]** that the analogous grafting operators $B^+_\delta$, with vertex decorations $\delta$, define Hochschild 1-cocycles.

The 1-cocycle condition for the operator $B^+_\delta$ then consists of the property that

$$\Delta B^+_\delta = (id \otimes B^+_\delta)\Delta + B^+_\delta \otimes 1. \tag{4.1}$$

LEMMA 4.1. *The operators $B^+_\delta$ on $\mathcal{H}^{nc}_{\text{flow},\mathcal{V}'}$, with $\delta \in \{\mathfrak{b}, \mathfrak{c}, \mathfrak{r}\}$, satisfy the 1-cocycle condition* (4.1).

PROOF. This case works exactly as in the renormalization setting, see **[2]**: we reproduce the argument here for the reader's convenience. For an element $x$ in a Hopf algebra $\mathcal{H}$ with coproduct

$$\Delta(x) = 1 \otimes x + x \otimes 1 + \sum x' \otimes x'',$$

where $x'$ and $x''$ denote lower degree terms, we use the notation

$$\tilde{\Delta}(x) := \sum x' \otimes x''.$$

One then sees easily that the cocycle condition (4.1) is equivalent to

$$\tilde{\Delta} B_\delta^+ = (id \otimes B_\delta^+)\tilde{\Delta} + id \otimes B_\delta^+(1), \tag{4.2}$$

where, as in the renormalization case, with $B_\delta^+(1) = v_\delta$ is a single vertex with assigned label $\delta$. The argument is then as in the original case, with the first term in the right hand side of (4.2) accounting for the admissible cuts where the root vertex remains attached to the $\rho_C(T)$ part of an admissible cut of $T$, and the second term counts the case where the admissible cut separates the root vertex completely. □

Notice that, for the cocycle condition to hold, we need to be able to assign the same type of label ($\mathfrak{b}$, $\mathfrak{c}$, or $\mathfrak{r}$) to vertices of arbitrary valence, hence the reason for including in the set of vertex decorations all the $k$-ary versions of composition and recursion and working with the version $\mathcal{H}^{nc}_{\text{flow},\mathcal{V}'}$ of the Hopf algebra of flow charts.

It is possible to define other interesting grafting operators in the Hopf algebras of flow charts, which, however, do not satisfy the cocycle condition. These can still be used to construct Dyson–Schwinger equations, but the set of solutions does not define a Hopf subalgebra. This will be discussed in an example in the following subsection.

**4.2. Systems of combinatorial Dyson–Schwinger equations.** Given a graded Hopf algebra $\mathcal{H}$, we consider the direct product $\overline{\mathcal{H}} = \prod_{n=0}^\infty \mathcal{H}_n$, whose elements we write as infinite sums $x = \sum_{n=0}^\infty x_n$, with $x_n \in \mathcal{H}_n$, endowed with the associative product and coproduct induced by those of $\mathcal{H}$. Following [**10**], (see also [**2**], [**34**]) given a formal power series

$$P(t) = \sum_{k=0}^\infty a_k t^k, \quad \text{with } a_0 = 1,$$

and a Hochschild 1-cocycle $B^+$ on the Hopf algebra $\mathcal{H}$, the associated Dyson–Schwinger equation is given by

$$X = B^+(P(X)). \tag{4.3}$$

One considers this as an equation in $\overline{\mathcal{H}}$ and interprets the infinite sum $P(X) = \sum_k a_k X^k$ as an element in $\overline{\mathcal{H}}$.

One should read (4.3) as a fixed point equation for the nonlinear transformation

$$X \mapsto B^+(P(X))$$

of $\overline{\mathcal{H}}$. Solutions to the fixed point equation are elements of $\overline{\mathcal{H}}$ that exhibit a self-similarity property with respect to the operation of first applying $X \mapsto P(X)$ (in the case of a polynomial, this operation heuristically replaces $X$ by a sum of multiple copies of itself weighted by the coefficients of the polynomial) and then grafting them together according to the cocycle $B^+$.

In fact, the equation (4.3) has a unique solution given by an element $X = \sum_k x_k$ with $x_k \in \mathcal{H}_k$ whose homogeneous components are determined inductively by the procedure

$$x_{n+1} = \sum_{k=1}^{n} \sum_{j_1+\cdots+j_k=n} a_k \, B^+(x_{j_1}\cdots x_{j_k}), \tag{4.4}$$

starting with $x_1$ given by $B^+(1)$, see Lemma 2 of [**2**] and Proposition 2 of [**10**].

Notice that we can still define the Dyson–Schwinger equation (4.3) for other types of grafting operators that do not necessarily satisfy the 1-cocycle condition (4.1), such as the examples discussed in the previous subsection. The existence and uniqueness of solutions, of the form (4.4) is also still valid. Where the 1-cocycle condition is crucially used is in showing that the associative subalgebra of $\mathcal{H}$ generated by the components $x_k$ of the solution is also a Hopf subalgebra, see Theorem 3 of [**2**].

4.2.1. *Systems of Dyson–Schwinger equations.* As discussed in the work of Foissy [**11**], in cases like our $\mathcal{H}^{nc}_{\text{flow},\mathcal{V}'}$, where one has different vertex labels $\delta \in \{\mathfrak{b}, \mathfrak{c}, \mathfrak{r}\}$, one can consider more complicated *systems* of Dyson–Schwinger equations, involving the grafting operators $B^+_\delta$ on $\mathcal{H}^{nc}_{\text{flow},\mathcal{V}'}$.

For $\delta \in \{\mathfrak{b}, \mathfrak{c}, r\}$, we consider the grafting cocycles $B^+_\delta$ on the Hopf algebra $\mathcal{H}^{nc}_{\text{flow},\mathcal{V}'}$, as above. We also consider data of three non-constant formal power series in three variables $X = (X_\delta)$

$$F_\delta(X) = \sum_{k_1,k_2,k_3} a^{(\delta)}_{k_1,k_2,k_3} X_{\mathfrak{b}}^{k_1} X_{\mathfrak{c}}^{k_2} X_{\mathfrak{r}}^{k_3}.$$

To these data we associate the system of Dyson–Schwinger equations as in [**11**]

$$X_\delta = B^+_\delta(F_\delta(X)). \tag{4.5}$$

As shown in Proposition 5 of [**11**], these systems of equations also have a unique solution of the form $X_\delta = \sum_\tau x_\tau \, \tau$, with the sum over all planar rooted trees with root decorated by $\delta$, with coefficients

$$x_\tau = \Big(\prod_{k=1}^{3} \frac{(\sum_{l=1}^{m_k} p_{\delta,l})!}{\prod_{l=1}^{m_k} p_{\delta,l}!}\Big) a^{(\delta)}_{\sum_{k=1}^3 p_{1,k}, \sum_{k=1}^3 p_{2,k}, \sum_{k=1}^3 p_{3,k}} x_{\tau_{1,1}}^{p_{1,1}} \cdots x_{\tau_{3,m_3}}^{p_{3,m_3}}, \tag{4.6}$$

when

$$\tau = B^+(\tau_{1,1}^{p_{1,1}} \cdots \tau_{1,m_1}^{p_{1,m_1}} \cdots \tau_{3,1}^{p_{3,1}} \cdots \tau_{3,m_3}^{p_{3,m_3}}).$$

**4.3. Dyson–Schwinger solutions: a computational perspective.** The solution $X_\delta = \sum_\tau x_\tau \tau$ of a Dyson–Schwinger equation $X_\delta = B^+_\delta(F_\delta(X))$ as above, is a formal infinite linear combination of templates $\tau$, with coefficients $x_\tau$ given by recursively defined computable functions of the coefficients $a^{(\delta)}_{k_1,k_2,k_3}$ of $F_\delta$. It takes an input $f^{\text{input}} = (f^{\text{input}}_\tau)_\tau$ consisting of a sequence of primitive recursive functions, where $f^{\text{input}}_\tau = (f^{\text{input}}_{\tau,1}, \dots, f^{\text{input}}_{\tau,n_\tau})$ are the $n_\tau$ inputs of a template $\tau$ given by a tree with $n_\tau$ leaves. It produces an output given by a formal linear combination $\sum_\tau x_\tau f^{\text{output}}_\tau$, where $f^{\text{output}}_\tau$ is the output of the template $\tau$ for the given input $f^{\text{input}}_\tau$. More precisely, to avoid dealing with infinite linear combinations, one can consider the individual components of the solution obtained by recursively solving the Dyson–Schwinger equation.

For simplicity, we look at how this works in a toy-model example of Dyson–Schwinger equation. Let us restrict to only one possible type of vertex labeling by composition $\mathfrak{c}$, and consider an equation of the form $X = 1 + B^+_{\mathfrak{c}}(X^2)$. The components $x_n$ of the solution are then obtained recursively as

$x_0 = 1, \quad x_1 =$  $, \quad x_2 =$  ,

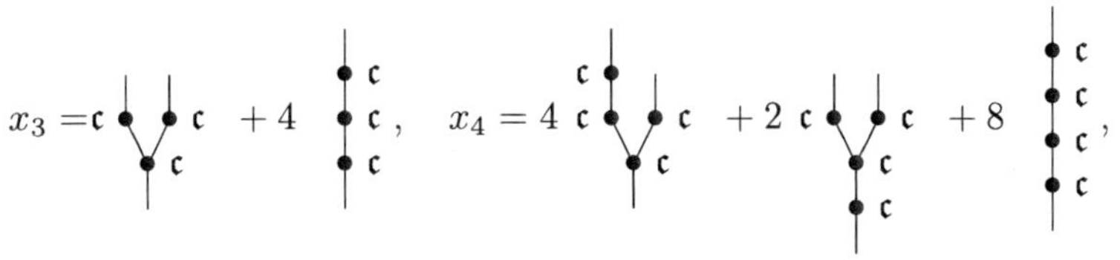


and so on. More realistically, one would have to count also templates with other possible labelings of vertices besides $\mathfrak{c}$, but for simplicity let us focus on this case. Then, for example, we interpret $x_3$ as a template that, for a given input $(f_1, f_2)$, consisting of a pair of primitive recursive functions computes the output function $6f_1 \circ f_2 + 8f_1$. Notice that the self-similar structure of the Dyson–Schwinger equation is not apparent when one looks at one individual component of the solution but only in the overall recursive relation $x_{n+1} = \sum_{k=0}^{n} B_{\mathfrak{c}}^{+}(x_k x_{n-k})$.

**4.4. Hopf subalgebras and Hopf ideals.** Bergbauer and Kreimer showed that, under the assumption that $B_{\delta}^{+}$ is a Hochschild cocycle, solutions of Dyson–Schwinger equations determine a sub-Hopf algebra. Namely, they considered Dyson–Schwinger equation in a Hopf algebra $\mathcal{H}$ of the form

$$X = 1 + \sum_{n=1}^{\infty} c_n \, B_{\delta}^{+}(X^{n+1}). \tag{4.7}$$

One then considers the associative algebra $\mathcal{A}$ defined as the *subalgebra* of $\mathcal{H}$ generated by the components $x_n$, with $n \geq 0$ of the unique solution of this Dyson–Schwinger equations. In Theorem 3 of [**2**], they showed that $\mathcal{A}$ is in fact a *sub-Hopf* algebra, by inductively using the cocycle condition (4.1) to see that

$$\Delta(x_n) = \sum_{k=0}^{n} \Pi_k^n \otimes x_k, \quad \text{where} \quad \Pi_k^n = \sum_{j_1+\cdots+j_{k+1}=n-k} x_{j_1} \cdots x_{j_{k+1}}. \tag{4.8}$$

This result was extended by Foissy to the more general form (4.3) of Dyson–Schwinger equations in [**10**] and to systems of Dyson–Schwinger equations in [**11**]. In Theorem 4 of [**10**] it is shown that, for an equation of the form (4.3), the subalgebra $\mathcal{A}$ spanned by the solutions is a Hopf subalgebra if and only if the formal series $P(t)$ satisfies the differential equation

$$(1 - \alpha\beta t)P'(t) = \alpha P(t), \quad \text{with} \quad P(0) = 1, \tag{4.9}$$

for some $\alpha, \beta \in \mathbb{K}$, again under the assumption that $B^+$ satisfies the cocycle condition (4.1). Similarly, in [**11**] combinatorial conditions and conditions on the multivariable series $F_i$ are identified that completely characterize when the solutions of that systems of Dyson–Schwinger equations generate a Hopf algebra, under the assumption that the $B^+$ satisfies the cocycle condition (4.1).

4.4.1. *Ideals of Dyson–Schwinger solutions.* It is natural to consider not only the associative *subalgebra* $\mathcal{A}$ spanned by the components of the solutions of Dyson–Schwinger equations, but also the *ideal* $\mathcal{I}$ spanned by the $(x_n)_{n\geq 1}$. As in the case of the subalgebra $\mathcal{A}$, it is also natural to ask for conditions that will ensure that it is a Hopf subalgebra, so with the ideal $\mathcal{I}$ of solutions it is natural to ask for conditions ensuring that it is a Hopf ideal, so that it makes sense to define a quotient Hopf algebra, playing the role of the ring of functions on the variety of solutions.

As an example, we consider here the Dyson–Schwinger equation (4.3), with $B^+ = B_{\mathfrak{b}}^{+}$ the grafting operator to a vertex labelled with the bracketing operation.

Assuming that parallel computations that do not feed into each other can be inputed in arbitrary order at the vertex, we can regard this grafting operator as descending to the commutative quotient $\mathcal{H}_{\text{flow},\mathcal{V}'}$ of the noncommutative Hopf algebra $\mathcal{H}^{nc}_{\text{flow},\mathcal{V}'}$. This heuristic rationale for passing to the commutative quotient does not take into account a proper definition of a notion of parallel processes, which would require considering the notion of leveled trees and, in the binary case, Hopf algebra structures related to permutohedra instead of associahedra. We refer the reader to [**14**], [**16**], [**22**].

Let $X = \sum_{n\geq 1} x_n$ be the unique solution of (4.3), obtained as in (4.4), with $x_1 = B^+_{\mathfrak{b}}(1) = v_{\mathfrak{b}}$ a single vertex labeled with the $\mathfrak{b}$ operation. We define $\mathcal{I}$ to be the ideal in the commutative algebra $\mathcal{H}_{\text{flow},\mathcal{V}'}$ generated by the $x_n$ with $n \geq 1$.

The same conditions that ensures that $\mathcal{A}$ is a Hopf subalgebra also ensure that $\mathcal{I}$ is a Hopf ideal. We verify it in the case of (4.7). The more general case of [**10**] is similar.

LEMMA 4.2. *The ideal $\mathcal{I}$ generated by the components $x_n$ with $n \geq 1$ of the solution of* (4.7) *is a Hopf ideal.*

PROOF. Elements of the ideal $\mathcal{I}$ in $\mathcal{H} = \mathcal{H}_{\text{flow},\mathcal{V}'}$ are finite sums $\sum_{m=1}^{M} h_m x_{k_m}$, with $h_m \in \mathcal{H}$ and $x_k$ the coordinates of the unique solution of (4.7). The condition that $\mathcal{I}$ is a Hopf ideal is that $\Delta(\mathcal{I}) \subset \mathcal{I} \otimes \mathcal{H} \oplus \mathcal{H} \otimes \mathcal{I}$. Using the formula (4.8) for the coproduct of the elements $x_k$, we see that we obtain a sum of terms of which the primitive part $1 \otimes x_k + x_k \otimes 1$ is in $\mathcal{H} \otimes \mathcal{I} \oplus \mathcal{I} \otimes \mathcal{H}$ and all the other terms $\tilde{\Delta}(x_k)$ are in $\mathcal{I} \otimes \mathcal{I}$. Then the coproducts $\Delta(h_m x_{k_m})$ will also be in $\mathcal{H} \otimes \mathcal{I} \oplus \mathcal{I} \otimes \mathcal{H}$. □

Notice that we can also work with the noncommutative Hopf algebra $\mathcal{H}^{nc}_{\text{flow},\mathcal{V}'}$ and the two sided ideal with elements of the form $\sum_m h_m x_{k_m} \ell_m$, with $h_m, \ell_m \in \mathcal{H}^{nc}_{\text{flow},\mathcal{V}'}$, but for the considerations that follow (see §4.5 below) it is more natural to work with the commutative quotient.

**4.5. Yanofsky's Galois theory of algorithms.** As we have seen above, one can pass to commutative quotients $\mathcal{H}_{\text{flow}}$ of the noncommutative Hopf algebras of flow charts. The meaning of passing to the commutative quotient $\mathcal{H}_{\text{flow}}$ is best expressed as follows: one can think of monomials in $\mathcal{H}^{nc}_{\text{flow}}$ as diagrams for parallel computations of a certain number of outputs (one for each planar tree in the forest). Imposing commutativity then corresponds to the reasonable expectation that if one has two parallel computations that do not feed one into the other, then computing them in either order will not affect the result. However, when one introduces operations that graft the output of one tree to the input of another, with $\mathfrak{c}$ and $\mathfrak{r}$ labels at the vertices, imposing the commutativity relation is no longer appropriate, as the result of grafting would not be well defined as a *planar* tree. In the case of grafting with $\mathfrak{c}$ label, we can still pass to the commutative quotient, as argued above.

4.5.1. *Commutative Hopf algebras and affine group schemes.* An advantage of working with commutative Hopf algebras is that they are dual to affine group schemes. Thus, the Hopf algebra $\mathcal{H}_{\text{flow}}$ determines an affine group scheme $G_{\text{flow}}$ such that, for any commutative $\mathbb{K}$-algebra $A$, one obtains a group $G_{\text{flow}}(A) = \text{Hom}(\mathcal{H}_{\text{flow}}, A)$, the set of homomorphisms of $\mathbb{K}$-algebras, with the product $(\phi_1 \star$

$\phi_2)(x) = \langle \phi_1 \otimes \phi_2, \Delta(x) \rangle$ dual to the coproduct and the inverse determined by the antipode.

4.5.2. *Galois groups of algorithms.* The idea elaborated by Yanofsky in the recent work **[33]**, of a Galois theory of algorithms, aims at considering all possible "reasonable sets of relations" on programs that would correspond to "implementing the same algorithm", viewing "algorithms" as an intermediate level between the labelled planar rooted trees and the primitive recursive functions they compute. Our point of view here is similar and directly inspired by the approach proposed by Yanofsky, with the main difference that we seek to implement relations at the level of the Hopf algebra $\mathcal{H}^{nc}_{\text{flow}}$ of flow charts.

The automorphism groups that arise in Yanofsky's setting are typically products of symmetric groups and the notion of equivalence relations considered does not take into account any additional structure on flow charts, such as Manin's Hopf algebra point of view. Moreover, a possible drawback of the current formulation of the approach of **[33]** lies in the fact that different symmetry groups can have the same orbit space and give rise to the same equivalence on algorithms. We propose instead to consider a different approach where equivalence relations on algorithms are imposed at the level of the Hopf algebra of flow charts.

Our proposal for an alternative approach to a Galois theory of algorithms is to take into account the Hopf algebra structure as part of the construction. By this we mean defining as "reasonable relations" between the planar rooted trees of $\mathcal{H}^{nc}_{\text{flow}}$ those that correspond to passing to quotient Hopf algebras. Each such quotient is determined by a Hopf ideal $\mathcal{I}$ in $\mathcal{H}^{nc}_{\text{flow}}$, the kernel of the map to the quotient Hopf algebra $\mathcal{H}_{\mathcal{I}}$.

In the commutative setting $\mathcal{H}_{\text{flow}}$ one can view, dually, the quotient Hopf algebras $\mathcal{H}_I$ as sub-group schemes $G_I$ of the affine group scheme of characters $G_{\text{flow}}$. The group schemes $G_I$ obtained in this way replace, in our approach, the basic objects of interest in Yanofsky's Galois theory of algorithms.

4.5.3. *Hopf ideals and Galois groups.* In light of Yanofsky's point of view, it is natural to consider ideals in $\mathcal{H}_{\text{flow}}$ generated by solutions of Dyson–Schwinger equations, and the resulting quotient Hopf algebras, rather than sub-Hopf algebras. The Hopf ideals and quotient Hopf algebras define, dually, the affine group schemes that provide our proposed notion of Galois groups of algorithms.

As a heuristic geometric interpretation, in the Hopf algebraic setting, we can think of the quotient $\mathcal{H}/\mathcal{I}$ of a Hopf algebra by a Hopf ideal as an analog of the algebro-geometric quotient $R/I$ of a polynomial ring $R$ by a prime ideal $I$, which provides the coordinate ring of the affine variety $V$ defined by $I$. In the case of the Hopf ideal $\mathcal{I}$ determined by the components of the solution of the Dyson–Schwinger equation, we can therefore think of $\mathcal{H}/\mathcal{I}$ as the "ring of function" of the variety of solutions. The associated group scheme $G_I$ consists of those symmetries in $G_{\text{flow}}$ that are compatible with the self-similarity structure described by the Dyson–Schwinger equation.

PROPOSITION 4.3. *A Dyson–Schwinger equation* (4.7) *in the commutative Hopf algebra* $\mathcal{H}_{\text{flow}}$ *determines a Galois group* $G_P \subset G_{\text{flow}}$.

This is simply a reformulation of Lemma 4.2, with our notion of what we consider to be an appropriate notion of Galois group. A similar result holds for

Dyson–Schwinger equations (4.3) with (4.9). While, on the one hand, our point of view takes into account more information, coming from the Hopf algebra structure, at another level it is less structured than Yanofsky's setting: in fact, we work here with the Hopf algebra where flow charts only have vertex decorations, so that the input functions (and the corresponding edge labelings through the graph) are not specified, while in Yanofsky's setting the underlying functions are assigned and preserved by the symmetries.

**4.6. A non-Hopf example with binary trees.** In the setting of Dyson–Schwinger equations described above, it is crucial that we consider the planar rooted trees in $\mathcal{H}^{nc}_{\text{flow},\mathcal{V}'}$ without external edges. In fact, this allows us to define the grafting operators $B^+_\delta$ as grafting any number of trees to the *same* root consisting of a single vertex decorated by $\delta$. If we reintroduce external edges, then the graphs consisting of a single vertex but with different numbers of incoming edges are viewed as different generators in the Hopf algebra and they no longer define a cocycle $B^+_\delta$.

To see this more explicitly, we focus on the case of binary trees, keeping track of external edges. Consider the following primitive elements in the Hopf algebra $\mathcal{H}^{nc,0}_{\text{flow},\mathcal{V}}$:

(4.10)

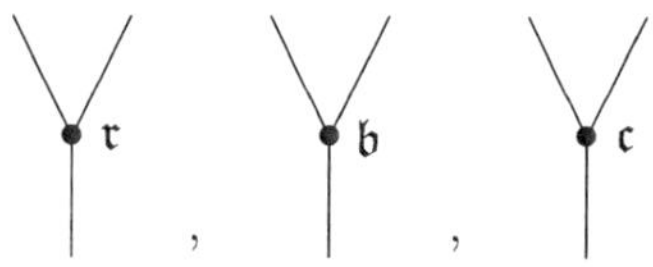

For each of these graphs $\tau$ one can define a grafting operation $B^+_\tau$ by setting the result equal to zero on monomials $T$ that cannot be grafted to $\tau$. This $B^+_\tau$ clearly does *not* satisfy the cocycle condition because one would have to have $B^+_\tau(1) = 0$, but then the expressions $\tilde{\Delta} B^+_\tau$ and $(id \otimes B^+_\tau)\tilde{\Delta}$, when applied to elements $T = \tau_1 \amalg \tau_2$ that are not in the kernel, would differ by a term corresponding to the cut that separates $\tau$ from $T$. On the other hand, if $B^+_\tau(1) = \tau$ as this case would require, then the cocycle condition fails on elements in the kernel of $B^+_\tau$.

Another way of defining a grafting operator $B^+_\tau$ with $\tau$ one of the trees (4.10) is to define, for a monomial $T = \tau_1 \amalg \cdots \amalg \tau_n$ the grafting $B^+_\tau(T)$ to be obtained by just gluing one of the outputs of $T$ (say, the first one) to one of the two inputs of $\tau$ (say, the first one), with the result of grafting taken to be $\tau$ itself if $T = \emptyset$. We use this choice in the explicit example below. Again, however, one sees that the grafting defined in this way does *not* satisfy the cocycle condition. In fact, consider the case of a monomial $T$ consisting of a forest with more than one component. Then only one of the components of $T$ is glued to $\tau$ by $B^+_\tau$. This means that, in the computation of the term $\tilde{\Delta} B^+_\tau$ there is not just one but several admissible cuts that separate $\tau$ from $T$, namely all the admissible cuts that include a cut of the edge used for grafting together with an arbitrary cut of the remaining components not connected to $\tau$. Thus, the difference between $\tilde{\Delta} B^+_\tau$ and $(id \otimes B^+_\tau)\tilde{\Delta}$ does not consist only of the term $id \otimes B^+_\delta(1) = id \otimes \tau$.

However, one can still consider a Dyson–Schwinger equation of the form

$$X = 1 + \sum_\delta B^+_{\tau_\delta}(X^2),$$

where the sum is on the three labels $\delta \in \{\mathfrak{b}, \mathfrak{c}, \mathfrak{r}\}$ of the vertex of the tree $\tau_\delta$ as in (4.10). This admits an explicit solution, as in the general case above, which are of the form

$$x_{n+1} = \sum_{k=0}^{n} \sum_{\delta} B^+_{\tau_\delta}(x_k x_{n-k}).$$

In $\overline{\mathcal{H}}^0_{\text{flow},\mathcal{V}}$, one can see that the first few terms are of the form $x_0 = I$ and

$x_1 =$ 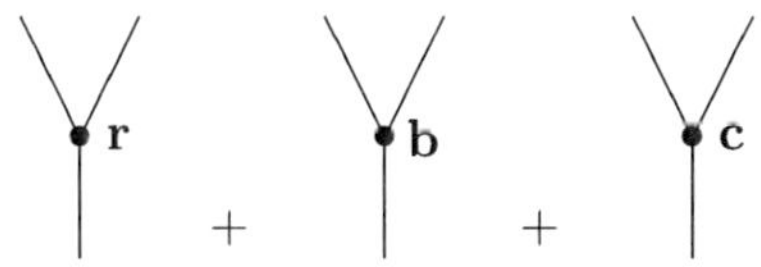


$$x_2 = \sum_a B^a_+(2x_1x_0)$$

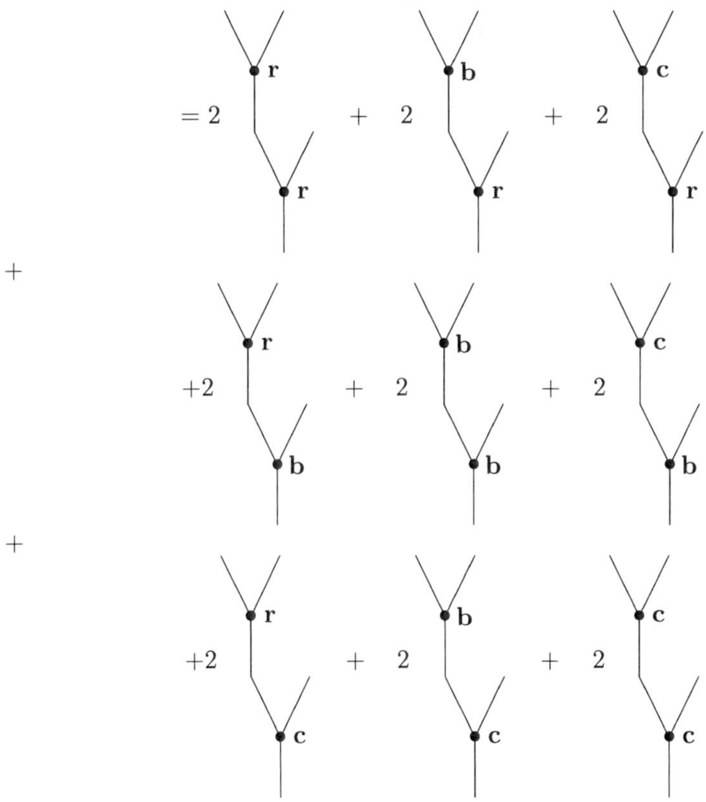


One can continue with the successive coefficients, such as $x_3 = \sum_a B^a_+(x_1^2 + 2x_2x_0)$, using the facts that for a vertex labeled by recursion, the order of inputs always matters, as well as for composition, while we can assume that order of inputs does not matter for bracketing as a consequence of the nature of parallel computation.

This example shows that, if we want to work with binary trees with external edges, which is a very natural choice from the point of view of computation theory, and we want to obtain a grafting operation that is non-trivial on monomials of arbitrary degree, we would have to extend the grafting to include, besides the primitive two-valent trees (4.10), also a choice of binary trees with larger numbers of input flags. This cannot be accommodated in the usual setting of combinatorial Dyson–Schwinger equations based on Hochschild cocycles in Hopf algebras. However, we shall see below that such generalizations exist naturally when one reformulates Dyson–Schwinger equations in the operadic setting.

## 5. Operadic viewpoint

Instead of using a Hopf algebraic setup to describe flow charts and primitive recursive functions, one can also adopt an operadic viewpoint, as suggested by Manin in [**27**].

Let us consider, as above, planar rooted trees $\tau$ with vertex labeling by the elementary operations of type $\{\mathfrak{b}, \mathfrak{c}, \mathfrak{r}\}$, oriented so that each tree has a single outgoing flag attached to the root vertex and a certain number of incoming flags.

DEFINITION 5.1. *We define the* operad of flow charts $\mathcal{O}_{\text{flow},\mathcal{V}'}$ *by setting* $\mathcal{O}(n)$ *to the* $\mathbb{K}$*-vector space spanned by labelled planar rooted trees with* $n$ *incoming flags and the operad composition operations*

$$\circ_{\mathcal{O}} : \mathcal{O}(n) \otimes \mathcal{O}(m_1) \otimes \cdots \otimes \mathcal{O}(m_n) \to \mathcal{O}(m_1 + \cdots + m_n)$$

*are given on generators* $\tau \otimes \tau_1 \otimes \cdots \otimes \tau_n$ *by grafting the output flag of the tree* $\tau_i$ *to the* $i$*-th input flag of the tree* $\tau$.

As pointed out in [**2**], Dyson–Schwinger equations admit a natural operadic interpretation. Namely, given a formal series $P(t) = 1 + \sum_{k=1}^{\infty} a_k t^k$ and a collection $\beta = (\beta_n)$ with $\beta_n \in \mathcal{O}(n)$, we consider the equation

$$X = \beta(P(X)), \tag{5.1}$$

where $X = \sum_k x_k$ is a formal sum with $x_k \in \mathcal{O}(k)$. In the right hand side we have $\beta(P(X))_1 = 1 + \beta_1 \circ x_1$, where 1 is the identity in $\mathcal{O}(1)$, and for $n \geq 2$

$$\beta(P(X))_n = \sum_{k=1}^{n} \sum_{j_1+\cdots+j_k=n} a_k \ \beta_k \circ (x_{j_1} \otimes \cdots \otimes x_{j_k}), \tag{5.2}$$

with $x_{j_1} \otimes \cdots \otimes x_{j_k} \in \mathcal{O}(j_1) \otimes \cdots \otimes \mathcal{O}(j_k)$, so that the composition $\beta_k \circ_{\mathcal{O}} (x_{j_1} \otimes \cdots \otimes x_{j_k}) \in \mathcal{O}(n)$, since $j_1 + \cdots + j_k = n$.

PROPOSITION 5.2. *For* $\mathcal{O} = \mathcal{O}_{\text{flow},\mathcal{V}'}$ *the operad of flow charts, if* $a_1\beta_1 \neq 1 \in \mathcal{O}(1)$*, the operadic Dyson–Schwinger equation* (5.1) *has a unique solution* $X \in \prod_{n\geq 1} \mathcal{O}(n)$ *given inductively by*

$$(1 - a_1\beta_1) \circ x_{n+1} = \sum_{k=2}^{n+1} \sum_{j_1+\cdots j_k=n+1} a_k \ \beta_k \circ (x_{j_1} \otimes \cdots \otimes x_{j_k}). \tag{5.3}$$

PROOF. First observe that in the operad of flow charts $\mathcal{O}(1)$ is one-dimensional, since it is spanned by the trees with a single vertex, one incoming and one outgoing flag, with the operations $\mathfrak{b}$, $\mathfrak{c}$ or $\mathfrak{r}$ at the vertex, but in the case of a single input, all of these operations are the identity, so that elements of $\mathcal{O}(1)$ are scalar multiples of the identity map, viewed as operations on primitive recursive functions. Thus, the element $\beta_1 \in \mathcal{O}(1)$ is a scalar multiple of the identity and, if we assume that $a_1\beta_1 \neq 1$ in $\mathcal{O}(1)$, then $1 - a_1\beta_1$ is invertible. Since we have $\beta(P(X))_1 = 1 + \beta_1 \circ x_1$

as the term in $\mathcal{O}(1)$ in the right hand side of (5.1), we obtain $(1 - a_1\beta_1)x_1 = 1$, which fixes $x_1 = (1 - a_1\beta_1)^{-1} \in \mathcal{O}(1)$. At the next step we have, from (5.2), $x_2 = a_1\beta_1 \circ x_2 + a_2\beta_2 \circ (x_1 \otimes x_1)$, which gives $x_2 = (1 - a_1\beta_1)^{-1}(a_2\beta_2 \circ (x_1 \otimes x_1))$, with $x_1$ as above. At the $(n+1)$-st step, $x_{n+1}$ is then determined uniquely by (5.3) in terms of the coefficients $a_k$ and the elements $\beta_k$ for $k = 1, \ldots, n+1$. □

Let then $\mathcal{O}_{\beta,P}(n)$ denote the $\mathbb{K}$-linear span of all the compositions $x_k \circ (x_{j_1} \otimes \cdots \otimes x_{j_k})$, for $k = 1, \ldots, n$ and $j_1 + \cdots + j_k = n$, with $x_k$ the coordinates of the solution $X = \sum_k x_k$ of the Dyson–Schwinger equation determined by $\beta = (\beta_n)$ and $P(t)$. The $\mathcal{O}_{\beta,P}(n)$ form a sub-operad under the composition maps induced from $\mathcal{O}(n)$.

For instance, for a choice of one of the operations $\delta \in \{\mathfrak{b}, \mathfrak{c}, \mathfrak{r}\}$, we can take $a_1 \neq 1$ and the element $\beta_k$ given by the tree with a single vertex marked by $\delta$, with $k$ incoming flags and one outgoing flag. This choice gives an operadic reformulation of the Hopf-theoretic Dyson–Schwinger equations with the cocycles $B^+_\delta$. However, we can now consider also more general operadic Dyson–Schwinger equations for different choices of $\beta = (\beta_n)$ that do not correspond to Hochschild cocycles in the Hopf algebra setting.

Restricting to the sub-operad $\mathcal{O}_{\beta,P}(n)$ generated by solutions of the Dyson–Schwinger equation should be regarded as considering those operations (formal combinations of flow charts) that satisfy a self-similarity property with respect to the transformation $X \mapsto \beta(P(X))$.

5.0.1. *The case of binary trees.* In particular, we can now revisit the example of the binary trees in this operadic setting. Let $\mathcal{O}^0_{\mathrm{flow},\mathcal{V}}(n)$ be the $\mathbb{K}$-linear space spanned by the planar *binary* trees with vertex labeling by the binary operations $\mathfrak{b} = \mathfrak{b}_2$, $\mathfrak{r}$, $\mathfrak{c}$. These form a sub-operad of $\mathcal{O}_{\mathrm{flow},\mathcal{V}'}$ with the induced composition maps.

For a fixed $\delta \in \{\mathfrak{b}, \mathfrak{r}, \mathfrak{c}\}$ let $\beta_{1,\delta}$ be the binary tree with a single vertex, two incoming and one outgoing flag, as in (4.10). We then consider a choice $\beta_n$ of a binary tree with $n$ input flags, for each $n \geq 1$, with a fixed choice of vertex labelings in $\{\mathfrak{b}, \mathfrak{r}, \mathfrak{c}\}$, not necessarily all of the same type. Each such choice of $\beta = (\beta_n)$, together with a choice of $P(t)$, now determines an operadic Dyson–Schwinger equation of the form (5.1), in the binary setting, with a corresponding solution as in Proposition 5.2.

5.0.2. *Systems of Dyson–Schwinger equations in operads.* By analogy to what happens in Hopf algebras, [**11**], one can pass from the case of a single Dyson–Schwinger equation to systems of Dyson–Schwinger equations. In the operadic setting a system of Dyson–Schwinger equations would be determined by the data of formal series $F_\delta(t_1, t_2, t_3) = \sum_{k_1,k_2,k_3} a^{(\delta)}_{k_1,k_2,k_3} t_1^{k_1} t_2^{k_2} t_3^{k_3}$ and of elements $\beta^{(\delta)} = (\beta^{(\delta)}_n)$, where for $\delta \in \{\mathfrak{b}, \mathfrak{c}, \mathfrak{r}\}$, the elements $\beta^{(\delta)}_n \in \mathcal{O}(n)$ are realized by rooted trees with root vertex marked by $\delta$. The operadic systems of Dyson–Schwinger equations are then of the form

$$X_\delta = \beta^{(\delta)}(F_\delta(X)).$$

The explicit form of the solutions is more cumbersome than in the case of a single equation in Proposition 5.2 above, and it follows a pattern similar to the derivation of the explicit solutions of Hopf theoretic systems of Dyson–Schwinger equations (4.5) as in [**11**]. These systems detect more elaborate forms of self-similarity in the operad of flow charts, involving different families of grafting operations simultaneously.

5.0.3. *Operads and Properads.* As mentioned in [**27**], one can extend the Hopf algebra of flow charts by considering, instead of labelled embedded rooted trees, more general labelled embedded acyclic graphs, namely graphs endowed with an acyclic orientation. Correspondingly, the operad of flow charts would be replaced by a properad, where the compositions extend from grafting output and input flags of trees to grafting outputs and inputs of acyclic graphs.

The notion of *properad* was introduced in [**31**] as an intermediate notion between operad and prop. Namely, a properad parameterizes operations with varying numbers of inputs and outputs that are labelled by connected acyclic graphs, while the operad case uses trees (varying number of inputs and a single output) and props allow for disconnected graphs.

More precisely, we consider the properad $\mathcal{P}_{\text{flow},\mathcal{V}'}$ where $\mathcal{P}(m,n)$ is the $\mathbb{K}$-vector space spanned by planar connected directed (acyclic) graphs with $m$ incoming flags and $n$ outgoing flags, with vertices decorated by operation that include the elementary $\mathfrak{b}$, $\mathfrak{c}$, $\mathfrak{r}$, with $m$ inputs and one output, but also include now a chosen set of additional operations with $m$ inputs and $n$ outputs. As we have seen in the case of extending the operations $\mathfrak{c}$ and $\mathfrak{r}$ from binary to $m$-ary inputs, in this case also such additional operations with $m$ inputs and $n$ outputs acting on primitive recursive functions can be decomposed in terms of the elementary operations: the resulting operations associated to vertices with $m$ incoming and $n$ outoing flags can be regarded as "macros", in the sense discussed in §2 of [**33**]. The properad has composition operations

$$\mathcal{P}(m,n)\otimes\mathcal{P}(j_1,k_1)\otimes\cdots\otimes\mathcal{P}(j_\ell,k_\ell)\to\mathcal{P}(j_1+\cdots+j_\ell,n),\quad\text{for}\quad k_1+\cdots+k_\ell=m.$$

5.0.4. *Dyson–Schwinger equations in properads.* We can define Dyson–Schwinger equations for properads, extending the operadic case considered above. Namely we consider again a choice of a formal power series $P(t)=1+\sum_k a_k t^k$ and a collection $\beta=(\beta_{m,n})$ of elements with $\beta_{m,n}\in\mathcal{P}(m,n)$. We then consider the equation

$$X=\beta(P(X)) \tag{5.4}$$

as in (5.1), where now the right hand side has $(m,n)$-component in $\mathcal{P}(m,n)$ given by

$$\beta(P(X))_{m,n}=\sum_{k=1}^{m}a_k\sum_{\substack{j_1+\cdots j_k=m\\ i_1+\cdots i_k=\ell}}\beta_{\ell,n}\circ(x_{j_1,i_1}\otimes\cdots\otimes x_{j_k,i_k}). \tag{5.5}$$

In order to construct solutions of the properad Dyson–Schwinger equations, we introduce transformations $\Lambda_n=\Lambda_n(a,\beta)$ with

$$\Lambda_n(a,\beta):\oplus_{k=1}^n\mathcal{P}(n,k)\to\oplus_{k=1}^n\mathcal{P}(n,k),\qquad\text{with}\qquad\Lambda_n(a,\beta)_{ij}=a_j\beta_{j,i}.$$

THEOREM 5.3. *If for all $n\geq 1$ the transformation $I-\Lambda_n(a,\beta)$ is invertible, with $I$ the identity on $\oplus_{k=1}^n\mathcal{P}(n,k)$, then the properad Dyson–Schwinger equation*

(5.4) *admits a unique solution, given by* $x_{1,1} = \Lambda_1^{-1}$, *and for* $m < n$ *by*

$$x_{m,n} = \sum_{k=1}^{m} a_k \beta_{k,n} \circ \left( \sum_{\ell=1}^{k} \sum_{\substack{j_1+\cdots+j_\ell=m \\ i_1+\cdots+i_\ell=k}} x_{j_1,i_1} \otimes \cdots \otimes x_{j_\ell,i_\ell} \right), \tag{5.6}$$

*while the remaining components* $x_{m,n}$ *with* $m \geq n$ *are determined by*

$$Y_n(x) = (I - \Lambda_n)^{-1} \Lambda_n V^{(n)}(x), \tag{5.7}$$

*where* $Y_n(x)^t = (x_{n,1}, \dots, x_{n,n})$ *and* $V^{(n)}(x)^t = (V^{(n)}(x)_j)_{j=1,\dots,n}$

$$V^{(n)}(x)_j = \sum_{k=2}^{n} \sum_{\substack{r_1+\cdots+r_k=n \\ s_1+\cdots+s_k=j}} x_{r_1,s_1} \otimes \cdots \otimes x_{r_k,s_k}.$$

PROOF. By (5.5) we see that we have $x_{1,1} = 1 + a_1\beta_{1,1} \circ x_{1,1}$, which can be solved for $x_{1,1}$ provided the invertibility of $1 - \Lambda_1 = 1 - a_1\beta_{1,1}$ holds. The components $x_{m,n}$ with $m < n$ are determined by (5.6), where the right hand side only involves components $x_{r,s}$ with $r < m$ and $s < n$. The equations for the remaining entries $x_{m,n}$ with $n \leq m$ break into systems of the form

$$(I - \Lambda_m) \begin{pmatrix} x_{m,1} \\ \vdots \\ x_{m,m} \end{pmatrix} = \sum_j (\Lambda_m)_{ij} \left( \sum_{k=2}^{m} \sum_{\substack{r_1+\cdots+r_k=m \\ s_1+\cdots+s_k=j}} x_{r_1,s_1} \otimes \cdots \otimes x_{r_k,s_k} \right)_j.$$

Provided that the invertibility condition for the transformation $I - \Lambda_m$ holds, these systems provide unique values for all the components $x_{m,n}$ with $n \leq m$, as a function of the components $x_{r,s}$ with $r < m$, that we have already been determined. □

Notice that one does not expect the invertibility of $I - \Lambda_n(a, \beta)$ to hold in general, as $\Lambda_n(a, \beta)$ involves compositions with the elements $\beta_{j,i}$. The simplest example in which one has invertibility can be obtained by taking $\beta_{m,n} = 0$ for $m \neq n$ and $I - \beta_{n,n}$ an invertible transformation in $\mathcal{P}(n, n)$.

As in the operad case, we can define $\mathcal{P}_{\beta,P}(m, n)$ to be the $\mathbb{K}$-vector spaced spanned by all the properad compositions

$$x_{k,n} \circ (x_{j_1,i_1} \otimes \cdots \otimes x_{j_\ell,i_\ell})$$

with $j_1 + \cdots + j_\ell = m$ and $i_1 + \cdots + i_\ell = k$ and where all the $x_{m,n}$ are components of the solution to the properad Dyson–Schwinger equation determined by the choice of $P(t)$ and of $\beta = (\beta_{m,n})$, with the invertibility condition of Theorem 5.3. The $\mathcal{P}_{\beta,P}(m, n)$ form a sub-properad of $\mathcal{P}_{\mathsf{flow},\mathcal{V}'}$ with the induced composition.

## 6. Renormalization of the halting problem

In [**25**], Manin adapted the formalism of perturbative renormalization to the halting problem in the theory of computation. To summarize Manin's approach, the basic idea is that one considers non-computable functions as an analog of divergent Feynman integrals. Renormalization consists of a procedure of extraction of finite values from divergences and in this setting Manin proposes that a suitable form of renormalization should lead to a procedure of extraction of a "computable part" from a non-computable function.

While this whole program is, at present, still in its early stages and some of the ideas involved are still speculative, we attempt here to give a sense of the general picture and formulate some of the main questions, as we see them. We will also discuss how Dyson–Schwinger equations may be accommodated in this general circle of ideas. While most of what is discussed in this section is more speculative than the previous parts of the paper, we regard this viewpoint as an important motivation for further developments.

**6.1. The halting problem: regularization and renormalization.** We describe here a setting for renormalization of the halting problem which is a minor elaboration on the procedure described in §3 of [**25**]. We do not add any significant new development beyond what is outlined as a possible program in Manin's work [**25**], but we will try to identify some specific questions that we hope to return to in future work.

6.1.1. *Partial recursive functions and Hopf algebras.* We first pass from the class of primitive recursive functions considered above to the larger class of partial recursive functions. These have a similar presentation as the one recalled in §2 above, with the same basic functions (successor, constants, and projections) and, in addition to the three elementary operations $\mathfrak{c}, \mathfrak{b}, \mathfrak{r}$ of composition, bracketing and recursion, an additional $\mu$ operation that assigns to an input function $f : \mathbb{N}^{n+1} \to \mathbb{N}$ assigns an output

$$h : \mathbb{N}^n \to \mathbb{N}, \quad h(x_1, \dots, x_n) = \min\{x_{n+1} \,|\, f(x_1, \dots, x_{n+1}) = 1\},$$

with domain

$$\mathcal{D}(h) = \{(x_1, \dots, x_n) \,|\, \exists x_{n+1} \geq 1 \,:\, f(x_1, \dots, x_{n+1}) = 1,$$

with $(x_1, \dots, x_n, k) \in \mathcal{D}(f), \forall k \leq x_{n+1}\}$. According to Church's thesis the partial recursive functions obtained in this way are exactly the semi-computable functions, namely those for which there exists a program that, for all $x \in \mathcal{D}(f)$, computes $f(x)$ and which either computes zero or runs for an infinite time when $x \notin \mathcal{D}(f)$, see §V of [**23**]. Notice that it is the presence of the additional operation $\mu$ that allows for the construction, starting from the basic functions, of functions that are only partially defined and semi-computable.

Correspondingly, we enrich the Hopf algebras of flow charts by additional vertex decorations by the $\mu$ operations. As in the case of the other operations, this requires defining a version of the $\mu$ operation for vertices of valence greater than two. For example, a possible choice may be to use the higher-arity bracketing operation and an application of $\mu$ coordinatewise. We shall still use the notations $\mathcal{H}^{nc}_{\text{flow},\mathcal{R}'}$ and $\mathcal{H}^{nc}_{\text{flow},\mathcal{V}'}$, for the resulting Hopf algebras with admissible decorations by partial recursive functions or with vertex decorations, respectively, and $\mathcal{H}_{\text{flow},\mathcal{R}'}$ and $\mathcal{H}_{\text{flow},\mathcal{V}'}$ for their commutative quotients.

6.1.2. *Partial recursive functions and the halting problem.* As in [**25**], we regard the halting problem as a particular instance of the general problem of recognizing whether a given $k \in \mathbb{N}$ belongs to the domain of definition $\mathcal{D}(f)$ of a partial recursive function $f$. The halting problem corresponds to the case where, given a partial recursive function $f$ and a program $P$ computing it, one considers the partial recursive function $t(P)$, the computation time, with domain $\mathcal{D}(t(P)) = \mathcal{D}(f)$, see [**25**], §3.1.(a) and §2.1, and also §IX of [**23**].

6.1.3. *Algebraic renormalization.* The basic algebraic formalism for perturbative renormalization, in its formulation as given in [**9**] (see also §5 of [**28**]), consists of an *algebraic Feynman rule*

$$\phi : \mathcal{H} \to \mathcal{B} \tag{6.1}$$

which is a *morphism of commutative algebras* from a *commutative Hopf algebra* $\mathcal{H}$ to a *Rota–Baxter algebra* $\mathcal{B}$ of weight $-1$.

Recall that a commutative Rota–Baxter algebra of weight $\lambda$ is a commutative, associative algebra $\mathcal{B}$ endowed with a linear operator $T$ on it that satisfies the identity

$$T(x)T(y) = T(xT(y)) + T(T(x)y) + \lambda T(xy). \tag{6.2}$$

The field of convergent Laurent series in one variable with the projection $T$ onto the polar part is an example of Rota–Baxter algebra of weight $\lambda = -1$, which is widely used in renormalization theory, [**3**], [**5**]. The Rota–Baxter operator $T$ of weight $-1$ determines a splitting of $\mathcal{B}$ into two commutative unital subalgebras, $\mathcal{B}_+ = (1-T)\mathcal{B}$ and $\mathcal{B}_-$ given by $T\mathcal{B}$ with a unit adjoined. The fact that $\mathcal{B}_\pm$ are indeed algebras and not just vector spaces follows from the Rota–Baxter identity (6.2).

The BPHZ renormalization of the algebraic Feynman rule (6.1) is then obtained by showing the existence of a multiplicative factorization

$$\phi = (\phi_- \circ S) \star \phi_+, \qquad \text{with} \qquad \phi_\pm : \mathcal{H} \to \mathcal{B}_\pm,$$

obtained inductively via the BPHZ preparation formula (see [**3**], [**9**])

(6.3)

$$\phi_-(x) = -T(\phi(x) + \sum \phi_-(x')\phi(x'')), \text{ and } \phi_+(x) = (1-T)(\phi(x) + \sum \phi_-(x')\phi(x'')),$$

where $x \in \mathcal{H}$, with $\Delta(x) = 1 \otimes x + x \otimes 1 + \sum x' \otimes x''$. The factorization is unique if normalized by $\epsilon_- \circ \phi_- = \epsilon$, where $\epsilon_- : \mathcal{B}_- \to \mathbb{C}$ is the augmentation map and $\epsilon$ is the counit of $\mathcal{H}$. Again, it is the Rota–Baxter property of $(\mathcal{B}, T)$ that ensures that $\phi_\pm$ obtained in this way are algebra homomorphisms. The homomorphism $\phi_+ : \mathcal{H} \to \mathcal{B}_+$ is the *renormalized* algebraic Feynman rule, while $\phi_- : \mathcal{H} \to \mathcal{B}_-$ is the *counterterm*, the divergence.

6.1.4. *An algebraic Feynman rule for the halting problem.* We recall here a proposal made by Manin in [**25**] of a possible construction of an algebraic Feynman rule in the setting of the halting problem.

We let $\mathcal{B}$ be the algebra of functions $\Phi : \mathbb{N}^m \to \mathcal{M}(D)$ from $\mathbb{N}^m$, for some $k$, to the algebra $\mathcal{M}(D)$ of analytic functions in the unit disk $D = \{z \in \mathbb{C} : |z| < 1\}$. Since the functions $\Phi$ can have different domains $\mathbb{N}^k$, the sums and products are defined by extending them to functions $\Phi(k_1, k_2, \ldots, k_n, \ldots)$ from a common domain $\mathbb{N}^\infty$, which depend on only finitely many variables, $\Phi = \Phi \circ \pi^\infty_m$, with $\pi^\infty_m : \mathbb{N}^\infty \to \mathbb{N}^m$. The Rota–Baxter operator $T$ on $\mathcal{B}$ is given by componentwise (for each $\underline{k}$) projection onto the polar part at $z = 1$. Thus, $\mathcal{B}_+$ consists of functions from $\mathbb{N}^\infty$ (depending only on finitely many $k_i$) to $\mathcal{M}(D)_+$, the algebra of meromorphic functions in $D$ that extend continuously at $z = 1$, while $\mathcal{B}_-$ consists of functions that are either constant or have a pole at $z = 1$.

We define algebraic Feynman rules for the Hopf algebras $\mathcal{H}_{\text{flow},\mathcal{R}'}$ and $\mathcal{H}_{\text{flow},\mathcal{V}'}$. In the case of $\mathcal{H}_{\text{flow},\mathcal{R}'}$, where both vertices and flags in the trees are labeled by admissible labelings, there is a well defined partial recursive function $f$ associated to each labelled rooted tree $\tau$ in $\mathcal{H}_{\text{flow},\mathcal{R}'}$, namely the function $f$ labeling the unique outgoing flag of $\tau$. Let us first assume that the partial recursive function $f$ has range in $\mathbb{N}$. We extend this $f : \mathbb{N}^m \to \mathbb{N}$ to a function $\bar{f} : \mathbb{N}^m \to \mathbb{Z}_{\geq 0}$ that computes $f(x)$ at $x \in \mathcal{D}(f)$ and takes value 0 at $x \notin \mathcal{D}(f)$.

We assign to a tree $\tau$ that computes $f$ an element in $\Phi_\tau(\underline{k}, z)$ in $\mathcal{B}$, defined (as in §3 of [**25**]) by

$$\Phi_\tau(\underline{k}, z) = \Phi(\underline{k}, f, z) := \sum_{n \geq 0} \frac{z^n}{(1 + n\bar{f}(\underline{k}))^2}. \tag{6.4}$$

The resulting function $\Phi_\tau(\underline{k}, \cdot) \in \mathcal{M}(D)$ has a pole at $z = 1$ iff $\underline{k} \notin \mathcal{D}(f)$. We extend this definition mutiplicatively to cases where the output of the flow chart $\tau$ is a partial recursive function $f : \mathbb{N}^k \to \mathbb{N}^\ell$, by setting $\Phi_\tau(\underline{k}, z) = \prod_{j=1}^{\ell} \Phi(\underline{k}, f_j, z)$. In this way, the resulting function has a pole at $z = 1$ if at least one of the $f_j$ does, that is, iff $\underline{k} \notin \mathcal{D}(f_1) \cap \cdots \cap \mathcal{D}(f_\ell) = \mathcal{D}(f)$, consistently with the above. We then extend the definition multiplicatively to the case of a monomial $\tau_1 \cdots \tau_n$ in $\mathcal{H}_{\text{flow},\mathcal{R}'}$ and additively to linear combinations of monomials. Thus, we obtain an algebra homomorphism $\Phi : \mathcal{H}_{\text{flow},\mathcal{R}'} \to \mathcal{B}$.

We write $\Phi(\underline{k}, f, z) := \Phi_f(\underline{k})(z)$ for the function $\Phi_f(\underline{k}) \in \mathcal{M}(D)$ of $z \in D$, associated to a given $\underline{k} \in \mathbb{N}^\infty$ and depending on the output $f$ of a given flow chart, with $f$ a partial recursive functions. If we denote by $\tau$ the decorated planar rooted tree describing the flowchart, we equivalently write $\Phi(\underline{k}, \tau, z)$ or $\Phi_{\underline{k},\tau}(z)$.

The specific form (6.4) of the proposed Feynman rule does not appear to be so important, as much as the two properties:

- The function $\Phi(\underline{k}, f, \cdot) \in \mathcal{M}(D)$ has a pole at $z = 1$ iff $\underline{k} \notin \mathcal{D}(f)$
- The function $\Phi(\underline{k}, f, \cdot)$ depends on $\underline{k}$ through the values $\bar{f}(\underline{k})$ and these values can be reconstructed from the function $\Phi(\underline{k}, f, z)$.

Any other choice of a function with these properties would be a suitable Feynman rule according to [**25**]. Thus, it would seem perhaps more convenient to consider the set of all such functions, possibly up to a suitable equivalence relation. We will discuss more below why it may be necessary to introduce an equivalence relation on Feynman rules. This has an analog in the quantum field theory setting, where the classification of divergences in terms of a Galois theory based on the Tannakian formalism was obtained in [**6**] precisely by introducing a suitable equivalence relation on algebraic Feynman rules. We will not develop this approach further in the present paper, but we hope to return to it elsewhere.

In Manin's idea a possible renormalization of the halting problem should provide a way of extracting a computable part from the halting problem for partial recursive functions that accounts for "subdivergences" created by subroutines within the flowcharts. Though, at present, how exactly this should be achieved is not yet fully understood, in the procedure described in [**25**] one would proceed by applying the BPHZ factorization (6.3) to the chosen algebraic Feynman rule. The result of factorization consists of two new algebra homomorphisms from the same Hopf algebra to the two parts of the target Rota–Baxter algebra. The main question is then the interpretation of both pieces in terms of computability and partial recursive functions.

When we apply the BPHZ procedure to $\Phi$ defined as in (6.4), we obtain

$$\Phi_-(\underline{k}, f_\tau, z) = -T(\Phi(\underline{k}, f_\tau, z) + \sum_C \Phi_-(\underline{k}, f_{\pi_C(\tau)}, z)\Phi(\underline{k}, f_{\rho_C(\tau)}, z)). \tag{6.5}$$

Because $f = f_\tau$ is the output function computed by the flow chart, that is, the label of the outgoing flag of $\tau$, we have $f_{\rho_C(\tau)} = f_\tau$, since these trees have the same root and outgoing flag with the same label. Thus, the expression above is simply of the form

$$\Phi_-(\underline{k}, f_\tau, z) = -T\left(\Phi(\underline{k}, f_\tau, z)(1 + \Phi_-(\underline{k}, \sum_C f_{\pi_C(\tau)}, z))\right). \tag{6.6}$$

This has the effect of considering not only the pole at $z = 1$ that is caused by $f_\tau$ itself, but also those that come from subdivergences created by all the partial recursive functions $f_{\pi_C(\tau)}$ computed in the intermediate steps of the computation performed by the flow chart $\tau$, which are the outputs of the "pruned parts" $\pi_C(\tau)$ of the admissible cuts on $\tau$.

This suggests that the main role of the negative piece of the Birkhoff factorization should be understood as a way of detecting the presence of a pole not only in the function $\Phi(\underline{k}, f, z)$ itself, but in the analogous functions associated to the results $f_{\pi_C(\tau)}$ and $f_{\rho_C(\tau)}$ computed by subroutines of the program $\tau$ that computes $f$.

The role of "subdivergences" in programs computing partial recursive functions is interesting, for example because one can have programs that compute total functions as output, but involve partial functions, so that the graph itself may be divergence-free and still contain subdivergences, as the next example (suggested to us by Joachim Kock) shows.

EXAMPLE 6.1. The identity function can be computed by a tree with one $\mathfrak{c}$ node and one $\mu$ node, as a composition of a successor function followed by a partial predecessor function, $\mu(|y + 1 - x|)$, undefined at 0 and $x - 1$ for $x > 0$.

Thus, replacing the polar part of $\Phi(\underline{k}, f, z)$ with the polar part $\Phi_-(\underline{k}, f_\tau, z)$ can be understood as a way to take into account all possible (and possibly hidden) subdivergences.

It is more difficult to obtain a good interpretation of the positive piece of the Birkhoff factorization. The analogy with quantum field theory suggests that it is not the value itself of $\Phi_+(\underline{k}, f_\tau, z)$ that matters. After all, even in the original physical setting, there is an intrinsic ambiguity in the renormalized values, coming from the action of the renormalization group. The main question of Manin's approach here appears to be the following (loosely formulated).

QUESTION 6.2. Is there a new function $f_{\text{ren}}$, which is primitive recursive, such that

$$\Phi_+(\underline{k}, f_\tau, z) = \Phi(\underline{k}, f_{\text{ren}}, z), \tag{6.7}$$

namely, is $\Phi_+$ equal to the original $\Phi$ of another function, this time primitive instead of partial recursive?

If one could answer this question positively, then one could think of the new function $f_{\text{ren}}$ as an "extraction of a computable piece" from the partial recursive function $f$ (and the program $\tau$ that computes it). Upon close inspection, it is easy to see that, as naively formulated in Question 6.2, it may not be possible in general to obtain such an $f_{\rm ren}$. However, one can perhaps formulate a better question of whether an identification (6.7) holds after considering the algebraic Feynman rules up to an equivalence relation, as with the gauge transformations on germs of meromorphic functions considered in the quantum field theory setting in [**6**]. We will not develop this issue further in the present paper, but we leave it as an open question.

6.1.5. *The case of* $\mathcal{H}_{\rm flow,\mathcal{V}'}$. We make here a sketch of a possible suggestion for a variation on the construction above, adapted to define an algebraic Feynman rule on $\mathcal{H}_{\rm flow,\mathcal{V}'}$, where the flags are not labelled and only vertices are. In this case, in order to define something like (6.4), we need a way of assigning inputs to the flow charts. A reasonable choice is to use the basic functions as input. We then define the algebraic Feynman rule as

$$\Phi(\underline{k},\tau,z) := \prod_{\sigma} \Phi(\underline{k}, f_{\tau,\sigma}, z), \tag{6.8}$$

where the product is over $\sigma$ ranging over the set of all possible functions $\sigma : E^{in}_{ext}(\tau) \to \{s, c, \pi\}$ that label the incoming external edges of $\tau$ by basic functions (successor, constant, or projection). The target of $\sigma$ contains all the possible basic functions (of type $s$, $c$, or $\pi$). Since this is an infinite set (there are infinitely many possible choices of projection functions $\pi^n_i$), if one wants to have a finite product in (6.8) one may have to restrict $\sigma$ by making a priori choices of smaller sets of basic functions as target, or else introduce further conditions on the class of functions $\Phi$ that would ensure convergence. The partial recursive function $f_{\tau,\sigma}$ is the output of the flow chart given by the tree $\tau$ with inputs assigned by $\sigma$. Each $\Phi(\underline{k}, f_{\tau,\sigma}, z)$ is computed as in (6.4), and (6.8) is extended to arbitrary elements of the Hopf algebra as before. In this setting, the function $\Phi_\tau(\underline{k}, z)$ has a pole at $z = 1$ iff there is at least a choice of a basic input $\sigma$ for which $\underline{k} \notin \mathcal{D}(f_{\tau,\sigma})$.

With this setting, the BPHZ formula for the algebraic Feynman rule (6.8) becomes more interesting than in the case of (6.6). We have again (6.5), in the form

$$\Phi_-(\underline{k},\tau,z) = -T(\Phi(\underline{k},\tau,z) + \sum_C \Phi_-(\underline{k},\pi_C(\tau),z)\Phi(\underline{k},\rho_C(\tau),z)), \tag{6.9}$$

with $\Phi(\underline{k},\rho_C(\tau),z)$ and $\Phi(\underline{k},\pi_C(\tau),z)$ again computed as in (6.8), but this time it is no longer true that $f_\tau = f_{\rho_C(\tau)}$ because on the tree $\rho_C(\tau)$ we are using the new inputs given by basic functions and not the input coming from the output of $\pi_C(\tau)$. The divergence (6.9) here combines the divergences of the $f_{\tau,\sigma}$ with divergences coming from partial recursive functions $f_{\pi_C(\tau),\sigma}$ and $f_{\rho_C(\tau),\sigma}$ that flowcharts $\pi_C(\tau)$ and $\rho_C(\tau)$ compute starting from inputs of basic functions.

**6.2. Dyson–Schwinger equations for the halting problem.** Finally, we tentatively propose a possible role for Dyson–Schwinger equations in the context of Manin's approach to the halting problem. These remarks are also, at this stage, very speculative, as they would depend for a more precise interpretation upon a better understanding of the main questions about the Birkhoff factorization discussed earlier in this section.

If we work with the Feynman rule described above on the Hopf algebra $\mathcal{H}_{\text{flow},\mathcal{V}'}$ with only vertex decorations, then we can readily consider Dyson–Schwinger equations as discussed in §4 above. However, if we work with the definition of the algebraic Feynman rule $\Phi$ described in (6.4) on the Hopf algebra $\mathcal{H}_{\text{flow},\mathcal{R}'}$ of flow charts with flag decorations, we need to extend the grafting operators appropriately that are needed to define Dyson–Schwinger equations from $\mathcal{H}_{\text{flow},\mathcal{V}'}$ to $\mathcal{H}_{\text{flow},\mathcal{V}'}$. This can be done by assigning as output the empty function $f = \emptyset$, mapped to the constant function $\Phi(\underline{k}, \emptyset, z) \equiv 1$, as the output of any tree $\tau$ that is obtained through a grafting where the flag labels don't match.

We can then consider a Dyson–Schwinger equation of the form (4.3), with $B^+ = \sum_\delta B_\delta^+$, with $\delta \in \{\mathfrak{b}, \mathfrak{c}, \mathfrak{r}, \mu\}$ or a more general system of Dyson–Schwinger equations (4.5), with the condition that the components $x_m$ of the unique solution span a Hopf subalgebra of $\mathcal{H}_{\text{flow},\mathcal{R}'}$. In such case, we can restrict the algebraic Feynman rule $\Phi$ constructed above to this Hopf subalgebra and still perform the BPHZ renormalization.

One can then reformulate the same questions that we have briefly discussed regarding Manin's approach to the halting problem, of how to interpret the resulting negative and positive parts of the Birkhoff factorization. In this case, the fact that one is restricting attention to the Hopf subalgebra generated by the components of the solution to the Dyson–Schwinger equation means that the negative part $\Phi_-(\underline{k}, f_{x_m}, z)$ (or $\Phi_-(\underline{k}, f_{x_m,\sigma}, z)$ in the case of $\mathcal{H}_{\text{flow},\mathcal{V}'}$) will now account for subdivergences that belong to the same Hopf subalgebra determined by the Dyson–Schwinger equation. This can be thought of, heuristically, as measuring the "amount of non-computability" that can be produced by sub-flow-charts that are characterized by a certain self-similarity property (defined by the Dyson–Schwinger equation). An analog in this setting of Question 6.2 on the interpretation of the positive part of the Birkhoff factorization can also be formulated in the Hopf subalgebra determined by the Dyson–Schwinger equation.

## Acknowledgment

The first author was supported for this project by the Summer Undergraduate Research Fellowship (SURF) program of Caltech, through a Herbert J. Ryser fellowship. The second author is partially supported by NSF grants DMS-0901221, DMS-1007207, DMS-1201512, and PHY-1205440. The second author acknowledges MSRI for hospitality and support. The authors are especially grateful to Joachim Kock for many helpful comments and suggestions that significantly improved the paper.

## References

[1] J. Baez and M. Stay, *Physics, topology, logic and computation: a Rosetta Stone*, New structures for physics, Lecture Notes in Phys., vol. 813, Springer, Heidelberg, 2011, pp. 95–172, DOI 10.1007/978-3-642-12821-9_2. MR2767046 (2012f:18015)

[2] Christoph Bergbauer and Dirk Kreimer, *Hopf algebras in renormalization theory: locality and Dyson-Schwinger equations from Hochschild cohomology*, Physics and number theory, IRMA Lect. Math. Theor. Phys., vol. 10, Eur. Math. Soc., Zürich, 2006, pp. 133–164, DOI 10.4171/028-1/4. MR2277758 (2007k:81138)

[3] Alain Connes and Dirk Kreimer, *Renormalization in quantum field theory and the Riemann-Hilbert problem. I. The Hopf algebra structure of graphs and the main theorem*, Comm. Math. Phys. **210** (2000), no. 1, 249–273, DOI 10.1007/s002200050779. MR1748177 (2002f:81070)

[4] A. Connes and D. Kreimer, *Insertion and elimination: the doubly infinite Lie algebra of Feynman graphs*, Ann. Henri Poincaré **3** (2002), no. 3, 411–433, DOI 10.1007/s00023-002-8622-9. MR1915297 (2003j:81118)
[5] Alain Connes and Matilde Marcolli, *Noncommutative geometry, quantum fields and motives*, American Mathematical Society Colloquium Publications, vol. 55, American Mathematical Society, Providence, RI; Hindustan Book Agency, New Delhi, 2008. MR2371808 (2009b:58015)
[6] Alain Connes and Matilde Marcolli, *Renormalization and motivic Galois theory*, Int. Math. Res. Not. **76** (2004), 4073–4091, DOI 10.1155/S1073792804143122. MR2109986 (2006b:81173)
[7] F. J. Dyson, *The S matrix in quantum electrodynamics*, Physical Rev. (2) **75** (1949), 1736–1755. MR0031388 (11,145b)
[8] Kurusch Ebrahimi-Fard, Dirk Kreimer, and Igor Mencattini, *On the insertion-elimination Lie algebra of Feynman graphs*, Lie theory and its applications in physics V, World Sci. Publ., River Edge, NJ, 2004, pp. 124–134, DOI 10.1142/9789812702562_0007. MR2172177
[9] Kurusch Ebrahimi-Fard, Li Guo, and Dirk Kreimer, *Integrable renormalization. II. The general case*, Ann. Henri Poincaré **6** (2005), no. 2, 369–395, DOI 10.1007/s00023-005-0211-2. MR2136196 (2006e:81161)
[10] Loïc Foissy, *Faà di Bruno subalgebras of the Hopf algebra of planar trees from combinatorial Dyson-Schwinger equations*, Adv. Math. **218** (2008), no. 1, 136–162, DOI 10.1016/j.aim.2007.12.003. MR2409411 (2009d:16057)
[11] Loïc Foissy, *Classification of systems of Dyson-Schwinger equations of the Hopf algebra of decorated rooted trees*, Adv. Math. **224** (2010), no. 5, 2094–2150, DOI 10.1016/j.aim.2010.01.024. MR2646118 (2011j:81220)
[12] Loïc Foissy, *Lie algebras associated to systems of Dyson-Schwinger equations*, Adv. Math. **226** (2011), no. 6, 4702–4730, DOI 10.1016/j.aim.2010.12.011. MR2775882 (2012d:16114)
[13] L. Foissy, *Les algèbres de Hopf des arbres enracinés décorés. I* (French, with English and French summaries), Bull. Sci. Math. **126** (2002), no. 3, 193–239, DOI 10.1016/S0007-4497(02)01108-9. MR1905177 (2003d:16049)
[14] Stefan Forcey, Aaron Lauve, and Frank Sottile, *New Hopf structures on binary trees* (English, with English and French summaries), 21st International Conference on Formal Power Series and Algebraic Combinatorics (FPSAC 2009), Discrete Math. Theor. Comput. Sci. Proc., AK, Assoc. Discrete Math. Theor. Comput. Sci., Nancy, 2009, pp. 411–420. MR2721530 (2012b:16100)
[15] Ralf Holtkamp, *Comparison of Hopf algebras on trees*, Arch. Math. (Basel) **80** (2003), no. 4, 368–383. MR1982837 (2004f:16067)
[16] Ralf Holtkamp, *Rooted trees appearing in products and co-products*, Combinatorics and physics, Contemp. Math., vol. 539, Amer. Math. Soc., Providence, RI, 2011, pp. 153–169, DOI 10.1090/conm/539/10633. MR2790307 (2012d:16115)
[17] Dirk Kreimer, *On the Hopf algebra structure of perturbative quantum field theories*, Adv. Theor. Math. Phys. **2** (1998), no. 2, 303–334. MR1633004 (99e:81156)
[18] Dirk Kreimer, *The core Hopf algebra*, Quanta of maths, Clay Math. Proc., vol. 11, Amer. Math. Soc., Providence, RI, 2010, pp. 313–321. MR2732056 (2012a:81104)
[19] Dirk Kreimer, *Dyson-Schwinger equations: from Hopf algebras to number theory*, Universality and renormalization, Fields Inst. Commun., vol. 50, Amer. Math. Soc., Providence, RI, 2007, pp. 225–248. MR2310307 (2008b:81178)
[20] Dirk Kreimer and Walter D. van Suijlekom, *Recursive relations in the core Hopf algebra*, Nuclear Phys. B **820** (2009), no. 3, 682–693, DOI 10.1016/j.nuclphysb.2009.04.025. MR2554483 (2011a:81161)
[21] Dirk Kreimer and Karen Yeats, *An étude in non-linear Dyson-Schwinger equations*, Nuclear Phys. B Proc. Suppl. **160** (2006), 116–121, DOI 10.1016/j.nuclphysbps.2006.09.036. MR2255485 (2007i:81162)
[22] Jean-Louis Loday and María O. Ronco, *Hopf algebra of the planar binary trees*, Adv. Math. **139** (1998), no. 2, 293–309, DOI 10.1006/aima.1998.1759. MR1654173 (99m:16063)
[23] Yu. I. Manin, *A course in mathematical logic for mathematicians*, 2nd ed., Graduate Texts in Mathematics, vol. 53, Springer, New York, 2010. Chapters I–VIII translated from the Russian by Neal Koblitz; With new chapters by Boris Zilber and the author. MR2562767 (2010k:03001)

[24] Yuri I. Manin, *Renormalization and computation I: motivation and background* (English, with English and French summaries), OPERADS 2009, Sémin. Congr., vol. 26, Soc. Math. France, Paris, 2013, pp. 181–222. MR3203373

[25] Yuri I. Manin, *Renormalisation and computation II: time cut-off and the Halting problem*, Math. Structures Comput. Sci. **22** (2012), no. 5, 729–751, DOI 10.1017/S0960129511000508. MR2981942

[26] Yuri I. Manin, *Infinities in quantum field theory and in classical computing: renormalization program*, Programs, proofs, processes, Lecture Notes in Comput. Sci., vol. 6158, Springer, Berlin, 2010, pp. 307–316, DOI 10.1007/978-3-642-13962-8_34. MR2678142 (2012g:81143)

[27] Yu.I. Manin, *Zipf's law and Levin's probability distributions*, arXiv:1301.0427.

[28] Matilde Marcolli, *Feynman motives*, World Scientific Publishing Co. Pte. Ltd., Hackensack, NJ, 2010. MR2604634 (2011g:81079)

[29] Julian Schwinger, *On the Green's functions of quantized fields. I, II*, Proc. Nat. Acad. Sci. U. S. A. **37** (1951), 452–455, 455–459. MR0045065 (13,520d)

[30] Walter D. van Suijlekom, *Renormalization of gauge fields: a Hopf algebra approach*, Comm. Math. Phys. **276** (2007), no. 3, 773–798, DOI 10.1007/s00220-007-0353-9. MR2350437 (2009d:81236)

[31] Bruno Vallette, *A Koszul duality for PROPs*, Trans. Amer. Math. Soc. **359** (2007), no. 10, 4865–4943, DOI 10.1090/S0002-9947-07-04182-7. MR2320654 (2008e:18020)

[32] Noson S. Yanofsky, *Towards a definition of an algorithm*, J. Logic Comput. **21** (2011), no. 2, 253–286, DOI 10.1093/logcom/exq016. MR2783414 (2012b:68295)

[33] N. Yanofsky, *Galois theory of algorithms*, arXiv:1011.0014.

[34] Karen Yeats, *Rearranging Dyson-Schwinger equations*, Mem. Amer. Math. Soc. **211** (2011), no. 995, x+82, DOI 10.1090/S0065-9266-2010-00612-4. With a foreword by Dirk Kreimer. MR2791969 (2012d:81239)

Physics Department, Caltech, 1200 E. California Blvd. Pasadena, California 91125
*E-mail address*: cdelaney@caltech.edu

Mathematics Department, Caltech, 1200 E. California Blvd. Pasadena, California 91125
*E-mail address*: matilde@caltech.edu

# Scattering amplitudes, Feynman integrals and multiple polylogarithms

Claude Duhr

ABSTRACT. We present a review of how ideas inspired by recent developments in number theory find applications in physics in the context of scattering amplitudes and Feynman integrals. In particular, we show how one can combine (and conjecturally extend) Goncharov's Hopf algebra on multiple polylogarithms by recent results by Brown on motivic multiple zeta values. These results can be used to derive in an effective way complicated relations among multiple polylogarithms. We conclude by illustrating the use of these concepts in various contexts related to the computation of scattering amplitudes and Feynman integrals.

## 1. Introduction

When computing theoretical predictions for physical observables in quantum field theories, a central role is played by the so-called scattering amplitudes, which, loosely speaking, encode the differential probability for a certain scattering process to happen. When computed in perturbation theory, scattering amplitudes can be expanded into a sum over Feynman diagrams which at each order involve Feynman graphs with a fixed number of loops. The loop diagrams are associated with Feynman integrals, and large classes of Feynman integrals are known to evaluate to very long and complicated combinations of (multiple) polylogarithms.

A lot of progress has been made over the last decade or so in the field of analytically computing scattering amplitudes and Feynman integrals. In particular, a lot of pioneering work was done in the context of the so-called $\mathcal{N} = 4$ Super Yang-Mills theory in the planar limit, a supersymmetric gauge theory which is known to enjoy various interesting mathematical properties. While $\mathcal{N} = 4$ Super Yang-Mills itself does not have any phenomenological relevance, it allows one to investigate in a clean environment the structure and the properties of scattering amplitudes which might not be apparent in more realistic theories. In particular, it was realised that scattering amplitudes in this theory are often characterized by very compact analytic expressions. Indeed, while scattering amplitudes are known to be generically expressible by combinations of thousands of special functions, loop amplitudes in $\mathcal{N} = 4$ Super Yang-Mills can often be written in a single line.

---

*Key words and phrases.* Feynman integrals, multiple polylogarithms.

The author is grateful to H. Gangl for collaboration on some of the topics presented in this paper. This work was supported by the ERC grant "IterQCD".

The first time such an unexpected simplicity was observed was in ref. [**1**], where the very long result for the two-loop six-point remainder function computed in refs. [**2**, **3**] was simplified to just a single line of classical polylogarithms. The main tool that made this simplification possible was the so-called *symbol map*, which allows one to associate to each multiple polylogarithm of weight $n$ with rational arguments an element in the $n$-fold tensor product over the group of rational functions. The merit of the symbol map lies in the fact that complicated and intricate functional equations between multiple polylogarithms are mapped, at least conjecturally, to simple algebraic relations among tensors. In other words, while it can be extremely complicated to prove a functional equation at the level of the functions, proving the corresponding relation among the tensors after applying the symbol map becomes an almost trivial exercise. The symbol map, however, is not invertible, and in particular its kernel contains all multiple zeta values, as well as $i\pi$.

The aim of this paper is to provide a concise review of the techniques developed in the context of Feynman integral computations based on ideas inspired by recent advances in number theory. In particular, we argue that by 'upgrading' the symbol map to a certain coaction defined on the comodule $\mathbb{Q}[i\pi] \otimes_{\mathbb{Q}} \mathcal{H}$, where $\mathcal{H}$ denotes Goncharov's Hopf algebra of multiple polylogarithms [**4**], we can eliminate almost all of the unwanted features of the symbol map. This construction is in fact a natural extension of recent work by Brown [**5**], and it allows one to derive functional equations among (large classes of) multiple polylogarithms with rational arguments which could not easily be obtained by traditional means. In a second part of the paper we show various examples where the knowledge of these functional equations can be helpful to perform explicit computations in quantum field theory. We note that the language used throughout this paper is the language used in physics: in particular, we sometimes do not make clear distinctions between the comodule $\mathbb{Q}[i\pi] \otimes_{\mathbb{Q}} \mathcal{H}$ and the Hopf algebra $\mathcal{H}$, and we will refer to the comodule as the 'extended' Hopf algebra. Furthermore, we do not make any distinction between zeta values and iterated integrals seen as complex numbers and their motivic counterparts. In addition, we assume throughout this paper that all transcendence conjectures for multiple zeta values are fulfilled.

The outline of this paper is as follows: In Section 2 we set the scene by briefly reviewing how Feynman integrals arise in the computation of scattering amplitudes. In Section 3 we discuss some mathematical properties of Feynman integrals and introduce the main actor in this work: the multiple polylogarithms. The symbol map and the Hopf algebra of multiple polylogarithms are introduced in Section 4, and we give some simple example how one can derive functional equations for polylogarithmic functions. Finally, we conclude this paper in Section 5 by showing three applications of the concepts introduced in Section 4 to the computation of Feynman integrals.

## 2. Scattering amplitudes and Feynman integrals

We start by giving a brief overview of how Feynman integrals arise in computations in quantum field theory. As the main focus of this work is to study Feynman integrals through the viewpoint of results from modern number theory, we will be short and only quote the main ideas in order to indicate how the results of subsequent sections connect to high-energy physics.

Quantum field theory allows one to make predictions for physically observable quantities in terms of probabilities. These probabilities can schematically by written as

$$\int d\Phi \, |\mathcal{A}(\{p_i\}, \{m_i\}; \alpha; D)|^2 \,, \tag{2.1}$$

where the integration measure $d\Phi$ is the so-called *phase space measure* whose exact form is irrelevant in the following. The integrand is the (square of the) *scattering amplitude* $\mathcal{A}$, whose computation will be the primary target of this paper. Generically, the scattering amplitude is a function of the momenta $p_i$ of the particles entering the scattering and the masses $m_i$ of the particles in the theory, as well as of a number $\alpha$, called the coupling constant, which parametrizes the 'strengths' of the interactions among the particles[1]. In addition, for reasons that will become clear below, we explicitly include the dependence of the scattering amplitude on the space-time dimension $D$. The functional dependence of the scattering amplitude on its arguments can be extremely complicated, and the functional form of $\mathcal{A}(\{p_i\}, \{m_i\}; \alpha; D)$ is in general unknown. In the case where the coupling constant $\alpha$ is not too large, we can approximate the scattering amplitude by its *perturbative expansion*,

$$\mathcal{A}(\{p_i\}, \{m_i\}; \alpha; D) = \sum_{L=0}^{\infty} \alpha^L \, \mathcal{A}^{(L)}(\{p_i\}, \{m_i\}; D) \,. \tag{2.2}$$

The coefficients $\mathcal{A}^{(L)}(\{p_i\}, \{m_i\}; D)$ that appear in the perturbative expansion can be computed order by order in perturbation theory as a sum over Feynman diagrams. In the following we assume that the reader is familiar with the basics of Feynman diagrams. We only recall that the exact form of the Feynman rules depends on the quantum field theory under consideration and that the coefficients $\mathcal{A}^{(L)}(\{p_i\}, \{m_i\}; D)$ receive contributions from Feynman diagrams with exactly $L$ loops. The coefficient corresponding to $L = 0$ only involves tree-level diagrams, and so the momenta of all internal lines in a Feynman diagram are fixed by momentum conservation once the external momenta $p_i$ are specified. Tree-level amplitudes therefore evaluate to rational functions of the external momenta $p_i$ and the masses $m_i$, and will not be considered any further.

Beyond tree-level, however, not all the momenta of all the internal lines in a Feynman diagram are fixed by the external momenta, and an $L$-loop diagram still depends on $L$ unspecified *loop momenta* $k_j$, $j \in \{1, \ldots, L\}$. Feynman rules then instruct us to integrate over the values of all loop momenta, which leads to the appearance of Feynman integrals of the form

$$\int \left( \prod_{j=1}^{L} \frac{d^D k_j}{i\pi^{D/2}} \right) \frac{\mathcal{N}(\{p_i\}, \{k_j\})}{(q_1^2 - m_1^2)^{\nu_1} \ldots (q_N^2 - m_N^2)^{\nu_N}} \,, \qquad N, \nu_j \in \mathbb{N} \,, \tag{2.3}$$

where $q_j$ are linear combinations of the external momenta $p_i$ and the internal momenta $k_j$, and the numerator $\mathcal{N}$ is a polynomial in the external and internal momenta. It is a well-known fact that, using the so-called *integration-by-parts identities* [**6, 7**], Feynman integrals with a non-trivial numerator structure can always be written as linear combinations with rational coefficients of Feynman integrals

[1] In general, there can be more than one coupling constant in the theory.

without non-trivial numerators, so without loss of generality we always assume $\mathcal{N}(\{p_i\},\{k_j\}) = 1$ in the following.

Feynman integrals, and thus also scattering amplitudes, are in general ill-defined, because they are plagued by divergences which can be classified into two categories:

(1) Ultraviolet (UV) divergences, which arise from integrations over large values of the loop momenta. In renormalizable theories these divergences can be removed consistently during the renormalization procedure to all orders in perturbation theory.
(2) Infrared (IR) divergences, which arise when the momenta of massless particles inside a loop become very small or collinear to each other. These divergences cannot be dealt with by renormalization, but they cancel in (inclusive-enough) physical observables among various contributions to a given order in perturbation theory.

Because of these divergences, Feynman integrals require regularization before they can be evaluated. The most prominent regularization scheme is the so-called *dimensional regularization*, where, loosely speaking, the space-time dimension $D$ is interpreted as a complex number $D$ and the divergences manifest themselves as poles in the complex $D$ plane. As we are primarily interested in four space-time dimensions, it proves convenient to introduce a parameter $\epsilon$ by $D = 4-2\epsilon$, and the poles of a dimensionally-regulated Feynman integral then arise as poles at $\epsilon = 0$. In other words, dimensionally-regulated Feynman integrals are interpreted as Laurent series around $\epsilon = 0$. The coefficients of the Laurent series are the main goal in Feynman integral computations.

To summarize, predictions for physical observables require the computation of scattering amplitudes, and in perturbation theory the computation of scattering is tightly connected to Feynman integrals. In the next section we will have a closer look at the mathematical properties of Feynman integrals and the state-of-the-art regarding their (analytic) computation.

## 3. Feynman integrals and multiple polylogarithms

In this section we will have a closer look at the analytic structure of Feynman integrals. From the discussion in the previous section it is clear that we do not need to consider Feynman integrals with non-trivial numerators. Feynman integrals without non-trivial numerators are precisely those that arise in scalar field theories, i.e., in the following we assume that the Feynman integrals we consider are represented by a graph $G$ whose edges correspond to the propagators in eq. (2.3).

We start by introducing the Feynman-parametrized version of the Feynman integral $I_G$ corresponding to a graph $G$ with $L$ loops and $N$ edges. One can show that

$$\begin{aligned} I_G &= e^{L\gamma_E\epsilon}\,(-1)^\nu\,\Gamma\left(\nu-L\frac{D}{2}\right)\int_0^\infty\prod_{j=1}^N\left(dx_j\,\frac{x_j^{\nu_j-1}}{\Gamma(\nu_j)}\right)\\ &\quad\times\delta\left(1-\sum_{j\in S}x_j\right)\,\frac{\mathcal{U}_G(\{x_i\})^{\nu-(L+1)D/2}}{(-\mathcal{F}_G(\{x_i\},\{p_i\cdot p_j\},\{m_i\}))^{\nu-LD/2}}, \end{aligned} \tag{3.1}$$

where $\nu = \sum_{i=1}^N \nu_i$ and $S$ is a non-empty subset of $\{1,\ldots,N\}$. Note that the value of the integral is independent of the choice of $S$ [**8**]. Furthermore, we have introduced the Dirac distribution $\delta(x)$ defined by

$$\int_{-\infty}^{+\infty} dx\,\delta(x)\,f(x) = f(0)\,. \tag{3.2}$$

The integrand contains the two polynomials $\mathcal{U}_G$ and $\mathcal{F}_G$, which can be constructed from the two Symanzik polynomials of the graph $G$. As the form of the polynomials is irrelevant for the rest of the paper, we do not discuss them any further, but we merely stress that their form is determined entirely by the topology of the graph $G$. Two comments are in order:

(1) Due to Lorentz invariance, scalar Feynman integrals can depend on the external moments only through Lorentz invariant products $p_i \cdot p_j$ and the internal masses $m_i^2$. We will refer to these quantities collectively as *scales* in the following.
(2) In eq. (3.1) we introduced the overall normalization factor $\exp(L\gamma_E\epsilon)$, where $\epsilon$ is the dimensional regulator and $\gamma_E = -\Gamma'(1)$ denotes the Euler-Mascheroni constant. The reason for this normalization lies in the fact that it removes all factors of $\gamma_E$ from the Laurent expansion of $I_G$. Moreover, it was shown in ref. [**9**] that, using this normalization, the coefficients of the Laurent series in $\epsilon$ evaluated for algebraic values of the scales are periods.

At one-loop, all Feynman integrals in $D = 4-2\epsilon$ dimensions can be expressed, up to terms of $\mathcal{O}(\epsilon)$ in the Laurent expansion, as linear combinations of Feynman integrals with at most four propagators. Analytic results are known for all Feynman integrals with at most four propagators in $D = 4-2\epsilon$ dimensions up to $\mathcal{O}(\epsilon^0)$, and in all cases the results can be written in terms of ordinary logarithms and dilogarithms,

$$\log z = \int_1^z \frac{dt}{t} \qquad \text{and} \qquad \mathrm{Li}_2(z) = -\int_0^z \frac{dt}{t}\log(1-t)\,. \tag{3.3}$$

For example, the so-called "one-mass box function",

$$I_{1m}(p_1,p_2,p_3,p_4;\epsilon) = e^{\gamma_E\epsilon}\int \frac{d^Dk}{i\pi^{D/2}}\frac{1}{k^2\,(k+p_1)^2\,(k+p_1+p_2)^2\,(k-p_4)^2}\,, \tag{3.4}$$

with $p_1^2 < 0$ and $p_i^2 = 0$ for $i \in \{2,3,4\}$, can be written in the form

$$\begin{aligned} I_{1m}(p_1,p_2,p_3,p_4;\epsilon) &= \frac{2r_\Gamma}{st}\frac{1}{\epsilon^2}\left[(-s)^{-\epsilon} + (-t)^{-\epsilon} - (-p_1^2)^{-\epsilon}\right] \\ &- \frac{2r_\Gamma}{st}\left[\mathrm{Li}_2\left(1-\frac{p_1^2}{s}\right) + \mathrm{Li}_2\left(1-\frac{p_1^2}{t}\right) + \frac{1}{2}\log^2\frac{s}{t} + \zeta_2\right] + \mathcal{O}(\epsilon)\,, \end{aligned} \tag{3.5}$$

where we defined $s \equiv (p_1+p_2)^2 < 0$ and $t \equiv (p_2+p_3)^2 < 0$ and

$$r_\Gamma = e^{\gamma_E\epsilon}\frac{\Gamma(1-\epsilon)^2\,\Gamma(1+\epsilon)}{\Gamma(1-2\epsilon)}\,. \tag{3.6}$$

Already at two-loop order, however, analytic results for Feynman integrals are in general unknown. For example, if we concentrate on two-loop Feynman graphs with four external legs in massless theories (which are the prototype Feynman integrals that contribute to a two-to-two scattering process), we currently only know the analytic expressions for the two-loop four-point functions where zero, one, two or all external lines are massive (see, e.g., ref. [**10**–**17**]). No analytic results are known in dimensional regularization for generic two-loop integrals with more than four external legs. The reason why Feynman integral computations beyond one-loop order are so difficult are manyfold:

(1) The order of the poles in $\epsilon$ increases with the loop number, and an $L$-loop Feynman integral may exhibit poles up to $\epsilon^{-2L}$.
(2) As is clear from eq. (3.1), both the number of integrations and the complexity of the integrand increases with the number of propagators.
(3) The functional dependence on the scales can quickly get very involved.
(4) Feynman integrals are not holomorphic functions of the scales, but they have a complicated branch cut structure. In particular, physical scattering processes usually require the careful analytic continuation of the Feynman integrals from regions where the functions are real to regions where they develop explicit imaginary parts. For example, while the one-mass box function in eq. (3.5) is real for $s, t, p_1^2 < 0$, it develops an imaginary part for $t < 0$ and $s, p_1^2 > 0$. It is this latter region that defines a physical scattering process.

In the rest of this paper we concentrate on a specific aspect of Feynman integral computations, namely the class of functions that arise in computations beyond one loop order.

It is well-known that large classes of interesting Feynman integrals can be expressed in terms of multiple polylogarithms[2]. Multiple polylogarithms are generalizations of the ordinary logarithm and the *classical* polylogarithms,

$$\mathrm{Li}_n(z) = \int_0^z \frac{dt}{t}\,\mathrm{Li}_{n-1}(t)\,. \tag{3.7}$$

They can be defined via the iterated integrals [**21**, **22**]

$$G(a_1,\ldots,a_n;z) = \int_0^z \frac{dt}{t-a_1}\,G(a_2,\ldots,a_n;t)\,, \qquad a_i \in \mathbb{C}\,, \tag{3.8}$$

with $G(x) = G(;x) = 1$, an exception being when $x = 0$ in which case we put $G(0) = 0$. In the special case where all the $a_i$'s are zero, we define, using the obvious vector notation $\vec{a}_n = \underbrace{(a,\ldots,a)}_{n}$, $a \in \mathbb{C}$,

$$G(\vec{0}_n;z) = \frac{1}{n!}\ln^n z\,. \tag{3.9}$$

The number $n$ of integrations in eq. (3.8) is called the *weight* of the multiple polylogarithm. We note at this point that throughout this paper we use the notation for multiple polylogarithms used in the physics literature. The relationship to the notation used in the mathematical literature is

$$G(a_1,\ldots,a_n;z) = I(0;a_n,\ldots,a_1;z)\,. \tag{3.10}$$

Multiple polylogarithms contain the classical polylogarithms as special cases,

$$G(\vec{0}_{n-1},1;z) = -\mathrm{Li}_n(z)\,. \tag{3.11}$$

---

[2]In some cases it is known that also elliptic functions can appear, see for example ref. [**18**–**20**].

Whenever they converge, multiple polylogarithms can equally well be represented [**21**] as multiple nested sums (e.g., for $|x_i| < 1$)

$$\text{Li}_{m_1,\ldots,m_k}(x_1,\ldots,x_k) = \sum_{0<n_1<n_2<\cdots<n_k} \frac{x_1^{n_1} x_2^{n_2} \cdots x_k^{n_k}}{n_1^{m_1} n_2^{m_2} \cdots n_k^{m_k}} . \tag{3.12}$$

The integral and series representations of multiple polylogarithms are related by

$$\text{Li}_{m_1,\ldots,m_k}(x_1,\ldots,x_k) = (-1)^k\, G\Bigg(\underbrace{0,\ldots,0}_{m_k-1},\frac{1}{x_k},\ldots,\underbrace{0,\ldots,0}_{m_1-1},\frac{1}{x_1\ldots x_k};1\Bigg) . \tag{3.13}$$

An important property of multiple polylogarithms (and of iterated integrals in general) is that they form a *shuffle algebra* [**23**], i.e., the product of two multiple polylogarithms of weight $n_1$ and $n_2$ can be written as a linear combination of multiple polylogarithms of weight $n_1 + n_2$,

$$G(a_1,\ldots,a_{n_1};z)\, G(a_{n_1+1},\ldots,a_{n_1+n_2};z) = \sum_{\sigma\in\Sigma(n_1,n_2)} G(a_{\sigma(1)},\ldots,a_{\sigma(n_1+n_2)};z), \tag{3.14}$$

where $\Sigma(n_1, n_2)$ denotes the set of all shuffles of $n_1 + n_2$ elements, i.e., the subset of the symmetric group $S_{n_1+n_2}$ defined by (cf. ref. [**24**], eq. (1.5.6))

$$\begin{aligned}\Sigma(n_1,n_2)\\ = \{\sigma\in S_{n_1+n_2} |\, \sigma^{-1}(1)<\ldots<\sigma^{-1}(n_1) \text{ and } \sigma^{-1}(n_1+1)<\ldots<\sigma^{-1}(n_1+n_2)\} .\end{aligned} \tag{3.15}$$

Many special cases of multiple polylogarithms have been (re)discovered by physicists in the computation of multi-loop Feynman integrals [**13, 25–29**]. Moreover, a folklore conjecture states that, if an $L$-loop Feynman integral can be expressed in terms of multiple polylogarithms, then the coefficient of $\epsilon^k$ in the Laurent expansion in the dimensional regulator involves multiple polylogarithms of weight at most $2L + k$.

Due to the importance of multiple polylogarithms in Feynman integral computations, it is important to understand the structure underlying this class of functions as well as possible. For example, analytic results for Feynman integrals are often characterized by very long and complicated expressions, involving thousands of multiple polylogarithms. The question whether a given expression can be simplified is thus of great importance. Such simplifications usually require the use of identities among multiple polylogarithms, which are often unknown in the literature. In the following section we introduce a way, inspired from ideas in modern number theory, to deal with, or rather circumvent, the complicated functional equations among multiple polylogarithms.

## 4. Functional equations for multiple polylogarithms

In this section we present a way to derive functional equations among (certain classes of) multiple polylogarithms. We concentrate in this section on purely mathematical aspects, and defer the applications to physics to the next sections. In order to present our method, we first need to introduce and review various concepts, which we will do in the following subsections.

**4.1. The symbol map.** We start by introducing the so-called symbol map [**1**, **24**, **30–32**]. We follow here the exposition of ref. [**1**], where the symbol was first introduced in the physics literature. Consider an iterated integral $F_k$ of weight $k$, defined recursively by

$$F_k = \int d\log R\, F_{k-1}\,, \tag{4.1}$$

where $R$ is a rational function. If the total differential of $F_k$ satisfies

$$dF_k = \sum_i F_{k-1,i}\, d\log R_i\,, \tag{4.2}$$

where $F_{k-1,i}$ are iterated integrals of weight $k-1$ and $R_i$ are rational functions, then we define the *symbol* of $F_k$ recursively by

$$\mathcal{S}(F_k) = \sum_i \mathcal{S}(F_{k-1,i}) \otimes R_i\,. \tag{4.3}$$

The symbol map $\mathcal{S}$ then assigns to each iterated integral of weight $k$ an element in the $k$-fold tensor product (over $\mathbb{Q}$) over the multiplicative group of rational functions. If we specialize to multiple polylogarithms, we have the differential equation [**22**],

$$dG(a_{n-1},\ldots,a_1;a_n) = \sum_{i=1}^{n-1} G(a_{n-1},\ldots,\hat{a}_i,\ldots,a_1;a_n)\, d\log\left(\frac{a_i-a_{i+1}}{a_i-a_{i-1}}\right), \tag{4.4}$$

where the hat indicates that the corresponding element is omitted. The symbol of a multiple polylogarithms can thus be defined recursively by

$$\mathcal{S}(G(a_{n-1},\ldots,a_1;a_n)) = \sum_{i=1}^{n-1} \mathcal{S}(G(a_{n-1},\ldots,\hat{a}_i,\ldots,a_1;a_n))\otimes\left(\frac{a_i-a_{i+1}}{a_i-a_{i-1}}\right). \tag{4.5}$$

For example, from

$$d\left(\frac{1}{n!}\log^n z\right) = \frac{1}{(n-1)!}\log^{n-1} z\, d\log z \text{ and } d\text{Li}_n(z) = \text{Li}_{n-1}(z)\, d\log z\,, \tag{4.6}$$

we conclude that

$$\mathcal{S}\left(\frac{1}{n!}\log^n z\right) = \underbrace{z\otimes\ldots\otimes z}_{n} \quad \text{and} \quad \mathcal{S}(\text{Li}_n(z)) = -[(1-z)\otimes\underbrace{z\otimes\ldots\otimes z}_{n-1}]\,, \tag{4.7}$$

where we explicitly put square brackets to indicate that the minus sign is to be interpreted as the coefficient of the tensor.

The map $\mathcal{S}$ is obviously linear. In addition, $\mathcal{S}$ is a shuffle algebra homomorphism,

$$\mathcal{S}(f\cdot g) = \mathcal{S}(f)\,\text{Ш}\,\mathcal{S}(g)\,, \tag{4.8}$$

where in the left-hand side the multiplication is given by ordinary multiplication of iterated integrals, while in the right-hand side multiplication is given by the shuffle product on tensors, defined by

$$(a_1\otimes\ldots\otimes a_{n_1})\,\text{Ш}\,(a_{n_1+1}\otimes\ldots\otimes a_{n_1+n_2}) = \sum_{\sigma\in\Sigma(n_1,n_2)} a_{\sigma(1)}\otimes\ldots\otimes a_{\sigma(n_1+n_2)}\,, \tag{4.9}$$

and extended by linearity.

Due to the fact that we work in the *multiplicative* group of rational functions, the usual linearity of the tensor product takes a rather unusual form,

$$\begin{aligned} \ldots \otimes (a \cdot b) \otimes \ldots &= \ldots \otimes a \otimes \ldots + \ldots \otimes b \otimes \ldots\,, \\ \ldots \otimes 1 \otimes \ldots &= 0\,. \end{aligned} \tag{4.10}$$

We conclude that, for every $k$-th root of unity $\rho_k$,

$$\ldots \otimes \rho_k \otimes \ldots = 0\,, \tag{4.11}$$

because

$$0 = \ldots \otimes 1 \otimes \ldots = k\,(\ldots \otimes \rho_k \otimes \ldots)\,. \tag{4.12}$$

The previous properties immediately imply that the symbol map must send to zero all terms proportional to multiple zeta values (MZVs) and $i\pi = \log(-1)$. MZVs can be defined as the values of multiple polylogarithm of the form $G(a_1, \ldots, a_n; 1)$, with $a_i \in \{0, 1\}$ (provided that the iterated integral converges). It is then easy to see that the symbol of all MZVs must vanish, for example,

$$\mathcal{S}(\zeta_n) = \mathcal{S}(\text{Li}_n(1)) = -[(1-1) \otimes \underbrace{1 \otimes \ldots \otimes 1}_{n-1}] = 0\,. \tag{4.13}$$

At this stage we want to emphasize a subtle point: from the definition of the symbol map we gave at the beginning of this section it seems that one should have $\ldots \otimes c \otimes \ldots = 0$, for every constant $c$, because $d \log c = 0$. However, it was shown in ref. [**32**] that the symbol of a multiple polylogarithm can be defined in an equivalent way using dissections of a certain decorated and rooted polygon attached to a multiple polylogarithm [**33**]. This latter definition of the symbol map is combinatorial in nature and so the distinction between $c$ being constant or not becomes immaterial. As the definition of the symbol map using dissections of rooted and decorated polygons is technically more involved, we do not give it here explicitly, but we refer to ref. [**32, 33**]. In the following we will always work with a definition of the symbol map where constants inside the tensor product are not put to zero.

There are two natural operations one can do on multiple polylogarithms: taking derivatives and taking discontinuities. Both of these operations have natural analogues in the tensor product over the group of rational functions. For simplicity, we discuss the case of rational functions depending on a single variable $x$. The generalization to more variables is straightforward. We start by defining two linear operators $D_x$ and $M_{x,a}$, $a \in \mathbb{C}$, acting on tensors by[3]

(4.14)
$$\begin{aligned} D_x(a_1(x) \otimes \ldots \otimes a_n(x)) &= (\partial_x \log a_n(x))\, a_1(x) \otimes \ldots \otimes a_{n-1}(x)\,, \\ M_{x,a}(a_1(x) \otimes \ldots \otimes a_n(x)) &= \begin{cases} a_2(x) \otimes \ldots \otimes a_n(x)\,, & \text{if } a_1(a) = 0 \text{ or } \infty\,, \\ 0\,, & \text{otherwise,} \end{cases} \end{aligned}$$

where $\partial_x$ denotes the usual partial derivative with respect to $x$. It is easy to check that $D_x$ and $M_{x,a}$ define derivations on the tensor product over the group of rational functions, i.e., they satisfy the Leibniz rule for the shuffle product on

[3]We note that, strictly speaking, we need to enlarge the coefficients of the tensors from rational numbers to rational functions for $D_x$ to be well-defined.

tensors. These operators correspond to taking derivatives or discontinuities at the level of functions. More precisely, we have

$$\begin{aligned} \mathcal{S}\left[\partial_x f(x)\right] &= D_x\, \mathcal{S}[f(x)]\,, \\ \mathcal{S}\left[\frac{1}{2\pi i}\text{Disc}_{x=a} f(x)\right] &= M_{x,a}\, \mathcal{S}[f(x)]\,, \end{aligned} \tag{4.15}$$

where $\text{Disc}_{x=a} f(x)$ denotes the discontinuity of the function $f(x)$ around $x = a$.

So far we have only discussed how to assign a tensor to a multiple polylogarithm, but we have not yet addressed the inverse question, i.e., whether for an arbitrary element $\xi$ from the tensor product over the multiplicative group of rational functions we can always find a function $f$ (e.g., a combination of multiple polylogarithms) such that $\xi = \mathcal{S}(f)$. It turns out that the answer is negative, but there is a necessary and sufficient condition to be satisfied by $\xi$ for such a function $f$ to exist: if we assume that $\xi$ has the form

$$\xi = \sum_{I=(i_1,\ldots,i_m)} c_I\, a_{i_1} \otimes \cdots \otimes a_{i_m}\,, \quad c_I \in \mathbb{Q}\,, \tag{4.16}$$

then there is a function $f$ whose symbol is $\xi$ if and only if the following *integrability condition* is satisfied [**24**, **34**],

$$\sum_{I=(i_1,\ldots,i_m)} c_I \left[d\log a_{i_j} \wedge d\log a_{i_{j+1}}\right] a_{i_1} \otimes \cdots \otimes \widehat{a}_{i_j} \otimes \widehat{a}_{i_{j+1}} \otimes \cdots \otimes a_{i_m} = 0\,, \tag{4.17}$$

where the hats indicate that we omit the corresponding factors in the tensor product and $\wedge$ denotes the usual wedge product on differential forms.

**4.2. Symbols and functional equations.** In this section we illustrate the use of the symbol map when proving or deriving functional equations among multiple polylogarithms. The basic idea is very simple: assume we are given two functions $F(x)$ and $G(x)$ (we assume for simplicity that the functions only depend on the single variable $x$, but the results are in fact independent on the number of variables), and assume we want to decide if these two functions are equal. In general, this can be a very complicated task, and it is often simpler to prove or disprove the identity at the level of the symbols, i.e., if the two functions $F(x)$ and $G(x)$ are equal, then their symbols must also be equal. As identities among symbols are essentially equivalent to identities among logarithms (see eq. (4.10)), proving identities among symbols is much easier then proving identities among functions.

Let us illustrate this on the following simple example: assume we are given the two functions

$$\begin{aligned} F(x) &= \text{Li}_3\left(\frac{1}{x}\right)\,, \\ G(x) &= \text{Li}_3(x) + \frac{1}{6}\log^3 x - \frac{i\pi}{2}\log^2 x - \frac{\pi^2}{3}\log x\,, \end{aligned} \tag{4.18}$$

where we assume $0 < x < 1$ and we have chosen a branch of the logarithm such that $\log(-1) = -i\pi$ . Note that this is nothing but the well-known inversion relation for the trilogarithm. While proving the equality at the level of the functions can be quite intricate, proving the equality at the level of the symbol is trivial. Indeed, we immediately get

$$\mathcal{S}(F(x)) = -(1-x) \otimes x \otimes x + x \otimes x \otimes x = \mathcal{S}(G(x))\,. \tag{4.19}$$

At this stage we have to stress that, despite the fact that we have established the equality of the symbols of $F(x)$ and $G(x)$, it would be premature to conclude that $F(x)$ and $G(x)$ are equal as functions. Indeed, as already mentioned, the symbol map is not invertible, which can immediately be seen from the fact that it maps to zero all MZVs. In general, we can at this stage only conclude that the two functions are equal, modulo elements in the kernel of the symbol map. We have already seen that the kernel of $\mathcal{S}$ necessarily contains all MZVs, as well as $i\pi$. The kernel of the symbol map however also contains more non-trivial elements, e.g., we have

$$\mathcal{S}\left[\mathrm{Li}_n\left(\frac{1}{2}\right)+\frac{(-1)^n}{n!}\log^n 2\right]=0\,,\qquad \forall n\in\mathbb{N}^\times\,. \tag{4.20}$$

In order to deal with (some of) the elements in the kernel of the symbol map, we need to introduce some more concepts, which will be done in the next subsection.

**4.3. The Hopf algebra of multiple polylogarithms.** In ref. [**4**] it was shown that multiple polylogarithms form a Hopf algebra. In the following we assume that the reader is familiar with the basic concepts of Hopf algebras, and we only recall the concepts used in the rest of this paper.

Let us denote by $\mathcal{H}$ the algebra of multiple polylogarithms, i.e., the vector space spanned by all multiple polylogarithms equipped with the shuffle product. The algebra is graded by the weight

$$\mathcal{H}=\bigoplus_{n=0}^{\infty}\mathcal{H}_n\,, \tag{4.21}$$

where by definition we put $\mathcal{H}_0=\mathbb{Q}$. In ref. [**4**] it was shown that $\mathcal{H}$ may be equipped with a coproduct, i.e., a linear map $\Delta:\mathcal{H}\to\mathcal{H}\otimes\mathcal{H}$ that respects the grading and and the multiplication (in other words, $\Delta$ is an algebra homomorphism),

$$\Delta(a\cdot b)=\Delta(a)\cdot\Delta(b)\,. \tag{4.22}$$

In addition, the coproduct $\Delta$ is coassociative, $(\mathrm{id}\otimes\Delta)\Delta=(\Delta\otimes\mathrm{id})\Delta$. All these properties are part of the defining axioms of a Hopf algebra, and indeed the algebra $\mathcal{H}$ equipped with the coproduct $\Delta$ forms a Hopf algebra. The explicit construction of $\Delta$ for multiple polylogarithms is rather involved, and we refer to ref. [**4, 22**] for details. Here we only quote two simple examples,

$$\begin{aligned}\Delta(\log x)&=1\otimes\log x+\log x\otimes 1\,,\\ \Delta(\mathrm{Li}_n(x))&=1\otimes\mathrm{Li}_n(x)+\mathrm{Li}_n(x)\otimes 1+\sum_{k=1}^{n-1}\mathrm{Li}_{n-k}(x)\otimes\frac{\log^k x}{k!}\,.\end{aligned} \tag{4.23}$$

In the following it will be useful to introduce some notation: we will denote the component of the iterated coproduct that takes values in $\mathcal{H}_{p_1}\otimes\ldots\otimes\mathcal{H}_{p_n}$ by $\Delta_{p_1,\ldots,p_n}$. It can then be checked that the symbol map introduced in Section 4.1 coincides with the maximal iteration of the coproduct, $\Delta_{1,\ldots,1}$. For example, from eq. (4.23) we immediately get

$$\Delta_{1,1,1}(\mathrm{Li}_3(x))=-\log(1-x)\otimes\log x\otimes\log x\,, \tag{4.24}$$

which agrees with the symbol of the trilogarithm, eq. (4.7), upon dropping the 'log' signs. A proof that the combinatorics of the maximal iteration of the coproduct agrees with the combinatorics of the symbol map can be found in ref. [**32**].

Having at our disposal the machinery of the Hopf algebra of multiple polylogarithms, and having identified the symbol map with the maximal iteration of the coproduct, we can try to remedy some of the problems we were facing in the previous section, namely that the symbol map did not allow us to determine the terms proportional to MZVs in eq. (4.18). Indeed, while the *maximal* iteration of the coproduct maps to zero all multiple zeta values, MZVs are not necessarily zero in other components. For example, from eq. (4.23) we immediately see that

$$\Delta(\zeta_n) = 1 \otimes \zeta_n + \zeta_n \otimes 1\,. \tag{4.25}$$

In other words, rather than working with $\Delta_{1,\ldots,1}$, we could for example work with the more 'refined' $\Delta_{2,1,\ldots,1}$. There is however still a caveat: the construction of the Hopf algebra $\mathcal{H}$ of ref. [4] requires to put to zero all $\zeta_{2n}$. An easy way to understand this is to consider for example the coproduct of $\zeta_4$,

$$\Delta(\zeta_4) = \Delta\left(\frac{2}{5}\zeta_2^2\right) = \frac{2}{5}\,\Delta(\zeta_2)^2 = 1\otimes\zeta_4 + \frac{4}{5}\zeta_2\otimes\zeta_2 + \zeta_4\otimes 1\,, \tag{4.26}$$

and so we have obtained a contradiction with eq. (4.25) unless '$\zeta_2 = 0$'. It thus seems at first glance that we have not gained much over the symbol approach. There is however a solution to this problem which we describe in the next subsection in the context of MZVs.

**4.4. Multiple zeta values.** Let us denote by $\mathcal{A}$ the Hopf algebra of MZVs. The coproduct is inherited from the coproduct on $\mathcal{H}$, and from the previous discussion we know that we have to work modulo $\zeta_2$. In ref. [**5**] it was shown that a consistent way to include $\zeta_2$ is to define the trivial comodule

$$\overline{\mathcal{A}} = \mathbb{Q}[\zeta_2]\otimes\mathcal{A}\,, \tag{4.27}$$

where the tensor product is taken over $\mathbb{Q}$. In other words, we consider elements of $\mathcal{A}$ whose coefficients are polynomials in $\zeta_2$ with rational coefficients. The coaction $\Delta : \overline{\mathcal{A}} \to \overline{\mathcal{A}}\otimes\mathcal{A}$ is, loosely speaking, defined from the coproduct on $\mathcal{A}$, augmented by the 'rule'

$$\Delta(\zeta_2) = \zeta_2\otimes 1\,. \tag{4.28}$$

In practice, this rule means that, after passing from $\mathcal{A}$ to $\overline{\mathcal{A}}$, we can solve the problem we faced at the end of the Section 4.3 by putting $\zeta_2$ to zero in all factors of the coproduct but the first. In the next section we show how we can extend the ideas of ref. [**5**] to multiple polylogarithms and we define a coaction which allows us to reconstruct almost all MZVs, thus obtaining an effective way to prove and derive functional equations among multiple polylogarithms.

**4.5. An 'extended Hopf algebra'.** Inspired by the ideas of ref. [**5**], it seems natural to define a comodule

$$\overline{\mathcal{H}} = \mathbb{Q}[\zeta_2]\otimes\mathcal{H}\,, \tag{4.29}$$

together with the coaction given by the coproduct on $\mathcal{H}$ and eq. (4.28). We will in the following take an additional step and define [**35**]

$$\overline{\mathcal{H}} = \mathbb{Q}[i\pi]\otimes\mathcal{H}\,, \tag{4.30}$$

together with the coaction given by the coproduct on $\mathcal{H}$ and

$$\Delta(i\pi) = i\pi\otimes 1\,. \tag{4.31}$$

In the following, by abuse of language, we will often refer to this comodule as the 'extended' Hopf algebra of multiple polylogarithms. The advantage of the definition (4.31) is that it allows not only to determine terms proportional to MZVs in functional equations, but also those terms proportional to $i\pi$. While we have no proof of our construction, we have used it in numerous applications, and we find that in all cases we obtain the correct results.

Armed with the concepts we have just introduced, we can effectively obtain a 'calculus' for deriving and proving functional equations, i.e., we can easily derive complicated functional equations for multiple polylogarithms that could not have been obtained easily by other means. We illustrate this technique on the example of the inversion relation of Section 4.2. More precisely, we will not only prove the equality of the functions $F(x)$ and $G(x)$, which was our goal in Section 4.2, but we will actually *derive* the function $G(x)$ just from the knowledge of the Hopf algebra of multiple polylogarithms and the inversion relations for the dilogarithms.

We start by acting with the symbol map on $F(x)$. We immediately obtain

$$\mathcal{S}(F(x)) = -(1-x)\otimes x\otimes x + x\otimes x\otimes x = \mathcal{S}\left[\mathrm{Li}_3(x) + \frac{1}{6}\log^3 x\right]. \tag{4.32}$$

This is all the information we can get out of the symbol map. We can next act with $\Delta_{1,1,1}$ on the difference,

$$\begin{aligned}\Delta_{1,1,1}\left[F(x) - \left(\mathrm{Li}_3(x) + \frac{1}{6}\log^3 x\right)\right] &= -i\pi\otimes\log x\otimes\log x\\ &= \Delta_{1,1,1}\left(-\frac{i\pi}{2}\log^2 x\right).\end{aligned} \tag{4.33}$$

We see that the iterated coproduct allows us to fix the imaginary part of $F(x)$. Following the same logic, the terms proportional to $\zeta_2$ can be determined by acting with $\Delta_{2,1}$ on the difference,

$$\begin{aligned}&\Delta_{2,1}\left[F(x) - \left(\mathrm{Li}_3(x) + \frac{1}{6}\log^3 x - \frac{i\pi}{2}\log^2 x\right)\right]\\ &= -\mathrm{Li}_2\left(\frac{1}{x}\right)\otimes\log x - \mathrm{Li}_2(x)\otimes\log x - \frac{1}{2}\log^2 x\otimes\log x + (i\pi\log x)\otimes\log x.\end{aligned} \tag{4.34}$$

At this stage we assume recursively that we have already determined the inversion relation for $\mathrm{Li}_2(x)$,

$$\mathrm{Li}_2\left(\frac{1}{x}\right) = -\mathrm{Li}_2(x) - \frac{1}{2}\log^2 x + i\pi\log x + \frac{\pi^2}{3}. \tag{4.35}$$

We then immediately obtain

$$\begin{aligned}\Delta_{2,1}\left[F(x) - \left(\mathrm{Li}_3(x) + \frac{1}{6}\log^3 x - \frac{i\pi}{2}\log^2 x\right)\right] &= -\frac{\pi^2}{3}\otimes\log x\\ &= \Delta_{2,1}\Big(-\frac{\pi^2}{3}\ln x\Big).\end{aligned} \tag{4.36}$$

We can thus conclude that

$$F(x) = \mathrm{Li}_3(x) + \frac{1}{6}\log^3 x - \frac{i\pi}{2}\log^2 x - \frac{\pi^2}{3}\log x + c_1\,\zeta_3 + c_2\,i\pi^3, \tag{4.37}$$

for some rational numbers $c_1$ and $c_2$. Note that we cannot fix these constants from the Hopf algebra alone. The reason for this is that $\zeta_3$ and $i\pi^3$ cannot be decomposed into objects of lower weight by the action of the coproduct,

$$\Delta(\zeta_3) = \zeta_3 \otimes 1 + 1 \otimes \zeta_3 \quad \text{and} \quad \Delta(i\pi^3) = i\pi^3 \otimes 1\,. \tag{4.38}$$

Note that this is similar to the findings of ref. [**5**], where it was shown that the coaction on $\overline{\mathcal{A}}$ introduced in the previous subsection can be used to decompose MZVs into a basis, and for each weight $w$ all the coefficients appearing in the basis are fixed uniquely, except for the coefficient of $\zeta_w$. Similarly, in our case we will not be able to determine the coefficients of pure powers of $i\pi$ as well as of elements that are *primitive* in $\mathcal{H}$, i.e., elements $\xi \in \mathcal{H}$ such that

$$\Delta(\xi) = 1 \otimes \xi + \xi \otimes 1\,. \tag{4.39}$$

These coefficients can however easily be determined by evaluating the function at a single point. In our example, we can evaluate eq. (4.37) at $x = 1$, and we find,

$$\zeta_3 = \zeta_3 + c_1\,\zeta_3 + c_2\,i\pi^3\,. \tag{4.40}$$

Thus we see that $c_1 = c_2 = 0$, and we immediately recover the expression for $G(x)$ in eq. (4.18).

We have given an example of how the comodule (4.30) allows us to derive functional equations among multiple polylogarithms. Although we have only discussed the simple example of a trilogarithm, we have used the techniques presented in this section in many circumstances to derive identities among multiple polylogarithms. In the next section we give various examples of how the ideas of this section can be used in Feynman integral computations.

## 5. The Hopf algebra of multiple polylogarithms and Feynman integrals

In this section we illustrate how the algebraic structures on multiple polylogarithms we reviewed in the previous section can be useful in the field of Feynman integral computations. We review several contexts in which the Hopf algebra of multiple polylogarithms can be used to substantially simplify the computation and/or the results. As applications of symbols in the context of Feynman integrals and scattering amplitudes are numerous by now, we stress that in the following we only give a small collection of possible applications, and we claim that this list is by no means exhaustive.

**5.1. Simplification of analytic results.** The first, and probably most obvious, application of the ideas presented in the previous section consists in trying to simplify the sometimes large and complicated results that arise from Feynman integral computations. Indeed, analytic results for two-loop multi-scale Feynman integrals and scattering amplitudes can often involve combinations of several thousands of multiple polylogarithms, and so the question whether a given expression can be rewritten in terms of 'simpler' functions and/or in a more compact form is highly relevant.

The first time such a striking simplification was achieved in ref. [**1**], where also the symbol map was introduced for the first time in the physics literature. In ref. [**2, 3**] the so-called two-loop six-point remainder function in the $\mathcal{N} = 4$ Super Yang-Mills theory was evaluated, and the result was expressed as a 17-page-long combination of multiple polylogarithms of uniform weight four. In ref. [**1**] it was shown that using the symbol map the result of ref. [**2**, **3**] can be rewritten as a single line of classical polylogarithms, thus exhibiting a simplicity in the answer which was not obvious from the original computation. While the $\mathcal{N} = 4$ Super Yang-Mills theory is a particular quantum field theory that is known to enjoy special mathematical properties, it was more recently shown in ref. [**35**] that, based on earlier observations at the level of the symbol [**36**], the analytic results for the helicity amplitudes for a Higgs boson plus three gluons computed in ref. [**37**] can be rewritten in a much more compact form that makes all the symmetries of the amplitudes manifest and that only involves classical polylogarithms.

In the remainder of this section we describe how such simplifications may be achieved. The basic idea is simple: we are looking for a functional equation among multiple polylogarithms that allows us to express the complicated expression we start with in terms of 'simpler' functions (e.g., classical polylogarithms or products of lower weight functions). We can try to apply the ideas of Section 4 to derive this functional equation. In the following we always assume that we are dealing with $\mathbb{Q}$-linear combinations of products of multiple polylogarithms of uniform weight; if not, we deal separately with the contributions of different weight and/or different rational function coefficients.

We first discuss several criteria that can be used to determine whether a simpler representation for a function exists, and then we turn to a procedure that allows to construct an ansatz for the space of functions in which the simpler representation can be found.

We have already seen that the coefficient of $\epsilon^k$ in the Laurent expansion of an $L$-loop Feynman diagram is conjectured to have weight at most $2L+k$. The constant term of a two-loop integral can thus have weight at most four. It is well-known (cf. ref. [**38**], §8.4.3, implicitly, as well as refs. [**39**, **40**]) that multiple polylogarithms of weight at most three can always be expressed in terms of classical polylogarithms, consistent with the fact that one-loop integrals in four space-time dimensions can always be expressed in terms of dilogarithms and ordinary logarithms. This statement is no longer true for multiple polylogarithms of weight four. In ref. [**1**] a necessary and sufficient condition, based on a conjecture in ref. [**41**], was presented which allows one to determine whether a given combination of multiple polylogarithms of weight four can be expressed in terms of classical polylogarithms only. To state this criterion, it is convenient to introduce a linear operator $\delta$ acting on tensors of weight four by

$$\delta(a_1 \otimes a_2 \otimes a_3 \otimes a_4) \equiv (a_1 \wedge a_2) \wedge (a_3 \wedge a_4)\,, \tag{5.1}$$

where we defined $a \wedge b \equiv a \otimes b - b \otimes a$. In terms of this operator, the aforementioned condition takes the simple form that a function $F$, given as a combination of multiple polylogarithms of weight four, can be expressed in terms of classical polylogarithms and ordinary logarithms only if and only if the symbol of $F$ lies in the kernel of $\delta$. Naturally, the symbols of the two-loop six-point remainder function of ref. [**2**, **3**] and the (weight four part of) the helicity amplitudes for a Higgs boson plus three gluons satisfy this criterion, explaining why a representation for these functions in terms of classical polylogarithms only exists.

While the operator $\delta$ allows one to determine whether a given function of weight four can be expressed in terms of classical polylogarithms of weight four, it is also useful to have a criterion to determine when a function can be written as linear combinations of products of lower weight functions only. It turns out that there is a projector that acts on symbols (and more generally on shuffle algebras) that achieves precisely this: the kernel of the projector is generated by all (non-constant) shuffle products. There is one projector $\Pi_w$ for each weight $w$, and it can be defined by $\Pi_1 = \text{id}$ and for $w \geq 2$ by [**23, 32, 42**]

$$\begin{aligned} &\Pi_w(a_1 \otimes \ldots \otimes a_w) \\ &\quad = \frac{w-1}{w}\left[\Pi_{w-1}(a_1 \otimes \ldots \otimes a_{w-1}) \otimes a_w - \Pi_{w-1}(a_2 \otimes \ldots \otimes a_w) \otimes a_1\right] . \end{aligned} \tag{5.2}$$

In other words, if the symbol of a function $F$ of weight $w$ vanishes under the action of the projector $\Pi_w$, then a representation of $F$ exists that involves only products of multiple polylogarithms of weight less than $w$.

Having discussed several criteria to determine whether a simpler representation for a given function exists, we now give an outline for how an ansatz for such a simpler function can be constructed in some cases. Once such an ansatz has been constructed, determining the coefficients of the functions that appear in the ansatz reduces to elementary linear algebra. In order to find such an ansatz, we need to determine the arguments of the polylogarithms that appear in the ansatz. In ref. [**32**] a procedure was described that allows one to determine, based on a given symbol, a natural set of rational functions that can appear as arguments of polylogarithms. We illustrate this procedure on the example of the classical polylogarithms, and refer to ref. [**32**] for the multiple polylogarithm case.

Let us assume that we are given the symbol $S$ of a function for which we know that a representation in terms of classical polylogarithms only exists. Without loss of generality we can assume that the entries that appear in the symbol are polynomials that are irreducible over $\mathbb{Z}$. Let us denote the set of these polynomials by $P_S$, and the elements of $P_S$ are denoted by $\pi_i$. From the elements of $P_S$ we can define a new set of polynomials $\overline{P}_S = P_S \cup P_S'$ (whose elements we denote in the following by $\overline{\pi}_i$), where $P_S'$ is the set of all irreducible polynomials (possibly including prime numbers as well) that appear in $\pi_i \pm \pi_j$ and $1 \pm \pi_i$, $\forall \pi_i, \pi_j \in \overline{P}_S$. The reason for introducing this larger set of polynomials stems from the fact that there might be cancellations between different multiple polylogarithms contributing to the symbols $S$. A natural ansatz for rational functions that may appear as arguments of polylogarithms whose symbols have its entries drawn from the set $\overline{P}_S$ is then

$$R^{\pm}_{n_1,\ldots,n_k} = \pm\overline{\pi}_1^{n_1} \ldots \overline{\pi}_k^{n_k} , \tag{5.3}$$

where $n_1, \ldots, n_k$ are integers. Let us denote the set of these functions by $\mathcal{R}_S$. Note that $\mathcal{R}_S$ is just the multiplicative span of the elements in $\overline{P}_S$, and as such it carries a natural group structure given by multiplication. The set $\mathcal{R}_S$ is infinite, and it turns out that indeed it is too large. To see why, if $R \in \mathcal{R}_S$ appears as an entry of a classical polylogarithm, then

$$\mathcal{S}(\text{Li}_n(R)) = -(1-R) \otimes R \otimes \ldots \otimes R , \tag{5.4}$$

and so if we want all the entries in the symbol of $\text{Li}_n(R)$ to be drawn from the set $\overline{P}_S$, we have to require that $1 - R$ admits a factorization into products of powers of elements in $\overline{P}_S$, or equivalently that $1 - R \in \mathcal{R}_S$. Let us denote the subset of $\mathcal{R}_S$ consisting of rational function with this property by

$$\mathcal{R}_S^{(1)} = \{R \in \mathcal{R}_S \mid 1 - R \in \mathcal{R}_S\} \subset \mathcal{R}_S . \tag{5.5}$$

The set $\mathcal{R}_S^{(1)}$ is the natural set of rational functions that may appear as arguments of classical polylogarithms whose symbols have only entries drawn from $\overline{P}_S$. We note that, empirically in all cases we have considered so far, the set $\mathcal{R}_S^{(1)}$ is no longer infinite, but it is rather small. In addition, it is easy to see that $\mathcal{R}_S^{(1)}$ does no longer carry a group structure. It comes however equipped with a natural action of the symmetric group $S_3$, defined for a rational function $R$ and rational functions $\sigma_i$ of one variable by

$$\begin{aligned}
&\sigma_1(R) = R\,, \quad \sigma_2(R) = 1 - R\,, \quad \sigma_3(R) = 1/R\,,\\
&\sigma_4(R) = 1/(1-R)\,, \quad \sigma_5(R) = 1 - 1/R\,, \quad \sigma_6(R) = R/(R-1)\,.
\end{aligned} \tag{5.6}$$

It is easy to see that the functions $\sigma_i$ (under composition of function) form a group isomorphic to $S_3$. In addition, whenever $R \in \mathcal{R}_S^{(1)}$, we have $\sigma_i(R) \in \mathcal{R}_S^{(1)}$, $\forall i$.

The aforementioned procedure was used in ref. [**32**] to construct arguments that can appear as arguments of classical polylogarithms arising from harmonic polylogarithms, as well as to construct the arguments of the classical polylogarithms of the representation of the two-loop helicity amplitudes for a Higgs boson plus three gluons derived in ref. [**35**]. Once the arguments have been determined for a function that can be expressed through classical polylogarithms only, it is easy to write down an ansatz for these classical polylogarithms and to determine their coefficients using elementary linear algebra.

**5.2. Symbolic integration of parametric integrals.** Parametric integrals, e.g., Feynman parameter integrals, are in general not easy to integrate. Not only are the integrals in general divergent and the singularities need to be carefully extracted, but even if they are finite the complexity of the integrands obtained by recursively integrating over the integration variables increases rapidly. In some special cases, however, it is possible to carry out this recursive integration procedure in terms of multiple polylogarithms one variable at the time. Loosely speaking, we need to require that at each integration step we can find an integration variable such that all denominators are linear in this variable. Indeed, if we can find such a variable, then it is easy to find a primitive of the integrand using eq. (3.8). A general necessary condition to determine whether this procedure terminates was given in ref. [**43**].

Even if we know that the integration terminates, we still need to carry out the integration explicitly. For example, let us consider a single term that may contribute during the recursive integration procedure,

$$\int_0^\infty \frac{dx_i}{x_i - a}\, G(\vec{b}; x_i) = G(a, \vec{b}; x_i)\Big|_{x_i=0}^{x_i=\infty}\,, \tag{5.7}$$

where $a$ and $\vec{b}$ may depend on the remaining integration variables. In order to proceed, we need to evaluate the primitive on the boundaries. The limit $x_i \to 0$ is trivial to compute (see eq. (3.12)). The limit $x_i \to \infty$ can be computed by letting $\bar{x}_i = 1/x_i$ and then computing the limit

$$\lim_{\bar{x}_i \to 0} G\left(a, \vec{b}; \frac{1}{\bar{x}_i}\right)\,. \tag{5.8}$$

In order to proceed, we need to know the inversion relations for the primitive, which are unknown in general. Furthermore, even if the inversion relations are known, the next integration variable may be 'hidden' inside the parameters $a$ and $\vec{b}$, and additional functional equations may be required to bring the integral (5.7) into a form where the next variable can be integrated out. In the rest of this section we show how the methods described in the previous section allow us to derive all the functional equations needed to integrate out all the variables one-by-one (in a given order).

We will illustrate the recursive integration of parametric integrals on the well-known example of the 'wheel with three spokes' in four dimensions, given by the integral

$$I_{W3S} = \int_0^\infty \left(\prod_{i=1}^5 dx_i\right) \frac{1}{\mathcal{U}^2}\,, \tag{5.9}$$

where the integrand is given by

$$\begin{aligned}\mathcal{U} &= x_1x_2 + x_1x_2x_3 + x_2x_3x_4 + x_2x_4 + x_1x_2x_5 + x_2x_3x_5 + x_2x_4x_5 + x_2x_5\\ &+ x_1x_3 + x_1x_4 + x_1x_3x_4 + x_3x_4 + x_1x_5 + x_3x_5 + x_1x_4x_5 + x_3x_4x_5\,.\end{aligned} \tag{5.10}$$

It can be checked that the integration variables can be integrated out one-by-one in the order $(x_1, x_2, x_3, x_4, x_5)$ [**43**]. The first three integrations are in fact trivial and only lead to ordinary logarithms. The remaining two integrations then give rise to polylogarithms. In order to proceed, we first write all the ordinary logarithms as multiple polylogarithms, e.g.,

$$\log(x_4 + x_5 + x_4x_5) = G(0; x_5) + G\left(-\frac{x_5}{1+x_5}; x_4\right)\,. \tag{5.11}$$

After this rewriting, the integral over $x_4$ is easy to compute. In the following we only discuss one single term that appears in the computation of $I_{W3S}$,

$$\int_0^\infty \frac{dx_4}{x_4 + x_5 + x_4x_5} G(-x_5; x_4) = \frac{1}{1+x_5} G\left(-\frac{x_5}{1+x_5}, -x_5; x_4\right)\Big|_{x_4=0}^{x_4=\infty}\,. \tag{5.12}$$

The primitive is easily shown to vanish at the lower boundary. In order to compute the limit $x_4 \to \infty$, we let $\bar{x}_4 = 1/x_4$ and compute the limit $\bar{x}_4 \to 0$. Using exactly the same technique as for the derivation of the inversion relation of the trilogarithm in Section 4.5, we obtain

$$\begin{aligned} G\left(-\frac{x_5}{1+x_5}, -x_5; \frac{1}{\bar{x}_4}\right) &= G(0; x_5)\, G(0; \bar{x}_4) + G(0, 0; \bar{x}_4)\\ - G\left(0, -\frac{1}{x_5}; \bar{x}_4\right) &- G\left(-\frac{1+x_5}{x_5}, 0; \bar{x}_4\right) + G\left(-\frac{1+x_5}{x_5}, -\frac{1}{x_5}; \bar{x}_4\right)\\ - G(0; x_5)\, G&\left(-\frac{1+x_5}{x_5}; \bar{x}_4\right) - G(-1, -1; x_5) + G(0, -1; x_5)\\ + G(0, 0; x_5)&\,.\end{aligned} \tag{5.13}$$

The limit is now trivial to take, and we get

$$\begin{aligned} G\left(-\frac{x_5}{1+x_5}, -x_5; x_4\right) &= G(0, 0; x_4) - G(0; x_5)\, G(0; x_4) - G(-1, -1; x_5)\\ &+ G(0, -1; x_5) + G(0, 0; x_5) + \mathcal{O}(1/x_4)\,.\end{aligned} \tag{5.14}$$

Note that the limit of this individual term contains logarithmic singularities. As the integral $I_{W3S}$ is convergent, these divergences must cancel in the sum over all contributions.

We can now repeat the same procedure for the last integration variable, and we immediately reproduce the well-known result

$$I_{W3S} = 6\,\zeta_3\,. \tag{5.15}$$

**5.3. Analytic continuation of Feynman integrals.** In this section we discuss how the 'extended' Hopf algebra of Section 4 can be used to analytically continue results for Feynman integrals. Indeed, it is well known that Feynman integrals are in general not holomorphic functions of the scales, but they may possess complicated branch cuts [**44–47**], and so the analytic expressions for Feynman integrals need to be carefully analytically continued from one region to another.

The discontinuities of a Feynman integral are most conveniently summarized through Cutkosky's rules [**47**], which state that the discontinuities of a Feynman integral are obtained by putting some of the propagators on shell. More precisely, for a Feynman integral of the type (2.3), the discontinuities arise by removing the principle value part of some of the propagators, which amounts to replacing a subset of the propagators in eq. (2.3) by positive energy $\delta$ functions,

$$\frac{1}{q_i^2-m_i^2+i\varepsilon} \to 2\pi i\,\delta_+(q_i^2-m_i^2) \equiv 2\pi i\,\delta(q_i^2-m_i^2)\,\Theta(q_i^0)\,, \tag{5.16}$$

where $\Theta$ denotes the Heaviside step function and $q_i^0$ is the energy component of the momentum $q_i$.

In the following we concentrate on massless theories, i.e., we only consider Feynman integrals without internal masses (the external legs may still be massive). In that case the branch cuts of a scattering amplitude or a Feynman integral extend between points where one of the scales is either zero or infinite, and it can be shown that the correct analytic continuation between regions where a given scale, say $s=(p_1+p_2)^2$, is positive or negative is obtained by assigning a small positive imaginary part to each scale, i.e., we have to perform the replacement $s \to s+i\varepsilon$, where $\varepsilon$ is small and positive. This amounts to choosing a branch for the logarithm such that

$$\log(-s-i\varepsilon) = \log|s| - i\pi\,\Theta(s)\,, \tag{5.17}$$

for each scale $s$. While the previous considerations allow one, at least in principal, to analytically continue a given Feynman integral from a region where, say, $s<0$ to the region where $s>0$, this procedure can become cumbersome for integrals whose analytic expressions involve complicated combinations of multiple polylogarithms depending on many variables. In the following we show how symbols and the 'extended' Hopf algebra allow one to solve this problem, at least for large classes of Feynman integrals. For simplicity, we will discuss the example of the one-mass one-loop box function given in eq. (3.5). This one-loop example already illustrates the main features of the method. For an application of this method to two-loop integrals we refer to ref. [**35**].

We start by making some general considerations about the symbols of Feynman integrals in massless theories. We have already argued that the integrals that appear in such theories have branch cuts extending only between points where one of the scales becomes zero or infinity. Furthermore, from eq. (4.15) it follows that the symbol of a function encodes the discontinuities of this function in its first entries. Combining these two statements, one immediately concludes that the first entries of the symbol of a massless Feynman integral must necessarily be a scale [**48**], i.e., a Lorentz invariant of the form $(p_{i_1}+\ldots+p_{i_k})^2$. For the one-loop box function this implies that the first entry can only be one of the three variables $s$, $t$ or $p_1^2$. This result is known as the *first entry condition* for Feynman integrals in the physics literature. Note that more complicated functions of $s$, $t$ and $p_1^2$ are allowed to appear inside the symbol starting from the second entry. From the first entry condition we immediately conclude that if we act with $\Delta_{1,n-1}$ on a Feynman integral that can be expressed in terms of multiple polylogarithms of weight $n$, the first entry must necessarily be an ordinary logarithm of a scale, for which the analytic continuation is given by eq. (5.17). In the following we will illustrate these results on the example of the one-mass one-loop box function.

For our purposes it is convenient to rewrite eq. (3.5) as

$$I_{1m}(p_1,p_2,p_3,p_4;\epsilon)=\frac{2r_\Gamma}{st}\left[\frac{1}{\epsilon^2}+\frac{1}{\epsilon}i_{-1}+i_0+\mathcal{O}(\epsilon)\right], \tag{5.18}$$

where we defined

$$\begin{aligned} i_{-1}&=\log(-p_1^2)-\log(-s)-\log(-t)\,,\\ i_0&=-\text{Li}_2\left(1-\frac{p_1^2}{s}\right)-\text{Li}_2\left(1-\frac{p_1^2}{t}\right)\\ &\quad-\frac{1}{2}\log^2(-p_1^2)+\log(-s)\log(-t)-\zeta_2\,. \end{aligned} \tag{5.19}$$

It is easy to see that the functions in eq. (5.19) are real for $s,t,p_1^2<0$. In physics applications it is often necessary to analytically continue the function to other regions, and in the following we will perform the analytic continuation to the region where $s>0$ and $t,p_1^2<0$. The logarithms in eq. (5.19) develop an imaginary part in this region, and we can use eq. (5.17) to write

$$\log(-s)=\log s-i\pi\,. \tag{5.20}$$

In addition, we also need to analytically continue the dilogarithms that appear in $i_0$. In the following we show how this can easily be achieved using the 'extended' Hopf algebra. To achieve this, we start by acting with $\Delta_{1,1}$ on $i_0$,

$$\begin{aligned} \Delta_{1,1}(i_0)&=-\log(-s)\otimes\log\left(1-\frac{p_1^2}{s}\right)+\log(-s)\otimes\log(-t)\\ &\quad+\log(-t)\otimes\log(-s)-\log(-t)\otimes\log\left(1-\frac{p_1^2}{t}\right)\\ &\quad+\log\left(-p_1^2\right)\otimes\log\left(1-\frac{p_1^2}{s}\right)+\log\left(-p_1^2\right)\otimes\log\left(1-\frac{p_1^2}{t}\right)\\ &\quad-\log\left(-p_1^2\right)\otimes\log\left(-p_1^2\right)\,. \end{aligned} \tag{5.21}$$

In the previous equation we have arranged the terms such as to make explicit the first entry condition satisfied by Feynman integrals. The advantage of the representation (5.21) is that, in the 'extended' Hopf algebra, the imaginary parts can easily be extracted from the first component of the coproduct, which itself is entirely determined by eq. (5.17). More precisely, we immediately see that in the region where $s > 0$ while $t, p_1^2 < 0$ we have

$$
\begin{aligned}
\Delta_{1,1}(i_0) &= -[\log s - i\pi] \otimes \log\left(1 - \frac{p_1^2}{s}\right) + [\log s - i\pi] \otimes \log(-t) \\
&\quad + \log(-t) \otimes \log(-s) - \log(-t) \otimes \log\left(1 - \frac{p_1^2}{t}\right) \\
&\quad + \log\left(-p_1^2\right) \otimes \log\left(1 - \frac{p_1^2}{s}\right) + \log\left(-p_1^2\right) \otimes \log\left(1 - \frac{p_1^2}{t}\right) \\
&\quad - \log\left(-p_1^2\right) \otimes \log\left(-p_1^2\right) \\
&= \Delta_{1,1}\Bigg[\mathrm{Li}_2\left(\frac{p_1^2}{s}\right) - \mathrm{Li}_2\left(1 - \frac{p_1^2}{t}\right) + \log\left(\frac{-p_1^2}{s}\right) \log\left(1 - \frac{p_1^2}{s}\right) \\
&\quad - \frac{1}{2}\log^2\left(-p_1^2\right) + \log s \log(-t) + i\pi \log\left(1 - \frac{p_1^2}{s}\right) - i\pi \log(-t)\Bigg].
\end{aligned}
\tag{5.22}
$$

The previous relation fixes the functional form of $i_0$ uniquely in the region where $s > 0$, up to a term proportional to $\zeta_2$, which cannot be determined from the coproduct alone. In other words, in the region where $s > 0$, we can write

$$
\begin{aligned}
i_0 &= \mathrm{Li}_2\left(\frac{p_1^2}{s}\right) - \mathrm{Li}_2\left(1 - \frac{p_1^2}{t}\right) + \log\left(\frac{-p_1^2}{s}\right) \log\left(1 - \frac{p_1^2}{s}\right) \\
&\quad - \frac{1}{2}\log^2\left(-p_1^2\right) + \log s \log(-t) + i\pi \log\left(1 - \frac{p_1^2}{s}\right) - i\pi \log(-t) + k\,\zeta_2 ,
\end{aligned}
\tag{5.23}
$$

for some rational number $k$. In the previous section we argued that this number can easily be determined by evaluating the function at a single point. For our purposes, we would like to 'match' the expressions (5.19) and (5.23) of the function at the branching point $s = 0$, but we have to face the problem that $i_0$ develops logarithmic divergences for $s = 0$. In the following we describe how the rational number $k$ can be uniquely determined.

We start by noting that (see, e.g., Proposition 25 of ref. [**43**]) for every multiple polylogarithm of the form $G(a_1, \ldots, a_n; z)$, where the $a_i$ are independent of $z$ and for every $z_0 \in \mathbb{C}$, there is a neighborhood $U$ of $z_0$ and holomorphic functions $f_i(z)$ on $U$, $i \in \{0, \ldots, n\}$ such that $\forall z \in U$

$$
G(a_1, \ldots, a_n; z) = \sum_{i=0}^{n} f_i(z) \log^i(z - z_0)\,.
\tag{5.24}
$$

In other words, around every point $z_0$ in the complex plane we can extract the discontinuities of a (combination of) multiple polylogarithm in terms of logarithms, and the coefficients of the polylogarithms are holomorphic function in a neighborhood of the branching point, and can thus be expanded into a power series around $z_0$. In our case, this implies that, rather than matching the expressions (5.19) and (5.23) at the branching point $s = 0$, which is divergent, we can equally well expand both expressions around $s$ to determine the rational number $k$. For example, on the one hand the expansion around $s = 0$ of eq. (5.19) reads,

(5.25)
$$\begin{aligned}&-\operatorname{Li}_2\left(1-\frac{p_1^2}{t}\right)-\log\left(-p_1^2\right)\log(-s)+\log(-s)\log(-t)+\frac{1}{2}\log^2(-s)+\mathcal{O}(s)\\&=-\operatorname{Li}_2\left(1-\frac{p_1^2}{t}\right)-\log\left(-p_1^2\right)\log s+\log s\log(-t)+\frac{1}{2}\log^2 s-3\zeta_2\\&-i\pi\log s+i\pi\log\left(-p_1^2\right)-i\pi\log(-t)+\mathcal{O}(s)\,.\end{aligned}$$

On the other hand, from eq. (5.23) we obtain

$$\begin{aligned}&-\operatorname{Li}_2\left(1-\frac{p_1^2}{t}\right)-\log\left(-p_1^2\right)\log s+\log s\log(-t)+\frac{1}{2}\log^2 s+(k-1)\zeta_2\\&+i\pi\log\left(-p_1^2\right)-i\pi\log s-i\pi\log(-t)+\mathcal{O}(s)\,.\end{aligned}\tag{5.26}$$

Comparing the two expansions, we immediately see that $k = -2$. The expression obtained in this way agrees with the known results in the physics literature. We emphasize that eq. (5.23) only involves (poly)logarithms that are *real* in the region where $s > 0$ and $t, p_1^2 < 0$.

Following exactly the same steps, we can analytically continue the one-mass box function to any region we like. For example, we could start from the region where $s > 0$ and $t, p_1^2 < 0$ and analytically continue the function to the region where both $s$ and $p_1^2$ are positive. We stress that although we have only illustrated this procedure on the example of a one-loop example written in terms of dilogarithms, the whole procedure is generic and extends to multiple polylogarithms in a straightforward way.

We conclude this section by discussing a technical point: In order to fix the coefficient of $\zeta_2$ after analytic continuation, we had to match the expansions of the function in a neighborhood of the branch point. While such a Taylor expansion is trivial in the case where only dilogarithms are involved, one might wonder how difficult this expansion is to obtain in the presence of more scales and/or multiple polylogarithms. It turns out that this problem can easily be solved in many cases by rewriting all the multiple polylogarithms as $G(a_1, \ldots, a_n; z)$, where $z$ denotes the variable in which we want to expand around the origin in the complex plane. Finding such a representation for the function is equivalent to the problem discussed in the previous section (where in that case $z$ corresponds to the variable $\bar{x}_4$). So, assuming we have found such a representation, we can immediately obtain the desired Taylor expansion using the series representation (3.12) of the multiple polylogarithms. To illustrate this, we return to the dilogarithm in the example considered above. It is not difficult to check that

$$\begin{aligned}\operatorname{Li}_2\left(1-\frac{p_1^2}{s}\right)&=G\left(-p_1^2,0;-s\right)-G(0,0;-s)-G\left(0;-p_1^2\right)G\left(-p_1^2;-s\right)\\&+G\left(0;-p_1^2\right)G(0;-s)-G\left(0,0;-p_1^2\right)-\zeta_2\,,\end{aligned}\tag{5.27}$$

and we can easily expand all multiple polylogarithms around $s = 0$ using eq. (3.12).

## 6. Conclusion

In this paper we have a given a review of how insight from recent advances in number theory have found their way into physics in the context of scattering amplitudes and Feynman integrals. We have argued that by extending Goncharov's

Hopf algebra of multiple polylogarithms to the comodule (4.30), we can derive very complicated functional equations among multiple polylogarithms. While we believe that this approach is interesting in its own right, the knowledge of functional equations among multiple polylogarithms is an important ingredient when computing multi-loop Feynman integrals. Not only do they allow in some cases to greatly simplify analytic results for scattering amplitudes, but also allow to derive analytic continuation formulas for Feynman integrals which are important when making predictions for physical observables. Indeed, while the analytic continuation of Feynman integrals has been considered a tedious task in the past, it is almost trivial to obtain the symbol of the discontinuity of a function [**48**, **49**], and there is good hope that in the future we might be able to perform the analytic continuation of Feynman integrals in a systematic way.

## References

[1] A. B. Goncharov, M. Spradlin, C. Vergu, and A. Volovich, *Classical polylogarithms for amplitudes and Wilson loops*, Phys. Rev. Lett. **105** (2010), no. 15, 151605, 4, DOI 10.1103/PhysRevLett.105.151605. MR2734443 (2011k:81206)

[2] Vittorio Del Duca, Claude Duhr, and Vladimir A. Smirnov, *An analytic result for the two-loop hexagon Wilson loop in $\mathcal{N} = 4$ SYM*, J. High Energy Phys. **3** (2010), 099, 17. MR2653448 (2011h:81155)

[3] Vittorio Del Duca, Claude Duhr, and Vladimir A. Smirnov, *The two-loop hexagon Wilson loop in $\mathcal{N} = 4$ SYM*, J. High Energy Phys. **5** (2010), 084, i, 119. MR2721123 (2011j:81235)

[4] A. B. Goncharov, *Galois symmetries of fundamental groupoids and noncommutative geometry*, Duke Math. J. **128** (2005), no. 2, 209–284, DOI 10.1215/S0012-7094-04-12822-2. MR2140264 (2007b:11094)

[5] Francis C. S. Brown, *On the decomposition of motivic multiple zeta values*, Galois-Teichmüller theory and arithmetic geometry, Adv. Stud. Pure Math., vol. 63, Math. Soc. Japan, Tokyo, 2012, pp. 31–58. MR3051238

[6] F. V. Tkachov, *A theorem on analytical calculability of 4-loop renormalization group functions*, Phys. Lett. B **100** (1981), no. 1, 65–68, DOI 10.1016/0370-2693(81)90288-4. MR607640 (82b:81030)

[7] K. G. Chetyrkin and F. V. Tkachov, *Integration by Parts: The Algorithm to Calculate beta Functions in 4 Loops,* Nucl. Phys. B **192** (1981) 159.

[8] H. Cheng and T. T. Wu, *Expanding Protons: Scattering At High-energies,* CAMBRIDGE, USA: MIT-PR. (1987) 285p

[9] Christian Bogner and Stefan Weinzierl, *Periods and Feynman integrals*, J. Math. Phys. **50** (2009), no. 4, 042302, 16, DOI 10.1063/1.3106041. MR2513973 (2010g:81110)

[10] N. I. Usyukina and A. I. Davydychev, *An Approach to the evaluation of three and four point ladder diagrams,* Phys. Lett. B **298** (1993) 363.

[11] V. A. Smirnov, *Analytical result for dimensionally regularized massless on shell double box,* Phys. Lett. B **460** (1999) 397 [hep-ph/9905323].

[12] J. B. Tausk, *Non-planar massless two-loop Feynman diagrams with four on-shell legs*, Phys. Lett. B **469** (1999), no. 1-4, 225–234, DOI 10.1016/S0370-2693(99)01277-0. MR1733866 (2000h:81073)

[13] T. Gehrmann and E. Remiddi, *Two loop master integrals for $\gamma^* \to 3$ jets: The Planar topologies,* Nucl. Phys. B **601** (2001) 248 [hep-ph/0008287].

[14] T. Gehrmann and E. Remiddi, *Two loop master integrals for $\gamma^* \to 3$ jets: The Nonplanar topologies,* Nucl. Phys. B **601** (2001) 287 [hep-ph/0101124].

[15] J. M. Henn, K. Melnikov and V. A. Smirnov, *Two-loop planar master integrals for the production of off-shell vector bosons in hadron collisions,* JHEP **1405** (2014) 090 [arXiv:1402.7078 [hep-ph]].

[16] F. Caola, J. M. Henn, K. Melnikov and V. A. Smirnov, *Non-planar master integrals for the production of two off-shell vector bosons in collisions of massless partons,* JHEP **1409** (2014) 043 [arXiv:1404.5590 [hep-ph]].

[17] C. G. Papadopoulos, D. Tommasini and C. Wever, *Two-loop Master Integrals with the Simplified Differential Equations approach,* JHEP **1501** (2015) 072 [arXiv:1409.6114 [hep-ph]].
[18] M. Caffo, H. Czyz, S. Laporta and E. Remiddi, *The Master differential equations for the two loop sunrise selfmass amplitudes,* Nuovo Cim. A **111** (1998) 365 [hep-th/9805118].
[19] S. Caron-Huot and K. J. Larsen, *Uniqueness of two-loop master contours,* JHEP **1210** (2012) 026 [arXiv:1205.0801 [hep-ph]].
[20] Dhritiman Nandan, Miguel F. Paulos, Marcus Spradlin, and Anastasia Volovich, *Star integrals, convolutions and simplices*, J. High Energy Phys. **5** (2013), 105, front matter+23. MR3080556
[21] A. B. Goncharov, *Multiple polylogarithms, cyclotomy and modular complexes*, Math. Res. Lett. **5** (1998), no. 4, 497–516, DOI 10.4310/MRL.1998.v5.n4.a7. MR1653320 (2000c:11108)
[22] A. B. Goncharov, *Multiple polylogarithms and mixed Tate motives,* (2001) [math/0103059v4].
[23] Rimhak Ree, *Lie elements and an algebra associated with shuffles*, Ann. of Math. (2) **68** (1958), 210–220. MR0100011 (20 #6447)
[24] Kuo Tsai Chen, *Iterated path integrals*, Bull. Amer. Math. Soc. **83** (1977), no. 5, 831–879. MR0454968 (56 #13210)
[25] E. Remiddi and J. A. M. Vermaseren, *Harmonic polylogarithms*, Internat. J. Modern Phys. A **15** (2000), no. 5, 725–754, DOI 10.1142/S0217751X00000367. MR1746652 (2001i:33003)
[26] A. I. Davydychev and M. Yu. Kalmykov, *Massive Feynman diagrams and inverse binomial sums*, Nuclear Phys. B **699** (2004), no. 1-2, 3–64, DOI 10.1016/j.nuclphysb.2004.08.020. MR2095786
[27] A. I. Davydychev and M. Yu. Kalmykov, *New results for the $\epsilon$-expansion of certain one-, two- and three-loop Feynman diagrams*, Nuclear Phys. B **605** (2001), no. 1-3, 266–318, DOI 10.1016/S0550-3213(01)00095-5. MR1845851 (2002d:81142)
[28] R. Bonciani, G. Degrassi, and A. Vicini, *On the generalized harmonic polylogarithms of one complex variable*, Comput. Phys. Comm. **182** (2011), no. 6, 1253–1264, DOI 10.1016/j.cpc.2011.02.011. MR2785326
[29] Jakob Ablinger, Johannes Blümlein, and Carsten Schneider, *Harmonic sums and polylogarithms generated by cyclotomic polynomials*, J. Math. Phys. **52** (2011), no. 10, 102301, 52, DOI 10.1063/1.3629472. MR2894586 (2012m:81057)
[30] Francis C. S. Brown, *Multiple zeta values and periods of moduli spaces $\overline{\mathfrak{M}}_{0,n}$* (English, with English and French summaries), Ann. Sci. Éc. Norm. Supér. (4) **42** (2009), no. 3, 371–489. MR2543329 (2010f:32013)
[31] A. B. Goncharov, *Explicit construction of characteristic classes*, I. M. Gel′fand Seminar, Adv. Soviet Math., vol. 16, Amer. Math. Soc., Providence, RI, 1993, pp. 169–210. MR1237830 (95c:57045)
[32] Claude Duhr, Herbert Gangl, and John R. Rhodes, *From polygons and symbols to polylogarithmic functions*, J. High Energy Phys. **10** (2012), 075, front matter + 77. MR3033860
[33] H. Gangl, A. B. Goncharov, and A. Levin, *Multiple polylogarithms, polygons, trees and algebraic cycles*, Algebraic geometry—Seattle 2005. Part 2, Proc. Sympos. Pure Math., vol. 80, Amer. Math. Soc., Providence, RI, 2009, pp. 547–593, DOI 10.1090/pspum/080.2/2483947. MR2483947 (2009m:11100)
[34] Francis C. S. Brown, *Multiple zeta values and periods of moduli spaces $\overline{\mathfrak{M}}_{0,n}$* (English, with English and French summaries), Ann. Sci. Éc. Norm. Supér. (4) **42** (2009), no. 3, 371–489. MR2543329 (2010f:32013)
[35] Claude Duhr, *Hopf algebras, coproducts and symbols: an application to Higgs boson amplitudes*, J. High Energy Phys. **8** (2012), 043, front matter + 45. MR3006955
[36] Andreas Brandhuber, Gabriele Travaglini, and Gang Yang, *Analytic two-loop form factors in $\mathcal{N} = 4$ SYM*, J. High Energy Phys. **5** (2012), 082, front matter+32. MR3042958
[37] T. Gehrmann, M. Jaquier, E. W. N. Glover and A. Koukoutsakis, *Two-Loop QCD Corrections to the Helicity Amplitudes for $H \to 3$ partons,* JHEP **1202** (2012) 056 [arXiv:1112.3554 [hep-ph]].
[38] Leonard Lewin, *Polylogarithms and associated functions*, North-Holland Publishing Co., New York-Amsterdam, 1981. With a foreword by A. J. Van der Poorten. MR618278 (83b:33019)
[39] Alexander Goncharov, *Volumes of hyperbolic manifolds and mixed Tate motives*, J. Amer. Math. Soc. **12** (1999), no. 2, 569–618, DOI 10.1090/S0894-0347-99-00293-3. MR1649192 (99i:19004)

[40] R. Kellerhals, *Volumes in hyperbolic 5-space*, Geom. Funct. Anal. **5** (1995), no. 4, 640–667, DOI 10.1007/BF01902056. MR1345017 (96g:52011)

[41] Alexander B. Goncharov, *Polylogarithms in arithmetic and geometry*, Proceedings of the International Congress of Mathematicians, Vol. 1, 2 (Zürich, 1994), Birkhäuser, Basel, 1995, pp. 374–387. MR1403938 (97h:19010)

[42] Gary Griffing, *Dual Lie elements and a derivation for the cofree coassociative coalgebra*, Proc. Amer. Math. Soc. **123** (1995), no. 11, 3269–3277, DOI 10.2307/2161065. MR1273493 (96h:16041)

[43] Francis Brown, *The massless higher-loop two-point function*, Comm. Math. Phys. **287** (2009), no. 3, 925–958, DOI 10.1007/s00220-009-0740-5. MR2486668 (2009k:81094)

[44] L. D. Landau, *On analytic properties of vertex parts in quantum field theory*, Nuclear Phys. **13** (1959), 181–192. MR0141405 (25 #4810)

[45] S. Mandelstam, *Determination of the pion-nucleon scattering amplitude from dispersion relations and unitarity. General theory*, Phys. Rev. (2) **112** (1958), 1344–1360. MR0098600 (20 #5057)

[46] Stanley Mandelstam, *Analytic properties of transition amplitudes in perturbation theory.*, Phys. Rev. (2) **115** (1959), 1741–1751. MR0109689 (22 #575)

[47] R. E. Cutkosky, *Singularities and discontinuities of Feynman amplitudes*, J. Mathematical Phys. **1** (1960), 429–433. MR0118351 (22 #9124)

[48] Davide Gaiotto, Juan Maldacena, Amit Sever, and Pedro Vieira, *Pulling the straps of polygons*, J. High Energy Phys. **12** (2011), 011, 41. MR2935648

[49] Lance J. Dixon, James M. Drummond, and Johannes M. Henn, *Bootstrapping the three-loop hexagon*, J. High Energy Phys. **11** (2011), 023, 34, DOI 10.1007/JHEP11(2011)023. MR2913300

INSTITUTE FOR THEORETICAL PHYSICS, ETH ZURICH, 8093 ZURICH, SWITZERLAND
*E-mail address*: duhrc@itp.phys.ethz.ch

# Equations D3 and spectral elliptic curves

Vasily Golyshev and Masha Vlasenko

ABSTRACT. We study modular determinantal differential equations of orders 2 and 3. We show that the expansion of the analytic solution of a non-degenerate modular equation of type D3 over the rational numbers with respect to the natural parameter coincides, under certain assumptions, with the $q$–expansion of the newform of its spectral elliptic curve and therefore possesses a multiplicativity property. We compute the complete list of D3 equations with this multiplicativity property and relate it to Zagier's list of non-degenerate modular D2 equations.

## 1. Introduction

Motivated by Apéry's proof of irrationality of $\zeta(3)$ and its analog, the result on the measure of irrationality of $\zeta(2)$, Zagier studies in [**Zag07**] the question of finding those triples of rational numbers $(A, B, \lambda)$ for which the sequence obtained by the recursive formula

$$(n+1)^2 u_{n+1} \;-\; (An^2 + An + \lambda)u_n \;+\; Bn^2 u_{n-1} \;=\; 0$$

starting with $u_0 = 1$ has all integer terms, i.e. $u_n \in \mathbb{Z}$. The generating function $\phi_0(t) = 1 + u_1 t + u_2 t^2 + \dots$ is the normalized analytic at $t = 0$ solution of the differential equation $\mathcal{L}\phi_0 = 0$ with

$$\mathcal{L}(t) \;=\; D^2 \;-\; t\,(AD^2 + AD + \lambda) \;+\; B\,t^2\,(D+1)^2\,, \tag{1.1}$$

where we use the notation $D = t\frac{d}{dt}$ throughout the paper. We will refer later on to (1.1) as the Beukers-Zagier differential operator since it also appeared in the work of Beukers [**Beu02**]. A table of respective triples $(A, B, \lambda)$ is obtained in [**Zag07**] by searching in a large range of values. It appears that all degenerate cases in the table, i.e. those with either $A^2 = 4B$ or $B = 0$ come as members of infinite families of triples $(A, B, \lambda)$ with $\phi_0(t) \in \mathbb{Z}[[t]]$. On the contrary, imposing the assumption

$$A^2 \neq 4B,\; B \neq 0 \tag{1.2}$$

one arrives at 14 "sporadic" cases with no obvious pattern. Remarkably, in all those sporadic cases the corresponding solution to (1.1) can be parametrized by modular forms. Namely, one can find a modular function $t(\tau)$ that vanishes at $\infty$ and a modular form $f(\tau)$ of weight 1 such that $\phi_0(t(\tau)) = f(\tau)$ for all $\tau$ in the upper half-plane with large enough imaginary part. These cases are listed in the table below.

| $A$ | $B$ | $\lambda$ | $t(\tau)$ | $f(\tau)$ | $u_0$ | $u_1$ | $u_2$ | $u_3$ | $u_4$ | $u_5$ |
|---|---|---|---|---|---|---|---|---|---|---|
| 7 | -8 | 2 | $\frac{1^3 6^9}{2^3 3^9}$ | $\frac{2^1 3^6}{1^2 6^3}$ | 1 | 2 | 10 | 56 | 346 | 2252 |
| 9 | 27 | 3 | $\frac{1^3 4^3 18^9}{2^9 9^3 36^3}$ | $\frac{2^9 3^1 12^1}{1^3 4^3 6^3}$ | 1 | 3 | 9 | 21 | 9 | -297 |
| 10 | 9 | 3 | $\frac{1^4 6^8}{2^8 3^4}$ | $\frac{2^6 3^1}{1^3 6^2}$ | 1 | 3 | 15 | 93 | 639 | 4653 |
| 11 | -1 | 3 | $q\prod_{n=1}^{\infty}(1-q^n)^{5(\frac{n}{5})}$ | $\left(\frac{1}{t(\tau)}\cdot\frac{5^5}{1^1}\right)^{1/2}$ | 1 | 3 | 19 | 147 | 1251 | 11253 |
| 12 | 32 | 4 | $\frac{1^4 4^2 8^4}{2^{10}}$ | $\frac{2^{10}}{1^4 4^4}$ | 1 | 4 | 20 | 112 | 676 | 4304 |
| 17 | 72 | 6 | $\frac{1^5 3^1 4^5 6^2 12^1}{2^{14}}$ | $\frac{2^{15} 3^2 12^2}{1^6 4^6 6^5}$ | 1 | 6 | 42 | 312 | 2394 | 18756 |
| 0 | -16 | 0 | $\frac{2^4 8^8}{4^{12}}$ | $\frac{4^{10}}{2^4 8^4}$ | 1 | 0 | 4 | 0 | 36 | 0 |

The products in the fourth and fifth columns stand for eta-products, e.g. $\frac{1^3 6^9}{2^3 3^9} = \frac{\eta(\tau)^3\eta(6\tau)^9}{\eta(2\tau)^3\eta(3\tau)^9}$. We use this notation throughout the paper. For each of the first 6 rows one should also consider $(\tilde{A}, \tilde{B}, \tilde{\lambda}) = (-A, B, -\lambda)$ with the corresponding $\tilde{u}_n = (-1)^n u_n$, $\tilde{t}(\tau) = -t(\tau)$, $\tilde{f}(\tau) = f(\tau)$. For the last row there is also a triple $(0, 16, 0)$ leading to $\tilde{u}_n = (-1)^{n/2} u_n$. In total, the table gives us 14 triples $(A, B, \lambda)$ satisfying (1.2) with $\phi_0(t) \in \mathbb{Z}[[t]]$. Zagier conjectures that there are no more such cases, or if there are, they will each have a modular parametrization.

We observe in this paper that the differential operator (1.1) satisfying the assumption (1.2) is a specific form of the so called determinantal differential operator of order 2. The necessary definitions and properties will be recalled in Section 2. We deal with determinantal differential operators of order 2 in Sections 3 and 4 and recover Zagier's list above in a new context. Then we proceed to obtain an analog of this list for determinantal differential operators of order 3, namely, the complete list of D3's that satisfy a multiplicativity property that we discuss later.

## 2. Determinantal differential equations

Determinantal differential equations of order $N$ were defined in [**GS07**]. A DN equation is obtained from an $(N+1)\times(N+1)$ matrix $A = (a_{ij})_{i,j=0}^{N}$ that satisfies

$$\begin{aligned} a_{ij} &= 0\,, \quad i-j>1 \\ a_{ij} &= 1\,, \quad i-j=1 \\ a_{ij} &= a_{N-j,N-i}\,, \quad i-j<1 \end{aligned} \tag{2.1}$$

The respective differential operator is then defined as

$$\mathcal{L}_{A,\infty}(z) = \det{}_{\text{right}}\left(\delta_{ij} z\frac{d}{dz} - a_{ij}\left(\frac{d}{dz}\right)^{j-i+1}\right)\left(\frac{d}{dz}\right)^{-1}$$

where $\delta_{ij}$ is the Kronecker symbol and $\det_{\text{right}}$ refers to the way of expanding the determinant of a matrix with non-commuting entries with respect to the rightmost column.

The matrix $A$ can be reconstructed from the coefficients of the differential operator ([**GS07**], Corollary 3.3). Assume in addition that all eigenvalues of $A$ are distinct. Then obviously $A$ is diagonalizable. In fact, for a matrix satisfying (2.1) the two conditions are equivalent: $A$ is diagonalizable if and only if all eigenvalues of $A$ are distinct. It follows immediately if one observes that $A$ cannot have an eigenvector whose last component is zero. According to Corollary 6.4 in [**GS07**] the singularities of the differential operator $\mathcal{L}_{A,\infty}(z)$ are regular singular points located at $\infty$ and the eigenvalues $\lambda_0, \dots, \lambda_N$ of $A$. Moreover, the differential equation has maximal unipotent monodromy at $z = \infty$ and the valuation of its analytic solution at $z = \infty$ is equal to 1. This motivates the following notation.

DEFINITION 2.1. A differential operator of order $N$ is of type $DN_{\infty,1}$ if it equals $\mathcal{L}_{A,\infty}(z)$ for some matrix $A$ satisfying (2.1).

We denote the characteristic polynomial of $A$ by $F(z) = \det(z - A)$ throughout the paper. It will be convenient to also use the variable $t = \frac{1}{z}$. Namely, consider the operator

$$\mathcal{L}_{A,0}(t) \;=\; (-1)^N \, \mathcal{L}_{A,\infty}\Big(\frac{1}{t}\Big) t\,.$$

The respective differential equation has maximal unipotent monodromy at $t = 0$ and the valuation of its analytic solution at this point equals 0.

DEFINITION 2.2. A differential operator of order $N$ is of type $DN_{0,0}$ if it equals $\mathcal{L}_{A,0}(t)$ for some matrix $A$ satisfying (2.1).

By DN we mean either $DN_{0,0}$ or $DN_{\infty,1}$ the case being clear from the context. Observe that the following operations with the defining matrices

$$A \mapsto A' = A + \varepsilon$$
$$A \mapsto A'' = \left(\lambda^{j-i+1} a_{ij}\right)$$

lead to the substitutions in the differential equations

$$\mathcal{L}_{A',\infty}(z) \;=\; \mathcal{L}_{A,\infty}(z - \varepsilon)\,, \qquad \mathcal{L}_{A',0}(t) \;=\; \mathcal{L}_{A,0}\Big(\frac{t}{1 - \varepsilon t}\Big)\,(1 - \varepsilon t) \tag{2.2}$$

and

$$\mathcal{L}_{A'',\infty}(z) \;=\; \lambda\,\mathcal{L}_{A,\infty}\Big(\frac{z}{\lambda}\Big)\,, \qquad \mathcal{L}_{A'',0}(t) \;=\; \mathcal{L}_{A,0}(\lambda t) \tag{2.3}$$

respectively.

## 3. The Beukers-Zagier equation as a D2 equation

Let us consider D2 equations in detail. According to our definitions one has

$$\mathcal{L}_{A,\infty}(z) \;=\; \det{}_{right} \begin{pmatrix} (z - a_{00})\frac{d}{dz} & -a_{01}\big(\frac{d}{dz}\big)^2 & -a_{02}\big(\frac{d}{dz}\big)^3 \\ -1 & (z - a_{11})\frac{d}{dz} & -a_{01}\big(\frac{d}{dz}\big)^2 \\ 0 & -1 & (z - a_{00})\frac{d}{dz} \end{pmatrix} \Big(\frac{d}{dz}\Big)^{-1}$$

$$= -a_{02}\Big(\frac{d}{dz}\Big)^2 - a_{01}\Big(\frac{d}{dz}\Big)^2(z - a_{00}) + (z - a_{00})\frac{d}{dz}\Big((z - a_{11})\frac{d}{dz}(z - a_{00}) - a_{01}\frac{d}{dz}\Big)$$

$$= F(z)\Big(\frac{d}{dz}\Big)^2 + F'(z)\,\frac{d}{dz} + (z - a_{00})$$

where

$$\begin{aligned} F(z) &= \det\bigl(z-A\bigr) = z^3 + \alpha_2 z^2 + \alpha_1 z + \alpha_0 \\ \alpha_2 &= -a_{11} - 2a_{00} \\ \alpha_1 &= 2a_{00}a_{11} + a_{00}^2 - 2a_{01} \\ \alpha_0 &= 2a_{00}a_{01} - a_{00}^2 a_{11} - a_{02}. \end{aligned}$$

A $D2_{\infty,1}$ differential operator is then any operator of the form

$$F(z)\left(\frac{d}{dz}\right)^2 + F'(z)\frac{d}{dz} + (z-\beta)$$

with a cubic monic polynomial with distinct roots $F(z) = z^3 + \alpha_2 z^2 + \alpha_1 z + \alpha_0$. One can recover the matrix $A$ from $\alpha_i$ and $\beta$ via

$$\begin{aligned} a_{00} &= \beta \\ a_{11} &= -2\beta - \alpha_2 \\ a_{01} &= -\frac{3}{2}\beta^2 - \beta\alpha_2 - \frac{1}{2}\alpha_1 \\ a_{02} &= -\beta^3 - \beta^2\alpha_2 - \beta a_1 - \alpha_0. \end{aligned}$$

The generic equation of type $D2_{0,0}$ is then

$$\begin{aligned} &F\Bigl(\frac{1}{t}\Bigr)\Bigl(-t^2\frac{d}{dt}\Bigr)^2 t + F'\Bigl(\frac{1}{t}\Bigr)\Bigl(-t^2\frac{d}{dt}\Bigr)t + \Bigl(\frac{1}{t}-\beta\Bigr)t \\ &= t^5 F\Bigl(\frac{1}{t}\Bigr)\Bigl(\frac{d}{dt}\Bigr)^2 + \Bigl(4t^4 F\Bigl(\frac{1}{t}\Bigr) - t^3 F'\Bigl(\frac{1}{t}\Bigr)\Bigr)\frac{d}{dt} + 2t^3 F\Bigl(\frac{1}{t}\Bigr) - t^2 F'\Bigl(\frac{1}{t}\Bigr) + 1 - \beta t \\ &= tG(t)\Bigl(\frac{d}{dt}\Bigr)^2 + tG'(t)\frac{d}{dt} + tH(t) \end{aligned}$$

with

$$\begin{aligned} G(t) &= t + \alpha_2 t^2 + \alpha_1 t^3 + \alpha_0 t^4 \\ H(t) &= -\beta + \alpha_1 t + 2\alpha_0 t^2. \end{aligned}$$

With the notation $D = t\frac{d}{dt}$ we can further rewrite it as

$$\begin{aligned} &(1+\alpha_2 t + \alpha_1 t^2 + \alpha_0 t^3)(D^2 - D) + (1 + 2\alpha_2 t + 3\alpha_1 t^2 + 4\alpha_0 t^3)D \\ &\qquad - \beta t + \alpha_1 t^2 + 2\alpha_0 t^3 \\ &= D^2 + t(\alpha_2 D^2 + \alpha_2 D - \beta) + \alpha_1 t^2 (D+1)^2 + \alpha_0 t^3 (D+1)(D+2). \end{aligned} \tag{3.1}$$

Notice that putting $\alpha_0 = 0$ we obtain precisely operator (1.1) with $A = -\alpha_2$, $B = \alpha_1$ and $\lambda = \beta$.

## 4. Modular D2 equations

Recall that a D2 differential equation depends on 4 parameters $(\alpha_2, \alpha_1, \alpha_0, \beta)$. It determines a local system of rank 2 over the base

$$\mathbb{P}^1(\mathbb{C}) \setminus \{\infty, \text{the roots of } z^3 + \alpha_2 z^2 + \alpha_1 z + \alpha_0\}.$$

Consider also the basis in the space of solutions of D2 near $t = 0$ which consists of normalized analytic and logarithmic solutions:

$$\begin{aligned}\phi_0(t) &= 1 + \beta t + \Big(-\frac{1}{2}\alpha_2\beta + \frac{1}{4}\beta^2 - \frac{1}{4}\alpha_1\Big)t^2 + \dots \\ \phi_1(t) &= \log t\,\phi_0(t) + \Big(-\alpha_2 - 2\beta\Big)t + \Big(\frac{1}{2}\alpha_2^2 + \frac{1}{2}\alpha_2\beta - \frac{3}{4}\beta^2 - \frac{1}{4}\alpha_1\Big)t^2 + \dots\end{aligned}$$

DEFINITION 4.1. We say that an equation D2 with parameters $(\alpha_2, \alpha_1, \alpha_0, \beta) \in \mathbb{Q}^4$ is modular if the analytic continuation of

$$\tau = \frac{1}{2\pi i}\frac{\phi_1(t)}{\phi_0(t)}$$

gives uniformization of the base by the upper halfplane with the group of deck transformations being a congruence subgroup of $\mathrm{SL}(2, \mathbb{Z})$ and the function $\tau \mapsto \phi_0(t(\tau))$ is a modular form of weight 1.

In this case $t(\tau)$ is a modular function whose $q$-expansion can be written explicitly. Indeed, inverting the series

$$q = \exp\Big(\frac{\phi_1(t)}{\phi_0(t)}\Big) = t + (-\alpha_2 - 2\beta)t^2 + (\alpha_2^2 + \frac{7}{2}\alpha_2\beta + \frac{13}{4}\beta^2 - \frac{1}{4}\alpha_1)t^3 + \dots$$

one gets

$$t = q + (\alpha_2 + 2\beta)q^2 + (\alpha_2^2 + \frac{9}{2}\alpha_2\beta + \frac{19}{4}\beta^2 + \frac{1}{4}\alpha_1)q^3 + \dots$$

Further, substituting this expansion into $\phi_0(t)$ one obtains

$$f = \phi_0(t(\tau)) = 1 + \beta q + (\frac{1}{2}\alpha_2\beta + \frac{9}{4}\beta^2 - \frac{1}{4}\alpha_1)q^2 + \dots$$

This must be a modular form of weight 1.

Put $Q = q^{\frac{1}{2}}$ and consider the series

$$t^{\frac{1}{2}}\sqrt{1 + \alpha_2 t + \alpha_1 t^2 + \alpha_0 t^3}\,\phi_0(t)^2 = \sum_{n=1}^{\infty} c_n Q^n \tag{4.1}$$

whose coefficients $c_n = c_n(\vec{\alpha}, \beta)$ can be determined explicitly as follows. One writes

$$Q = \exp\Big(\frac{1}{2}\frac{\phi_1(t)}{\phi_0(t)}\Big) = t^{\frac{1}{2}}\Big(1 + (-\frac{1}{2}\alpha_2 - \beta)t + (\frac{3}{8}\alpha_2^2 + \frac{5}{4}\alpha_2\beta + \frac{9}{8}\beta^2 - \frac{1}{8}\alpha_1)t^2 + \dots\Big)$$

and inverts this series in order to get

$$t^{\frac{1}{2}} = Q\Big(1 + (\frac{1}{2}\alpha_2 + \beta)Q^2 + (\frac{3}{8}\alpha_2^2 + \frac{7}{4}\alpha_2\beta + \frac{15}{8}\beta^2 + \frac{1}{8}\alpha_1)Q^4 + \dots\Big)$$

which can be substituted into the left-hand side of (4.1). We have

$$\begin{aligned} c_1 &= 1 \\ c_2 &= 0 \\ c_3 &= \alpha_2 + 3\beta \\ c_4 &= 0 \\ c_5 &= \alpha_2^2 + \frac{25}{4}\alpha_2\beta + \frac{75}{8}\beta^2 + \frac{1}{8}\alpha_1 \\ &\dots \end{aligned} \tag{4.2}$$

It is not hard to see that all even coefficients in fact vanish.

THEOREM 4.2. *Assume we have a nondegenerate modular D2 equations with parameters $\alpha_2, \alpha_1, \alpha_0, \beta \in \mathbb{Q}$ . If the modular form $\sum_{n=1}^{\infty} c_n q^n$ of weight 2 with $c_n$ defined by (4.1) is a newform then*

$$L(s) \;=\; \sum_{n=1}^{\infty} \frac{c_n}{n^s}$$

*is the L-function of the elliptic curve*

$$y^2 = z^3 + \alpha_2 z^2 + \alpha_1 z + \alpha_0 \,. \tag{4.3}$$

By being a newform we mean that the modular form belongs to the subspace on newforms of certain level. We do not require it to be a Hecke eigenform a priori; rather, the Hecke–eigen property is a consequence of the theorem. In particular, we have the following

COROLLARY 4.3. *If $\sum_{n=1}^{\infty} c_n q^n$ is a newform, then the coefficients $c_n$ are multiplicative, i.e.*

$$c_{mn}(\vec{\alpha},\beta) \;=\; c_m(\vec{\alpha},\beta)\,\cdot\, c_n(\vec{\alpha},\beta) \tag{4.4}$$

*as soon as $m$ and $n$ are coprime.*

We will solve equations (4.4) with respect to the parameters $\alpha_2, \alpha_1, \alpha_0, \beta$ later in this section. It appears that modulo a certain transformation which preserves both the sequence $\{c_n; n \geq 1\}$ and the $L$-function of (4.3) there are finitely many cases.

The proof of Theorem 4.2 will rely on the following result.

**Theorem** (Atkin & Swinnerton-Dyer congruences, Theorem 4 in [**ASD71**]) *Let $p \neq 2, 3$, and let $y^2 = z^3 + Bz + C$ be an elliptic curve over $\mathbb{Z}_p$ with good reduction. Choose a local parameter at 0 so that $z = \xi^{-2} + \sum_{n=-1}^{\infty} d_n \xi^n$ and $y = \xi^{-3} + \dots$ are the respective expansions , and write*

$$-\frac{1}{2}\frac{dz}{y} \;=\; \Big(\sum_{n=1}^{\infty} c_n \xi^n\Big)\frac{d\xi}{\xi}\,.$$

*If $B, C, d_n, c_n$ are p-adic integers, then*

$$c_{np} \;-\; a_p\, c_n \;+\; p\, c_{\frac{n}{p}} \;\equiv\; 0 \mod p^{\mathrm{ord}_p(n)+1} \tag{4.5}$$

*where*

$$a_p \;=\; -\sum_{m=0}^{p-1}\Big(\frac{m^3 + Bm + C}{p}\Big)\,.$$

Notice that this theorem can be applied to an elliptic curve defined over $\mathbb{Q}$ with good reduction at $p$ as soon as the coefficients $B$, $C$, $d_n$ and $c_n$ do not contain $p$ in their denominators. Moreover, $a_p = p + 1 - \#E(\mathbb{F}_p)$ is then the $p$-th coefficient of the $L$-function of this elliptic curve.

PROOF OF THEOREM 4.2. Let $a_n$, $n \geq 1$ be the coefficients of the $L$-function $L(s) = \sum_n \frac{a_n}{n^s}$ of the elliptic curve (4.3). One can check that

$$\phi_1'(t)\phi_0(t) - \phi_1(t)\phi_0'(t) \;=\; \frac{1}{t(1+\alpha_2 t + \alpha_1 t^2 + \alpha_0 t^3)} \;=\; t^{-4} F\Big(\frac{1}{t}\Big)^{-1},$$

hence

$$\sum_{n=1}^{\infty} c_n(\vec{\alpha},\beta)Q^n \frac{dQ}{Q} = t^2 F\Big(\frac{1}{t}\Big)^{\frac{1}{2}} \phi_0(t)^2\, d\Big(\frac{1}{2}\frac{\phi_1(t)}{\phi_0(t)}\Big)$$

$$= \frac{1}{2}t^2 F\Big(\frac{1}{t}\Big)^{\frac{1}{2}} \Big(\phi_1'(t)\phi_0(t) - \phi_1(t)\phi_0'(t)\Big)\, dt = \frac{1}{2}t^{-2} F\Big(\frac{1}{t}\Big)^{-\frac{1}{2}} dt = -\frac{1}{2}\frac{dz}{y}$$

where we substitute $z = 1/t$, $y^2 = F(z)$. This is a holomorphic differential on the curve (4.3), and since $Q \sim t^{\frac{1}{2}}$ for small $t$ we conclude that $Q$ is a local parameter on the curve near the origin. Moreover, $z \sim Q^{-2}$ and $y \sim Q^{-3}$ and therefore the theorem of Atkin and Swinnerton-Dyer would be applicable for every prime $p$ not dividing the conductor of the curve as soon as all $c_n$ and $d_n$ defined from the expansion $z = \frac{1}{t} = Q^{-2} + \sum_{n=-1}^{\infty} d_n Q^n$ do not contain $p$ in their denominators. First we show that this is indeed the case for all but finitely many primes $p$ using the assumption of modularity.

We have $Q(t(\tau)) = q^{\frac{1}{2}}$. Looking at (4.1) we see that $\sum_{n=1}^{\infty} c_n q^n$ is the $q$-expansion of the modular form $\Big[t^{\frac{1}{2}}\sqrt{1+\alpha_2 t + \alpha_1 t^2 + \alpha_0 t^3} f^2\Big](2\tau)$ of weight 2. It follows that possibly after multiplication by an integer all $c_n$ become integers simultaneously. The same holds for $d_n$ since $z(\tau)$ is a modular function. Therefore for all but finitely many prime numbers $p$ we have congruences (4.5). Another consequence of modularity of $\sum_{n=1}^{\infty} c_n q^n$ is that

$$c_n = o(n)\,, \qquad n \to \infty\,. \tag{4.6}$$

Our next step is to show that (4.5) together with (4.6) imply that $c_n = a_n$ for all $n$ not divisible by a finite set of primes. Since $a_n = o(n^{\frac{1}{2}+\varepsilon})$ for any $\varepsilon > 0$ and $c_n = o(n)$ there is a number $N$ such that

$$\Big|\frac{c_n}{n}\Big| < \frac{1}{2}, \quad \Big|\frac{a_n}{n}\Big| < \frac{1}{2}$$

for all $n > N$. Obviously we can assume that (4.5) is true for all $p > N$ increasing $N$ if necessary. From (4.5) with $n = 1$ we get

$$c_p \equiv a_p \mod p\,.$$

Since for all $p > N$ also $|c_p|, |a_p| < \frac{p}{2}$ we conclude that $c_p = a_p$. Suppose $p > N$ and we have proved that $c_{p^m} = a_{p^m}$ for all $m \leq M$. Since

$$a_{p^{M+1}} - a_p a_{p^M} + p a_{p^{M-1}} = 0$$

$$c_{p^{M+1}} - a_p c_{p^M} + p c_{p^{M-1}} \equiv 0 \mod p^{M+1}$$

we conclude that

$$c_{p^{M+1}} \equiv a_{p^{M+1}} \mod p^{M+1}\,.$$

Therefore $c_{p^{M+1}} = a_{p^{M+1}}$ because $p^{M+1} > N$ again. It follows now by induction that $c_{p^m} = a_{p^m}$ for all $m \geq 0$ and $p > N$. Our next step is to show that $c_n = a_n$ for all $n$ not divisible by finite number of primes $p \leq N$. Let $n$ be such a number and suppose that for every proper divisor $n'|n$ we have already proved that $a_{n'} = c_{n'}$. By (4.5) with a prime divisor $p|n$ and $n' = \frac{n}{p}$ instead of $n$ we have

$$c_n - a_p c_{n'} + p c_{\frac{n'}{p}} = c_n - a_n \equiv 0 \mod p^m$$

where $m = \mathrm{ord}_p(n)$. Since this is true for every prime dividing $n$ we conclude that $c_n \equiv a_n \mod n$, and therefore $c_n = a_n$ again by our estimate.

Consider both newforms $\sum_n c_n q^n$ and $\sum_n a_n q^n$ on the intersection of the corresponding congruence subgroups which is again a congruence subgroup. Since $c_n = a_n$ for all $n$ not divisible by primes form a certain finite set, it follows that both forms have the same eigenvalues for infinitely many Hecke operators. Therefore by multiplicity one theorem [**AL70**] these forms are just equal, and our theorem is proved. □

Now we can substitute the polynomials (4.2) into the multiplicativity relations (4.4) and solve the resulting equations. We do not expect finitely many solutions because the shifts (2.2) preserve modularity. Under the shift $A \mapsto A' = A + \varepsilon$ the parameters become

$$(\alpha_2', \alpha_1', \alpha_0', \beta') = (\alpha_2 - 3\varepsilon, 3\varepsilon^2 - 2\alpha_2\varepsilon + \alpha_1, -\varepsilon^3 + \alpha_2\varepsilon^2 - \alpha_1\varepsilon + \alpha_0, \beta + \varepsilon).$$

Zagier's choice $\alpha_0 = 0$ is not natural from this point of view because one can make $\alpha_0 = 0$ only if $F(z)$ has a rational root. It is more natural to choose the equation with $\beta = 0$ as a unique representative of the orbit of the shifts.

Solving the first few relations with Gröbner bases (we used computer algebra system [**MAGMA**])

$$\begin{aligned} c_6 &= c_2 \cdot c_3\,, & c_{10} &= c_2 \cdot c_5\,, & c_{12} &= c_4 \cdot c_3\,, & c_{14} &= c_2 \cdot c_7\,, \\ c_{15} &= c_3 \cdot c_5\,, & c_{18} &= c_2 \cdot c_9\,, & c_{21} &= c_3 \cdot c_7\,, & c_{22} &= c_2 \cdot c_{11} \end{aligned} \tag{4.7}$$

we reduce drastically the domain of our search. Namely, we obtain 8 points $(\alpha_2, \alpha_1, \alpha_0)$ plus two one-parametric families $(0, 0, \alpha_0)$ and $(0, \alpha_1, 0)$ as the solutions to the incomplete system of multiplicativity equations above. Once this has been established, we analyze the families separately by specializing $(\alpha_2, \alpha_1, \alpha_0)$ to $(0, 0, \alpha_0)$ resp. $(0, \alpha_1, 0)$, thereby reducing the number of variables. This, in turn, enables us to consider more multiplicativity relations involving about 200 further coefficients. This new, heavily overdetermined system turns out to have finitely many solutions. The results are given in the table below.

| $\alpha_2$ | $\alpha_1$ | $\alpha_0$ | $F(z)$ | $(A, B, \lambda)$ |
|---|---|---|---|---|
| 1 | 0 | 0 | $z^2(z+1)$ | $(-1,0,0), (2,1,1)$ |
| -1 | 0 | 0 | $z^2(z-1)$ | $(1,0,0), (-2,1,-1)$ |
| -4 | -80 | -192 | $(z-12)(z+4)^2$ | $(-32,256,-12), (16,0,4)$ |
| 4 | -80 | 192 | $(z+12)(z-4)^2$ | $(32,256,12), (-16,0,-4)$ |
| -2 | -40 | -75 | $(z+3)(z^2-5z-25)$ | $(11,-1,3)$ |
| 2 | -40 | 75 | $(z-3)(z^2+5z-25)$ | $(-11,-1,-3)$ |
| -1 | -24 | -36 | $(z-6)(z+2)(z+3)$ | $(-17,72,-6), (7,-8,2), (10,9,3)$ |
| 1 | -24 | 36 | $(z+6)(z-2)(z-3)$ | $(17,72,6), (-7,-8,-2), (-10,9,-3)$ |
| 0 | $16\zeta_4$ | 0 | $z(z^2-16\zeta_4)$ | $(0,\pm16,0), (12,32,4), (-12,32,-4)$ |
| 0 | 0 | $27\zeta_6$ | $z^3-27\zeta_6$ | $(9,27,3), (-9,27,-3)$ |

The first three columns contain all solutions of a few first multiplicativity equations with $\beta = 0$. The fourth column shows decomposition of the respective polynomial $F(z) = z^3 + \alpha_2 z^2 + \alpha_1 z + \alpha_0$ over rational numbers. Polynomials in the first 4 rows appear to have multiple roots meaning that the respective differential operator is degenerate. In the last column we shift the differential operator by various rational roots of $F(z)$ taken with negative sign in order to obtain operators with $\alpha_0 = 0$, the respective values of the parameters $(A, B, \lambda) = (-\alpha_2, \alpha_1, \beta)$ being listed. For example, in the first row $(A, B, \lambda) = (-1, 0, 0)$ and $(2, 1, 1)$ correspond to $(\alpha_2,\alpha_1,\beta)=(1,0,0)$ itself and $(-2,1,1)$, where the last triple is obtained from the first one via (4.7) with $\varepsilon=1$. In the fifth row, shifting $(-2,-40,-75)$ by $\varepsilon=3$ we obtain $(-11,-1,3)$, and there are no more triples in the right-most column because $F(z)$ has only one rational root. In the seventh row, $F(z)$ has three rational roots and triples on the right correspond to the shifts with $\varepsilon=-6$, 2 and 3 respectively. Nondegenerate D2 equations are listed in the last 6 rows, and triples on the right give us precisely Zagier's table. One can easily check at this point that in each case the statement of Theorem 4.2 holds, and therefore the respective $(\alpha_2,\alpha_1,\alpha_0)$ indeed solve all multiplicativity equations. The triples corresponding to the degenerate differential equations from the first four rows can be found in **[Zag07]** as #1, #3, #19 and #11. On the other hand, the degenerate triples #14, #20 and #25 are also modular but do not appear on our list.

## 5. Differential equations of type D3

Our goal in this section is to write the generic form of a D3 equation by making exactly the same steps as in Section 3 but now with $N=3$. We get

$$\mathcal{L}_{A,\infty}=\det{}_{\text{right}}\begin{pmatrix}(z-a_{00})\frac{d}{dz} & -a_{01}\left(\frac{d}{dz}\right)^2 & -a_{02}\left(\frac{d}{dz}\right)^3 & -a_{03}\left(\frac{d}{dz}\right)^4\\ -1 & (z-a_{11})\frac{d}{dz} & -a_{12}\left(\frac{d}{dz}\right)^2 & -a_{02}\left(\frac{d}{dz}\right)^3\\ 0 & -1 & (z-a_{11})\frac{d}{dz} & -a_{01}\left(\frac{d}{dz}\right)^2\\ 0 & 0 & -1 & (z-a_{00})\frac{d}{dz}\end{pmatrix}\left(\frac{d}{dz}\right)^{-1}$$

$$=F(z)\left(\frac{d}{dz}\right)^3+\frac{3}{2}F'(z)\left(\frac{d}{dz}\right)^2+\left(\frac{1}{2}F''(z)+G(z)\right)\frac{d}{dz}+\frac{1}{2}G'(z)$$

with

$$\begin{aligned}F(z)&=\det(z-A)=z^4+\alpha_3z^3+\alpha_2z^2+\alpha_1z+\alpha_0\\ \alpha_3&=-2a_{11}-2a_{00}\\ \alpha_2&=4a_{00}a_{11}+a_{00}^2+a_{11}^2-2a_{01}-a_{12}\\ \alpha_1&=-2a_{02}-2a_{11}a_{00}^2+2a_{00}(a_{01}-a_{11}^2+a_{12})+2a_{11}a_{01}\\ \alpha_0&=-a_{03}+2a_{00}a_{02}+(a_{11}^2-a_{12})a_{00}^2-2a_{11}a_{01}a_{00}+a_{01}^2,\\ G(z)&=z^2+\beta_1z+\beta_0\\ \beta_1&=-2a_{00}\\ \beta_0&=2a_{00}a_{11}-a_{11}^2-2a_{01}+a_{12}.\end{aligned}$$

Recall that a $D3_{\infty,1}$ differential equation is called non–degenerate whenever the roots of $F(z)$ are distinct. Notice that our order 3 differential operator is the symmetric square of the order 2 operator

$$F(z)\left(\frac{d}{dz}\right)^2+\frac{1}{2}F'(z)\frac{d}{dz}+\frac{1}{4}G(z).$$

We also compute the generic $D3_{0,0}$. We have

$$-F\left(\frac{1}{t}\right)\left(-t^2\frac{d}{dt}\right)^3t-\frac{3}{2}F'\left(\frac{1}{t}\right)\left(-t^2\frac{d}{dt}\right)^2t-\left(\frac{1}{2}F''\left(\frac{1}{t}\right)+G\left(\frac{1}{t}\right)\right)\left(-t^2\frac{d}{dt}\right)t-\frac{1}{2}G'\left(\frac{1}{t}\right)t$$
$$=t\left[H(t)\left(\frac{d}{dt}\right)^3+\frac{3}{2}H'(t)\left(\frac{d}{dt}\right)^2+\left(\frac{1}{2}H''(t)+U(t)\right)\frac{d}{dt}+\frac{1}{2}U'(t)\right]$$

with

$$H(t) = t^6F\Big(\frac{1}{t}\Big) = t^2 + \alpha_3t^3 + \alpha_2t^4 + \alpha_1t^5 + \alpha_0t^6$$
$$U(t) = t^2G\Big(\frac{1}{t}\Big) + 3t^4F\Big(\frac{1}{t}\Big) - t^3F'\Big(\frac{1}{t}\Big) = 3\alpha_0t^4 + 2\alpha_1t^3 + (\alpha_2+\beta_0)t^2 + \beta_1 t.$$

Finally, this differential operator can be written as

$$\begin{aligned} & D^3 + t\big(D+\frac{1}{2}\big)\big(\alpha_3(D^2+D) + \beta_1\big) \\ & + t^2(D+1)\big(\alpha_2(D+1)^2 + \beta_0\big) \\ & + \alpha_1t^3(D+2)\big(D+\frac{3}{2}\big)(D+1) \\ & + a_0t^4(D+3)(D+2)(D+1). \end{aligned} \tag{5.1}$$

## 6. Nondegenerate modular D3 equations

In this section we will prove the analog of Theorem 4.2 for D3 equations. In order to state it we first associate to such an equation an appropriate elliptic curve. Recall that $F(z) = \det\big(z-A\big)$ has distinct roots, so the discriminant of $F$ is nonzero. For a $D3$ equation we have $F(z) = z^4 + \alpha_3z^3 + \alpha_2z^2 + \alpha_1z + \alpha_0$. Consider the curve

$$w^2 = z^4 + \alpha_3z^3 + \alpha_2z^2 + \alpha_1z + \alpha_0. \tag{6.1}$$

Put it into the Weierstrass form.

DEFINITION 6.1. The spectral elliptic curve of a D3 equation is

$$\begin{aligned} y^2 = x^3 &+ \Big(\alpha_1\alpha_3 - \frac{1}{3}\alpha_2^2 - 4\alpha_0\Big)x \\ &+ \Big(\alpha_0\alpha_3^2 - \frac{1}{3}\alpha_1\alpha_2\alpha_3 + \frac{2}{27}\alpha_2^3 - \frac{8}{3}\alpha_0\alpha_2 + \alpha_1^2\Big). \end{aligned} \tag{6.2}$$

It is indeed an elliptic curve because the discriminant of the cubic polynomial in the right-hand side is equal to the discriminant of $F(z)$ (as a function of $\alpha_i$), and therefore the right-hand side has 3 distinct roots.

LEMMA 6.2. *Curves* (6.1) *and* (6.2) *are birational over the splitting field of the polynomial* $F(z)$. *Moreover, the holomorphic differential* $\frac{dx}{y}$ *on the spectral elliptic curve transforms into* $-\frac{dz}{w}$ *on* (6.1) *under this birational equivalence.*

PROOF. Let $F(z_0) = 0$. Then

$$w^2 = (z-z_0)^4 + \tilde\alpha_3(z-z_0)^3 + \tilde\alpha_2(z-z_0)^2 + \tilde\alpha_1(z-z_0)$$

with $\tilde\alpha_i = \frac{1}{i!}F^{(i)}(z_0)$, and

$$\Big(\frac{\tilde\alpha_1\, w}{(z-z_0)^2}\Big)^2 = \tilde\alpha_1^2 + \tilde\alpha_1\tilde\alpha_3\Big(\frac{\tilde\alpha_1}{z-z_0}\Big) + \tilde\alpha_2\Big(\frac{\tilde\alpha_1}{z-z_0}\Big)^2 + \Big(\frac{\tilde\alpha_1}{z-z_0}\Big)^3.$$

Hence the variables

$$x = \frac{\tilde\alpha_1}{z-z_0} + \frac{\tilde\alpha_2}{3}, \quad y = \frac{\tilde\alpha_1\, w}{(z-z_0)^2} \tag{6.3}$$

satisfy $y^2 = x^3 + Bx + C$ with

$$\begin{aligned} B &= \tilde{\alpha}_1\tilde{\alpha}_3 - \frac{\tilde{\alpha}_2^2}{3} = \alpha_1\alpha_3 - \frac{1}{3}\alpha_2^2 - 4\alpha_0\,, \\ C &= \tilde{\alpha}_1^2 + \frac{2}{27}\tilde{\alpha}_2^3 - \frac{1}{3}\tilde{\alpha}_1\tilde{\alpha}_2\tilde{\alpha}_3 = \alpha_0\alpha_3^2 - \frac{1}{3}\alpha_1\alpha_2\alpha_3 + \frac{2}{27}\alpha_2^3 - \frac{8}{3}\alpha_0\alpha_2 + \alpha_1^2\,. \end{aligned}$$

The equality of the differentials follows immediately. □

Consider the normalized analytic and logarithmic solutions of D3 near $t = 0$:

$$\begin{aligned} \phi_0(t) &= 1 - \frac{1}{2}\beta_1 t + \Big(\frac{3}{16}\alpha_3\beta_1 + \frac{3}{32}\beta_1^2 - \frac{1}{8}\alpha_2 - \frac{1}{8}\beta_0\Big)t^2 + \dots \\ \phi_1(t) &= \log t\,\phi_0(t) + \psi(t) = \log t\,\phi_0(t) + \Big(-\frac{1}{2}\alpha_3 + \frac{1}{2}\beta_1\Big)t + \dots \end{aligned} \tag{6.4}$$

Again, the 6 parameters naturally split into two groups. Parameters $\alpha_3, \alpha_2, \alpha_1, \alpha_0$ determine the base

$$\mathbb{P}^1(\mathbb{C}) \setminus \{\infty, \text{the roots of } z^4 + \alpha_3 z^3 + \alpha_2 z^2 + \alpha_1 z + \alpha_0\}$$

and also the spectral elliptic curve (6.2). Of course the solutions (6.4) and the respective local system of rank 3 depend also on the remaining parameters $\beta_1$ and $\beta_0$.

DEFINITION 6.3. We say that an equation D3 with $\alpha_3, \alpha_2, \alpha_1, \alpha_0, \beta_1, \beta_0 \in \mathbb{Q}$ is modular if the analytic continuation of

$$\tau = \frac{1}{2\pi i}\frac{\phi_1(t)}{\phi_0(t)} \tag{6.5}$$

gives uniformization of the base by the upper halfplane with the group of deck transformations being a congruence subgroup of $\mathrm{SL}(2,\mathbb{Z})$ and the function $\tau \mapsto \phi_0(t(\tau))$ is a modular form of weight 2.

Consider the power series

$$q(t) = \exp\Big(\frac{\phi_1(t)}{\phi_0(t)}\Big) = t\exp\Big(\frac{\psi(t)}{\phi_0(t)}\Big) = t + \Big(-\frac{1}{2}\alpha_3 + \frac{1}{2}\beta_1\Big)t^2 + \dots$$

We can invert this expansion in order to write $t$ as a power series in $q$, and for accessory values of parameters this must be then the $q$-expansion of a modular function. Analogously, $\phi_0(t)$ written as a power series in $q$ will be the $q$-expansion of a modular form of weight 2 since the local system is of rank 3. We will be specifically interested in the coefficients of the modular form

$$t\,\phi_0(t) = \sum_{n=1}^{\infty} c_n q^n\,, \tag{6.6}$$

where $c_n = c_n(\vec{\alpha}, \vec{\beta})$. These coefficients can be computed explicitly as polynomials in the initial parameters

$$\begin{aligned} c_1 &= 1 \\ c_2 &= \frac{1}{2}\alpha_3 - \beta_1 \\ c_3 &= \frac{3}{16}\alpha_3^2 - \frac{27}{32}\alpha_3\beta_1 + \frac{57}{64}\beta_1^2 + \frac{1}{16}\alpha_2 - \frac{3}{16}\beta_0 \\ &\dots \end{aligned} \tag{6.7}$$

THEOREM 6.4. *Assume we have a non-degenerate modular D3 equation with parameters $\alpha_3, \alpha_2, \alpha_1, \alpha_0, \beta_1, \beta_0$. If the modular form (6.6) of weight 2 is a newform then*

$$\sum_{n=1}^{\infty} \frac{c_n}{n^s}$$

*is the L-function of the spectral elliptic curve (6.2).*

PROOF. We will use the same method as for D2. Let $a_n, n \geq 1$ be the coefficients of the $L$-function $L(s) = \sum_n a_n n^{-s}$ of the spectral elliptic curve (6.2). One can check that

$$\begin{aligned} t\,\phi_0(t)\,\frac{dq}{q} &= t\,\phi_0(t)\,d\log q = t\frac{\phi_1'(t)\phi_0(t) - \phi_1(t)\phi_0'(t)}{\phi_0(t)}dt \\ &= F\Big(\frac{1}{t}\Big)^{-\frac{1}{2}} t^{-2}\,dt = -\frac{dz}{w} \end{aligned} \tag{6.8}$$

where we substitute $z = 1/t$, $w^2 = F(z)$. Now using the birational transformation (6.3) we see that $t$ (hence also $q$) is a local parameter on the spectral elliptic curve near the point $P = \left(\frac{1}{3}\tilde{\alpha}_2, \tilde{\alpha}_1\right)$ where $\tilde{\alpha}_i = \frac{1}{i!}F^{(i)}(z_0)$ and $z_0$ is a chosen root of $F(z) = 0$ as in the proof of Lemma 6.2. Since our differential equation is modular, composition of modular uniformization with the birational transform gives a map from a modular curve to the spectral elliptic curve. The preimages of both points $\pm P = \left(\frac{1}{3}\tilde{\alpha}_2, \pm\tilde{\alpha}_1\right)$ are cusps because $t$ is zero there and this is a cuspidal value as we know. Therefore by the Manin-Drinfeld theorem their difference $P-(-P) = 2P$ and hence also $P$ is a point of finite order. Let us write the spectral curve (6.2) as $y^2 = x^3 + Bx + C$. The differential $-\frac{dz}{w}$ in (6.8) transforms to $\frac{dx}{y}$ according to Lemma 6.2. One can find an isogenous curve $\tilde{y}^2 = \tilde{x}^3 + B\tilde{x} + C$ where $P$ is mapped to the origin and the differential is mapped to $-\frac{1}{2}\frac{d\tilde{x}}{\tilde{y}}$. Then $q$ is a local parameter on the latter curve near the origin, and the $L$-function of this curve is again $L(s) = \sum_n a_n n^{-s}$ since isogenous curves have equal $L$-functions. One can check that since the expansion $-\frac{1}{2}\frac{d\tilde{x}}{\tilde{y}} = \left(\sum_{n=1}^{\infty} c_n q^n\right)\frac{dq}{q}$ starts with $c_1 = 1$ then $\tilde{x} \sim q^{-2}$. All the coefficients $c_n$ in (6.6) and $d_n$ in the expansion $\tilde{x} = q^{-2} + \sum_{n=-1}^{\infty} d_n q^n$ do not contain $p$ in denominators for all but finitely many primes $p$ because these are $q$-expansions of a modular form of weight 2 and modular function respectively. The rest of the proof goes exactly like in Theorem 4.2. Namely, one has Atkin and Swinnerton–Dyer congruences for all but finitely many primes and together with $c_n = o(n)$, which is another consequence of modularity, this implies that $a_n = c_n$

for all $n$ non divisible by a finite set of primes. The normalized newforms $\sum_n a_n q^n$ and $\sum_n c_n$ are then in one Hecke-eigenspace and therefore are equal by multiplicity one theorem for the space of newforms. □

By this theorem, if $\sum_{n=1}^{\infty} c_n q^n$ is a newform then the coefficients $c_n$ are the coefficients of the $L$-function of an elliptic curve, and we have the following consequence.

COROLLARY 6.5. *If $\sum_{n=1}^{\infty} c_n q^n$ is a newform, then the coefficients $c_n$ are multiplicative, i.e.*

$$c_{mn}(\vec{\alpha},\vec{\beta}) = c_m(\vec{\alpha},\vec{\beta}) \cdot c_n(\vec{\alpha},\vec{\beta}) \tag{6.9}$$

*as soon as $m$ and $n$ are coprime.*

Let us solve equations (6.9) for the parameters $(\vec{\alpha},\vec{\beta})$.

LEMMA 6.6. *A D3 equation has the following properties with respect to the shift* (2.2)*:*

(i) *the parameters change according to the rule*

$$\begin{aligned}&(\alpha_3',\alpha_2',\alpha_1',\alpha_0')\\ &\quad= (\alpha_3-4\varepsilon,\alpha_2-3\varepsilon\alpha_3+6\varepsilon^2,\alpha_1-2\varepsilon\alpha_2+3\varepsilon^2\alpha_3-4\varepsilon^3,\alpha_0-\varepsilon\alpha_1+\varepsilon^2\alpha_2-\varepsilon^3\alpha_3+\varepsilon^4)\\ &(\beta_1',\beta_0') = (\beta_1-2\varepsilon,\beta_0-\beta_1\varepsilon+\varepsilon^2)\end{aligned}$$

(ii) *the coefficients of the spectral elliptic curve* (6.2) *do not change*
(iii) *all $c_n$ in* (6.7) *do not change*
(iv) *modular differential equations transform into modular ones (assuming $\varepsilon \in \mathbb{Q}$)*

PROOF. Indeed, since for $A' = A+\varepsilon$ one has $\mathcal{L}_{A',\infty}(z) = \mathcal{L}_{A,\infty}(z-\varepsilon)$ the formulas in (i) follow from

$$\begin{aligned}&z^4+\alpha_3'z^3+\alpha_2'z^2+\alpha_1'z+\alpha_0'\\ &\qquad= (z-\varepsilon)^4+\alpha_3(z-\varepsilon)^3+\alpha_2(z-\varepsilon)^2+\alpha_1(z-\varepsilon)+\alpha_0,\\ &z^2+\beta_1'z+\beta_0' = (z-\varepsilon)^2+\beta_1(z-\varepsilon)+\beta_0\,.\end{aligned}$$

Part (ii) follows by a tedious computation. Next, from $\mathcal{L}_{A',0}(t) = \mathcal{L}_{A,0}\left(\frac{t}{1-\varepsilon t}\right)(1-\varepsilon t)$ we conclude that Frobenius bases must simply transform as $\phi_i'(t) = (1-\varepsilon t)\phi_i\left(\frac{t}{1-\varepsilon t}\right)$ for $i=0,1,2$. Therefore the uniformization maps (6.5) differ by the transformation $t \mapsto \frac{t}{1-\varepsilon t}$ of the base, from where (iii) and (iv) follow immediately. □

It follows from Lemma 6.6 that it suffices to solve equations (6.9) for a single representative of every orbit of shifts. There are two natural choices, $\alpha_3 = 0$ and $\beta_1 = 0$. We will use the latter one. It appears that the system of equations (6.9) has finitely many solutions with $\beta_1 = 0$. We will give the complete list later, but first we list the rational solutions that correspond to nondegenerate D3, i.e. such that the roots of $F(z)$ are all distinct. There are exactly 18 of them. We list the $\alpha_i$'s In the table below. These determine the spectral elliptic curve (6.2) which we denote by $\mathcal{E}$. Then we give its $j$-invariant, its level $N$ and the newform $g(\tau)$ of level $N$ whose Mellin transform is the $L$-function of $\mathcal{E}$. We give the value of $\beta_0$ in the last column.

| $\alpha_3$ | $\alpha_2$ | $\alpha_1$ | $\alpha_0$ | $j(\mathcal{E})$ | $N(\mathcal{E})$ | $g_{\mathcal{E}}(\tau)$ | $\beta_0$ |
|---|---|---|---|---|---|---|---|
| 0 | -44 | 0 | -16 | $-\frac{20720464}{15625}$ | 20 | $\mathbf{2^2 \cdot 10^2}$ | -4 |
| 0 | 44 | 0 | -16 | | 80 | $\mathbf{4^6 \cdot 20^6/2^2 \cdot 8^2 \cdot 10^2 \cdot 40^2}$ | 4 |
| 0 | -28 | 0 | -128 | $\frac{207646}{6561}$ | 24 | $\mathbf{2 \cdot 4 \cdot 6 \cdot 12}$ | -4 |
| 0 | 28 | 0 | -128 | | 48 | $\mathbf{4^4 \cdot 12^4/2 \cdot 6 \cdot 8 \cdot 24}$ | 4 |
| 0 | -40 | 0 | 144 | $\frac{35152}{9}$ | 24 | $\mathbf{2 \cdot 4 \cdot 6 \cdot 12}$ | -8 |
| 0 | 40 | 0 | 144 | | 48 | $\mathbf{4^4 \cdot 12^4/2 \cdot 6 \cdot 8 \cdot 24}$ | 8 |
| 0 | -68 | 0 | 1152 | $\frac{3065617154}{9}$ | 48 | $\mathbf{4^4 \cdot 12^4/2 \cdot 6 \cdot 8 \cdot 24}$ | -28 |
| 0 | 68 | 0 | 1152 | | 24 | $\mathbf{2 \cdot 4 \cdot 6 \cdot 12}$ | 28 |
| 0 | -48 | 0 | 512 | 287496 | 32 | $\mathbf{4^2 \cdot 8^2}$ | -16 |
| 0 | 48 | 0 | 512 | | 32 | $\mathbf{4^2 \cdot 8^2}$ | 16 |
| 0 | 0 | 0 | -256 | 1728 | 32 | $\mathbf{4^2 \cdot 8^2}$ | 0 |
| 0 | 0 | 0 | 256 | | 64 | $\mathbf{8^8/4^2 \cdot 16^2}$ | 0 |
| 0 | -36 | 0 | 432 | 54000 | 144 | $\mathbf{12^{12}/6^4 \cdot 24^4}$ | -12 |
| 0 | 36 | 0 | 432 | | 36 | $\mathbf{6^4}$ | 12 |
| -4 | -88 | -300 | -304 | $-\frac{122023936}{161051}$ | 11 | $\mathbf{1^2 \cdot 11^2}$ | -8 |
| 0 | 0 | -108 | 0 | 0 | 27 | $\mathbf{3^2 \cdot 9^2}$ | 0 |
| -2 | -43 | -156 | -216 | $\frac{4733169839}{3515625}$ | 15 | $\mathbf{1 \cdot 3 \cdot 5 \cdot 15}$ | -5 |
| -2 | -59 | -136 | -80 | $\frac{4956477625}{941192}$ | 14 | $\mathbf{1 \cdot 2 \cdot 7 \cdot 14}$ | -5 |

In the proof of the Theorem 6.4 we have constructed the point $P = \left(\frac{1}{3}\tilde{\alpha}_2, \tilde{\alpha}_1\right)$ of finite order on the spectral elliptic curve. The order of $P$ is 4 in the first 14 cases and for the last 4 rows it is 5,3,8 and 3 respectively.

## 7. All solutions of the multiplicativity equations for D3

The goal of this section is to list all the solutions to the multiplicativity equations. We have obtained them via Gröbner bases with the aid of the computer algebra system [**MAGMA**], doing computations over several finite fields and lifting solutions afterwards. Apart from the non–degenerate cases which we listed in the previous section, there are solutions defined over number fields and also solutions with degenerate $F(z)$. In order to list them in an efficient way we consider the twists $A \mapsto A'' = \left(\lambda^{j-i+1} a_{ij}\right)$ which lead to the simple variable change in the differential equation $\mathcal{L}_{A'',0}(t) = \mathcal{L}_{A,0}(\lambda t)$.

LEMMA 7.1. *A D3 equation has the following properties with respect to the twist* $\mathcal{L}_{A'',0}(t) = \mathcal{L}_{A,0}(\lambda t)$*:*

(i) *the parameters transform according to the rule*

$$\vec{\alpha}'' = (\lambda\alpha_3, \lambda^2\alpha_2, \lambda^3\alpha_1, \lambda^4\alpha_0), \quad \vec{\beta}'' = (\lambda\beta_1, \lambda^2\beta_0)$$

(ii) *the spectral curve transforms via*

$$y^2 = x^3 + \lambda^4 Bx + \lambda^6 C$$

(iii) *the coefficients* $c_n(\vec{\alpha}, \vec{\beta})$ *transform via* $c''_n = \lambda^{n-1} c_n$

(iv) *for the function* $\tau(t) = \frac{1}{2\pi i}\frac{\phi_1(t)}{\phi_0(t)}$*, one has*

$$\tau'' = \tau - \frac{\log \lambda}{2\pi i},$$

The proof is straightforward. The way the solutions transform under the twists described in (iii) and (iv) of Lemma 7.1 shows that only finitely many twists are possible for a given D3 that preserve the multiplicativity property, and the only twists possible are those by roots of unity. We give the complete list of solutions to (6.9) below. We list only one representative in every family of twists and give all possible twists in the right-most column. We start with nondegenerate cases. These have already been given in the previous section up to the twists by roots of unity.

| $\alpha_3$ | $\alpha_2$ | $\alpha_1$ | $\alpha_0$ | $j(\mathcal{E})$ | $N(\mathcal{E})$ | $g_{\mathcal{E}}(\tau)$ | $\beta_0$ | twists |
|---|---|---|---|---|---|---|---|---|
| 0 | -44 | 0 | -16 | $-\frac{20720464}{15625}$ | 20 | $\mathbf{2^2 \cdot 10^2}$ | -4 | $\lambda^2 = \pm 1$ |
| 0 | -28 | 0 | -128 | $\frac{207646}{6561}$ | 24 | $\mathbf{2 \cdot 4 \cdot 6 \cdot 12}$ | -4 | $\lambda^2 = \pm 1$ |
| 0 | -40 | 0 | 144 | $\frac{35152}{9}$ | 24 | $\mathbf{2 \cdot 4 \cdot 6 \cdot 12}$ | -8 | $\lambda^2 = \pm 1$ |
| 0 | 68 | 0 | 1152 | $\frac{3065617154}{9}$ | 24 | $\mathbf{2 \cdot 4 \cdot 6 \cdot 12}$ | 28 | $\lambda^2 = \pm 1$ |
| 0 | 48 | 0 | 512 | 287496 | 32 | $\mathbf{4^2 \cdot 8^2}$ | 16 | $\lambda^{16} = 1$ |
| 0 | 0 | 0 | -256 | 1728 | 32 | $\mathbf{4^2 \cdot 8^2}$ | 0 | $\lambda^{16} = 1$ |
| 0 | 36 | 0 | 432 | 54000 | 36 | $\mathbf{6^4}$ | 12 | $\lambda^{36} = 1$ |
| -4 | -88 | -300 | -304 | $-\frac{122023936}{161051}$ | 11 | $\mathbf{1^2 \cdot 11^2}$ | -8 | $\lambda = \pm 1$ |
| 0 | 0 | -108 | 0 | 0 | 27 | $\mathbf{3^2 \cdot 9^2}$ | 0 | $\lambda^{18} = 1$ |
| -2 | -43 | -156 | -216 | $\frac{4733169839}{3515625}$ | 15 | $\mathbf{1 \cdot 3 \cdot 5 \cdot 15}$ | -5 | $\lambda = \pm 1$ |
| -2 | -59 | -136 | -80 | $\frac{4956477625}{941192}$ | 14 | $\mathbf{1 \cdot 2 \cdot 7 \cdot 14}$ | -5 | $\lambda = \pm 1$ |

In addition, there is one more nondegenerate solution over $\mathbb{Q}(\sqrt{5})$

| $\alpha_3$ | $\alpha_2$ | $\alpha_1$ | $\alpha_0$ | $j(\mathcal{E})$ | $\beta_0$ | twists |
|---|---|---|---|---|---|---|
| 0 | $22 - 30\sqrt{5}$ | 0 | $1000 - 440\sqrt{5}$ | $\notin \mathbb{Q}$ | $18 - 10\sqrt{5}$ | $\lambda^2 = \pm 1$ |

which appears to give the same modular form $t(\tau)\phi_0(\tau) = \mathbf{2^2 \cdot 10^2}$ as in the first row of the above table.

The solutions of (6.9) with degenerate polynomial $F(z) = z^4 + \alpha_3 z^3 + \alpha_2 z^2 + \alpha_1 z + \alpha_0$ are given in the table below. Remarkably, some of them are still modular.

| $\alpha_3$ | $\alpha_2$ | $\alpha_1$ | $\alpha_0$ | $\beta_0$ | $t(\tau)\phi_0(\tau)$ | twists |
|---|---|---|---|---|---|---|
| 0 | 0 | 0 | 0 | 0 | $q$ | — |
| 0 | 4 | 0 | 0 | -4 | $q/(1-q^2)$ | $\lambda^2 = \pm 1$ |
| 0 | -8 | 0 | 16 | -8 | $q/(1-q^2)$ | $\lambda^2 = \pm 1$ |
| 0 | 128 | 0 | 4096 | 64 | $\mathbf{2^4 \cdot 8^4/4^4}$ | $\lambda^2 = \pm 1$ |
| 0 | 64 | 0 | 0 | 0 | $\mathbf{4^8/2^4}$ | $\lambda^2 = \pm 1$ |
| 4 | 0 | 0 | 0 | 0 | $q/(1-q)^2$ | $\lambda = \pm 1$ |
| 2 | 1 | 0 | 0 | -1 | $q/(1-q)$ | $\lambda = \pm 1$ |
| -2 | -3 | 0 | 0 | 3 | $q/(1+q+q^2)$ | $\lambda = \pm 1$ |
| -6 | -135 | -540 | -648 | -9 | $\mathbf{1^3 \cdot 9^3/3^2}$ | $\lambda = \pm 1$ |
| 4 | -80 | 192 | 0 | -16 | $\mathbf{2^4 \cdot 6^4/1^2 \cdot 3^2}$ | $\lambda = \pm 1$ |
| 2 | 9 | -216 | 432 | -9 | $\mathbf{3^3 \cdot 6^3/1 \cdot 2}$ | $\lambda = \pm 1$ |
| 2 | -55 | -100 | 1000 | -25 | $\mathbf{5^5/1}$ | $\lambda = \pm 1$ |
| 8 | -176 | 768 | -1024 | -16 | $\mathbf{2^{10} \cdot 8^4/1^4 \cdot 4^6}$ | $\lambda = \pm 1$ |

## 8. From D2's to D3's

Let $\phi_0(t) = 1 + u_1 t + u_2 t^2 + \dots$ be the solution of the differential equation (3.1) with $\alpha_0 = 0$. As we already mentioned, this is exactly the class of D2's considered by Zagier in [**Zag07**] where his parameters $A, B, \lambda$ are our $-\alpha_2, \alpha_1, \beta$ correspondingly. One then has

$$(n+1)^2 u_{n+1} + (\alpha_2 n^2 + \alpha_2 n - \beta) u_n + \alpha_1 n^2 u_{n-1} = 0\,,$$

and we observe that the sequence $u'_n = \binom{2n}{n} u_n$ satisfies

$$(n+1)^3 u_{n+1} + 2(2n+1)(\alpha_2 n^2 + \alpha_2 n - \beta) u_n + 4\alpha_1 (2n+1)(2n-1) n u_{n-1} = 0\,.$$

In other words, $\phi'_0(t) = \sum_{n=0}^{\infty} \binom{2n}{n} u_n t^n$ is a solution of

$$\begin{aligned} D^3 \;&+\; t\big(D + \tfrac{1}{2}\big)\big(4\alpha_2(D^2 + D) \;-\; 4\beta\big) \\ &+\; t^2 (D+1)\big(16\alpha_1 (D+1)^2 \;-\; 4\alpha_1\big) \end{aligned}$$

which turns out to be a D3 equation with parameters

$$(\alpha'_3, \alpha'_2, \alpha'_1, \alpha'_0) \;=\; (4\alpha_2, 16\alpha_1, 0, 0)\,, \quad (\beta'_1, \beta'_0) = (-4\beta, -4\alpha_1)\,.$$

This equation is degenerate as its symbol has double roots. Passing to $\phi''_0(t) = \sum_{n=0}^{\infty} \binom{2n}{n} u_n t^{2n}$, which is a solution of

$$D^3 \;+\; t^2 (D+1)\big(4\alpha_2 (D+1)^2 - 16\beta - 4\alpha_2\big) \;+\; 16\alpha_1 t^4 (D+3)(D+2)(D+1)\,,$$

we thus come to a D3 equation whose parameters are

$$(\alpha''_3, \alpha''_2, \alpha''_1, \alpha''_0) \;=\; (0, 4\alpha_2, 0, 16a_1)\,, \quad (\beta''_1, \beta''_0) = (0, -16\beta - 4\alpha_2)\,.$$

This D3 is nondegenerate if the initial D2 was nondegenerate. Indeed, the symbol $F''(z) = z^4 + 4\alpha_2 z^2 + 16\alpha_1$ has 4 distinct roots if and only if $\alpha_1 \neq 0$ and $\alpha_2^2 \neq 4\alpha_1$, which are exactly the conditions for the roots of the symbol $F(z) = z^3 + \alpha_2 z^2 + \alpha_1 z$ of the initial D2 to be all distinct. Therefore we have a map from nondegenerate D2's with $\alpha_0 = 0$ to nondegenerate D3's with $\alpha_3 = \alpha_1 = \beta_1 = 0$ given by

$$\begin{aligned} \vec{\alpha} &= (\alpha_2, \alpha_1, 0) & & \vec{\alpha} = (0, 4\alpha_2, 0, 16\alpha_1) \\ & & \mapsto & \\ \vec{\beta} &= (\beta) & & \vec{\beta} = (0, -16\beta - 4\alpha_2). \end{aligned} \tag{8.1}$$

This map is obviously bijective. The analytic solution at $t = 0$ transforms as

$$\sum_{n=0}^{\infty} u_n\, t^n \quad \mapsto \quad \sum_{n=0}^{\infty} \binom{2n}{n} u_n\, t^{2n}\,.$$

We find that under the map (8.1) Zagier's triples all go into D3's from our list of accessory equations and, moreover, they exhaust all D3's in our list with $\alpha_3 = \alpha_1 = 0$. There 14 cases on both lists, below we show half of them as we already did in Section 1. Namely, for every D2 listed in the table with the parameters $(\alpha_2, \alpha_1, \alpha_0)$ there is also $(-\alpha_2, \alpha_1, -\alpha_0)$. We denote the spectral elliptic curves of the D2 and D3 differential equations by $\mathcal{E}_2$ and $\mathcal{E}_3$ respectively.

| D2 | | | | | D3 | | | | |
|---|---|---|---|---|---|---|---|---|---|
| $\alpha_2$ | $\alpha_1$ | $\beta$ | $j(\mathcal{E}_2)$ | $N(\mathcal{E}_2)$ | $\alpha_2$ | $\alpha_0$ | $\beta_0$ | $j(\mathcal{E}_3)$ | $N(\mathcal{E}_3)$ |
| -7 | -8 | 2 | $\frac{1556068}{81}$ | 24 | -28 | -128 | -4 | $\frac{207646}{6561}$ | 24 |
| -9 | 27 | 3 | 0 | 144 | -36 | 432 | -12 | 54000 | 144 |
| -10 | 9 | 3 | $\frac{1556068}{81}$ | 24 | -40 | 144 | -8 | $\frac{35152}{9}$ | 24 |
| -11 | -1 | 3 | $\frac{488095744}{125}$ | 20 | -44 | -16 | -4 | $-\frac{20720464}{15625}$ | 20 |
| -12 | 32 | 4 | 1728 | 32 | -48 | 512 | -16 | 287496 | 32 |
| -17 | 72 | 6 | $\frac{1556068}{81}$ | 48 | -68 | 1152 | -28 | $\frac{3065617154}{9}$ | 48 |
| 0 | -16 | 0 | 1728 | 32 | 0 | -256 | 0 | 1728 | 32 |

The spectral elliptic curves appear to be non–isomorphic in general but their levels coincide. In fact, the respective spectral curves are isogenous, and therefore their $L$-functions are equal. Indeed, the spectral curve of D2

$$\mathcal{E}_2 : \quad y^2 \;=\; z^3 + \alpha_2 z^2 + \alpha_1 z$$

is isogenous to the elliptic curve

$$\mathcal{E}_2' : \quad y^2 \;=\; z^3 - 2\alpha_2 z^2 + (\alpha_2^2 - 4\alpha_1)z\,,$$

the isogeny of degree 2 given by

$$\begin{gathered}\mathcal{E}_2 \to \mathcal{E}_2' \\ (z,y) \mapsto \Big(\frac{y^2}{z^2}, \frac{y(\alpha_1 - z^2)}{z^2}\Big)\end{gathered}$$

(see Example 4.5 in [**Sil09**]). This latter curve is in turn isomorphic to the spectral elliptic curve of the respective D3

$$\mathcal{E}_3 : \quad y^2 \;=\; z^3 + (-\frac{16}{3}\alpha_2^2 - 64\alpha_1)z + (\frac{128}{27}\alpha_2^3 - \frac{512}{3}\alpha_1\alpha_2)$$

with the map

$$\begin{gathered}\mathcal{E}_2' \to \mathcal{E}_3 \\ (z,y) \mapsto \Big(\frac{1}{4}z + \frac{2}{3}\alpha_2, \frac{1}{8}y\Big)\,.\end{gathered}$$

**Remark.** One could argue that an alternative and perhaps more natural way of constructing a 3–rd order equation out of an equation of type D2 would be via the passage to the symmetric square. In fact, the symmetric square will be related to the Hadamard product considered above by an algebraic transformation of Clausen type. We refer the reader to [**AvSZ11**] for Clausen–type identities that appear in the context of Calabi–Yau differential equations of order 4.

## References


[Zag07] Don Zagier, *Integral solutions of Apéry-like recurrence equations*, Groups and symmetries, CRM Proc. Lecture Notes, vol. 47, Amer. Math. Soc., Providence, RI, 2009, pp. 349–366. MR2500571 (2010h:11069)

[Beu02] Frits Beukers, *On Dwork's accessory parameter problem*, Math. Z. **241** (2002), no. 2, 425–444, DOI 10.1007/s00209-002-0424-8. MR1935494 (2003i:12013)

[GS07] Vasily Golyshev and Jan Stienstra, *Fuchsian equations of type DN*, Commun. Number Theory Phys. **1** (2007), no. 2, 323–346, DOI 10.4310/CNTP.2007.v1.n2.a3. MR2346574 (2008g:16039)

[ASD71] A. O. L. Atkin and H. P. F. Swinnerton-Dyer, *Modular forms on noncongruence subgroups*, Combinatorics (Proc. Sympos. Pure Math., Vol. XIX, Univ. California, Los Angeles, Calif., 1968), Amer. Math. Soc., Providence, R.I., 1971, pp. 1–25. MR0337781 (49 #2550)

[Sil09] Joseph H. Silverman, *The arithmetic of elliptic curves*, 2nd ed., Graduate Texts in Mathematics, vol. 106, Springer, Dordrecht, 2009. MR2514094 (2010i:11005)

[MAGMA] Wieb Bosma, John Cannon, and Catherine Playoust, *The Magma algebra system. I. The user language*, J. Symbolic Comput. **24** (1997), no. 3-4, 235–265, DOI 10.1006/jsco.1996.0125. Computational algebra and number theory (London, 1993). MR1484478

[AL70] A. O. L. Atkin and J. Lehner, *Hecke operators on* $\Gamma_0(m)$, Math. Ann. **185** (1970), 134–160. MR0268123 (42 #3022)

[AvSZ11] Gert Almkvist, Duco van Straten, and Wadim Zudilin, *Generalizations of Clausen's formula and algebraic transformations of Calabi-Yau differential equations*, Proc. Edinb. Math. Soc. (2) **54** (2011), no. 2, 273–295, DOI 10.1017/S0013091509000959. MR2794653 (2012e:33016)

[AvSZ06] Gert Almkvist and Wadim Zudilin, *Differential equations, mirror maps and zeta values*, Mirror symmetry. V, AMS/IP Stud. Adv. Math., vol. 38, Amer. Math. Soc., Providence, RI, 2006, pp. 481–515. MR2282972 (2008j:14073)



Algebra and Number Theory Sector, Institute for Information Transmission Problems, B. Karetny 19, Moscow 127994, Russia

*E-mail address*: golyshev@mccme.ru

School of Mathematical Sciences, University College Dublin, Belfield, Dublin 4, Ireland

*E-mail address*: masha.vlasenko@gmail.com

# Quantum fields, periods and algebraic geometry

Dirk Kreimer

ABSTRACT. We discuss how basic notions of graph theory and associated graph polynomials define questions for algebraic geometry, with an emphasis given to an analysis of the structure of Feynman rules as determined by those graph polynomials as well as algebraic structures of graphs. In particular, we discuss the appearance of renormalization scheme independent periods in quantum field theory.

## 1. Introduction

In this contribution, we want to review work concerning the structure of local renormalizable quantum field theories. Our emphasis will be to exhibit the most recent developments by way of example, and in particular to stress that at the time of writing we witness two simultaneous developments: a better understanding of the algebro-geometric underpinning of field theory in four dimensions of space time, and also as a consequence the emergence of computational approaches which surpass the hitherto established state of the art.

## 2. Graphs and algebras

**2.1. Wheels in wheels.** It is the purpose of this section to completely analyse an example. We choose wheels with three or four spokes, inserted at most once into each other. Results for them are available by methods which were recently developed [**6**–**8**, **21**] and which are presented elsewhere [**18**, **19**].

We consider the free commutative $\mathbb{Q}$-algebra generated by a sole generator in degree zero, $\mathbb{I}$, which serves as a unit for the algebra. In degree three we put

$\Gamma_3 =$ 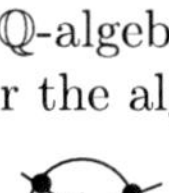 ,

whilst the only generator in degree four is

$\Gamma_4 =$ 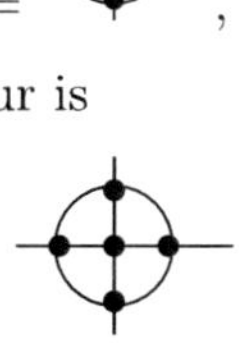 .

2010 *Mathematics Subject Classification.* Primary 81T15.

*Key words and phrases.* Quantum fields, Feynman rules, periods.

Author supported by the Alexander von Humboldt Foundation and the BMBF through an Alexander von Humboldt Professorship.

In degree six we have $\Gamma_3 \times \Gamma_3$ and

$\Gamma_{33} =$  ,

whilst in degree seven we have $\Gamma_3 \times \Gamma_4$ and

$\Gamma_{43} =$ 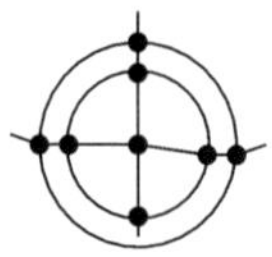 , $\Gamma_{34a} =$ 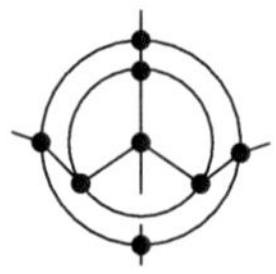 , $\Gamma_{34b} =$ 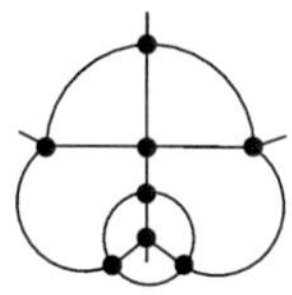 .

Note that $\Gamma_{34a}, \Gamma_{34b}$ are the only two different topologies we can obtain by replacing one of the five vertices of $\Gamma_4$ by $\Gamma_3$. The four vertices of $\Gamma_4$ which are connected to an external momentum all give $\Gamma_{34b}$ (modulo permutations of edge labels), whilst inserting at the internal vertex of $\Gamma_4$ gives $\Gamma_{34a}$.

Finally, in degree eight we only consider $\Gamma_4 \times \Gamma_4$ and insertion at the internal vertex:

$\Gamma_{44} =$ 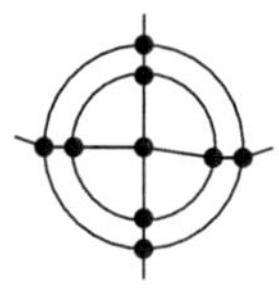 .

At higher degrees, we only allow products of the generators listed so far.

We make this into a bi-algebra by setting $\Delta(\mathbb{I}) = \mathbb{I} \otimes \mathbb{I}$ and

$$\Delta\Gamma_3 = \Gamma_3 \otimes \mathbb{I} + \mathbb{I} \otimes \Gamma_3,$$
$$\Delta\Gamma_4 = \Gamma_4 \otimes \mathbb{I} + \mathbb{I} \otimes \Gamma_4,$$
$$\Delta\Gamma_{33} = \Gamma_{33} \otimes \mathbb{I} + \mathbb{I} \otimes \Gamma_{33} + \Gamma_3 \otimes \Gamma_3,$$
$$\Delta\Gamma_{44} = \Gamma_{44} \otimes \mathbb{I} + \mathbb{I} \otimes \Gamma_{44} + \Gamma_4 \otimes \Gamma_4,$$
$$\Delta\Gamma_{43} = \Gamma_{43} \otimes \mathbb{I} + \mathbb{I} \otimes \Gamma_{43} + \Gamma_4 \otimes \Gamma_3,$$
$$\Delta\Gamma_{34a} = \Gamma_{34a} \otimes \mathbb{I} + \mathbb{I} \otimes \Gamma_{34a} + \Gamma_3 \otimes \Gamma_4,$$
$$\Delta\Gamma_{34b} = \Gamma_{34b} \otimes \mathbb{I} + \mathbb{I} \otimes \Gamma_{34b} + \Gamma_3 \otimes \Gamma_4,$$

and $\Delta(h_1 \times h_2) = \Delta(h_1) \times \Delta(h_2)$.

We get a Hopf algebra by setting $S(\mathbb{I}) = \mathbb{I}$, and recursively $S(h) = -m_H(S \otimes P)\Delta$, with $P$ the projection onto elements of positive degree, i.e. the augmentation ideal.

Define two maps into the augmentation ideal

$$B_+^3 : H \to PH, \text{ and } B_+^4 : H \to PH$$

by

$$B_+^3(\mathbb{I}) = \Gamma_3, \; B_+^4(\mathbb{I}) = \Gamma_4,$$
$$B_+^3(\Gamma_3) = \Gamma_{33}, \; B_+^3(\Gamma_4) = \Gamma_{43},$$
$$B_+^4(\Gamma_4) = \Gamma_{44}, \; B_+^4(\Gamma_3) = \frac{1}{2}\left(\Gamma_{34a} + \Gamma_{34b}\right),$$

and $B_+^i(h) = 0$, $i \in \{3, 4\}$, else.

Then $\forall h \in \{\mathbb{I}, \Gamma_3, \Gamma_4\}$ and $i \in \{3,4\}$,

$$\Delta B_+^i(h) = B_+^i(h) \otimes \mathbb{I} + (\mathrm{id} \otimes B_+^i)\Delta(h),$$

which ensures that these maps behave as Hochschild one-cocycles in the examples below.

REMARK 2.1. Effectively, we are working in a Hopf algebra of graphs generated and co-generated by $\Gamma_3$ and $\Gamma_4$, a quotient of the full Hopf algebra of graphs. Note that, for example,

$$\Delta B_+^4(\Gamma_{44}) = 0 \neq B_+^4(\Gamma_{44}) \otimes \mathbb{I} + (\mathrm{id} \otimes B_+^4)\Delta(\Gamma_{44}) = \Gamma_4 \otimes \Gamma_{44}.$$

This is a consequence of restricting to a finite Hopf algebra. It poses no problems for our applications below in this finite Hopf algebra.

Now, let $\alpha_i : H \to \mathbb{C}$ be algebra maps, and let $b\alpha_i : H \to H$ be defined by

$$b\alpha_i(h) = m(\mathrm{id} \otimes \alpha_i)\Delta(h) - \alpha_i(h)\mathbb{I}.$$

Then $b\alpha_i(\mathbb{I}) = 0$ and $b\alpha_i(\Gamma_j) = \alpha_i(\mathbb{I})\Gamma_j$, $\forall i, j \in \{3,4\}$.

REMARK 2.2. Were the $B_+^i$ to provide Hochschild one-cocycles, the $b\alpha_i$ would provide co-boundaries.

REMARK 2.3. We choose wheels in wheels as an example as results for them are on the brink of computability at the moment. The methods of Francis Brown [**6**, **7**] combined with [**8**] allow to compute the period provided by $\Gamma_{33}$ [**18**, **19**], whilst the periods from a symmetric combination of graphs $s_{34}$ as defined below are realistically in reach by this approch -and this approach only, it seems-, and the eight-loop $\Gamma_{44}$ period remains a challenge.

**2.2. Co-radical filtration.** Note that there is an obvious co-radical filtration and associated grading here, given by the kernels of $\sigma^{\star j}$, with $\sigma := S \star P = m(S \otimes P)\Delta$, i.e. using projections into the augmentation ideal combined with the co-product (see [**8**]).

We find in grading one the primitives $\Gamma_3, \Gamma_4$ and, more interestingly, the primitive elements

$$p_{33} := 2\Gamma_{33} - \Gamma_3 \times \Gamma_3,\ p_{44} := 2\Gamma_{44} - \Gamma_4 \times \Gamma_4,$$

and in particular

$$p_{34} := \overbrace{\frac{1}{2}\Gamma_{34a} + \frac{1}{2}\Gamma_{34b} + \Gamma_{43}}^{=:s_{34}} - \Gamma_3 \times \Gamma_4,$$

which also defines the co-symmetric $s_{34}$ and

$$p_{a-b} := \Gamma_{34a} - \Gamma_{34b}.$$

They are all linear combinations of elements in filtration two which combine to give primitive elements in the Hopf algebra, hence of co-radical degree 1. Note that $s_{34}$ is a co-symmetric element (of co-radical degree two) in the Hopf algebra, which is the reason why we can subtract the commutative product $\Gamma_3 \times \Gamma_4$ to obtain a primitive element.

Let us also define a co-antisymmetric element in degree two,

$$c_{34} = \frac{1}{2}\Gamma_{34a} + \frac{1}{2}\Gamma_{34b} - \Gamma_{43}.$$

Then, its reduced co-product $\Delta' := (P \otimes P)\Delta$ delivers

$$\Delta' c_{34} = \Gamma_3 \otimes \Gamma_4 - \Gamma_4 \otimes \Gamma_3,$$

an element which indeed changes sign when we swap the elements on the lhs and rhs of the tensor product, contrary to

$$\Delta' s_{34} = \Gamma_3 \otimes \Gamma_4 + \Gamma_4 \otimes \Gamma_3.$$

**2.3. Lie algebra aspects.** Let us now consider the Lie algebra $\mathcal{L}$ with generators $Z$ which are Kronecker-dual to the Hopf algebra generators $h$. Its bracket is determined by

$$\langle Z_i \otimes Z_j - Z_j \otimes Z_i, \Delta(h)\rangle = \langle [Z_i, Z_j], h\rangle,\ h \in H.$$

Here, $\langle Z_a, \Gamma_b\rangle = \delta_{ab}$ is the Kronecker pairing between elements $Z_a \in \mathcal{L}$ and $\Gamma_b \in H$, and $a, b$ range over the set of subscripts $3, 4, 33, 44, 34a, \ldots$ used to denote the graphs.

Consider also the corresponding universal enveloping algebra

$$U(\mathcal{L}) = \mathbb{Q}\mathbb{I} \oplus \mathcal{L} \oplus (\mathcal{L} \otimes_S \mathcal{L}) \oplus \cdots.$$

Here, $\otimes_S$ denotes the symmetrized tensor-product, and $U(\mathcal{L})$ can be identified, albeit non-canonically, with the symmetric tensor algebra of $\mathcal{L}$.

The Lie algebra $\mathcal{L}$ itself has a (descending) lower central series decomposition:

$$\mathcal{L}_1 := \mathcal{L}, \mathcal{L}_k := [\mathcal{L}, \mathcal{L}_{k-1}],\ k > 1.$$

The co-product of an element in $H$ is not co-commutative. It pays to decompose images of $\Delta'$ and its iterations into symmetric and antisymmetric parts.

The idea on which we elaborate in the following is to map elements in the Hopf algebra to elements in the above universal enveloping algebra of its dual Lie algebra, taking some extra information from physics: we will soon see that Feynman rules assign to Hopf algebra elements polynomials in a variable $L$, bounded by the co-radical degree, which respects a decomposition into co-symmetric and co-antisymmetric terms in the Hopf algebra which is particularly illuminating in comparison with the universal enveloping algebra.

Concretely, let us consider the following map (extended by linearity) $\sigma : H \to U(\mathcal{L})$. We start with primitive elements $\Gamma_3, \Gamma_4, p_{33}, p_{44}, p_{34}, p_{a-b}$, which, as we will see, all evaluate under the Feynman rules to terms linear in $L$:

$$\sigma(\Gamma_i) = Z_i \in \mathcal{L}_1 \subset U(\mathcal{L}),\ i \in 3, 4,$$

$$\sigma(p_{ii}) = \overbrace{Z_{ii}}^{\in \mathcal{L}_1}\ i \in 3, 4,$$

$$\sigma(s_{34}) = \overbrace{Z_{s_{34}}}^{\in \mathcal{L}_1},$$

$$\sigma(p_{a-b}) = \overbrace{Z_{p_{a-b}}}^{\in \mathcal{L}_1}.$$

Note that under $\sigma$ these primitives have images $\in \mathcal{L}_1$, but $\notin \mathcal{L}_2$.

Let us now consider non-primitive elements. As we will see under the Feynman rules, the next two examples give polynomials quadratic in $L$. This is reflected in $\sigma$:

$$\sigma(\Gamma_{ii}) = \overbrace{Z_{ii}}^{\in\mathcal{L}_1} + \overbrace{\frac{1}{2} Z_i \otimes Z_i}^{\in\mathcal{L}_1\otimes_S\mathcal{L}_1},\ i \in 3, 4,$$

$$\sigma(s_{34}) = \overbrace{Z_{s_{34}}}^{\in\mathcal{L}_1} + \overbrace{Z_3 \otimes Z_4 + Z_4 \otimes Z_3}^{\in\mathcal{L}_1\otimes_S\mathcal{L}_1},$$

Note that the second symmetric tensor power shows up here, reflecting the $L^2$ term in the Feynman rules.

Finally, we have the co-antisymmetric element. It is of co-radical degree two, but is linear in $L$ under the Feynman rules. We map

$$\sigma(c_{34}) = [Z_3, Z_4] \in \mathcal{L}_2,\ [Z_3, Z_4] \notin \mathcal{L}_3,$$

with

$$[Z_3, Z_4] = \frac{1}{2} Z_{34a} + \frac{1}{2} Z_{34b} - Z_{43}.$$

Note that the second symmetric tensor power does not show up here due to the co-antisymmetry of $c_{34}$. Nicely, the Feynman rules play along.

All others evaluations of $\sigma$ follow by linearity.

REMARK 2.4. The fact that the Dynkin operator $S \star Y = m(S \otimes Y)\Delta$, -with $Y$ the grading operator multiplying a Hopf algebra element of co-radical degree $k$ by $k$-, of $H$ vanishes on products very much suggests to construct $\sigma$ as above. The fact that it maps pre-images $\sigma^{-1}$ of co-symmetric elements in $\mathcal{L}_1$ to primitive elements of $H$ motivates to look at the lower central series of $\mathcal{L}$ for the co-antisymmetric elements. Also, note that pre-images of co-symmetric elements can be generated from $\mathbb{I}$ through shuffles of one-cocycles, for example $(B_+^3 B_+^4 + B_+^4 B_+^3)(\mathbb{I}) = s_{34}$.

## 3. Feynman Rules

We now give the Feynman rules for Hopf algebra elements, next study them in examples provided by our small Hopf algebra, and discuss the induced Feynman rules on the Lie side at the end.

Feynman rules on the Hopf algebra side are provided for scalar fields from the two Symanzik polynomials, together with the above Hopf algebra structure. For gauge fields, a third polynomial [**15**] allows to obtain the Feynman rules for gauge theory from the scalar field rules [**12**]. We follow [**8, 9, 12**].

**3.1. The first Kirchhoff polynomial $\psi_\Gamma$.** For the first Kirchhoff polynomial consider the short exact sequence

$$0 \to H^1 \to \mathbb{Q}^E \overset{\partial}{\frown\!\!\!\to} \mathbb{Q}^{V,0} \to 0. \tag{3.1}$$

Here, $H^1$ is provided by a chosen basis for the independent loops of a graph $\Gamma$. $E = |E^\Gamma|$ is the number of edges and $V = |V^\Gamma|$ the number of vertices, so $\mathbb{Q}^E$ is an $E$-dimensional $\mathbb{Q}$-vectorspace generated by the edges, similar $\mathbb{Q}^{V,0}$ for the vertices with a side constraint setting the sum of all vertices to zero.

Consider the matrix (see [**2, 3**])

$$N_0 \equiv (N_0)_{ij} = \sum_{e \in l_i \cap l_j} A_e,$$

for $l_i, l_j \in H^1$.

Define the *first Kirchhoff polynomial* as the determinant

$$\psi_\Gamma := |N_0|.$$

PROPOSITION 3.1. ([**10**], see also [**2**, Prop.2.2]) The first Kirchhoff polynomial can be written as

$$\psi_\Gamma = \sum_T \prod_{e \notin T} A_e$$

where the sum on the right is over spanning trees $T$ of $\Gamma$.

**3.2. The second Kirchhoff polynomial $\phi_\Gamma$ and $|N|_{\mathrm{Pf}}$.** To each edge $e$ we assign an auxiliary four-vector $\xi_e$.

Let then $\sigma^i$, $i \in 1,2,3$ be the three Pauli matrices, and $\sigma^0 = \mathbb{I}_{2\times 2}$ the unit matrix.

For the second Kirchhoff polynomial, augment the matrix $N_0$ to a new matrix $N$ in the following way:

(1) Assign to each edge $e$ a quaternion

$$\mathbf{q}_e := \xi_{e,0}\sigma^0 - i\sum_{j=1}^{3} \xi_{e,j}\sigma^j,$$

so that $\xi_e^2 \mathbb{I}_{2\times 2} = \mathbf{q}_e \bar{\mathbf{q}}_e$, and to the loop $l_i$, the quaternion

$$u_i = \sum_{e \in l_i} A_e \mathbf{q}_e.$$

(2) Consider the column vector $u = (u_i)$ and the conjugated transposed row vector $\bar{u}$. Augment $u$ as the rightmost column vector to $M$, and $\bar{u}$ as the bottom row vector.

(3) Add a new diagonal entry at the bottom right $\sum_e \mathbf{q}_e \bar{\mathbf{q}}_e A_e$.

Note that by momentum conservation, to each vertex, we assign a momentum $\xi_v$, and a corresponding quaternion $\mathbf{q}_v$.

REMARK 3.2. Note that we use that we work in four dimensions of space-time, by rewriting the momentum four-vectors in a quaternionic basis. This strictly four-dimensional approach can be extended to twistors [**1**].

The matrix $N$ has a well-defined Pfaffian determinant (see [**3**]) with a remarkable form obtained for generic $\xi_e$ and hence generic $\xi_v$:

LEMMA 3.3. ([**3**, Eq.3.12])

$$|N|_{\mathrm{Pf}} = -\sum_{T_1 \cup T_2} \left( \sum_{e \notin T_1 \cup T_2} \tau(e)\xi_e \right)^2 \prod_{e \notin T_1 \cup T_2} A_e,$$

where $\tau(e)$ is $+1$ if $e$ is oriented from $T_1$ to $T_2$ and $-1$ else.

Here, $T_1, T_2$ are two trees such that their union contains all vertices of the graph, i.e. $T_1 \cup T_2$ is a spanning 2-tree.

Note that $|N|_{\mathrm{Pf}} = |N|_{\mathrm{Pf}}(\{\xi_v\})$ is a function of all $\xi_v$, $v \in \Gamma^{[0]}$. From the viewpoint of graph theory, this is the natural polynomial. It gives the second Symanzik polynomial upon setting the $\xi_e$ in accordance with the external momenta $p_e$:

$$Q : \xi_e \to +p_e.$$

REMARK 3.4. Adding to the second Symanzik polynomial a term $\psi_\Gamma \sum_{e\in\Gamma} A_e m_e^2$ allows to treat masses $m_e$.

REMARK 3.5. For $\gamma \subset \Gamma$ a non-trivial subgraph, and $\kappa \in \{\phi, \psi\}$ we have almost factorization: $\kappa_\Gamma = \kappa_{\Gamma/\gamma}\psi_\gamma + R^\kappa_{\Gamma,\gamma}$, with the remainders $R^\kappa_{\Gamma,\gamma}$ homogeneous polynomials of higher degree in the sub-graph variables than $\psi_\gamma$.

**3.3. The unrenormalized integrand.** In Schwinger parametric form, the unrenormalized Feynman amplitude $\mathcal{I}_\Gamma$ (omitting trivial overall factors of powers of $\pi$ and such) comes from an integrand $I_\Gamma$

$$\mathcal{I}_\Gamma = \int \underbrace{\frac{e^{-\frac{\phi_\Gamma}{\psi_\Gamma}}}{\psi_\Gamma^2}}_{I_\Gamma} \prod_e dA_e. \tag{3.2}$$

This form gets modified if we allow for spin and other such complications. An exhaustive study of how to obtain gauge theory amplitudes from such an integrand is given in [**12**].

REMARK 3.6. A regularized integrand can be obtained by raising the denominator $1/\psi_\Gamma^2$ to a noninteger power (dimensional regularization), or multiplication by non-integer powers of edge variables, together with suitable $\Gamma$-functions (analytic regularization). The latter suffices to treat the Mellin transforms as used for example in [**23**] and discussed below.

**3.4. The renormalized integrand.** We can render the integrand $I_\Gamma$ integrable wrt to the domain $\sigma_\Gamma$ prescribed by parametric integration by a suitable sum over forests. We define

$$I_\Gamma^R := \sum_{f\in\mathcal{F}_\Gamma} (-1)^{|f|} I_{\Gamma/f} I_f^0,$$

where for $f = \bigcup_i \gamma_i$, $I_f = \prod_i I_{\gamma_i}$ and the superscript $^0$ indicates that kinematic variables are specified according to renormalization conditions.

The formula for $I_\Gamma^R$ is correct as long as all sub-graphs are overall log-divergent, the necessary correction terms in the general case are given in [**8**]. In our examples below, we can always identify the one log-divergent subgraph -if any- with the unique non-trivial forest.

**3.5. The renormalized integral.** Accompanying this integrand is the renormalized result which can be written projectively:

$$\Phi^R(\Gamma) := \int_{\mathbb{P}^{|E_I^\Gamma|}(\mathbb{R}_+)} \sum_{f\in\mathcal{F}_\Gamma} (-1)^{|f|} \frac{\ln \frac{\phi_{\Gamma/f}\psi_f + \phi_f^0 \psi_{\Gamma/f}}{\phi^0_{\Gamma/f}\psi_f + \phi_f^0 \psi_{\Gamma/f}}}{\psi^2_{\Gamma/f}\psi_f^2} \Omega_\Gamma,$$

for notation see [**8, 9**] or [**12**]. Let us just mention that for the domain of integration we will abbreviate from now on

$$\mathbb{P}^{|E_I^\Gamma|}(\mathbb{R}_+) = \mathbb{P}_\Gamma.$$

Note that this is a well-defined integral obtained from the use of the forest formula. It is obtained without using an intermediate regulator. It is well-suited to analyse the mathematical structure of perturbative contributions to Green functions.

Also, combining this approach with [**12**], it furnishes a reference point against which to check in a situation where intermediate regulators are spoiling the symmetries of the theory.

Below, we will shortly compare the structure of this integrand to the appaearance of analytic regulators provided by anomalous dimensions of quantum fields, wich then define Mellin transforms for the primitives in the Hopf algebra.

**3.6. Scales and Angles.** Feynman graphs have their external edges labelled by momenta, and internal edges labelled by masses.

Renormalized Feynman rules above are therefore functions of scalar products $Q_i \cdot Q_j$ and mass-squares $m_e^2$. Equivalently, upon defining a positive definite linear combination $S$ of such variables, we can write them as functions of such a scale $S$, and angles $\Theta_{ij} := Q_i \cdot Q_j/S$, $\Theta_e := m_e^2/S$. We use $S^0, \Theta_{ij}^0, \Theta_e^0$ to specify scale and angles for the renormalization point. A graph which furnishes only a single scalar product $Q \cdot Q$ as a scale is denoted a 1-scale graph.

Isolating short-distance singularities in 1-scale sub-graphs has many advantages, including a systematic separation of angles and scales, and a clean approach to the renormalization group as well as an identification of the freedom provided by exact terms in the Hochschild cohomology, as we discuss below, see also [**11, 16, 17**].

Following [**8**], we have the decomposition

THEOREM 3.7.

$$\Phi^R(S/S^0, \{\Theta, \Theta^0\}) = \Phi_{\text{fin}}^{-1}(\{\Theta^0\}) \star \Phi_{\text{1-s}}^R(S/S^0) \star \Phi_{\text{fin}}(\{\Theta\}).$$

Here, the angle-dependent Feynman rules $\Phi_{\text{fin}}$ are computed by eliminating short-distance singularities through the comparison, via the Hopf algebra, with 1-scale graphs evaluated at the same scale as the initial graphs, while the 1-scale Feynman rules $\Phi_{\text{1-s}}^R(S/S_0)$ eliminate short-distance singularities by renormalizing 1-scale graphs at a reference scale $S_0$.

REMARK 3.8. Feynman rules in parametric renormalization allow to treat the computation of Feynman graphs as a problem of algebraic geometry, analysing the structure of two kinds of homogeneous polynomials [**4, 6–8**].

REMARK 3.9. The fact that it is basically the denominator structure which determines the computability of Feynman graphs in parametric renormalization makes this approach very efficient in computing periods in the $\epsilon$-expansion of regularized integrands.

REMARK 3.10. We assume throughout that angles and scales are such that we are off any infrared singularities, for example by off-shell external momenta. The latter would not be cured by the forest sums which eliminate short-distance singularities.

## 4. Examples

**4.1. Overall finite graphs.** From now on, we write $\phi_\Gamma = \phi_\Gamma(\Theta), \phi_\Gamma^0 \equiv \phi_\Gamma(\Theta_0)$. For a 1-scale graph $\Gamma$, we let $\Gamma^\bullet$ be the graph where the two external vertices of $\Gamma$ are identified. One has $\phi_\Gamma = \psi_{\Gamma^\bullet}$.

Assume we are considering a superficially convergent graph $\Gamma$ free of subdivergences. For example, a graph $\Gamma$ in four dimensions of space time on $n > 2|\Gamma|$ edges delivers the integrable form

$$\frac{1}{S^{n-2|\Gamma|}}\int\frac{1}{\psi^2}\left(\frac{\psi}{\phi(\Theta)}\right)^{n-2|\Gamma|}\Omega_\Gamma.$$

This is polynomial in the scale dependence, while the angle dependence is dilogarithmic for good reasons [**3**].

Inserting logarithmic subdivergences, we get the integrable form (it is integrable as long as external momenta are off-shell such that no infrared singularities arise)

$$\frac{1}{S^{n-2|\Gamma|}}\int_{\mathbb{P}_\Gamma}\sum_f(-1)^{|f|}\frac{1}{\psi^2_{\Gamma/f}\psi^2_f}\left(\frac{\psi_{\Gamma/f}\psi_f}{\phi_{\Gamma/f}\psi_f+\phi^0_f\psi_{\Gamma/f}}\right)^{n-2|\Gamma|}\Omega_\Gamma.$$

Note that $\phi_\emptyset=0,\psi_\emptyset=1$.

REMARK 4.1. Note that for the logarithmic divergent case $n=2|\Gamma|$, we got a logarithm in the numerator of the renormalized integrand, reflecting the superficial degree of divergence zero. In the convergent case, the above power of $n-2|\Gamma|$ is then reflecting the superficial degree of convergence $2(n-2|\Gamma|)$.

**4.2. Primitive graphs.** Consider now a logarithmically divergent graph without sub-divergences, $L=\ln S/S^0$. Then,

$$\Phi_R(\Gamma)=c^1_\Gamma L+c^0_\Gamma(\Theta,\Theta_0).$$

We have

$$c^1_\Gamma=\int_{\mathbb{P}(\Gamma)}\frac{\Omega_\Gamma}{\psi^2_\Gamma},$$

$$c^0_\Gamma(\Theta,\Theta_0)=\int_{\mathbb{P}(\Gamma)}\frac{\ln\frac{\phi_\Gamma}{\phi^0_\Gamma}\Omega_\Gamma}{\psi^2_\Gamma}.$$

The finite part $c^0_\Gamma(\Theta,\Theta_0)$ can equivalently be expressed in the form of overall finite graphs. Let $P_e$ be the propagator at edge $e$, $P^0_e$ the same propagator, but with its external momenta evaluated as prescribed by the renormalization condition. Then,

$$\frac{1}{P_e}-\frac{1}{P^0_e}=\frac{P^0_e-P_e}{P_eP^0_e},$$

where internal loop momenta in edge $e$ drop out in the difference $P^0_e-P_e$.

By telescoping we can extend to products of propagators provided by graphs, and hence express the finite part of an overall logarithmically divergent graph as an overall convergent graph, which is an element of a larger Hopf algebra provided by general Feynman integrals.

**4.3. Structure of a graph with a sub-divergence.** Consider $\Gamma=\Gamma_{43}$ say, as a generic example. We have $\Delta'(\Gamma)=\Gamma_4\otimes\Gamma_3$.

Then

$$\Phi_R(\Gamma)=c^2_\Gamma L^2+c^1_\Gamma(\Theta,\Theta_0)L+c^0_\Gamma(\Theta,\Theta_0).$$

We have

$$\Phi_R(\Gamma_{43})=\int_{\mathbb{P}_\Gamma}\left(\frac{\ln\frac{\frac{S}{S_0}\phi_\Gamma}{\phi^0_\Gamma}}{\psi^2_\Gamma}-\frac{\ln\frac{\frac{S}{S_0}\phi_{\Gamma_3}\psi_{\Gamma_4}+\phi^0_{\Gamma_4}\psi_{\Gamma_3}}{\phi^0_{\Gamma_3}\psi_{\Gamma_4}+\phi^0_{\Gamma_4}\psi_{\Gamma_3}}}{\psi^2_{\Gamma_4}\psi^2_{\Gamma_3}}\right)\Omega_\Gamma.$$

We then have for the scale independent part

$$c_\Gamma^0(\Theta,\Theta_0) = \int_{\mathbb{P}_\Gamma} \left( \frac{\ln \frac{\phi_\Gamma}{\phi_\Gamma^0}}{\psi_\Gamma^2} - \frac{\ln \frac{\phi_{\Gamma_3}\psi_{\Gamma_4}+\phi_{\Gamma_4}^0\psi_{\Gamma_3}}{\phi_{\Gamma_3}^0\psi_{\Gamma_4}+\phi_{\Gamma_4}^0\psi_{\Gamma_3}}}{\psi_{\Gamma_4}^2\psi_{\Gamma_3}^2} \right) \Omega_\Gamma,$$

and for the term linear in $L$:

$$c_\Gamma^1(\Theta,\Theta_0) = \int_{\mathbb{P}_\Gamma} \left( \frac{1}{\psi_\Gamma^2} - \frac{\phi_{\Gamma_3}\psi_{\Gamma_4}}{\psi_{\Gamma_4}^2\psi_{\Gamma_3}^2\left[\phi_{\Gamma_3}\psi_{\Gamma_4}+\phi_{\Gamma_4}^0\psi_{\Gamma_3}\right]} \right) \Omega_\Gamma. \tag{4.1}$$

The term quadratic in $L$ gives

$$c_\Gamma^2 = \int_{\mathbb{P}_\Gamma} \left( \frac{\phi_{\Gamma_3}\psi_{\Gamma_4}\psi_{\Gamma_3}\phi_{\Gamma_4}^0}{\psi_{\Gamma_4}^2\psi_{\Gamma_3}^2\left[\phi_{\Gamma_3}\psi_{\Gamma_4}+\phi_{\Gamma_4}^0\psi_{\Gamma_3}\right]^2} \right) \Omega_\Gamma.$$

Scaling out from the edge variables of the subgraph one of its variables $\lambda$ say- and integrating it, so that $\Omega_\Gamma \to \Omega_{\Gamma_3} \wedge \Omega_{\Gamma_4} \wedge d\lambda$ (a careful treatment of such changes of variables is in [**8**]) gives us

$$\begin{aligned} c_\Gamma^2 &= \int_{\mathbb{P}_{\Gamma_3}\times\mathbb{P}_{\Gamma_4}} \left( \int_0^\infty \frac{\phi_{\Gamma_3}(\Theta)\psi_{\Gamma_4}\psi_{\Gamma_3}\phi_{\Gamma_4}(\Theta_0)}{\psi_{\Gamma_4}^2\psi_{\Gamma_3}^2\left[\phi_{\Gamma_3}(\Theta)\psi_{\Gamma_4}+\lambda\phi_{\Gamma_4}(\Theta_0)\psi_{\Gamma_3}\right]^2} d\lambda \right) \Omega_{\Gamma_3}\wedge\Omega_{\Gamma_4} \\ &= \int_{\mathbb{P}_{\Gamma_3}} \frac{1}{\psi_{\Gamma_3}^2}\Omega_{\Gamma_3} \int_{\mathbb{P}_{\Gamma_4}} \frac{1}{\psi_{\Gamma_4}^2}\Omega_{\Gamma_4}, \end{aligned}$$

which fully exhibits the desired factorization.

One easily checks that $\partial_L^k$ vanishes for $k$ greater than the co-radical degree.

**4.4. Periods from insertion places.** Let us now consider the primitive $p_{a-b}$. The two graphs involved are distinguished only by the insertion place into which we insert the subgraph $\Gamma_3$. From the previous result is it evident that for $p_{a-b}$ we could at most find up to a linear term in $L$

$$\Phi_R(p_{a-b}) = c_{p_{a-b}}^1 L + c_{p_{a-b}}^0(\Theta,\Theta^0).$$

We find for this scheme-independent -and hence well-defined- period

$$c_{p_{a-b}}^1 = \int_{\mathbb{P}_{\Gamma_{34}}} \left( \frac{1}{\psi_{\Gamma_{34a}}^2} - \frac{1}{\psi_{\Gamma_{34b}}^2} \right) \Omega_{\Gamma_{34}},$$

where $\mathbb{P}_{\Gamma_{34}}$ and $\Omega_{\Gamma_{34}}$ are obviously independent of the insertion place.

Note that the difference is completely governed by $R^\psi_{\Gamma_{34a},\Gamma_3}$ as compared to $R^\psi_{\Gamma_{34b},\Gamma_3}$, while for the term constant in $L$ we also need to consider $R^\phi_{\Gamma_{34a},\Gamma_3}$ as compared to $R^\phi_{\Gamma_{34b},\Gamma_3}$.

From now on we discard the constant terms in $L$, which we regard as originating from overall convergent integrals.

**4.5. Periods for co-commutative elements.** Next, let us look at $s_{34}$ which is of co-radical degree two. Clearly,

$$\Phi_R(s_{34}) = c_3^1 c_4^1 L^2 + c_{s_{34}}^1.$$

In general, $c_{s_{34}}^1$ is not a period but rather a complicated function of $\Theta, \Theta^0$.

We now assume that we subtract at $\Theta = \Theta_0$.

$c^1_{s34}$ could then still be a function of the angles $\Theta$. Instead, it is a constant, as is immediately clear by using Eq.(4.1). This constant is a period (associated to $p_{34}$) which is now known to us using the methods of [**19**], see [**20**].

$$\begin{aligned} c^1_{s34} &= \int_{\mathbb{P}_\Gamma} \Bigg( \frac{1}{2}\frac{1}{\psi^2_{\Gamma_{34a}}} + \frac{1}{2}\frac{1}{\psi^2_{\Gamma_{34b}}} + \frac{1}{\psi^2_{\Gamma_{43}}} \\ &\quad - \frac{\phi_{\Gamma_3}\psi_{\Gamma_4} + \phi_{\Gamma_4}\psi_{\Gamma_3}}{\psi^2_{\Gamma_4}\psi^2_{\Gamma_3}\left[\phi_{\Gamma_3}\psi_{\Gamma_4} + \phi_{\Gamma_4}\psi_{\Gamma_3}\right]} \Bigg) \Omega_\Gamma \\ &= \int_{\mathbb{P}_\Gamma} \left( \frac{1}{2}\frac{1}{\psi^2_{\Gamma_{34a}}} + \frac{1}{2}\frac{1}{\psi^2_{\Gamma_{34b}}} + \frac{1}{\psi^2_{\Gamma_{43}}} - \frac{1}{\psi^2_{\Gamma_4}\psi^2_{\Gamma_3}} \right) \Omega_\Gamma, \end{aligned} \tag{4.2}$$

where the notation $\mathbb{P}_\Gamma, \Omega_\Gamma$ is justified, as edges can be consistently labeled in all terms. Note that the step above from the first to the second line follows as we have $\phi_\Gamma(\Theta_0) = \phi_\Gamma(\Theta) = \phi_\Gamma, \forall \Gamma$.

**4.6. Angle dependence in commutators.** For anti-cocommutative elements like $c_{34}$ angle dependence remains, even if we set $\Theta = \Theta^0$. In such a setting, we find

$$\Phi_R(c_{34}) = c^1_{34}(\Theta)L.$$

$$\begin{aligned} c^1_{34}(\Theta) &= \int_{\mathbb{P}_\Gamma} \Bigg( \frac{1}{2}\frac{1}{\psi^2_{\Gamma_{34a}}} + \frac{1}{2}\frac{1}{\psi^2_{\Gamma_{34b}}} - \frac{1}{\psi^2_{\Gamma_{43}}} \\ &\quad - \frac{\phi_{\Gamma_4}\psi_{\Gamma_3} - \phi_{\Gamma_3}\psi_{\Gamma_4}}{\psi^2_{\Gamma_4}\psi^2_{\Gamma_3}\left[\phi_{\Gamma_3}\psi_{\Gamma_4} + \phi_{\Gamma_4}\psi_{\Gamma_3}\right]} \Bigg) \Omega_\Gamma. \end{aligned} \tag{4.3}$$

In this way, when renormalizing at unchanged scattering angles, angle dependence is relegated to anti-cocommutativity.

**4.7. 1-scale sub-graphs vs $\Phi_{1-s}$.** In [**8**] scale and angle dependence were separated using 1-scale renormalized Feynman rules $\Phi^R_{1-s}$. These are massless Feynman rules which act by choosing two distinct vertices for each subdivergent graph $\gamma \subset \Gamma$ and evaluating the counterterms for this subgraph treating it as a 1-scale graph $\gamma_2$, so that we have $\phi_\gamma = \psi_{\gamma^\bullet}$.

Also, $\Gamma$ itself allows external momenta only at two distinct vertices.

As discussed in [**8**], see also [**9**] for a detailed example, we can enlarge the set of graphs to be considered by graphs $G_2$ say so that $\Gamma/\gamma = G_2/g_2$, and

$$\phi_{\Gamma/\gamma}\psi_{g_2} + \psi_{\gamma^\bullet}\psi_{\Gamma/\gamma}$$

is the two-vertex join of co- and subgraph [**8**]. In $G_2$, edges which connect $G_2 - g_2$ to $g_2$ all originate from the two distinct vertices chosen in $\gamma$. The reader will have to consider [**8**] for precise definitions.

Feynman rules for graphs $G_2$ are the canonical $\Phi_R$ subtracting at $S = S_0$, and for example for $\Gamma = \Gamma_{43}$,

$$\Phi_R(G_2) - \Phi^R_{1-s}(\Gamma) = \int_{\mathbb{P}_\Gamma} \left( \frac{1}{\psi^2_{G_2}} - \frac{1}{\psi^2_\Gamma} \right) \Omega_\Gamma,$$

gives us another period.

$G_2$ in this example is the graph

$G_2 =$ 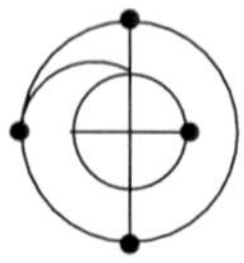 ,

with the understanding that momenta are zero at two of its four marked external vertices when acted upon by $\Phi^R_{1-s}$. Its wheel with four spokes subgraph is rendered 1-scale upon enforcing a five-valent vertex and hence must be treated in a suitably enlarged Hopf algebra.

The general case is studied in [**8**] in great detail.

**4.8. Mellin transforms and 1-scale subgraphs.** Graphs of the form $G_2$ can be computed by defining a suitable Mellin transform [**13, 14, 22**]. This holds even if the co-graph is not 1-scale, the important fact being that the subgraph is.

This Mellin transform is defined by raising a quadric $Q(e)$ for an internal edge $e$, or a linear combination of such quadrics, to a non-integer power

$$\frac{1}{Q(e)} \to \frac{1}{Q(e)^{1+\rho}},$$

in a cograph which has no subdivergences. This defines a Mellin transform (staying in the above example)

$$M_{\Gamma_3}(\rho, L) = e^{-\rho L} f_3(\rho),$$

where $f_3(\rho)$ has a first order pole in $\rho$ at zero with residue $6\zeta(3) = c^1_{\Gamma_3}$.

Also, $f_3(\rho) = f_3(1-\rho)$. We set

$$f_3(\rho) = \frac{6\zeta(3)}{\rho(1-\rho)}(1 + d^1_3(\Theta)\rho + \mathcal{O}(\rho^2)).$$

We can compute $\Phi_R(G_2)$ as

$$\Phi_R(G_2) = -\overbrace{c^1_{\Gamma_4}}^{20\zeta(5)}\partial_\rho(e^{-\rho L} - 1) f(\rho).$$

One hence finds that $c^2_{G_2} = 60\zeta(3)\zeta(5)$ and $c^1_{G_2} = 120\zeta(3)\zeta(5) - \zeta(5)d^1_3(\Theta)$.

Can we confirm this structure from the parametric approach?

We first note that

$$|\psi_{G_2/\Gamma_4}| = |\phi_{G_2/\Gamma_4}| - 1,$$

$$|\psi_{G_2/\Gamma_4}| = |\psi_{G_2-\Gamma_4}| + 1.$$

Returning to affine coordinates, scaling out a subgraph variable and integrating it, using $R^{\psi}_{G_2,\Gamma_4} = \psi_{\Gamma_4}\psi_{G_2-\Gamma_4}$ and returning to $\mathbb{P}_{\Gamma_3} \times \mathbb{P}_{\Gamma_4}$ delivers

$$c^1_{G_2} = \underbrace{\int_{\mathbb{P}_{\Gamma_4}} \frac{1}{\psi^2_{\Gamma_4}}\Omega_{\Gamma_4}}_{20\zeta_5} \underbrace{\int_{\mathbb{P}_{\Gamma_3}} \frac{1}{\psi^2_{\gamma_3}}\left(1 - \ln\frac{\psi_{G_2-\Gamma_4}\phi_{\Gamma_3}}{\psi^2_{\Gamma_3}}\right)\Omega_{\Gamma_3}}_{6\zeta_3 - d^1_3}, \tag{4.4}$$

as desired.

**4.9. Well-defined periods from the dual of commutators.** Finally, a further angle independent period is furnished by the 1-scale version $c_{34,1-s}$ of the anti-cocommutative $c_{34}$.

$$c^1_{c_{34,1-s}} = \int_{\mathbb{P}_\Gamma} \left( \frac{1}{\psi^2_{\Gamma_{43}}} - \frac{1}{2\psi^2_{\Gamma_{34a}}} - \frac{1}{2\psi^2_{\Gamma_{34b}}} - \frac{\psi_{\Gamma_3^\bullet}\psi_{\Gamma_4} - \psi_{\Gamma_4^\bullet}\psi_{\Gamma_3}}{\psi^2_{\Gamma_3}\psi^2_{\Gamma_4}(\psi_{\Gamma_3^\bullet}\psi_{\Gamma_4} + \psi_{\Gamma_4^\bullet}\psi_{\Gamma_3})} \right) \Omega_\Gamma,$$

where it is understood that all graphs have their subgraphs as 1-scale subgraphs as in $G_2$. By the previous result, this can be decomposed into two separate projective integrals.

**4.10. The role of exact co-boundaries.** We have seen that when a subgraph $\gamma$ is 1-scale, the evaluation of the full graph factorizes the period $c^1_\gamma$. This is the crucial fact which allows to use co-boundaries to alter the Taylor coefficients of Mellin transforms [**11, 16, 17**].

For example, with $B_+^3$ now effecting a 1-scale insertion, $B_+^3(\Gamma_4) = G_2$ and $b\alpha_3(\Gamma_4) = \alpha_3(\mathbb{I})\Gamma_4$:

$$\Phi_R((B_+^3 + b\alpha_3)(\Gamma_4) = \Phi_R(\Gamma_2) + 20\zeta(5)\alpha_3(\mathbb{I})L,$$

where we are free to choose $\alpha_3(\mathbb{I})$ to modify $d^1_3(\Theta)$, a useful fact in light of the manipulations in [**22, 23**].

## 5. Feynman rules from a Lie viewpoint

The map $\sigma : H \to U(\mathcal{L})$ can be combined with projectors $T_k$ into the $k$-th symmetric tensorpower of $\mathcal{L}$. Let then

$$\sigma_k := T_k \circ \sigma.$$

Then, $\sigma_1$ takes values in $\mathcal{L}$, $\sigma_2$ takes values in $\mathcal{L} \otimes_s \mathcal{L}$, and so on.

The map $\sigma_1 : H \to \mathcal{L}$ is such that an element $h \in H$ in the $k$-th co-radical filtration (so that $\Delta'^k(h) \neq 0$) has contributions in $\mathcal{L}_k$ at most, for example the co-radical degree two $c_{34}$ fulfils this bound as it maps to $[Z_3, Z_4] \in \mathcal{L}_2$.

In general, a co-radical degree $k$ element has a non-vanishing component in $\mathcal{L}_k$ if and only if $\Delta'^k(h)$ contains corresponding anti-symmetric elements.

The symmetric parts in $\Delta'^k(h)$ map under $\sigma_1$ to an element $l_1(h) \in \mathcal{L}_1$ say, $l_1(h) \notin \mathcal{L}_2$, so that the Dynkin operator $S \star Y$ maps the pre-image $\sigma_1^{-1}(l_1)$ to a primitive element in $H$,

$$\Phi_R(S \star Y\sigma_1^{-1}(l_1(h))) = c^1_{l_1} L,$$

while all other terms in $c^1_h$ come from the pre-images of elements in $\mathcal{L}_k, k > 1$.

Finally, pre-images of $\sigma_k$ provide the contributions to order $L^k$ similarly, in full accordance with the co-radical filtration and the renormalization group [**8**].

## Acknowledgments

Foremost, I want to thank David Broadhurst, with whom my interest in the periods emerging in QFT started some twenty years ago [**5**]. Spencer Bloch helped to uncover the mathematics behind it, as did Francis Brown, Christian Bogner, Alain Connes, Dzmitry Doryn, Hélène Esnault, Erik Panzer, Oliver Schnetz, and others. It is a pleasure to thank James Drummond for hospitality at CERN, Geneva, Feb 24-March 01 2013, as well as the IHES, where parts of this paper were written.

## References


[1] S. Bloch, talk at Spring School: Feynman Graphs and Motives, Bingen, march 2013, http://www.sfb45.de/events/spring-school-feynman-graphs-and-motives.

[2] Spencer Bloch, Hélène Esnault, and Dirk Kreimer, *On motives associated to graph polynomials*, Comm. Math. Phys. **267** (2006), no. 1, 181–225, DOI 10.1007/s00220-006-0040-2. MR2238909 (2008a:81074)

[3] Spencer Bloch and Dirk Kreimer, *Feynman amplitudes and Landau singularities for one-loop graphs*, Commun. Number Theory Phys. **4** (2010), no. 4, 709–753, DOI 10.4310/CNTP.2010.v4.n4.a4. MR2793425 (2012k:81104)

[4] C. Bogner and M. Lüders, *Multiple polylogarithms and linearly reducible Feynman graphs,* arXiv:1302.6215 [hep-ph], these proceedings.

[5] D. J. Broadhurst and D. Kreimer, *Knots and numbers in $\phi^4$ theory to 7 loops and beyond*, Internat. J. Modern Phys. C **6** (1995), no. 4, 519–524, DOI 10.1142/S012918319500037X. MR1352337 (97a:81143)

[6] F. Brown, *On the periods of some Feynman integrals*, arXiv:0910.0114v1, (2009), 1-69.

[7] Francis Brown, *The massless higher-loop two-point function*, Comm. Math. Phys. **287** (2009), no. 3, 925–958, DOI 10.1007/s00220-009-0740-5. MR2486668 (2009k:81094)

[8] Francis Brown and Dirk Kreimer, *Angles, scales and parametric renormalization*, Lett. Math. Phys. **103** (2013), no. 9, 933–1007, DOI 10.1007/s11005-013-0625-6. MR3077961

[9] F. Brown and D. Kreimer, *Decomposing Feynman rules,* arXiv:1212.3923 [hep-th], Proceedings of Science (LL2012)049.

[10] G. Kirchhoff, *Über die Auflösung der Gleichungen, auf welche man bei der Untersuchung der linearen Vertheilung galvanischer Ströme geführt wird*, Annalen der Physik und Chemie 72 no. 12 (1847), 497-508.

[11] D. Kreimer and E. Panzer, *Renormalization and Mellin transforms,* arXiv:1207.6321 [hep-th], Proceedings Linz July 2012, to appear.

[12] Dirk Kreimer, Matthias Sars, and Walter D. van Suijlekom, *Quantization of gauge fields, graph polynomials and graph homology*, Ann. Physics **336** (2013), 180–222, DOI 10.1016/j.aop.2013.04.019. MR3086033

[13] Dirk Kreimer and Karen Yeats, *Recursion and growth estimates in renormalizable quantum field theory*, Comm. Math. Phys. **279** (2008), no. 2, 401–427, DOI 10.1007/s00220-008-0431-7. MR2383593 (2009g:81124)

[14] Dirk Kreimer and Karen Yeats, *An étude in non-linear Dyson-Schwinger equations*, Nuclear Phys. B Proc. Suppl. **160** (2006), 116–121, DOI 10.1016/j.nuclphysbps.2006.09.036. MR2255485 (2007i:81162)

[15] Dirk Kreimer and Karen Yeats, *Properties of the corolla polynomial of a 3-regular graph*, Electron. J. Combin. **20** (2013), no. 1, Paper 41, 12. MR3035051

[16] E. Panzer, *Hopf algebraic Renormalization of Kreimer's toy model,* arXiv:1202.3552 [math.QA], Master Thesis at Humboldt University.

[17] E. Panzer, *Renormalization, Hopf algebras and Mellin transforms* , http://www.mathematik.hu-berlin.de/ maphy/panzer.pdf, these proceedings.

[18] E. Panzer, talk at Spring School: Feynman Graphs and Motives, Bingen, March 2013, http://www.sfb45.de/events/spring-school-feynman-graphs-and-motives.

[19] Erik Panzer, *On the analytic computation of massless propagators in dimensional regularization*, Nuclear Phys. B **874** (2013), no. 2, 567–593, DOI 10.1016/j.nuclphysb.2013.05.025. MR3079414

[20] E. Panzer, *Feynman Integrals and Hyperlogarithms*, http://www2.mathematik.hu-berlin.de/~kreimer/wp-content/uploads/Panzerphd.pdf. PhD thesis. Humboldt University, 2014.

[21] O. Schnetz, *Graphical functions and single-valued multiple polylogarithms*, arXiv:1302.6445.

[22] Karen Yeats, *Rearranging Dyson-Schwinger equations*, Mem. Amer. Math. Soc. **211** (2011), no. 995, x+82, DOI 10.1090/S0065-9266-2010-00612-4. With a foreword by Dirk Kreimer. MR2791969 (2012d:81239)

[23] K. Yeats, *Some combinatorial interpretations in perturbative quantum field theory,* arXiv:1302.0080 [math-ph], these proceedings.

Dept. of Physics and Dept. of Mathematics, Humboldt University, Unter den Linden 6, 10099 Berlin, Germany

*E-mail address*: kreimer@physik.hu-berlin.de

# Renormalization, Hopf algebras and Mellin transforms

Erik Panzer

ABSTRACT. This article aims to give a short introduction into Hopf-algebraic aspects of renormalization, enjoying growing attention for more than a decade by now. As most available literature is concerned with the minimal subtraction scheme, we like to point out properties of the kinematic subtraction scheme which is also widely used in physics (under the names of MOM or BPHZ).

In particular we relate renormalized Feynman rules $\phi_{\mathrm{R}}$ in this scheme to the universal property of the Hopf algebra $H_R$ of rooted trees, exhibiting a refined renormalization group equation which is equivalent to $\phi_{\mathrm{R}} : H_R \to \mathbb{K}[x]$ being a morphism of Hopf algebras to the polynomials in one indeterminate.

Upon introduction of analytic regularization this results in efficient combinatorial recursions to calculate $\phi_{\mathrm{R}}$ in terms of the Mellin transform. We find that different Feynman rules are related by a distinguished class of Hopf algebra automorphisms of $H_R$ that arise naturally from Hochschild cohomology.

Also we recall the known results for the minimal subtraction scheme and shed light on the interrelationship of both schemes.

Finally we incorporate combinatorial Dyson-Schwinger equations to study the effects of renormalization on the physical meaningful correlation functions. This yields a precise formulation of the equivalence of the two different renormalization prescriptions mentioned before and allows for non-perturbative definitions of quantum field theories in special cases.

## Motivation: The renormalization problem

Suppose we want to assign a value to the logarithmically divergent integral $\phi_s(\bullet) := \int_0^\infty \frac{\mathrm{d}x}{x+s}$, which we associate to the tree $\bullet$. Observing the (absolutely) integrable difference

$$\int_0^\infty \left[\frac{\mathrm{d}x}{x+s} - \frac{\mathrm{d}x}{x+\mu}\right] = -\ln\frac{s}{\mu} =: -\ell \tag{0.1}$$

allows for the definition of $\phi_{\mathrm{R},s}(\bullet) := \phi_s(\bullet) - \phi_\mu(\bullet) = -\ln\frac{s}{\mu} = -\ell$, which we call the *renormalized* value of the expression $\phi_s(\bullet)$. We need to choose the *renormalization point* $\mu$ to fix the constant not determined by (0.1). This natural *renormalization scheme* given by subtraction at a *reference scale* $s \mapsto \mu$ is commonly employed in quantum field theory (where similar divergent expressions occur as we briefly describe in section 7) and will be called *kinematic subtraction scheme* in the sequel.

2010 *Mathematics Subject Classification.* Primary 81T15.

I thank the participants and organizers of the 2012 ICMAT workshop "Periods and Motives" for creating a memorable event and I appreciate very much the opportunity I had to give a talk, which resulted in this review.

When we apply the same idea to multi-dimensional integrals, we have to take care of *subdivergences* as occurring for example in the expression

$$\phi_s\left(\underset{\bullet\;\bullet}{\bullet}\right) := \int_0^\infty \frac{\mathrm{d}x}{x+s}\left[\int_0^\infty \frac{\mathrm{d}y}{x+y}\cdot\int_0^\infty \frac{\mathrm{d}z}{x+z}\right] = \int_0^\infty \frac{\mathrm{d}x}{x+s}\left[\phi_x(\bullet)\right]^2. \tag{0.2}$$

A single subtraction at $s=\mu$ is insufficient as the sub integrals over $y$ and $z$ remain divergent. This problem is circumvented by applying renormalization to these first:

$$\begin{aligned}\phi_{\mathrm{R},s}\left(\underset{\bullet\;\bullet}{\bullet}\right) &:= \int_0^\infty \left[\frac{\mathrm{d}x}{x+s}-\frac{\mathrm{d}x}{x+\mu}\right]\left\{\int_0^\infty\left[\frac{\mathrm{d}y}{x+y}-\frac{\mathrm{d}y}{\mu+y}\right]\cdot\int_0^\infty\left[\frac{\mathrm{d}z}{x+z}-\frac{\mathrm{d}z}{\mu+z}\right]\right\}\\ &= \int_0^\infty\left[\frac{\mathrm{d}x}{x+s}-\frac{\mathrm{d}x}{x+\mu}\right]\left[\phi_{\mathrm{R},x}(\bullet)\right]^2 = \int_0^\infty\left[\frac{\mathrm{d}x}{x+s}-\frac{\mathrm{d}x}{x+\mu}\right]\left(\ln\frac{x}{\mu}\right)^2 = -\frac{\ell^3}{3}-\frac{\pi^2}{3}\ell.\end{aligned} \tag{0.3}$$

We want to summarize how this procedure is formulated in terms of Hopf algebras, study under which conditions it can be applied and reveal the main properties of the resulting maps $\phi_{\mathrm{R},s}$. In particular we will show that they are morphisms of Hopf algebras, taking values in the polynomials in $\ell$.

For a quick start, we prove this analytically in section 2, along the ideas [**26**] originating from quantum field theory. Section 3 exploits an artificial regulator to rederive the same results in a setup more common to the original literature dealing with dimensional regularization and minimal subtraction. Along this way we take the time to recall the common algebraic techniques and contrast both methods.

After this construction of renormalized Feynman rules, we study their algebraic properties in section 4 focusing on the renormalization group. Together with the Mellin transform we can derive compact recursion relations, allowing for efficient combinatorial calculations.

At this point we turn towards the *minimal subtraction scheme* in section 5. We summarize the known results and particularly relate the different realizations of the renormalization group equations in the two schemes, developing the duality between the concepts of *finiteness* in the subtraction scheme and *locality* in minimal subtraction.

Section 6 is devoted to Dyson-Schwinger equations, which link the combinatorics of the Hopf algebra to the physically meaningful correlation functions. In particular we observe how the change of renormalization scheme is equivalent to a redefinition of the coupling constant, proving the renormalization group equation in its physical form.

Finally we comment on the necessary modifications for generalizations of the model in different directions, like the presence of multiple parameters or higher degrees of divergence.

For reference and convenience of the reader, we collected the required features of the Hopf algebras $H_R$ of rooted trees and $\mathbb{K}[x]$ of polynomials in the appendix. We also added a collection of well-known results on the *Dynkin operator* $S\star Y$ which plays a prominent role in sections 3 and 5 when we use a regulator.

## 1. Notations and preliminaries

The essential structure behind perturbative renormalization is the Hopf algebra as discovered in [**23**]. As the literature grew comprehensive already, we content ourselves with fixing notation and recommend [**30, 32**] for extended accounts of these concepts with a particular focus on their application to renormalization.

**1.1. Hopf algebras.** Throughout we consider associative, coassociative, commutative, unital and counital Bialgebras $(H, m, u, \Delta, \varepsilon)$ given a connected ($H_0 = \mathbb{K} \cdot \mathbb{1}$) grading $H = \bigoplus_{n \geq 0} H_n$. For homogeneous $0 \neq x \in H_n$, write $|x| := n$ while the induced *grading operator* $Y \in \operatorname{End}(H), x \mapsto Yx := |x| \cdot x$ exponentiates to the one-parameter group $\mathbb{K} \ni t \mapsto \theta_t$ of Hopf algebra automorphisms

$$(1.1)\quad \theta_t := \exp(tY) = \sum_{n \in \mathbb{N}_0} \frac{(tY)^n}{n!}, \quad \forall n \in \mathbb{N}_0: \quad H_n \ni x \mapsto \theta_t(x) = e^{t|x|}x = e^{nt}x.$$

Algebras $(\mathcal{A}, m_{\mathcal{A}}, u_{\mathcal{A}})$ are unital, associative and commutative, giving rise to the associative *convolution product* on $\operatorname{Hom}(H, \mathcal{A})$ with unit given by $e := u_{\mathcal{A}} \circ \varepsilon$:

$$\operatorname{Hom}(H, \mathcal{A}) \ni \phi, \psi \mapsto \phi \star \psi := m_{\mathcal{A}} \circ (\phi \otimes \psi) \circ \Delta \in \operatorname{Hom}(H, \mathcal{A}).$$

As $H = \mathbb{K} \cdot \mathbb{1} \oplus \ker \varepsilon = \operatorname{im} u \oplus \ker \varepsilon$ splits into the scalars and the *augmentation ideal* $\ker \varepsilon$, we obtain a projection $P := \operatorname{id} - u \circ \varepsilon\colon\ H \twoheadrightarrow \ker \varepsilon$ and use Sweedler's [**33**] notation $\Delta(x) = \sum_x x_1 \otimes x_2$ and $\widetilde{\Delta}(x) = \sum_x x' \otimes x''$ to abbreviate the reduced coproduct $\widetilde{\Delta} := \Delta - \mathbb{1} \otimes \operatorname{id} - \operatorname{id} \otimes \mathbb{1}$. The connectedness implies:

(1) Under $\star$, the *characters* (morphisms of unital algebras) form a group $G^H_{\mathcal{A}} := \{\phi \in \operatorname{Hom}(H, \mathcal{A})\colon\ \phi \circ u = u_{\mathcal{A}} \text{ and } \phi \circ m = m_{\mathcal{A}} \circ (\phi \otimes \phi)\}$.
(2) These biject along $\exp_\star\colon\ \mathfrak{g}^H_{\mathcal{A}} \to G^H_{\mathcal{A}}$ with inverse $\log_\star\colon\ G^H_{\mathcal{A}} \to \mathfrak{g}^H_{\mathcal{A}}$ to the *infinitesimal characters* $\mathfrak{g}^H_{\mathcal{A}} := \{\phi \in \operatorname{Hom}(H, \mathcal{A})\colon\ \phi \circ m = \phi \otimes e + e \otimes \phi\}$, using the pointwise finite series

$$(1.2)\qquad \exp_\star(\phi) := \sum_{n \in \mathbb{N}_0} \frac{\phi^{\star n}}{n!} \quad \text{and} \quad \log_\star(\phi) := \sum_{n \in \mathbb{N}} \frac{(-1)^{n+1}}{n} (\phi - e)^{\star n}.$$

(3) The unique inverse $S := \operatorname{id}^{\star -1} \in G^H_H$ is called *antipode* and reveals $H$ as Hopf algebra. For all characters $\phi \in G^H_{\mathcal{A}}$ we have $\phi^{\star -1} = \phi \circ S$.

In general we assume the ground field $\mathbb{K}$ to be $\mathbb{R}$ or $\mathbb{C}$, though the reader will easily recognize that the majority of results allows for more generality (often characteristic zero suffices). Note that by $\operatorname{Hom}(\cdot, \cdot)$ and $\operatorname{End}(\cdot)$ we always denote $\mathbb{K}$-linear maps and explicitly spell out if more structure should enjoy preservation by a morphism. Finally, we write $\operatorname{lin} M$ for the linear span of $M$.

**1.2. Hochschild cohomology.** The Hochschild cochain complex [**1, 8, 32**] we associate to $H$ contains the functionals $H' = \operatorname{Hom}(H, \mathbb{K})$ as zero-cochains. We will only consider one-cocycles $L \in \mathrm{HZ}^1_\varepsilon(H) \subset \operatorname{End}(H)$ which are defined to solve $\Delta \circ L = (\operatorname{id} \otimes L) \circ \Delta + L \otimes \mathbb{1}$. The differential of this complex then becomes

$$(1.3)\qquad \delta : H' \to \mathrm{HZ}^1_\varepsilon(H), \alpha \mapsto \delta\alpha := (\operatorname{id} \otimes \alpha) \circ \Delta - u \circ \alpha \in \mathrm{HB}^1_\varepsilon(H) := \delta(H')$$

and defines the coboundaries $\mathrm{HB}^1_\varepsilon(H)$ and thus the first cohomology group given by $\mathrm{HH}^1_\varepsilon(H) := \mathrm{HZ}^1_\varepsilon(H)/\mathrm{HB}^1_\varepsilon(H)$. Note the elementary

LEMMA 1.1. *Cocycles $L \in HZ^1_\varepsilon(H)$ map into the augmentation ideal* $\operatorname{im} L \subseteq \ker \varepsilon$ *and* $L(\mathbb{1}) \in \operatorname{Prim}(H) := \ker \widetilde{\Delta}$ *is primitive. The map* $HH^1_\varepsilon(H) \to \operatorname{Prim}(H)$, $[L] = L + HB^1_\varepsilon(H) \mapsto L(\mathbb{1})$ *is well-defined since* $\delta\alpha(\mathbb{1}) = 0$ *for all* $\alpha \in H'$.

## 2. Finiteness of renormalization by kinematic subtraction

Originally, perturbative quantum field theory assigns (divergent) expressions to combinatorial objects called *Feynman graphs*, as we will comment on in section 7. However the Hopf algebra $H_R$ of rooted trees summarized in appendix A suffices to encode the structure of subdivergences [**8**, **9**, **23**] such that we can focus on *Feynman rules* of the form $\phi : H_R \to \mathcal{A}$ as above. The *target algebra* $\mathcal{A}$ has to sustain divergent expressions which only become finite after we accomplished the renormalization. Therefore we consider $\mathcal{A}$ as the integrands (differential forms) which for convenience we nevertheless write as integrals, keeping in mind that we do not evaluate them.

Guided by the examples (0.1) and (0.2) we make

DEFINITION 2.1. *By virtue of* A.3 *let* $\phi \in G_{\mathcal{A}}^{H_R}$ *be the character fixed through*

$$\phi_s\left(B_+(w)\right) := \int \frac{\mathrm{d}\zeta}{s} f\left(\frac{\zeta}{s}\right) \phi_\zeta(w) \quad \textit{for any forest} \quad w \in H_R, \tag{2.1}$$

*where the function* $f$ *generalizes our choice of* $f(\zeta) = \frac{1}{1+\zeta}$ *in the introduction and will be dictated by the Feynman rules in a physical application. Excluding infrared divergences by requiring* $f$ *to be bounded on* $[0, \infty)$, *we restrict our study to* ultraviolet[1] *divergences at* $\zeta \to \infty$.

Observe how each node of a tree corresponds to an integration of $f$ times the function $\phi_\zeta(w)$ given by its children, so (2.1) ensures that all information about subdivergences of these Feynman rules $\phi$ is encoded in the coproduct of $H_R$.

EXAMPLE 2.2. *Looking at* (0.2), $\widetilde{\Delta}\left(\boldsymbol{\Lambda}\right) = 2\bullet \otimes \mathbf{I} + \bullet\bullet \otimes \bullet$ *informs us about:*

(1) *Two individual subdivergences* $\int \frac{\mathrm{d}y}{x+y} = \phi_x(\bullet) = \int \frac{\mathrm{d}z}{x+z}$ *of the type* $\phi(\bullet)$ *nested inside the outer integrals* $\int_0^\infty \frac{\mathrm{d}x}{x+s} \int_0^\infty \frac{\mathrm{d}z}{x+z}$ *and* $\int_0^\infty \frac{\mathrm{d}x}{x+s} \int_0^\infty \frac{\mathrm{d}y}{x+y}$ *respectively, each being of type* $\phi\left(\mathbf{I}\right)$.

(2) *One subdivergence* $\int \frac{\mathrm{d}y}{x+y} \cdot \int \frac{\mathrm{d}z}{x+z}$ *of the kind* $\phi(\bullet\bullet)$ *(when* $y$ *and* $z$ *approach* $\infty$ *jointly) inside the outer integration* $\phi_s(\bullet) = \int_0^\infty \frac{\mathrm{d}x}{x+s}$.

**2.1. Subtraction scheme.** Note that the integrands $\phi$ depend on a free *external parameter* $s > 0$ which models a physical mass or momentum. Our goal is to replace $\phi$ by integrable integrands to achieve well-defined functions of $s$. As exemplified in the introduction, this can be achieved by subtracting the integrand at $s \mapsto \mu$. Renormalizing the subdivergences first as in (0.3) motivates

DEFINITION 2.3. *Fixing a renormalization point* $\mu > 0$ *we define the character* $\phi_R : H_R \to \mathcal{A}$ *(again as an instance of* A.3*) by requiring*

$$\phi_{R,s}(B_+(w)) := \int \mathrm{d}\zeta \left[\frac{f(\frac{\zeta}{s})}{s} - \frac{f(\frac{\zeta}{\mu})}{\mu}\right] \phi_{R,\zeta}(w) \quad \textit{for any} \quad w \in H_R. \tag{2.2}$$

To actually achieve finiteness this way we need to constrain the growth of $f(\zeta)$ at $\zeta \to \infty$ to be not worse than $\zeta^{-1}$, corresponding to a *logarithmic divergence* in

[1] physically $\zeta$ corresponds to a momentum, so the limit $\zeta \to \infty$ means high energies

THEOREM 2.4. *Let $f$ be a bounded, piecewise differentiable function on $[0,\infty)$ with $f(\zeta) \in \mathcal{O}(\zeta^{-1})$, so in particular $c_{-1} := \lim_{\zeta\to\infty}[\zeta f(\zeta)]$ exists. If furthermore*

$$f(\zeta) - \frac{c_{-1}}{\zeta},\, f(\zeta) + \zeta f'(\zeta) \in \mathcal{O}\left(\zeta^{-1-\varepsilon}\right) \quad \textit{for some} \quad \varepsilon > 0, \tag{2.3}$$

*then for any forest $w \in H_R$ the integral $\phi_{R,s}(w)$ is absolutely convergent and evaluates to a polynomial $\phi_{R,s}(w) \in \mathbb{K}[\ell]$ in $\ell := \ln\frac{s}{\mu}$.*

We remark that these requirements on $f$ could be relaxed[2], but they are already fulfilled for physical cases of logarithmic divergent Feynman graphs[3]. In particular (2.3) holds for all rational functions $f \in \mathcal{O}(\zeta^{-1}) \cap \mathbb{K}(\zeta)$ without poles in $[0,\infty)$.

PROOF. By definition 2.3 we have $\phi_R(\mathbb{1}) = 1$ and $\phi_R(ww') = \phi_R(w)\cdot\phi_R(w')$ and can therefore proceed inductively. We may thus assume the statement to be true for some element $w \in H_R$ and only need to prove it for the tree $t = B_+(w)$. But then the difference in brackets in (2.2) falls of like $\zeta^{-1-\varepsilon}$ from (2.3) while $\phi_{R,\zeta}(w)$ only grows like $\ln^N\zeta$ for the degree $N$ of $\phi_R(w)$. Hence (2.2) is absolutely convergent (the logarithmic singularities $\ln^N\zeta$ at $\zeta\to 0$ are integrable anyway) and thus $\phi_{R,s}(B_+(w))$ finite.

By (2.3) we can also interchange integration with the partial derivative $\partial_\ell$ in

$$-\partial_\ell\phi_{R,s}(t) = \int_0^\infty \mathrm{d}\zeta \left[\frac{f(\frac{\zeta}{s})}{s} + \frac{\zeta}{s}\frac{f'(\frac{\zeta}{s})}{s}\right]\phi_{R,\zeta}(w) = \int_0^\infty \mathrm{d}\zeta\,[f(\zeta)+\zeta f'(\zeta)]\,\phi_{R,\zeta s}(w).$$

Exploiting that $\phi_{R,\zeta s}(w)$ is polynomial in $\ln\frac{\zeta s}{\mu} = \ln\zeta + \ell$ we can evaluate

$$\int_0^\infty \mathrm{d}\zeta\,[f(\zeta)+\zeta f'(\zeta)]\,(\ln\zeta+\ell)^n = \sum_{i=0}^{n}\binom{n}{i}\ell^{n-i}c_{i-1}(-1)^i i! = \sum_{i=0}^{\infty} c_{i-1}\,(-\partial_\ell)^i\,\ell^n \tag{2.4}$$

upon defining the constants (which are *periods* [**20**] for algebraic functions $f$)

$$c_{n-1} := \int_0^\infty \mathrm{d}\zeta\,[f(\zeta)+\zeta f'(\zeta)]\,\frac{(-\ln\zeta)^n}{n!} \qquad \text{for any} \qquad n\in\mathbb{N}_0. \tag{2.5}$$

Thus linearity shows $\partial_{-\ell}\phi_{R,s}(t) \in \mathbb{K}[\ell]$ and we merely have to integrate once. □

Not only did we achieve our goal of renormalization, but we found an explicit recursion (2.4) determining $\phi_R$ completely using the universal property A.3 in

COROLLARY 2.5. *The constants $c_\cdot$ of (2.5) determine the renormalized Feynman rules $\phi_R \in G^{H_R}_{\mathbb{K}[\ell]}$ completely through the universal property* A.3 *by*

$$\phi_R \circ B_+ = P\circ F(-\partial_\ell)\circ\phi_R, \quad \textit{where} \quad F(-\partial_\ell) := \sum_{n\geq -1} c_n(-\partial_\ell)^n \in \mathrm{End}\left(\mathbb{K}[\ell]\right). \tag{2.6}$$

*For convenience we write here $(-\partial_\ell)^{-1} := -\int_0$ for the integral operator. Recall that the projection $P: \mathbb{K}[\ell] \twoheadrightarrow \ell\,\mathbb{K}[\ell]$ annihilates any constants.*

---

[2]for example, $f(\zeta)+\zeta f'(\zeta) \in \mathcal{O}(\zeta^{-1-\varepsilon})$ already follows from $f(\zeta) - \frac{c_{-1}}{\zeta} \in \mathcal{O}(\zeta^{-1-\varepsilon})$ using the theorem of L'Hôpital

[3]these include the vertex graphs of QED and $\varphi^4$-theory in four dimensions of space-time

Example 3.9 explicitly shows how this recursion works in detail.

In section 4 we will see that (2.6) implies the *renormalization group* upon realizing that $P \circ F(-\partial_\ell) \in \mathrm{HZ}^1_\varepsilon(\mathbb{K}[\ell])$ is a Hochschild-1-cocycle. But before let us review the

**2.2. Algebraic renormalization process.** Renormalization of a character $\phi \in G^H_{\mathcal{A}}$ can be described as a *Birkhoff decomposition* into the *renormalized* $\phi_{\mathrm{R}} := \phi_+ \in G^H_{\mathcal{A}}$ and the *counterterms* $\phi_- \in G^H_{\mathcal{A}}$ subject to the conditions that

$$\phi = \phi_-^{\star -1} \star \phi_+ \quad \text{and} \quad \phi_\pm(\ker \varepsilon) \subseteq \mathcal{A}_\pm. \tag{2.7}$$

It depends on a splitting $\mathcal{A} = \mathcal{A}_+ \oplus \mathcal{A}_-$ of the target algebra, determining the *renormalization scheme* which we identify with the corresponding projection $R\colon \mathcal{A} \twoheadrightarrow \mathcal{A}_-$ along $\mathcal{A}_+$.

THEOREM 2.6 ([**9**,**30**,**32**]). *A unique Birkhoff decomposition* (2.7) *exists given that $R$ is a Rota-Baxter map, meaning*

$$m \circ (R \otimes R) = R \circ m \circ [R \otimes \mathrm{id} + \mathrm{id} \otimes R - \mathrm{id} \otimes \mathrm{id}]. \tag{2.8}$$

*On the augmentation ideal* $\ker \varepsilon$ *it may be computed inductively by*

$$\phi_-(x) = -R \circ \bar{\phi}(x) \qquad \textit{and} \qquad \phi_+(x) = (\mathrm{id} - R) \circ \bar{\phi}(x), \tag{2.9}$$

*using the* Bogoliubov character $\bar{\phi}$ *(also* $\bar{R}$-operation*) which is defined as*

$$\bar{\phi}(x) := \phi(x) + \sum_x \phi_-(x')\phi(x'') = \phi_+(x) - \phi_-(x). \tag{2.10}$$

DEFINITION 2.7. *The* kinematic subtraction scheme $R_\mu$ *by evaluation at* $s \mapsto \mu$ *is defined as*

$$\mathrm{End}(\mathcal{A}) \ni R_\mu := \mathrm{ev}_\mu = \left(\mathcal{A} \ni f \mapsto f|_{s=\mu}\right) \tag{2.11}$$

*and splits $\mathcal{A}$ into* $\operatorname{im} R_\mu = \mathcal{A}_-$ *($s$-independent integrals) and* $\ker R_\mu = \mathcal{A}_+$*, those integrals that vanish at $s = \mu$.*

As $R_\mu$ is a character of $\mathcal{A}$, it not only fulfills (2.8) and we obtain a unique Birkhoff decomposition, but also the recursion (2.9) simplifies a lot to just

$$\phi_- = R_\mu \circ \phi \circ S = \phi_\mu \circ S = \phi_\mu^{\star -1} \qquad \text{and} \qquad \phi_+ = \phi_\mu^{\star -1} \star \phi_s. \tag{2.12}$$

EXAMPLE 2.8. *In accordance with* (0.1) *we find*

$$\phi_{R,s}(\bullet) = \phi_{+,s}(\bullet) = \phi_-(\bullet) + \phi_s(\bullet) = \int_0^\infty \left[ -\left.\frac{\mathrm{d}x}{x+s}\right|_{s \mapsto \mu} + \frac{\mathrm{d}x}{x+s} \right],$$

*and* $\bar{\phi}(\Lambda) \underset{(2.9)}{=} \phi_s(\Lambda) + 2\phi_-(\bullet)\,\phi_s(:) + \phi_-(\bullet\bullet)\,\phi_s(\bullet)$ *indeed agrees with* (0.3) *using* $\phi_R(\Lambda) = (\mathrm{id} - R_\mu)\bar{\phi}(\Lambda)$ *after rearranging the terms*[4]

$$\bar{\phi}(\Lambda) = \int_0^\infty \mathrm{d}x \int_0^\infty \mathrm{d}y \int_0^\infty \mathrm{d}z \left[ \frac{1}{(s+x)(x+y)(x+z)} - \frac{1}{\mu+y}\frac{1}{(s+x)(x+z)} - \frac{1}{\mu+z}\frac{1}{(s+x)(x+y)} + \frac{1}{\mu+y}\frac{1}{\mu+z}\frac{1}{s+x} \right].$$

[4]Note that we need to track the correspondence of variables and nodes.

We remark that the recursion (2.9) makes explicit reference to the divergent counterterms $\phi_-$. In (2.2) we anticipated the much more practical formula resulting from the special structure A.3 of the Feynman rules $\phi$ of (2.1) in

THEOREM 2.9. *Let the character* $\phi : H_R \to \mathcal{A}$ *be subject to* $\phi \circ B_+ = L \circ \phi$ *for some* $L \in \mathrm{End}(\mathcal{A})$ *and the renormalization scheme* $R \in \mathrm{End}(\mathcal{A})$ *such that it ensures*

$$L \circ m_{\mathcal{A}} \circ (\phi_- \otimes \mathrm{id}) = m_{\mathcal{A}} \circ (\phi_- \otimes L), \tag{2.13}$$

*which means linearity of* $L$ *over the counterterms. Then we have*

$$\bar{\phi} \circ B_+ = L \circ \phi_+ \qquad \text{and therefore} \qquad \phi_+ \circ B_+ = (\mathrm{id} - R) \circ L \circ \phi_+ \tag{2.14}$$

PROOF. This is a straightforward consequence of the cocycle property of $B_+$:

$$\begin{aligned}\bar{\phi} \circ B_+ &= (\phi_- \star \phi - \phi_-) \circ B_+ = m_{\mathcal{A}} \circ (\phi_- \otimes \phi) \circ [(\mathrm{id} \otimes B_+) \circ \Delta + B_+ \otimes \mathbb{1}] - \phi_- \circ B_+ \\ &= \phi_- \star (\phi \circ B_+) = \phi_- \star (L \circ \phi) \underset{(2.13)}{=} L \circ (\phi_- \star \phi) = L \circ \phi_+.\end{aligned}$$

□

As for $R_\mu$ the counterterms $\phi_-(x) \in \mathcal{A}_-$ are independent of $s$, they separate from the integration in (2.1) and (2.13) is fulfilled indeed. This is a general feature of quantum field theories: The counterterms to not depend on any external variables[5].

The significance of (2.14) lies in the expression of the renormalized $\phi_+(t)$ for a tree $t = B_+(w)$ only in terms of the renormalized value $\phi_+(w)$. This allows for inductive proofs like 2.4 on properties of $\phi_{\mathrm{R}} = \phi_+$, without having to consider the unrenormalized Feynman rules or their counterterms (both of which are divergent) at all.

Summarizing, we proved in 2.4 that for any forest $w \in H_R$, the expression $\phi_+(w) \in \mathcal{A}_+$ is actually integrable and may be directly written as a convergent integral using (2.2).

## 3. Regularization and Mellin transforms

A technique often applied prior the renormalization is the introduction of a *regulator* to assign finite values also to divergent expressions. Popular methods usually either alter the domain of integration:

(1) Confine integrations to the bounded interval $[0, \Lambda]$ for a *cut-off* $\Lambda > 0$. Then all integrals converge but acquire a dependence on $\Lambda$, which will in general diverge in the *physical limit* $\Lambda \to \infty$ resembling the original situation. After renormalization however, this limit will be finite.
(2) Variations of mixed Hodge structures [**3**] also vary the chain of integration to avoid singularities.

or modify the integrand:

(3) Choose an *analytic regulator* $0 < z < 1$ and replace each $\int_0^\infty \mathrm{d}x$ with $\int_0^\infty x^{-z}\mathrm{d}x$. This increases the decay of the integrand at $x \to \infty$ and we again get finite results which depend on $z$. As for the cut-off, these typically diverge in the physical limit $z \to 0$, unless we renormalize.

[5]Even if the divergence of a Feynman graph does depend on external momenta as happens for higher degrees of divergence, this dependence is only polynomial and stripped off by extracting the individual coefficients. In the Hopf algebra this can be encoded with *external structures* which are given by distributions in [**9**]. So in any case, $\phi_-$ maps to constants.

(4) *Dimensional regularization* is similar in introducing a complex parameter $z \neq 0$ associated to a shifted dimension $D = 4 - 2z$ of space-time. It is tailor made for Feynman integrals in quantum field theory and we refer to [**7**] for its definition and examples.

We study the analytic regularization in detail, as it allows for the simplest algebraic description: Due to the regulator all integrals converge and give functions of both $s$ and $z$ that lie in the target algebra $\mathcal{A} = \mathbb{K}[z^{-1}, z]][s^{-z}]$ of Laurent series in $z$ as we shall see in proposition 3.2.

DEFINITION 3.1. *The analytically regularized Feynman rules* ${}_z\phi \in G_{\mathcal{A}}^{H_R}$ *are given through the universal property* A.3 *by requiring*

$$ {}_z\phi_s \circ B_+ = \int_0^\infty \frac{f(\frac{\zeta}{s})\zeta^{-z}}{s}\, {}_z\phi_\zeta \, \mathrm{d}\zeta = \int_0^\infty f(\zeta)(s\zeta)^{-z}\, {}_z\phi_{s\zeta}\, \mathrm{d}\zeta. \tag{3.1}$$

All these integrals can conveniently be evaluated in terms of the coefficients $c_n$ of the *Mellin transform*[6]

$$F(z) := \int_0^\infty f(\zeta)\zeta^{-z}\, \mathrm{d}\zeta = \sum_{n=-1}^{\infty} c_n z^n \in z^{-1}\mathbb{K}[[z]], \tag{3.2}$$

which we already encountered in (2.6): Indeed, a partial integration proves that

$$\begin{aligned} c_{n-1} n! &\underset{(2.5)}{=} \int_0^\infty \mathrm{d}\zeta \left[f(\zeta) + \zeta f'(\zeta)\right] (-\ln\zeta)^n = \left.\frac{\partial^n}{\partial z^n}\right|_{z=0} \int_0^\infty \mathrm{d}\zeta \left[f(\zeta) + \zeta f'(\zeta)\right] \zeta^{-z} \\ &= \left.\frac{\partial^n}{\partial z^n}\right|_{z=0} \left\{ \left[f(\zeta)\zeta^{1-z}\right]_{\zeta=0}^{\infty} + \int_0^\infty \mathrm{d}\zeta \left[f(\zeta) - (1-z) f(\zeta)\right]\zeta^{-z} \right\} = \left.\frac{\partial^n}{\partial z^n}\right|_{z=0} \{zF(z)\}. \end{aligned}$$

PROPOSITION 3.2. *For any forest* $w \in \mathcal{F}$ *we have (called* BPHZ model *in* [**4**]*)*

$$ {}_z\phi_s(w) = s^{-z|w|} \prod_{v \in V(w)} F(z\, |w_v|). \tag{3.3}$$

PROOF. As both sides of (3.3) are clearly multiplicative, it is enough to inductively assume the claim for a forest $w \in \mathcal{F}$ and prove it for the tree $t = B_+(w)$:

$$\begin{aligned} {}_z\phi_s(t) &\underset{(3.1)}{=} \int_0^\infty (s\zeta)^{-z} f(\zeta)\, {}_z\phi_{s\zeta}(w)\, \mathrm{d}\zeta = \int_0^\infty (s\zeta)^{-z} f(\zeta) (s\zeta)^{-z|w|} \prod_{v\in V(w)} F(z\,|w_v|)\, \mathrm{d}\zeta \\ &= s^{-z|B_+(w)|} \left[ \prod_{v \in V(w)} F(z\,|w_v|) \right] F(z\,|B_+(w)|) = s^{-z|t|} \prod_{v\in V(t)} F(z\,|t_v|). \qquad \square \end{aligned}$$

EXAMPLE 3.3. *Using* (3.3), *we can directly write down the Feynman rules like* ${}_z\phi_s(\bullet) = s^{-z}F(z)$, ${}_z\phi_s$ () $= s^{-2z}F(z)F(2z)$ *and* ${}_z\phi_s$ () $= s^{-3z}[F(z)]^2F(3z)$.

Many examples (choices of $F$) are discussed in [**4**], the particular case of the one-loop propagator graph $\gamma$ of Yukawa theory is in [**5**] and for scalar Yukawa theory in six dimensions one has $F(z) = \frac{1}{z(1-z)(2-z)(3-z)}$ as in [**32**]. Already noted in [**24**], the highest order pole of ${}_z\phi_s(w)$ is independent of $s$ and just the tree factorial

$$ {}_z\phi_s(w) \in s^{-z|w|} \prod_{v\in V(w)} \left\{ \tfrac{c_{-1}}{z|w_v|} + \mathbb{K}[[z]] \right\} \underset{(A.4)}{\subset} \frac{1}{w!} \left(\tfrac{c_{-1}}{z}\right)^{|w|} + z^{1-|w|}\mathbb{K}[[z]]. \tag{3.4}$$

---

[6] Conditions (2.3) suffice to prove that $F(z)$ is a Laurent series of this form.

**3.1. Finiteness.** Using (3.3) and (2.12) we can quickly write down explicitly the values of the renormalized Feynman rules like in

EXAMPLE 3.4. *We find* ${}_z\phi_{R,s}(\bullet) = (s^{-z} - \mu^{-z}) F(z)$ *and* $S\left(\overset{\bullet}{\underset{\bullet}{|}}\right) = -\overset{\bullet}{\underset{\bullet}{|}} + \bullet\bullet$ *gives*

$$ {}_z\phi_{R,s}\left(\overset{\bullet}{\underset{\bullet}{|}}\right) = \left(s^{-2z} - \mu^{-2z}\right) F(z)F(2z) - \left(s^{-z} - \mu^{-z}\right)\mu^{-z}F^2(z). \tag{3.5}$$

As the *physical limit* $z \to 0$ reconstructs the original (unregularized) Feynman rules (2.1), the finiteness of theorem 2.4 is equivalent (by Lebesgue's theorem on dominated convergence) to the existence of the limit

$$\phi_{\mathrm{R}} := \lim_{z\to 0} {}_z\phi_R. \tag{3.6}$$

COROLLARY 3.5. *The renormalized regularized Feynman rules are holomorphic, that is they map into* $\operatorname{im}\left({}_z\phi_{R,s}\right) \subset \mathbb{K}[[z]]$.

EXAMPLE 3.6. *Indeed we find* ${}_z\phi_{R,s}(\bullet) \in -c_{-1}\ln\frac{s}{\mu} + z\mathbb{K}[[z]]$. *For (3.5) check*

$$\phi_R\left(\overset{\bullet}{\underset{\bullet}{|}}\right) \underset{(3.6)}{=} \lim_{z\to 0}\left\{-\left[-z\ln\tfrac{s}{\mu} + \tfrac{z^2}{2}\left(\ln^2 s + 2\ln s\ln\mu - 3\ln^2\mu\right)\right]\cdot\left[\tfrac{c_{-1}^2}{z^2} + 2\tfrac{c_{-1}c_0}{z}\right]\right.$$
$$\left. + \left[-2z\ln\tfrac{s}{\mu} + 2z^2\left(\ln^2 s - \ln^2\mu\right)\right]\cdot\left[\tfrac{c_{-1}^2}{2z^2} + \tfrac{3c_0c_{-1}}{2z}\right]\right\} = \frac{c_{-1}^2}{2}\ln^2\tfrac{s}{\mu} - c_{-1}c_0\ln\tfrac{s}{\mu},$$

*where all poles in $z$ perfectly cancel.*

Observe that we proved the by now purely combinatorial statement 3.5 of the cancellation of all pole terms in ${}_z\phi_R$ analytically by estimates on the asymptotic growths in theorem 2.4. As we absorbed all analytic input of the integrands in $F(z) \in z^{-1}\mathbb{K}[[z]]$ in the series (3.3) we can also give a completely combinatorial proof as we shall do in lemma 3.8.

Note that the analytic regularization yields a very simple dependence on the parameter $s$: Setting now $\mathcal{A} := \mathbb{C}[z^{-1}, z]]$ and ${}_z\phi := {}_z\phi_1 = {}_z\phi|_{s=1} \in G_{\mathcal{A}}^{H_R}$, (3.3) fixes the scale dependence ${}_z\phi_s = {}_z\phi \circ \theta_{-z\ln s}$ completely through the grading, see also [**11, 30**]. Therefore we can write

$$ {}_z\phi_{R,s} = {}_z\phi_\mu^{\star -1} \star {}_z\phi_s = {}_z\phi \circ \left[(S\circ\theta_{-z\ln\mu})\star\theta_{-z\ln s}\right] = {}_z\phi\circ\left(S\star\theta_{-z\ln\frac{s}{\mu}}\right)\circ\theta_{-z\ln\mu} \tag{3.7}$$

and characterize the finiteness of the physical limit (3.6) in

PROPOSITION 3.7. *For any character* ${}_z\phi \in G_{\mathcal{A}}^{H_R}$, *the following are equivalent:*

(1) *The physical limit* $\phi_R := \lim_{z\to 0} {}_z\phi_R$ *exists*
(2) *For any* $\ell \in \mathbb{K}$, ${}_z\phi^{\star -1} \star ({}_z\phi\circ\theta_{-\ell z}) = {}_z\phi\circ(S\star\theta_{-\ell z})$ *maps into* $\mathbb{C}[[z]]$.
(3) *For every* $n \in \mathbb{N}_0$, ${}_z\phi^{\star -1}\star({}_z\phi\circ Y^n) = {}_z\phi\circ(S\star Y^n)$ *maps into* $z^{-n}\mathbb{C}[[z]]$.
(4) ${}_z\phi^{\star -1}\star({}_z\phi\circ Y) = {}_z\phi\circ(S\star Y)$ *maps into* $\frac{1}{z}\mathbb{C}[[z]]$, *equivalently the limit* $\lim\limits_{z\to 0} {}_z\phi^{\star -1}\star({}_z\phi\circ zY)$ *exists.*

PROOF. From (3.7), (1) $\Leftrightarrow$ (2) is just composition with the holomorphic $\theta_{-z\ln\mu}$ or $\theta_{z\ln\mu} = \theta_{-z\ln\mu}^{-1}$ while (2) $\Leftrightarrow$ (3) merely expands $\theta_{-\ell z} = \sum_{n\geq 0}\frac{(-\ell zY)^n}{n!}$. It remains to prove (4) $\Rightarrow$ (3) inductively with

$$ {}_z\phi\circ\left(S\star Y^{n+1}\right) = {}_z\phi\circ(S\star Y^n)\circ Y + \left[{}_z\phi\circ(S\star Y)\right]\star\left[{}_z\phi\circ(S\star Y^n)\right],$$

exploiting $(S\circ Y)\star\mathrm{id} = -S\star Y$ in the formula ($\alpha$ arbitrary)

$$S\star(\alpha\circ Y) - (S\star\alpha)\circ Y = -(S\circ Y)\star\alpha = -\left[(S\circ Y)\star\mathrm{id}\right]\star S\star\alpha = S\star Y\star S\star\alpha. \qquad \square$$

LEMMA 3.8. *Let* ${}_z\phi \in G_{\mathcal{A}}^{H_R}$ *be the character defined by* (A.2) *with*

$$(3.8)\quad {}_z\phi \circ B_+(w) = {}_z\phi(w) \cdot F(z\,|B_+(w)|) \qquad \textit{for any fixed} \qquad F(z) \in z^{-1}\mathbb{K}[[z]].$$

*Then* ${}_z\phi$ *fulfills the conditions of proposition* 3.7. *In particular,* $\operatorname{im}({}_z\phi_R) \subseteq \mathbb{K}[[z]]$ *allows the finite physical limit* $\phi_R = \lim_{z\to 0} {}_z\phi_R \subseteq \mathbb{K}[\ell, c_\cdot]$ *taking values in the polynomials in* $\ell = \ln\frac{s}{\mu}$ *and the coefficients* $c_n$ *of the series* $F(z)$.

PROOF. We show (2) of 3.7 inductively along the grading of $H_R$. So let it be true on $H_{R,m}$, then by the multiplicativity of ${}_z\phi \circ (S \star \theta_{-z\ell})$ it holds for all products in $H_{R,m+1}$ and we only need consider trees $t = B_+(w)$ for some $w \in H_{R,m}$. For any $k \in \mathbb{N}$ observe holomorphy of $\partial^k_{-\ell}\big|_{\ell=0}{}_z\phi \circ (S \star \theta_{-z\ell})$ through

$$\begin{aligned}
{}_z\phi \circ (S \star [zY]^k)(t) &\underset{(A.1)}{=} {}_z\phi \circ \left\{S \star ([zY]^k \circ B_+)\right\}(w) \\
&= {}_z\phi^{\star -1} \star ({}_z\phi \circ B_+ \circ [z(Y + \mathrm{id})]^{k+1})(w) \\
&\underset{(3.8)}{=} \sum_{n\geq -1} c_n \cdot {}_z\phi \circ \left\{S \star [z(Y+\mathrm{id})]^{n+k}\right\}(w)
\end{aligned}$$

$$(3.9)\qquad = \sum_{n\geq -1} c_n \sum_{j=0}^{n+k} \binom{n+k}{j} z^{n+k-j}\, \partial^j_{-\ell}\Big|_{\ell=0} {}_z\phi \circ (S \star \theta_{-z\ell})(w) \in \mathbb{K}[[z]],$$

while for $k = 0$ we use $S \star [zY]^0 = S \star \mathrm{id} = e$ and $e \circ B_+ = 0$. □

**3.2. Feynman rule recursion from Mellin transforms.** In fact this serves an alternative prove of the recursion (2.6), as in the physical limit $z \to 0$ only the contributions of $j = n + k$ in (3.9) survive:

(3.10)

$$\begin{aligned}
\phi_{\mathrm{R}} \circ B_+ &\underset{(3.7)}{=} \sum_{k\in\mathbb{N}} \frac{(-\ell)^k}{k!} \lim_{z\to 0} {}_z\phi \circ (S \star [zY]^k) \circ B_+ \underset{(3.9)}{=} \sum_{\substack{k\in\mathbb{N}\\ n\geq -1}} c_n \frac{(-\ell)^k}{k!} \left[\partial^{n+k}_{-\ell} \phi_{\mathrm{R}}\right]_{\ell=0} \\
&= P \circ F(-\partial_\ell) \circ \phi_{\mathrm{R}}.
\end{aligned}$$

Recall that $P = \mathrm{id} - \mathrm{ev}_0 : \mathbb{K}[\ell] \twoheadrightarrow \ker\varepsilon = \ell\,\mathbb{K}[\ell]$ projects out the constant terms and we defined $\partial_\ell^{-1} := \int_0$. This delivers an efficient recursion to calculate $\phi_{\mathrm{R}}$ combinatorially in terms of the Mellin transform coefficients $c_\cdot$ without any need for series expansions in $z$ as in example 3.6 or integrations like in (2.2):

EXAMPLE 3.9. *Applying* (2.6) *we can reproduce example* 3.6 *as*

$$\begin{aligned}
\phi_R(\bullet) &= \phi_R \circ B_+(\mathbb{1}) = [P \circ F(-\partial_\ell)]\left(\phi_R(\mathbb{1})\right) = P\left(-c_{-1}\textstyle\int_0 1 + c_0\right) = -c_{-1}\ell \\
\phi_R\left(\begin{smallmatrix}\bullet\\ |\\ \bullet\end{smallmatrix}\right) &= \phi_R \circ B_+(\bullet) = [P \circ F(-\partial_\ell)]\left(\phi_R(\bullet)\right) \\
&= P\left(\left[-c_{-1}\textstyle\int_0 + c_0 - c_1\partial_\ell\right]\left(-c_{-1}\ell\right)\right) = c_{-1}^2\frac{\ell^2}{2} - c_{-1}c_0\,\ell, \\
\phi_R\left(\begin{smallmatrix} & \bullet & \\ \bullet & & \bullet\end{smallmatrix}\right) &= \phi_R \circ B_+(\bullet\bullet) = [P \circ F(-\partial_\ell)]\left(\phi_R(\bullet\bullet)\right) = P \circ \left[-c_{-1}\textstyle\int_0 + c_0 - c_1\partial_\ell\right]\left\{\left(-c_{-1}\ell\right)^2\right\} \\
&= -c_{-1}^3\frac{\ell^3}{3} + c_{-1}^2c_0\,\ell^2 - 2c_{-1}^2c_1\,\ell.
\end{aligned}$$

*Here we can substitute* $c_{-1} = 1$, $c_0 = 0$ *and* $c_1 = \zeta(2) = \frac{\pi^2}{6}$ *to finally verify* (0.3) *from the introduction, where the choice* $f(\zeta) = \frac{1}{1+\zeta}$ *results in the beta function*

$$F(z) \underset{(3.2)}{=} B(z, 1-z) = \Gamma(z)\Gamma(1-z) = \frac{\pi}{\sin(\pi z)} \in z^{-1} + \frac{\pi^2}{6} z + \frac{7\pi^4}{360} z^3 + \mathcal{O}\left(z^5\right).$$

COROLLARY 3.10. *As in* $F(-\partial_\ell)$ *only* $-c_{-1}\int_0$ *increases the degree in* $\ell$, *the highest order contribution (called* leading log*) of* $\phi_{\mathrm{R}}$ *is the tree factorial we already saw in* (3.4)*: For any forest* $w \in \mathcal{F}$,

$$\phi_R(w) \in \left[{}^{-c_{-1}\int_0}\right]_{\rho(w)} + \mathcal{O}\left(x^{|w|-1}\right) \underset{\text{A.5}}{=} \frac{\left(-c_{-1}x\right)^{|w|}}{w!} + \mathbb{K}[x]_{<|w|}. \tag{3.11}$$

## 4. Hopf algebra morphisms and the renormalization group

From now on we identify $\phi_{\mathrm{R}}\colon H_R \to \mathbb{K}[x]$ with the polynomials that evaluate to the renormalized Feynman rules $\phi_{\mathrm{R},s} = \mathrm{ev}_\ell \circ \phi_{\mathrm{R}}$ at $x \mapsto \ell = \ln\frac{s}{\mu}$. In (2.6) and (3.10) we independently proved

COROLLARY 4.1. *As* $P \circ F(-\partial_x) \in HZ^1_\varepsilon(\mathbb{K}[x])$ *in* (2.6) *is a Hochschild-1-cocycle by* (B.1) *and* (B.3), *theorem* A.3 *implies that* $\phi_R\colon H_R \to \mathbb{K}[x]$ *is a morphism of Hopf algebras.*

Therefore $\Delta \circ \phi_{\mathrm{R}} = (\phi_{\mathrm{R}} \otimes \phi_{\mathrm{R}}) \circ \Delta$ and the induced map $G_{\mathbb{K}}^{\mathbb{K}[x]} \to G_{\mathbb{K}}^{H_R}$ given by $\mathrm{ev}_\ell \mapsto \phi_{\mathrm{R}}|_\ell := \mathrm{ev}_\ell \circ \phi_{\mathrm{R}}$ becomes a morphism of groups, implying

COROLLARY 4.2. *Using* (B.3) *we obtain the* renormalization group equation *(called* Chen's lemma *in* [**24**]*)*

$$\phi_R|_\ell \star \phi_R|_{\ell'} = \phi_R|_{\ell+\ell'}, \quad \textit{for any} \quad \ell, \ell' \in \mathbb{K}. \tag{4.1}$$

Before we obtain the generator of this one-parameter group in 4.4, note how this result imposes non-trivial relations between individual trees like

$$\begin{aligned} \phi_{\mathrm{R},\ell} \star \phi_{\mathrm{R},\ell'}\left(\begin{smallmatrix}\bullet\\ \vert\\ \bullet\end{smallmatrix}\right) &\underset{(4.1)}{=} \phi_{\mathrm{R},\ell}\left(\begin{smallmatrix}\bullet\\ \vert\\ \bullet\end{smallmatrix}\right) + \phi_{\mathrm{R},\ell}(\bullet)\,\phi_{\mathrm{R},\ell'}(\bullet) + \phi_{\mathrm{R},\ell'}\left(\begin{smallmatrix}\bullet\\ \vert\\ \bullet\end{smallmatrix}\right) \\ &\underset{(3.9)}{=} c_{-1}^2 \frac{\ell^2 + \ell'^2}{2} - c_{-1}c_0\left(\ell + \ell'\right) + c_{-1}^2 \ell\ell' \underset{(3.9)}{=} \phi_{\mathrm{R},\ell+\ell'}\left(\begin{smallmatrix}\bullet\\ \vert\\ \bullet\end{smallmatrix}\right). \end{aligned}$$

PROPOSITION 4.3. *Let* $\phi\colon H \to \mathbb{K}[x]$ *be a morphism of bialgebras*[7], *then* $\log_\star \phi$ *is precisely the monomial linear in* $x$*:*

$$\log_\star \phi = x \cdot \left[\partial_0 \circ \phi\right]. \tag{4.2}$$

PROOF. Letting $\phi\colon C \to H$ and $\psi\colon H \to \mathcal{A}$ denote morphisms of coalgebras and algebras, exploiting $\left(\psi \circ \phi - u_{\mathcal{A}} \circ \varepsilon_C\right)^{\star n} = \psi \circ \left(\phi - u_H \circ \varepsilon_H\right)^{\star n} = \left(\psi - u_{\mathcal{A}} \circ \varepsilon_H\right)^{\star n} \circ \phi$ in (1.2) proves $(\log_\star \psi) \circ \phi = \log_\star(\psi \circ \phi) = \psi \circ \log_\star \phi$. Now set $\psi = \mathrm{ev}_a$ and use lemma B.5. □

DEFINITION 4.4. *The* anomalous dimension $\gamma$ *of* $\phi_R$ *is the infinitesimal character*

$$H_R' \supset \mathfrak{g}_{\mathbb{K}}^{H_R} \ni \gamma := -\partial_0 \circ \phi_R \underset{(4.2)}{=} -\frac{1}{x}\log_\star \phi_R. \tag{4.3}$$

---

[7]This already implies $\phi$ to be a morphism of Hopf algebras since $H$ is connected.

*It completely determines all higher powers of $x$ by means of*

$$\phi_R = \exp_\star(-x\cdot\gamma) = \sum_{n\in\mathbb{N}_0} \frac{\gamma^{\star n}}{n!}(-x)^n. \tag{4.4}$$

EXAMPLE 4.5. *Reading off* $\gamma(\bullet) = c_{-1}$, $\gamma\big(\overset{\bullet}{\underset{\bullet}{|}}\big) = c_{-1}c_0$ *and* $\gamma\big(\underset{\bullet\,\bullet}{\overset{\bullet}{\wedge}}\big) = 2c_{-1}^2c_1$ *from the example 3.9 above, (4.4) correctly determines the higher powers of $x$ through*

$$\phi_R\big(\overset{\bullet}{\underset{\bullet}{|}}\big) \underset{(1.2)}{=} \left[e - x\gamma + x^2\frac{\gamma\star\gamma}{2}\right]\big(\overset{\bullet}{\underset{\bullet}{|}}\big) = 0 - x\gamma\big(\overset{\bullet}{\underset{\bullet}{|}}\big) + x^2\frac{\gamma^2(\bullet)}{2} = -c_{-1}c_0\,x + c_{-1}^2\frac{x^2}{2},$$

$$\begin{aligned}\phi_R\big(\underset{\bullet\,\bullet}{\overset{\bullet}{\wedge}}\big) &= 0 - x\gamma\big(\underset{\bullet\,\bullet}{\overset{\bullet}{\wedge}}\big) + x^2\frac{\gamma\otimes\gamma}{2}\big(2\bullet\otimes\overset{\bullet}{\underset{\bullet}{|}} + \bullet\bullet\otimes\bullet\big) - x^3\frac{\gamma\otimes\gamma\otimes\gamma}{6}(2\bullet\otimes\bullet\otimes\bullet)\\ &= -\gamma^3(\bullet)\frac{x^3}{3} + x^2\gamma(\bullet)\gamma\big(\overset{\bullet}{\underset{\bullet}{|}}\big) - 2c_{-1}^2c_1\,x = -c_{-1}^3\frac{x^3}{3} + c_{-1}^2c_0\,x^2 - 2c_{-1}^2c_1\,x.\end{aligned}$$

Note how the fragment $\bullet\bullet\otimes\bullet$ of $\Delta\big(\underset{\bullet\,\bullet}{\overset{\bullet}{\wedge}}\big)$ does not contribute to the quadratic terms $\frac{x^2}{2}\gamma\star\gamma$, as $\gamma$ vanishes on products. We will exploit this in (6.7) of section 6.1 on Dyson-Schwinger equations.

EXAMPLE 4.6. *In the leading*-log *case* (A.3) *we read off* $\partial_0\circ\varphi = Z_\bullet \in \mathfrak{g}_{\mathbb{K}}^{H_R}$ *where* $Z_\bullet(w) := \delta_{w,\bullet}$. *Comparing* $\varphi = \exp_\star(xZ_\bullet)$ *with* (A.3) *shows* $|w|! = w!\cdot Z_\bullet^{\star|w|}(w)$, *implying the following combinatorial relation among tree factorials noted in* [**24**]:

$$\frac{|w|}{w!} = \frac{1}{(|w|-1)!}\sum_w Z_\bullet(w_1)Z_\bullet^{\star|w|-1}(w_2) = \sum_{w:\ w_1=\bullet}\frac{1}{|w_2|!}Z_\bullet^{\star|w_2|}(w_2) = \sum_{w:\ w_1=\bullet}\frac{1}{w_2!}.$$

*This Sweedler sum picks only the parts with $w_1 = \bullet$ of the coproduct, which means that we sum over all leaves (nodes without children) $w_1$ of $w$ and $w_2$ is obtained by cutting off this leaf.*

**4.1. The regularized viewpoint.** We can obtain these results also by exploiting the regulator as in [**10**]:

LEMMA 4.7. *For* ${}_z\phi \in G_{\mathcal{A}}^H$ *subject to 3.7, the* anomalous dimension

$$\gamma := -\partial_\ell|_0\phi_R = \lim_{z\to 0} z\cdot{}_z\phi\circ(S\star Y) = \operatorname{Res}{}_z\phi\circ(S\star Y) \in \mathfrak{g}_{\mathbb{K}}^H \tag{4.5}$$

*is the residue (coefficient of $\frac{1}{z}$) of ${}_z\phi\circ(S\star Y)\in z^{-1}\mathbb{K}[[z]]$ and fulfills*

$$-\frac{\partial}{\partial\ell}\phi_R = \gamma\star\phi_R, \quad \textit{therefore} \quad \phi_R = \exp_\star(-\ell\gamma)\star\phi_R|_{\ell=0} = \exp_\star(-\ell\gamma). \tag{4.6}$$

PROOF. Proposition C.2 renders $\gamma\in\mathfrak{g}_{\mathbb{K}}^H$ immediate. We employ the coderivation property $\Delta\circ Y = (\mathrm{id}\otimes Y + Y\otimes\mathrm{id})\circ\Delta$ and $S\circ Y = -S\star Y\star S$ in

$$\begin{aligned}&-\frac{\partial}{\partial\ell}\lim_{z\to0}{}_z\phi\circ(S\star\theta_{-z\ell}) = \lim_{z\to0} z\cdot{}_z\phi\circ(S\star[\theta_{-z\ell}\circ Y])\\ &= \lim_{z\to0}\left\{z\cdot{}_z\phi\circ(S\star\theta_{-z\ell})\circ Y\right\} + \lim_{z\to0}\left\{z\cdot{}_z\phi\circ(S\star Y)\star{}_z\phi\circ(S\star\theta_{-z\ell})\right\}.\end{aligned}$$

The first term vanishes by the existence of $\lim_{z\to0}{}_z\phi\circ(S\star\theta_{-z\ell})$, while the second factorizes as desired. It remains to observe $\phi_{\mathrm{R}}|_{\ell=0} = {}_z\phi\circ(S\star\mathrm{id}) = {}_z\phi\circ e = e$. □

Clearly we can easily rewrite this in the form of (4.1) since

$$\phi_{\mathrm{R},\ell} \star \phi_{\mathrm{R},\ell'} = \exp_\star(-\ell\gamma) \star \exp_\star(-\ell'\gamma) = \exp_\star(-(\ell+\ell')\gamma) = \phi_{\mathrm{R},\ell+\ell'}. \tag{4.7}$$

Note how this reasoning fails if $\phi_{\mathrm{R}}$ disrespects the coproduct: Then

$$\log_\star \phi_{\mathrm{R}} = \sum_{n\in\mathbb{N}} \frac{\gamma_n}{n!} x^n \tag{4.8}$$

would contain higher powers in $x$ and a family $\gamma_n \in H'$ of functionals. As these do not necessarily commute under $\star$, also $\mathrm{ev}_\ell \circ \log_\star(\phi_{\mathrm{R}})$ and $\mathrm{ev}_{\ell'} \circ \log_\star(\phi_{\mathrm{R}})$ do not commute such that (4.7) is not applicable.

While the renormalization group allows us to reduce all computations to the linear terms $\gamma$, in our setup (3.1) we can give a simple recursion for $\gamma$ itself in term of the Mellin transform coefficients in

Corollary 4.8. *From* $\gamma \circ B_+ \underset{(2.6)}{=} -\partial_0 \circ F(-\partial_\ell) \circ \phi_R \underset{(4.4)}{=} \mathrm{ev}_0 \circ [zF(z)]_{-\partial_x} \circ \exp_\star(-x\gamma)$ *we obtain the inductive formula* $\gamma \circ B_+ = \sum_{n\in\mathbb{N}_0} c_{n-1}\gamma^{\star n} = [zF(z)]_{z\mapsto\gamma}$.

As $\gamma \in \mathfrak{g}^H_{\mathbb{K}}$ vanishes on products, evaluating it on trees is all we need such that 4.8 is sufficient to determine $\gamma$.

Example 4.9. *Starting with* $\gamma(\bullet) = c_{-1}\varepsilon(\mathbb{1}) = c_{-1}$ *we can recursively calculate*

$$\begin{aligned}
\gamma\left(\substack{\bullet\\|\\\bullet}\right) &= c_{-1}\varepsilon(\bullet) + c_0\gamma(\bullet) = c_{-1}c_0,\\
\gamma\left(\substack{\bullet\\|\\\bullet\\|\\\bullet}\right) &= c_{-1}\varepsilon\left(\substack{\bullet\\|\\\bullet}\right) + c_0\gamma\left(\substack{\bullet\\|\\\bullet}\right) + c_1\gamma\star\gamma\left(\substack{\bullet\\|\\\bullet}\right) = c_{-1}c_0^2 + c_1[\gamma(\bullet)]^2 = c_{-1}c_0^2 + c_{-1}^2c_1,\\
\gamma\left(\substack{\bullet\\ \bigwedge \\\bullet\ \ \bullet}\right) &= c_{-1}\varepsilon(\bullet\bullet) + c_0\gamma(\bullet\bullet) + c_1\gamma\star\gamma(\bullet\bullet) = 2c_1[\gamma(\bullet)]^2 = 2c_{-1}^2c_1 \quad \textit{and so on.}
\end{aligned}$$

## 5. Locality, finiteness and minimal subtraction

In the presence of a regulator $z$, considering a general character ${}_z\phi \in G^H_{\mathcal{A}}$ for $\mathcal{A} = \mathbb{K}[z^{-1}, z]]$ and scale dependence fixed by ${}_z\phi_s = {}_z\phi \circ \theta_{-z\ell}$ as in section 3 naturally leads to a different idea of renormalization in

Definition 5.1. *The* minimal subtraction scheme $R_{MS}$ *splits* $\mathcal{A} = \mathcal{A}_- \oplus \mathcal{A}_+$ *by projection on the poles* $\mathcal{A}_- := z^{-1}\mathbb{K}[z^{-1}]$ *along the holomorphic* $\mathcal{A}_+ := \mathbb{K}[[z]]$.

*One easily checks* (2.8) *and therefore obtains a unique Birkhoff decomposition* ${}_z\phi_+ = {}_z\phi_- \star ({}_z\phi \circ \theta_{-z\ell})$*, with physical limit* $\phi_+ := \lim_{z\to 0} {}_z\phi_+$.

This setup is the starting point in **[10]** and subject to many articles like **[11, 30]**. Note that this scheme does not specify a subtraction point $\mu$, but we included $\mu$ as an arbitrary scale inside $\ell = \ln\frac{s}{\mu}$ when we agreed on ${}_z\phi_s = {}_z\phi \circ \theta_{-z\ell}$. Physically this is necessary to obtain the dimensionless argument $\frac{s}{\mu}$ in the logarithm instead of expressions like $\ln s$ alone, as $s$ is a quantity carrying a unit (typically momentum or energy). In any case, replacing $\ln s$ by $\ell$ is nothing but a rescaling of $s$.

This renormalization scheme $R_{\mathrm{MS}}$ and the resulting Birkhoff decomposition differ from the kinematic subtraction scheme $R_\mu$, compare example 3.9 with

EXAMPLE 5.2. *For ${}_z\phi$ arising from (3.8), minimal subtraction yields*

$$
\begin{aligned}
{}_z\phi_+(\bullet) &= (\mathrm{id} - R_{MS})\,{}_z\phi_s(\bullet) = (\mathrm{id} - R_{MS})e^{-z\ell}F(z) = e^{-z\ell}F(z) - \frac{c_{-1}}{z} \\
\phi_+(\bullet) &= \lim_{z\to 0} {}_z\phi_+(\bullet) = c_0 - c_{-1}\ell \qquad (5.1)\\
{}_z\phi_+\left(\substack{\bullet\\ \bullet}\right) &= (\mathrm{id} - R_{MS})\left[{}_z\phi_s\left(\substack{\bullet\\ \bullet}\right) + {}_z\phi_-(\bullet)\,{}_z\phi_s(\bullet)\right] \\
&= (\mathrm{id} - R_{MS})\left[e^{-2z\ell}F(z)F(2z) - \frac{c_{-1}}{z}e^{-z\ell}F(z)\right] \\
&= e^{-2z\ell}F(z)F(2z) - \frac{c_{-1}}{z}e^{-z\ell}F(z) + \frac{c_{-1}^2}{2z^2} - \frac{c_{-1}c_0}{2z} \\
\phi_+\left(\substack{\bullet\\ \bullet}\right) &= \frac{c_{-1}^2}{2}\ell^2 - 2c_{-1}c_0\ell + c_0^2 + \frac{3}{2}c_{-1}c_1. \qquad (5.2)
\end{aligned}
$$

Note that by this choice of $\mathcal{A}_+ = \mathbb{K}[[z]]$, the finiteness of $\phi_+ := \lim_{z\to 0} {}_z\phi_+$ is automatic such that we can finitely renormalize any ${}_z\phi \in G_{\mathcal{A}}^{H}$ using $R_{\mathrm{MS}}$. This seems preferable considering that the kinematic subtraction scheme only yields finite results under the conditions of proposition 3.7. However, the physics of local field theory requires *local* counterterms that are constants, independent of the external parameters (in our setup $s$)[8].

DEFINITION 5.3 ([**10**]). *A Feynman rule ${}_z\phi \in G_{\mathcal{A}}^{H_R}$ is called* local *iff the minimal subtraction counterterm ${}_z\phi_{-,s} = ({}_z\phi \circ \theta_{-z\ell})_-$ is independent of $\ell \in \mathbb{K}$.*

By definition 2.7, counterterms in the kinematic subtraction scheme are $s$-independent a priori. For the minimal subtraction scheme $R_{\mathrm{MS}}$, locality is a true condition and the study and characterization of local Feynman rules in this setting is a main theme of [**10, 30**]. It is therefore illuminating to find

PROPOSITION 5.4. *Locality of ${}_z\phi \in G_{\mathcal{A}}^{H_R}$ in the $R_{MS}$ scheme is equivalent to the finiteness conditions of proposition 3.7 in the kinematic subtraction scheme.*

PROOF. Given condition 3.7 (2), ${}_z\phi_+ \star [{}_z\phi^{\star -1} \star ({}_z\phi \circ \theta_{-z\ell})]$ maps to $\mathbb{K}[[z]]$ so

$$
{}_z\phi \circ \theta_{-z\ell} = {}_z\phi_-^{\star -1} \star \left\{{}_z\phi_+ \star \left[{}_z\phi^{\star -1} \star ({}_z\phi \circ \theta_{-z\ell})\right]\right\}
$$

is a Birkhoff decomposition and its uniqueness implies the locality $({}_z\phi \circ \theta_{-z\ell})_- = {}_z\phi_-$. Conversely, for local ${}_z\phi$ we have ${}_z\phi_- = ({}_z\phi \circ \theta_{-z\ell})_-$ wherefore

$$
{}_z\phi_+^{\star -1} \star ({}_z\phi \circ \theta_{-z\ell})_+ = {}_z\phi^{\star -1} \star {}_z\phi_-^{\star -1} \star ({}_z\phi \circ \theta_{-z\ell})_- \star ({}_z\phi \circ \theta_{-z\ell}) = {}_z\phi^{\star -1} \star ({}_z\phi \circ \theta_{-z\ell})
$$

shows 3.7 (2) as the left hand side maps to $\mathcal{A}_+ = \mathbb{K}[[z]]$. □

We have seen how algebraically the problems of finiteness in the kinematic subtraction scheme and locality in minimal subtraction coincide. Both finite renormalization and local counterterms are simultaneously only achieved under the conditions of proposition 3.7, no matter which of the schemes $\{R_\mu, R_{\mathrm{MS}}\}$ is chosen.

---

[8]See also section 7: Counterterms may depend polynomially on parameters, but not logarithmically as might in general happen for $\operatorname{im}\left({}_z\phi_{-,s}\right) \subseteq z^{-1}\mathbb{K}[z^{-1}, \ell]$. In practice one actually defines *form-factors* as the coefficients of these polynomials and therefore become indeed completely independent of kinematics.

**5.1. Renormalization group.** We again identify $\phi_+$ with the polynomials in $\mathbb{K}[x]$ such that $\phi_{+,s} = \mathrm{ev}_\ell \circ \phi_+$, but in contrast to $\phi_{\mathrm{R}}$ from the kinematic subtraction scheme, these feature constant terms $\phi_{+,\mu} = \phi_+\big|_{x=0} = \mathrm{ev}_0 \circ \phi_+ = \varepsilon \circ \phi_+$ as we observed in the examples 5.2. Therefore the renormalization group equation (4.4) can not hold, but is instead replaced by

COROLLARY 5.5. *For local* ${}_z\phi \in G^H_{\mathcal{A}}$*, the* beta functional [**10**]

$$\beta := z \cdot {}_z\phi_-^{\star -1} \circ (S \star Y) = -\operatorname{Res} \circ {}_z\phi_- \circ Y = -\operatorname{Res} \circ {}_z\phi_- \circ (S \star Y) \in \mathfrak{g}^H_{\mathbb{K}} \tag{5.3}$$

*completely dictates the scale dependence of the physical limit* $\phi_+$ *through*

$$\phi_{+,s} = \lim_{z\to 0} ({}_z\phi \circ \theta_{-z\ell})_+ = \exp_\star(-\ell\beta) \star \big(\varepsilon \circ \phi_+\big). \tag{5.4}$$

PROOF. As ${}_z\phi_-^{\star -1}$ is local by lemma 5.6, proposition 3.7 allows to invoke 5.4 which proves the finiteness of (5.3): $\operatorname{im}\big({}_z\phi_-^{\star -1} \circ (S \star Y)\big) \subseteq z^{-1}\mathbb{K}[[z]] \cap \mathcal{A}_- = \mathbb{K}\cdot z^{-1}$,

$$\beta = \lim_{z\to 0} \left[z \cdot {}_z\phi_-^{\star -1} \circ (S \star Y)\right] = z \cdot {}_z\phi_-^{\star -1} \circ (S \star Y) \in \mathfrak{g}^H_{\mathbb{K}} \tag{5.5}$$

converges and (4.4) applies to give $\lim_{z\to 0} {}_z\phi_-^{\star -1} \circ (S \star \theta_{-z\ell}) = \exp_\star(-\ell\beta)$. Inserting this into the Birkhoff decomposition (here we set ${}_z\phi_+ = {}_z\phi_{+,\mu} = {}_z\phi_+\big|_{\ell=0}$)

$$\begin{aligned} {}_z\phi_{+,s} &= ({}_z\phi \circ \theta_{-z\ell})_- \star ({}_z\phi \circ \theta_{-z\ell}) = {}_z\phi_- \star \left[({}_z\phi_-^{\star -1} \star {}_z\phi_+) \circ \theta_{-z\ell}\right] \\ &= \left[{}_z\phi_+^{\star -1} \circ (S \star \theta_{-z\ell})\right] \star \left({}_z\phi_+ \circ \theta_{-z\ell}\right) \end{aligned}$$

and exploiting $\lim_{z\to 0}\left({}_z\phi_+ \circ \theta_{-z\ell}\right) = \lim_{z\to 0} {}_z\phi_+ = \phi_+\big|_{\ell=0}$ shows (5.4). The relations in (5.3) follow from $S(w) = -w \mod (\ker\varepsilon)^2$, $S \star Y(w) = Y(w) \mod (\ker\varepsilon)^2$ and the fact that $\operatorname{Res} \circ {}_z\phi_- \in \mathfrak{g}^H_{\mathbb{K}}$ vanishes on products, because for any $w, w' \in \ker\varepsilon$

$${}_z\phi_-(w \cdot w') = {}_z\phi_-(w) \cdot {}_z\phi_-(w') \in \mathcal{A}_-^2 = z^{-2}\mathbb{K}[[z]]$$

has no pole of first order. □

LEMMA 5.6. *Let* ${}_z\phi \in G^H_{\mathcal{A}}$ *be local, then* ${}_z\phi_-^{\star -1} \in G^H_{\mathcal{A}}$ *is local as well.*

PROOF. As in [**30**], from ${}_z\phi_+, {}_z\phi_{+,s} \in G^H_{\mathcal{A}_+}$ we deduce that

$${}_z\phi_-^{\star -1} \circ \theta_{-z\ell} = \left({}_z\phi \star {}_z\phi_+^{\star -1}\right) \circ \theta_{-z\ell} = {}_z\phi_s \star \left({}_z\phi_+^{\star -1} \circ \theta_{-z\ell}\right) = {}_z\phi_-^{\star -1} \star {}_z\phi_{+,s} \star \left({}_z\phi_+^{\star -1} \circ \theta_{-z\ell}\right)$$

is a Birkhoff decomposition and read off $\left({}_z\phi_-^{\star -1} \circ \theta_{-z\ell}\right)_- = {}_z\phi_-$ by uniqueness. □

In the minimal subtraction scheme we can rephrase the renormalization group (5.4) as expressing all $z$-poles of the counterterms ${}_z\phi_- : H \to \mathbb{K}[z^{-1}]$ in terms of the first order poles only [**10**]. This comes about as $\beta$ captures all information on the character ${}_z\phi_-^{\star -1} \in \mathfrak{g}^H_{\mathcal{A}}$ since $\operatorname{im}(S \star Y)$ generates the full Hopf algebra by C.3. For clarity we shall demonstrate this in

EXAMPLE 5.7. *First we take example* 5.2 *to read off the counterterms*

$$\begin{aligned} {}_z\phi_-^{\star -1}(\bullet) &= {}_z\phi_-(-\bullet) = -R_{MS}\left[-e^{-z\ell}F(z)\right] = \frac{c_{-1}}{z} \\ {}_z\phi_-^{\star -1}\left(\begin{smallmatrix}\bullet\\ \vert\\ \bullet\end{smallmatrix}\right) &= {}_z\phi_-\left(-\begin{smallmatrix}\bullet\\ \vert\\ \bullet\end{smallmatrix} + \bullet\bullet\right) = -\frac{c_{-1}^2}{2z^2} + \frac{c_{-1}c_0}{2z} + \left[\frac{c_{-1}}{z}\right]^2 = \frac{c_{-1}^2}{2z^2} + \frac{c_{-1}c_0}{2z} \end{aligned}$$

*and apply* (5.3) *the get hold of the residues* $\beta(\bullet) \underset{(5.3)}{=} z \cdot {}_z\phi_-^{\star -1}(\bullet) = c_{-1}$ *as well as*

$$\beta\left(\begin{smallmatrix}\bullet\\ \vert\\ \bullet\end{smallmatrix}\right) \underset{(5.3)}{=} z \cdot {}_z\phi_-^{\star -1}\left(2\begin{smallmatrix}\bullet\\ \vert\\ \bullet\end{smallmatrix} - \bullet\bullet\right) = c_{-1}c_0.$$

*Observe how for* $S\star Y\left(\overset{\bullet}{\underset{\bullet}{|}}\right)=2\overset{\bullet}{\underset{\bullet}{|}}-\bullet\bullet$ *we indeed obtained only a first order pole in* ${}_z\phi_-^{\star-1}$*, contrary to* $\overset{\bullet}{\underset{\bullet}{|}}$ *itself entailing a second order pole as well. This one we can now predict using* (C.4) *applied to* ${}_z\phi_-^{\star-1}\circ(S\star Y)=\frac{\beta}{z}$*:*

$$
\begin{aligned}
{}_z\phi_-^{\star-1}\left(\overset{\bullet}{\underset{\bullet}{|}}\right) &= \left\{e+\frac{\beta\circ Y^{-1}}{z}+\frac{\left[(\beta\circ Y^{-1})\star\beta\right]\circ Y^{-1}}{z^2}\right\}\left(\overset{\bullet}{\underset{\bullet}{|}}\right)\\
&= \frac{\beta\left(\frac{1}{2}\overset{\bullet}{\underset{\bullet}{|}}\right)}{z}+\frac{\left[\beta(\bullet)\right]^2}{2z^2}=\frac{c_{-1}c_0}{2z}+\frac{c_{-1}^2}{2z^2}.
\end{aligned}
$$

**5.2. Relating $R_\mu$ with $R_{\mathrm{MS}}$.** Though both schemes seem so different, already [4] exploited their relationship given in

LEMMA 5.8. *For local* ${}_z\phi\in G_{\mathcal{A}}^H$*, the scale dependence of* ${}_z\phi_{+,s}$ *(in the* $R_{MS}$ *scheme) is dictated by* ${}_z\phi_{R,s}$ *(kinematic subtraction scheme) through*

$$
{}_z\phi_{+,s}=\left(R_\mu\circ{}_z\phi_{+,s}\right)\star{}_z\phi_{R,s}. \tag{5.6}
$$

PROOF. Locality of the counterterms ${}_z\phi_-$ implies $R_\mu\circ{}_z\phi_-={}_z\phi_-$, hence

$$
\begin{aligned}
\left(R_\mu\circ{}_z\phi_{+,s}\right)\star{}_z\phi_{R,s} &= \left[R_\mu\circ\left({}_z\phi_-\star{}_z\phi_s\right)\right]\star\left(R_\mu\circ{}_z\phi_s\right)^{\star-1}\star{}_z\phi_s\\
&= \left[R_\mu\circ\left({}_z\phi_-\star{}_z\phi_s\star{}_z\phi_s^{\star-1}\right)\right]\star{}_z\phi_s=\left(R_\mu\circ{}_z\phi_-\right)\star{}_z\phi_s={}_z\phi_{+,s}. \qquad\square
\end{aligned}
$$

Note how $R_\mu\circ{}_z\phi_{+,s}={}_z\phi_+\big|_{\ell=0}$ reduces to the constants $\varepsilon\circ\phi_+\in G_{\mathbb{K}}^H$ in the physical limit. There (5.6) takes the form of

COROLLARY 5.9. *The characters* $\phi_R,\phi_+: H_R\to\mathbb{K}[x]$ *fulfill the relations*

$$
\phi_+=\left(\varepsilon\circ\phi_+\right)\star\phi_R,\quad\textit{equivalently}\quad\Delta\circ\phi_+=\left(\phi_+\otimes\phi_R\right)\circ\Delta. \tag{5.7}
$$

EXAMPLE 5.10. *After reading off the constants* $\varepsilon\circ\phi_+(\bullet)=c_0$ *and* $\varepsilon\circ\phi_+\left(\overset{\bullet}{\underset{\bullet}{|}}\right)=c_0^2+\frac{3}{2}c_{-1}c_1$*, we can verify* (5.7) *against the examples* 5.2 *making use of* 3.9*:*

$$
\begin{aligned}
\phi_+(\bullet) &= \varepsilon\circ\phi_+(\bullet)+\phi_R(\bullet)=c_0-c_{-1}\ell\\
\phi_+\left(\overset{\bullet}{\underset{\bullet}{|}}\right) &= \varepsilon\circ\phi_+\left(\overset{\bullet}{\underset{\bullet}{|}}\right)+\left[\varepsilon\circ\phi_+(\bullet)\right]\cdot\phi_R(\bullet)+\phi_R\left(\overset{\bullet}{\underset{\bullet}{|}}\right)\\
&= c_0^2+\frac{3}{2}c_{-1}c_1-c_0c_{-1}\ell+\frac{c_{-1}^2}{2}\ell^2-c_0c_{-1}\ell=\frac{c_{-1}^2}{2}\ell^2-2c_0c_{-1}\ell+c_0^2+\frac{3}{2}c_{-1}c_1.
\end{aligned}
$$

COROLLARY 5.11. *Inserting both* (5.4) *and* (4.4) *into* (5.7) *reveals*

$$
\beta\star\left(\varepsilon\circ\phi_+\right)=\left(\varepsilon\circ\phi_+\right)\star\gamma. \tag{5.8}
$$

Hence $\beta$ and $\gamma$ differ only by conjugation with the character $\varepsilon\circ\phi_+\in G_{\mathbb{K}}^H$ and therefore in particular agree on any cocommutative elements of $H$.

EXAMPLE 5.12. *For the rooted trees* $H_R$*, the cocommutative elements include the* ladders $B_+^n(\mathbb{1})$*. In the examples* 5.7 *and* 4.9 *we explicitly checked the first two cases of* $n\in\{1,2\}$*:* $\beta(\bullet)=c_{-1}=\gamma(\bullet)$ *as well as* $\beta\left(\overset{\bullet}{\underset{\bullet}{|}}\right)=c_{-1}c_0=\gamma\left(\overset{\bullet}{\underset{\bullet}{|}}\right)$.

## 6. Dyson-Schwinger equations and correlation functions

Until now we considered the renormalized Feynman rules $\phi_{\mathrm{R}}$ on their own, but these form only one ingredient to quantum field theory. The counterpart is the *perturbation series* $X(g)$ they are being applied to.

EXAMPLE 6.1 ([**32**]). *In* Yukawa theory, *the propagation of a fermion is the superposition of infinitely many possible interactions with a scalar field each of which is represented by a Feynman diagram. Among these are contributions like*

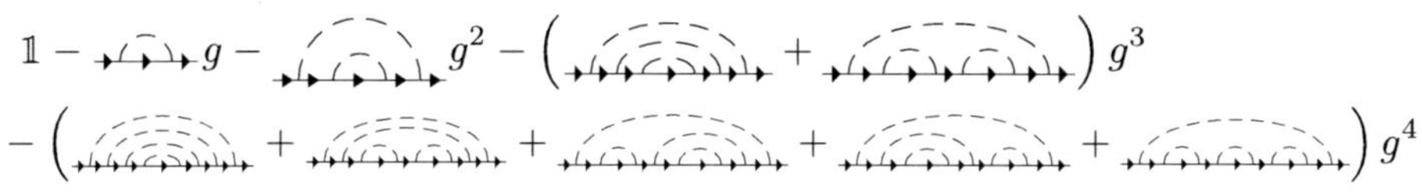

$$+\mathcal{O}\left(g^5\right).$$

*The Feynman rules map each graph to an individual amplitude, but physically these are not distinguishable and need to be all summed up. Further, a* coupling constant $g$ *takes the strength of an interaction into account.*

DEFINITION 6.2. *A* perturbation series *is a formal power series*

$$X(g)=\sum_{n\in\mathbb{N}_0} x_n g^n\in H_R[[g]] \quad \textit{with} \quad x_0=\mathbb{1} \tag{6.1}$$

*taking values in the Hopf algebra $H_R$ of rooted trees and indexed by the* coupling constant $g$. *Evaluation of the renormalized Feynman rules $\phi_R\in G^{H_R}_{\mathbb{K}[\ell]}$ on $X(g)$ delivers the* correlation function

$$G(g):=\phi_R\circ X(g)=\sum_{n\in\mathbb{N}_0}\phi_R(x_n)g^n\in\mathbb{K}[\ell][[g]], \tag{6.2}$$

*while the* physical anomalous dimension *is* $\widetilde{\gamma}(g):=\gamma\circ X(g)\underset{(4.3)}{=}-\partial_\ell|_0 G(g)\in\mathbb{K}[[g]]$.

The crucial property of perturbation series is the possibility of insertions: In the above example, we started with the *primitive*[9] graph 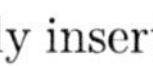 and iteratively inserted it as a subdivergence into itself. Though the full perturbation series contains many more graphs, this illustrates how $X(g)$ may efficiently be described by means of recursive insertions. Those are represented by Hochschild-1-cocycles motivating

DEFINITION 6.3. *To a parameter $\kappa\in\mathbb{K}$ and a family of cocycles $B_\cdot:\mathbb{N}\to HZ^1_\varepsilon(H_R)$ we associate the* combinatorial Dyson-Schwinger equation[10]

$$X(g)=\mathbb{1}+\sum_{n\in\mathbb{N}} g^n B_n\left(X^{1+n\kappa}(g)\right). \tag{6.3}$$

This type of equations is folklore in physics, but had not been cast into its pure algebraic form before [**1**]. Referring to [**17**] we recall the main results in

LEMMA 6.4. *As* perturbation series (6.1), *the equation* (6.3) *allows a unique solution which is determined recursively by*

$$x_k=\sum_{0\le m+n\le k}\binom{1+\kappa n}{m}B_n\left(\sum_{\substack{i_1+\ldots+i_m+n=k\\ i_1,\ldots,i_m\ge 1}} x_{i_1}\cdots x_{i_m}.\right) \tag{6.4}$$

*Most importantly, these coefficients generate a Hopf subalgebra $H_X:=\langle\{x_n:\ n\in\mathbb{N}\}\rangle$ (isomorphic to the Fàa di Bruno Hopf algebra when $\kappa\neq 0$). Explicitly we find*

$$\Delta X(g)=\sum_{n\in\mathbb{N}_0}[X(g)]^{1+n\kappa}\otimes g^n x_n\in(H_R\otimes H_R)[[g]]. \tag{6.5}$$

[9]That means it is free of subdivergences.

[10]As $x_0=\mathbb{1}$, for arbitrary $p$ we define $[X(g)]^p:=\sum_{n\in\mathbb{N}_0}\binom{p}{n}[X(g)-\mathbb{1}]^n\in H_R[[g]]$.

We learn that the solution of (6.3) has a very special property: The coproduct $\Delta x_n \in H_X \otimes H_X$ can be expressed by coefficients $x_\cdot$ alone, with (6.5) serving an explicit formula. Before we exploit this information on $X$ let us give some examples.

EXAMPLE 6.5. *When we set* $\kappa = 0$, $\Delta X(g) = X(g) \otimes X(g)$ *is group-like such that* $\Delta x_n = \sum_{i+j=n} x_i \otimes x_j$. *The corresponding Dyson-Schwinger equation* $X(g) = \mathbb{1} + B_+(X(g))$ *is linear and generates the cocommutative ladders* $x_n = B_+^n(\mathbb{1})$.

Recall that we take the trees in $H_R$ as substitute for Feynman graphs, each node representing an insertion into some other graph.

EXAMPLE 6.6. *In* [**5**,**32**] *we find the equation* $X(g) = \mathbb{1} - gB_+\left(\frac{1}{X(g)}\right)$ *featuring* $\kappa = -2$ *which corresponds to the propagator example* 6.1. *The solution sums all trees with the factor counting the number of distinct ordered embeddings:*

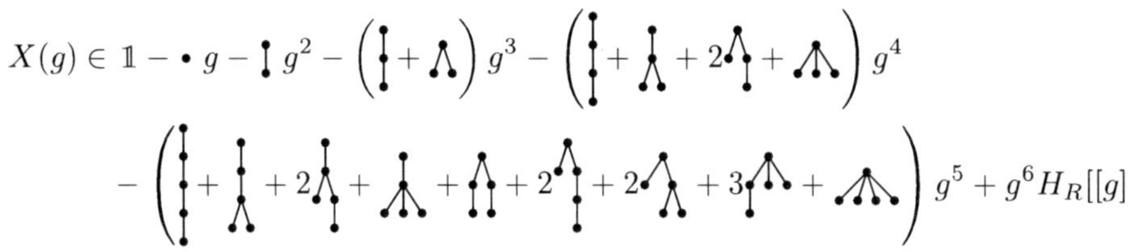

*The first factor of two arises from the different embeddings*  *and* 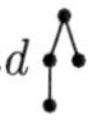 *and correctly accounts for the fact that these two shall represent different graphs (though they are the same elements in* $H_R$*):* 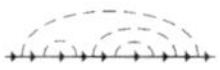 *and* 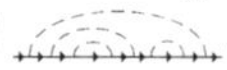.

EXAMPLE 6.7. *Let us consider the tree factorial Feynman rules* $\varphi$ *from* (A.3) *applied to the above series. The anomalous dimension* $\widetilde{\gamma}(g) \underset{4.6}{=} -Z_\bullet \circ X(g) = g$ *is linear, while the first terms of the correlation function become*

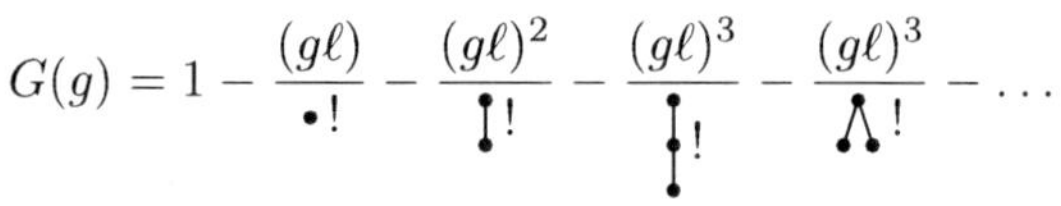

$$= 1 - g\ell - \frac{1}{2}(g\ell)^2 - \frac{1}{2}(g\ell)^3 + \mathcal{O}\left((g\ell)^4\right).$$

**6.1. Propagator coupling duality.** The Hopf subalgebra of the perturbation series allows to calculate convolutions in

LEMMA 6.8. *Let* $\psi \in \mathfrak{g}_{\mathcal{A}}^{H_R}$ *denote an infinitesimal character,* $\Psi \in G_{\mathcal{A}}^{H_R}$ *a character and* $\lambda \in \operatorname{Hom}(H_R, \mathcal{A})$ *a linear map. Then we obtain*

$$(\Psi \star \lambda) \circ X(g) = [\Psi \circ X(g)] \cdot \lambda \circ X\left(g\,[\Psi \circ X(g)]^\kappa\right) \tag{6.6}$$

$$:= [\Psi \circ X(g)] \cdot \sum_{n \in \mathbb{N}_0} \lambda(x_n) \cdot \left(g\,[\Psi \circ X(g)]^\kappa\right)^n \in \mathcal{A}[[g]]$$

$$(\psi \star \lambda) \circ X(g) = [\psi \circ X(g)] \cdot (\mathrm{id} + \kappa g \partial_g)\,[\lambda \circ X(g)] \in \mathcal{A}[[g]]. \tag{6.7}$$

PROOF. These are immediate consequences of lemma 6.4, for (6.7) consider

$$\psi\left([X(g)]^{1+n\kappa}\right) \cdot g^n = \sum_{i \in \mathbb{N}_0} \binom{1+n\kappa}{i} \psi\left([X(g) - \mathbb{1}]^i\right) g^n = \psi\left(X(g) - \mathbb{1}\right) \cdot (1 + n\kappa) g^n.$$

□

By combining (6.5) with the renormalization group equation we can calculate the correlation function out of the knowledge of $\widetilde{\gamma}$ only:

EXAMPLE 6.9. *Continuing example 6.6, we can calculate the* $\propto \ell^2$*-term* $Z_\bullet^{\star 2}(X(g)) = -g(1-2g\partial_g)(-g) = -g^2$ *and all further convolution products*

$$Z_\bullet^{\star n+1}(X(g)) \underset{(6.7)}{=} -g^{n+1}(2n-1)(2n-3)\cdots(1) = -g^{n+1}\frac{(2n)!}{2^n n!},$$

*proving* $\varphi(x_{n+1}) = -2^{-n}C_n\ell^{n+1}$ *with the* Catalan numbers $C_n$ *already noted in* [**29**]. *Combining their known generating function* $2g\sum_{n\in\mathbb{N}_0} g^n C_n = 1-\sqrt{1-4g}$ *with* $\varphi = \exp_\star(-\ell Z_\bullet)$ *allows us to completely determine the correlation function as* $G(g) = \sqrt{1-2g\ell}$.

COROLLARY 6.10. *As* $\phi_R$ *is a morphism of Hopf algebras by 4.1, for any* $\ell, \ell' \in \mathbb{K}$ *we can factorize the correlation function at* $\ell + \ell'$ *in two different ways*

$$G_{\ell+\ell'}(g) = (\phi_{R,\ell}\star\phi_{R,\ell'})\circ X(g) \underset{(6.6)}{=} G_\ell(g)\cdot G_{\ell'}\left[gG_\ell^\kappa(g)\right] = G_{\ell'}(g)\cdot G_\ell\left[gG_{\ell'}^\kappa(g)\right]. \tag{6.8}$$

At this point we like to briefly highlight the non-analytic nature of perturbative quantum field theory. The correlation function $G(g)$ at $g \neq 0$ (6.2) is a physical object that can in principle be measured through experiment. It is only by the nature of the perturbative method we apply that we are merely able to calculate the formal series expansion (6.2) of $G(g)$ around $g = 0$ with the help of the Feynman rules. The main issue is that in the interesting cases, the function $G(g)$ is not analytic at this point and the series (6.2) has zero radius of convergence[11].

However, in this perturbative approach we just deduced the functional equations (6.8) for the formal series. Therefore it is natural to impose these on the true correlation functions, such that we gain a non-perturbative handle on quantum field theory. We will continue to stress similar examples in this section.

But first observe how (6.8) takes the infinitesimal form

COROLLARY 6.11. *With the help of* $-\frac{\mathrm{d}}{\mathrm{d}x}\phi_R = \gamma\star\phi_R = \phi_R\star\gamma$ *or by differentiating* (6.8) *with respect to* $\ell'$ *at zero we find the differential equations*

$$G_\ell(g)\cdot\widetilde{\gamma}\left[gG_\ell^\kappa(g)\right] \underset{(6.6)}{=} -\partial_\ell G_\ell(g) \underset{(6.7)}{=} \widetilde{\gamma}(g)\cdot(1+\kappa g\partial_g)\,G_\ell(g). \tag{6.9}$$

The first of these equations generalizes the *propagator coupling duality* observed in [**5**, **29**]. For any fixed coupling $g$, it expresses the correlation function as the solution of the first order ordinary differential equation

$$-\frac{\mathrm{d}}{\mathrm{d}\ell}\ln G_\ell(g) = \widetilde{\gamma}\left[ge^{\kappa\ln G_\ell(g)}\right] \quad \text{with initial condition} \quad \ln G_0(g) = 0. \tag{6.10}$$

Note how this equation reconstructs $G_\ell(g)$ completely only from the input $\widetilde{\gamma}(g)$. This demonstrates the power of the renormalization group: Though $G$ depends on $g$ and $\ell$, after imposing (4.1) only a one-dimensional degree of freedom is left. As before, (6.10) serves a non-perturbative relation and need not be restrained to the perturbative series alone.

EXAMPLE 6.12. *The* leading-log *expansion takes only the highest power of* $\ell$ *in each* $g$*-order, so* $\widetilde{\gamma}(g) = cg^n$ *is a monomial for some* $c \in \mathbb{K}$, $n \in \mathbb{N}$ *(otherwise*

[11] Hence the example 6.9 is still far away from quantum field theory as its correlation function is analytic at $g \to 0$.

*different g-powers would mix for a given order in $\ell$). In this case* (6.10) *integrates to*

$$G_{leading\text{-}\log}(g) = \left[1 + cn\kappa\ell g^n\right]^{-\frac{1}{n\kappa}}. \tag{6.11}$$

*As a special case we recover example* 6.7 *for $n = c = 1$ and $\kappa = -2$.*

EXAMPLE 6.13. *For the linear Dyson-Schwinger equation $\kappa = 0$,* (6.8) *states $G_{\ell+\ell'}(g) = G_\ell(g) \cdot G_{\ell'}(g)$ which is solved by the* scaling solution $G_\ell(g) = e^{-\ell\widetilde{\gamma}(g)}$ *of* (6.10)*, well-known from* [**27**].

EXAMPLE 6.14. *The physical situation of vertex insertions as in* [**2**] *corresponds to $\kappa = 1$ and $G_{\ell+\ell'}(g) = G_{\ell'}(g) \cdot G_\ell\big[\widetilde{G}_{\ell'}(g)\big]$ can be interpreted as the running of the coupling constant $\widetilde{G} := g \cdot G$: A change in scale by $\ell'$ is (up to a multiplicative constant) equivalent to replacing the coupling $g$ by $\widetilde{G}_{\ell'}(g)$.*

**6.2. Running coupling.** The idea of this last example 6.14 leads us to another form of the renormalization group equation, common to the physics literature like (7.3.15) and (7.3.21) in [**7**]. We introduce the *$\beta$-function*[12] $\beta(g) := -\kappa g\widetilde{\gamma}(g)$ and the *running coupling* $g(\mu)$ as the solution of

$$\mu\frac{\mathrm{d}}{\mathrm{d}\mu}g(\mu) = \beta\big(g(\mu)\big), \quad \text{so} \quad \mu\frac{\mathrm{d}}{\mathrm{d}\mu}G\left(g(\mu), \ln\tfrac{s}{\mu}\right) \underset{(6.9)}{=} \widetilde{\gamma}\big(g(\mu)\big)G\left(g(\mu), \ln\tfrac{s}{\mu}\right). \tag{6.12}$$

Integration over $\mu$ results in a relation of the correlation functions corresponding to different choices of the renormalization point $\mu$:

$$\begin{aligned} G\left(g(\mu_2), \ln\tfrac{s}{\mu_2}\right) &= G\left(g(\mu_1), \ln\tfrac{s}{\mu_1}\right) \cdot \exp\left[\int_{\mu_1}^{\mu_2} \widetilde{\gamma}\big(g(\mu)\big)\tfrac{\mathrm{d}\mu}{\mu}\right] \\ &\underset{(6.12)}{=} G\left(g(\mu_1), \ln\tfrac{s}{\mu_1}\right) \cdot \left[\frac{g(\mu_2)}{g(\mu_1)}\right]^{-\frac{1}{\kappa}}. \end{aligned}$$

This result is important from a conceptual point of view: To achieve renormalization, we introduced a parameter $\mu$ that is completely arbitrary, yet the shape of the correlation function is a measurable quantity wherefore it clearly has to be insensitive to the choice of $\mu$.

Indeed we see that a change from $\mu_1$ to $\mu_2$ affects the correlation function only by a constant overall factor and a redefinition of the coupling constant (which itself is a parameter of the theory). Hence the physical content of $G$ is left invariant of the choice of renormalization point.

Similar to (6.10) we can express $G$ through a differential equation involving the running coupling $g(s)$ after choosing $\mu_1 = s$:

$$G_\ell(g) = \left[\frac{g}{g(s)}\right]^{-\frac{1}{\kappa}}, \quad \text{with } g(s) \text{ subject to} \quad \ell = \ln\frac{s}{\mu} = \int_g^{g(s)} \frac{\mathrm{d}g'}{\beta(g')}. \tag{6.13}$$

**6.3. Relation to Mellin transforms.** So far we exploited the renormalization group equation (4.4) and the Hopf subalgebra (6.5) of perturbation series. Now we like to take the special structure (2.6) of the Feynman rules from (2.2) into account. As different cocycles[13] represent different Feynman integrals, we allow for a

---

[12]This should not be confused with the $\beta$-functional of (5.3), though both are related.

[13]This is most easily considered in the Hopf algebra of decorated rooted trees (section A.2) where each $B_n$ inserts into a node of decoration $n$.

family $F_{\cdot} : \mathbb{N} \to z^{-1}\mathbb{K}[[z]]$ of Mellin transforms such that $\phi_{\mathrm{R}} \circ B_n = P \circ F_n(-\partial_\ell) \circ \phi_{\mathrm{R}}$. Applying this to (6.3) results in

$$G_\ell(g) \underset{(6.3)}{=} 1 + \sum_{n \in \mathbb{N}} g^n \phi_{\mathrm{R}} \circ B_n \left( X(g)^{1+n\kappa} \right) \underset{(2.6)}{=} 1 + P \circ \sum_{n \in \mathbb{N}} g^n F_n \left( -\partial_\ell \right) G_\ell(g)^{1+n\kappa}$$

and taking a derivative brings us to the differential equation of

COROLLARY 6.15. *The power series $G_\ell(g) \in \mathbb{K}[\ell][[g]]$ is fully determined by*

$$\partial_{-\ell} G_\ell \left( g \right) \underset{(2.6)}{=} \sum_{n \in \mathbb{N}} g^n \left[ z F_n(z) \right]_{z=-\partial_\ell} \left( G_\ell(g)^{1+n\kappa} \right) \quad \textit{and} \quad G_\ell(0) = 1. \tag{6.14}$$

Though (6.14) is always well defined for the formal power series (since the growing powers of $g$ allow only finitely many contributions in each order), the differential operator $[zF(z)]_{z=-\partial_\ell}$ can be of infinite order which might hinder a non-perturbative interpretation of this equation. However we can proceed in a couple of interesting cases, allowing us to construct the full correlation function or anomalous dimension (and not just individual terms of the perturbation series):

EXAMPLE 6.16. *Consider a single cocycle $F_k(z) = F(z)\delta_{k,n}$ and choose $F(z) = \frac{c_{-1}}{z}$. Then* (6.14) *reproduces the leading-log example* (6.11) *as it becomes*

$$\partial_{-\ell} G_\ell(g) = g^n c_{-1} G_\ell(g)^{1+n\kappa}. \tag{6.15}$$

More generally, for rational $F(z) = \frac{p(z)}{q(z)} \in \mathbb{K}(z)$ with polynomials $p(z), q(z) \in \mathbb{K}[z]$ we can apply $q(-\partial_\ell)$ on both sides of (6.14) resulting in a finite order differential equation

$$q(-\partial_\ell) G_\ell(g) = g^n p(-\partial_\ell) G_\ell(g)^{1+n\kappa}. \tag{6.16}$$

Enjoying this situation we can directly interpret it non-perturbatively (extending the algebraic $\partial_\ell \in \mathrm{End}(\mathbb{K}[\ell])$ to the analytic differential operator).

EXAMPLE 6.17. *The single Mellin transform $F(z) = \frac{1}{z(1-z)}$ in the propagator-type equation ($\kappa = -2$ as in example 6.6) occurs in the first approximation to quantum electrodynamics and the Yukawa theory. The equation*

$$\frac{g}{G_\ell(g)} = \partial_{-\ell} \left( 1 - \partial_{-\ell} \right) G_\ell(g) \underset{(6.9)}{=} \widetilde{\gamma}(g) \left( 1 - 2g\partial_g \right) \left[ 1 - \widetilde{\gamma}(g) \left( 1 - 2g\partial_g \right) \right] G_\ell(g),$$

*evaluates at $\ell = 0$ to the compact form*

$$g = \widetilde{\gamma}(g) - \widetilde{\gamma}(g)(1 - 2g\partial_g)\widetilde{\gamma}(g). \tag{6.17}$$

*We stress how this equation determines the anomalous dimension non-perturbatively, in fact it can be expressed in terms of the complementary error function as analyzed in* [**5**, **36**]. *Reference* [**34**] *is devoted to a detailed study of this type of equations and also solves the case $\kappa = 1$ with the help of the Lambert $W$ function.*

**6.4. Variations of Mellin transforms.** Consider a change of the Mellin transform $F$ to a different $\widetilde{F}$ that keeps $c_{-1}$ fixed but is free to alter the other coefficients $c_n$ with $n \in \mathbb{N}_0$. Then by (B.2) the difference

$$\delta\alpha := P \circ \left[ \widetilde{F} - F \right]_{-\partial_\ell} \in \mathrm{HB}^1_\varepsilon(\mathbb{K}[\ell]) \tag{6.18}$$

is a Hochschild-1-coboundary and (A.6) shows that we can relate the two resulting renormalized Feynman rules $\phi_{\mathrm{R}}$ and $\widetilde{\phi_{\mathrm{R}}}$ by composition with a distinguished Hopf algebra automorphism:

$$\widetilde{\phi_{\mathrm{R}}} \underset{(2.6)}{=} {}^{P\circ\widetilde{F}(-\partial_\ell)}\rho = {}^{P\circ F(-\partial_\ell)+\delta\alpha}\rho \underset{(A.6)}{=} {}^{P\circ F(-\partial_\ell)}\rho \circ {}^{\left[\alpha\circ {}^{P\circ F(-\partial_\ell)}\rho\right]}\chi \underset{(2.6)}{=} \phi_{\mathrm{R}} \circ {}^{\left[\alpha\circ\phi_{\mathrm{R}}\right]}\chi.$$

This accentuates that the Feynman rules coming from (2.2) obey even more structure than just the renormalization group (4.4): The origin from the Mellin transform poses restrictions on the generator $\gamma$ of $\phi_{\mathrm{R}}$. For illustration consider how example 4.9 implies

$$\gamma(\bullet)\cdot\left[2\gamma\left(\mathbf{I}_3\right) - \gamma\left(\Lambda\right)\right] = 2c_{-1}^2 c_0^2 = 2\left[\gamma\left(\mathbf{I}\right)\right]^2. \tag{6.19}$$

EXAMPLE 6.18. *Assume that $c_{-1} = -1$, then we can relates $\phi_R$ to the tree factorial character $\varphi = \int_0 \rho$: The difference in Mellin transforms is*

$$\delta\alpha = P\circ\sum_{n\in\mathbb{N}_0} c_n\partial_{-x}^n \in HB_{\varepsilon}^1(\mathbb{K}[x]), \quad \textit{therefore} \quad \alpha \underset{(B.2)}{=} \varepsilon\circ\sum_{n\in\mathbb{N}_0} c_n\partial_{-x}^n \in \mathbb{K}[x]' \tag{6.20}$$

*such that we find* $\alpha\circ\varphi(w) = \alpha\left(\frac{x^{|w|}}{w!}\right) = (-1)^{|w|}\frac{|w|!}{w!}c_{|w|}$. *Now we can verify*

$$\phi_R(\bullet) = x = \varphi(\bullet) \underset{A.8}{=} \varphi\circ{}^{\alpha\circ\varphi}\chi(\bullet), \phi_R(\mathbf{I}) = \frac{x^2}{2} + c_0 x = \varphi\left\{\mathbf{I} + \eta(1)\bullet\right\} \underset{A.8}{=} \varphi\circ{}^{\alpha\circ\varphi}\chi(\mathbf{I}),$$

$$\phi_R\left(\mathbf{I}_3\right) = \frac{x^3}{6} + x^2 c_0 + x(c_0^2 - c_1) = \varphi\left\{\mathbf{I}_3 + 2c_0\mathbf{I} + \left[c_0^2 - c_1\right]\bullet\right\} \underset{A.8}{=} \varphi\circ{}^{\alpha\circ\varphi}\chi\left(\mathbf{I}_3\right) \textit{ and}$$

$$\phi_R(\Lambda) = \frac{x^3}{3} + c_0\cdot x^2 - 2c_1\cdot x = \varphi\left\{\Lambda + c_0\bullet\bullet - 2c_1\bullet\right\} \underset{A.8}{=} \varphi\circ{}^{\alpha\circ\varphi}\chi(\Lambda).$$

Having related different Feynman rules by such automorphisms, we may ask how the actual correlation functions are influenced by such a change. Here we can observe a kind of rigidity of the Dyson-Schwinger equation in

COROLLARY 6.19. *The new correlation function $\phi_R\circ X = \varphi\circ\widetilde{X}$ equals the original Feynman rules $\varphi$ applied to a modified perturbation series $\widetilde{X}(g)$, fulfilling an equivalent Dyson-Schwinger equation that merely differs in the cocycles by coboundaries. By* (A.7) *the leading logs coincide and explicitly*

$$\widetilde{X}(g) := {}^{\alpha\circ\varphi}\chi\circ X(g) = \mathbb{1} + \sum_{n\in\mathbb{N}} g^n\left(B_n + \delta\alpha_n\right)\left(\widetilde{X}(g)^{1+n\kappa}\right).$$

This follows directly through application of ${}^{\alpha\circ\varphi}\chi$ to the original Dyson-Schwinger equation. Since we consider now many cocycles $B_n$, each of which corresponding to a different Mellin transform $F_n$, also the functionals $\alpha_n$ are now indexed by $n$.

**6.5. Minimal subtraction.** In section 6.2 we already understood precisely why the arbitrary choice of the renormalization point $\mu$ in the kinematic subtraction scheme does not influence the physical interpretation. Now we can find an equivalent relation for the minimal subtraction scheme in

COROLLARY 6.20. *Applying (6.6) to (5.6) or (5.7) expresses the correlation function $G_{\text{MS}}$ of the $R_{MS}$-scheme in terms of $G$ in the kinematic subtraction scheme by a redefinition of the coupling constant and an overall factor:*

$$G_{\text{MS},\ell}(g) = G_{\text{MS},0}(g) \cdot G_\ell\Big(g \cdot [G_{\text{MS},0}(g)]^\kappa\Big). \tag{6.21}$$

## 7. Extensions towards quantum field theory

**7.1. Feynman graphs and Feynman integrals.** In the formulation of perturbative quantum field theory, the Hopf algebra $H_R$ of rooted trees is replaced by the Hopf algebra $H_{FG}$ of Feynman graphs [**9**]. Most importantly, it features insertion operators that act like Hochschild-1-cocycles on the relevant subspace of $H_{FG}$. Under the Feynman rules, these result in (divergent) sub integrals just as in (2.1) and the coproduct of $H_{FG}$ again mirrors the structure of subdivergences by definition.

The integrals are multi-dimensional and entail algebraic functions as integrands. For massive theories, a *Wick rotation* to Euclidean space-time disposes of all singularities in these integrands such that divergences only occur from the integrations at infinity. These may be renormalized by suitable subtractions again, though one has to face two new issues: Multiple parameters and higher degrees of divergence.

Before addressing these we remark that dimensional regularization [**7**] can be used to introduce a regulator $z \in \mathbb{C} \setminus \{0\}$ like in section 3 and assign a Laurent series ${}_z\phi$ in $z$ to every Feynman graph. Then dimensional analysis reveals a scale dependence of the form

$${}_z\phi = {}_z\phi|_{s=\tilde{s}} \circ \theta_{-z\ell}, \tag{7.1}$$

where we employ the grading $Y$ by *loop number* of $H_{FG}$ and $\ell = \ln \frac{s}{\tilde{s}}$ encodes the ratio of simultaneous rescaling of all dimensionful parameters. Therefore the techniques of section 3 become available in quantum field theory as shown for the first time in [**10**].

**7.2. Multiple parameters.** Correlation functions of quantum field theory typically depend on many variables, namely the masses of internal particles and the momenta of external particles. For illustration consider a logarithmic divergence with two parameters $(s, t)$: We can still renormalize

$$\phi_{\text{R},(s,t)} = \int_0^\infty \left[\frac{x\,\mathrm{d}x}{(x+s)(x+t)} - \frac{x\,\mathrm{d}x}{(x+\tilde{s})(x+\tilde{t})}\right] = \frac{s\ln s - t\ln t}{t-s} - \frac{\tilde{s}\ln\tilde{s} - \tilde{t}\ln\tilde{t}}{\tilde{t}-\tilde{s}} \tag{7.2}$$

by a single subtraction at a reference point $(\tilde{s}, \tilde{t})$ in the parameter space. However, the function is no longer a plain logarithm in a single variable $\ell$. In fact the dependence of correlation functions on the parameters becomes indeed extremely complicated and is only fully understood for the simplest (one-loop) Feynman graphs or slightly better in special situations like massless or supersymmetric theories, with on-shell conditions or in space-time dimensions different from four.

Crucially though the fundamental properties of a theory are described by asymptotic behavior, and the renormalization group still persists: If we rescale all parameters simultaneously by a factor $e^\ell$, then (7.2) simplifies drastically to $\phi_{\text{R},(\tilde{s}\cdot e^\ell, \tilde{t}\cdot e^\ell)} = -\ell$.

Analogously to the single scale case we considered, the presence of subdivergences requests additional subtractions generating richer dependence on $\ell$, which nevertheless stays polynomial throughout.

Explicitly, encode all parameters as multiples of a distinguished *scale* $s$ and dimensionless ratios $\theta \in \Theta$ called *angles* and choose a renormalization point $(\tilde{s}, \tilde{\Theta})$ for the subtraction scheme $R_{(\tilde{s},\tilde{\Theta})}$ evaluating $(s, \Theta) \mapsto (\tilde{s}, \tilde{\Theta})$. Then we can state (proof is provided in [**28**] for Feynman integrals) the replacement for (4.4) as

THEOREM 7.1. *The renormalized Feynman rules* $\phi_R = \phi_R|_{\Theta=\tilde{\Theta}} \star \phi_R|_{s=\tilde{s}}$ *factorize into the angle-dependent part* $\phi_R|_{s=\tilde{s}}$ *(independent of* $\ell$ *and only a function of* $\Theta$ *and* $\tilde{\Theta}$*) and the scale-dependence* $\phi_R|_{\Theta=\tilde{\Theta}}$. *The later depends only on* $\ell = \frac{s}{\tilde{s}}$ *(and the fixed renormalization point angles* $\tilde{\Theta}$*) and defines a morphism of Hopf algebras*

$$\phi_R|_{\Theta=\tilde{\Theta}} = \exp_\star(-\ell\mathcal{P}): \quad H \to \mathbb{K}[\ell]. \tag{7.3}$$

*Its generator* $\mathcal{P} \in \mathfrak{g}_{\mathbb{K}}^H$, *commonly called* period, *is given by* $\mathcal{P} := -\partial_\ell|_{\ell=0}\phi_R|_{\Theta=\tilde{\Theta}}$.

We also recommend [**6**] for a different decomposition and detailed analysis of the angle- and scale dependence.

**7.3. Higher degrees of divergence.** So far we restricted ourselves to logarithmic divergences only. Recall from (0.1) that in this case though $\phi$ itself diverges, the derivative $\frac{\partial}{\partial s}\phi$ is convergent (by differentiating the integrand we obtain an integrable form).

In general one defines the *superficial degree of divergence* sdd by simple power counting of the integrand $f$ such that $f(\zeta) \in \mathcal{O}\left(\zeta^{\operatorname{sdd}-n}\right)$, where $n$ counts the number of variables that we integrate over and the asymptotics are to be understood as all of these variables approaching $\infty$ jointly (for rational functions $f$, sdd is plainly the degree of the numerator minus the degree of the denominator, less the number of variables in the integral).

In this situation we find that any derivative $\frac{\partial^{\operatorname{sdd}+k}}{\partial s^{\operatorname{sdd}+k}}\phi$ for $k \in \mathbb{N}$ is convergent. Hence we can renormalize and keep these derivatives intact by subtracting a polynomial in $\mathbb{K}[s]_{\leq \operatorname{sdd}}$ of degree $\leq$ sdd. In the logarithmic case, this freedom is precisely a single constant we parametrized by $\mu$ so far.

This renormalization scheme is common practice in quantum field theory under the name BPHZ and the involved analytic estimates on the integrands necessary to prove the finiteness (as we did in theorem 2.4 in the simple setup of definition 2.1) have been worked out in [**35, 37**]. Variants exist for massless theories as well, while minimal subtraction (in connection with dimensional regularization) is particularly popular to handle gauge theories.

Subtractions of different polynomials in the external parameters are encoded into the Hopf algebra $H_{FG}$ by adjunction of auxiliary marked vertices and specification of *external structures* [**9**], also called *form factor decomposition.*

As we apply different subtractions depending on the value of sdd for each individual graph, the algebraic formulation of section 2.2 needs to be relaxed to encompass *Rota-Baxter families* [**11**]. The *exponential renormalization* introduced in [**12**] allows for even more general renormalization schemes and further explicitly captures the idea of order-by-order renormalization through counterterms common in physics. It also shows the role of reparametrizations of the coupling constant we commented on in section 6 and we further recommend [**13**] focussing on BPHZ.

We close by only briefly mentioning the increasing freedom in the Feynman rules coming along with growing degrees of divergence. If the divergences can attain arbitrarily high degrees, infinitely many subtraction terms are necessary in the renormalization process and thus generate as many constants (like our subtraction point $\mu$) to be fixed. Such a theory is unfortunately called *unrenormalizable*: Though it *is* renormalizable, it loses any predictive power due to the infinity of unknown constants.

Contrary, *renormalizable* theories adhere to an upper bound on the degree of divergences that occur. Therefore only finitely many parameters (*renormalization conditions*) have to be fixed through measurements, whereafter all other processes may in principle be predicted.

The divergences of scalar field theory together with their renormalization according to BPHZ is for example elaborated on in [**6**] in the parametric representation, though most textbooks contain at least a basic account of this theme.

**7.4. Overlapping divergences.** Historically, the possibility of overlapping (compared to nested and disjoint) subdivergences in Feynman graphs caused difficulties in proofs of renormalizability. However the forest formula and its Hopf algebraic counterparts (the coproduct and the antipode) as for example formulated in [**9**] encompass overlapping divergences seamlessly, without the need of a special separate treatment.

The article [**25**] explains how this is achieved and [**22**] constructs an alternative Hopf algebra to address these structures and clarifies their equivalence.

Also note that the so-called *multi-scale renormalization* avoids this issue altogether, as is explained in [**21**] with particular emphasis on its Hopf-algebraic formulation and the relations between these different Hopf algebras.

**7.5. Systems of Dyson-Schwinger equations.** Until now we only considered a single Dyson-Schwinger equation in section 6, though quantum field theorys typically involve different types of fields (like fermions, photons or scalars) and also a variety of couplings (*vertices*). Each of those is represented by an (mostly) infinite series over Feynman graphs and these series may be inserted into each other in many ways.

These changes can be incorporated combinatorially by considering systems of Dyson-Schwinger equations as for example studied in detail in [**18, 19**].

## 8. Summary

We reviewed renormalization of logarithmic ultraviolet divergences in the Hopf algebraic framework working with the rooted trees $H_R$. In the kinematic subtraction scheme we arrived at the same renormalized Feynman rules $\phi_{\mathrm{R}}$ either by direct integration or with an analytic regulator being present.

After renormalization, the physical limit revealed a very special structure as being a Hopf algebra morphism $\phi_{\mathrm{R}}\colon H_R \to \mathbb{K}[x]$. This is the *renormalization group property* and reduces the full character down to the linear terms $\gamma$ only.

The minimal subtraction scheme does not allow for such a simple description: Though we could obtain the scale dependence, the constant terms in this scheme are not as easily understood.

All along the case of Feynman rules that can be described by the Mellin transform is special in that it gives simple explicit recursions for the renormalization process.

In section 6 studying the correlation functions, we understood the physical equivalence of two different renormalization schemes and the renormalization group. We hinted at how these lead to possible non-perturbative formulations.

Hochschild cohomology appeared ubiquitously in form of the universal property of rooted trees, governing most constructions we made. Its importance lies not only in the concept and power of Dyson-Schwinger equations alone, but also in the induced automorphisms (A.5) of $H_R$ which help to understand variations of Feynman rules coming from different Mellin transforms.

## Appendix A. The Hopf algebra of rooted trees

Loïc Foissy studied rooted trees in depth and we only mention his thesis [**15**,**16**] and in particular the article [**14**] which discloses the structure of $H_R$ as a free shuffle algebra. Here we restrict to introduce the notions relevant to understand this article and in particular elaborate on the universal property.

As an algebra, $H_R = \mathbb{K}[\mathcal{T}]$ is free commutative[14] generated by the *rooted trees* $\mathcal{T}$ and spanned by their disjoint unions (products) called *rooted forests* $\mathcal{F}$:

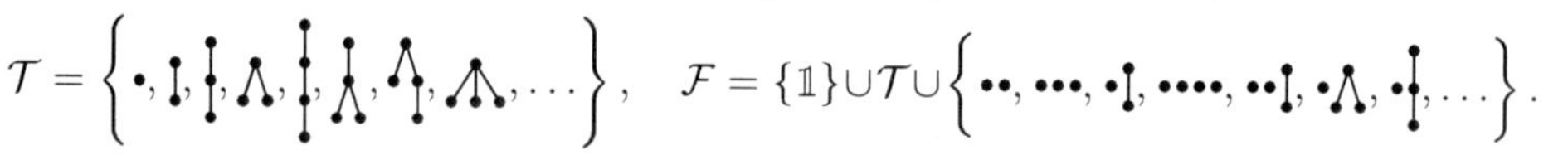

Every $w \in \mathcal{F}$ is just the monomial $w = \prod_{t\in\pi_0(w)} t$ of its multiset of tree components $\pi_0(w)$, while $\mathbb{1}$ denotes the empty forest. The number $|w| := |V(w)|$ of nodes $V(w)$ induces the grading $H_{R,n} = \operatorname{lin} \mathcal{F}_n$ where $\mathcal{F}_n := \{w \in \mathcal{F}\colon\ |w| = n\}$.

Definition A.1. *The (linear)* grafting operator $B_+ \in \operatorname{End}(H_R)$ *attaches all trees of a forest to a new root, for example* $B_+(\mathbb{1}) = \bullet$, $B_+(\bullet) =$  *and* $B_+(\bullet\bullet) =$ .

$B_+$ is homogenous of degree one and restricts to a bijection $B_+\colon\ \mathcal{F} \to \mathcal{T}$. The coproduct $\Delta$ is defined to make $B_+$ a cocycle by requiring

$$\Delta \circ B_+ = B_+ \otimes \mathbb{1} + (\operatorname{id} \otimes B_+) \circ \Delta. \tag{A.1}$$

Lemma A.2. $B_+(\mathbb{1}) = \bullet \neq 0$ *implies that* $0 \neq [B_+] \in HH^1_\varepsilon(H_R)$ *is non-trivial.*

$H_R$ is characterized through the *universal property* (theorem 2 in [**8**]) of

Theorem A.3. *For an algebra $\mathcal{A}$ and $L \in \operatorname{End}(\mathcal{A})$ there exists a unique morphism ${}^L\rho\colon H_R \to \mathcal{A}$ of unital algebras such that*

$$ {}^L\rho \circ B_+ = L \circ {}^L\rho, \quad \textit{equivalently} \quad \begin{array}{ccc} H_R & \xrightarrow{{}^L\rho} & \mathcal{A} \\ {\scriptstyle B_+}\downarrow & & \downarrow{\scriptstyle L} \\ H_R & \xrightarrow[{}^L\rho]{} & \mathcal{A} \end{array} \quad \textit{commutes.} \tag{A.2}$$

*In case of a bialgebra $\mathcal{A}$ and a cocycle $L \in HZ^1_\varepsilon(\mathcal{A})$, ${}^L\rho$ is a morphism of bialgebras and even of Hopf algebras when $\mathcal{A}$ is Hopf.*

---

[14] We consider *unordered* trees  =  and forests  = , sometimes called *non-planar*.

Note that ${}^{L}\rho = e$ trivializes if $L \in \mathrm{HB}^1_\varepsilon(\mathcal{A})$ is a coboundary.

EXAMPLE A.4. *The action of ${}^{L}\rho$ is plainly replacement of $B_+$ by $L$:*

$$ {}^{L}\rho\left(\Lambda - 3\bullet\right) = {}^{L}\rho\left\{B_+\left([B_+(\mathbb{1})]^2\right) - 3B_+(\mathbb{1})\right\} = L\left([L(\mathbb{1}_{\mathcal{A}})]^2\right) - 3L(\mathbb{1}_{\mathcal{A}}). $$

EXAMPLE A.5. *The cocycle $\int_0 \in HZ^1_\varepsilon(\mathbb{K}[x])$ of appendix* B *induces the character*

$$ \text{(A.3)} \qquad \varphi := {}^{\int_0}\rho \in G^{H_R}_{\mathbb{K}[x]} \quad \textit{mapping any forest } w \in \mathcal{F} \textit{ to } \quad \varphi(w) = \frac{x^{|w|}}{w!}, \quad \textit{using} $$

DEFINITION A.6. *The* tree factorial $(\cdot)! \in G^{H_R}_{\mathbb{K}}$ *is for any $w \in \mathcal{F}$ given by*

$$ \text{(A.4)} \qquad [B_+(w)]! = w! \cdot |B_+(w)|, \quad \textit{equivalently} \quad w! = \prod_{v \in V(w)} |w_v|. $$

[15]

**A.1. Automorphisms of $H_R$.** Applying the universal property to $H_R$ itself, adding coboundaries to $B_+$ leads to

DEFINITION A.7. *For any $\alpha \in H_R'$, theorem* A.3 *defines the Hopf algebra morphism*

$$ \text{(A.5)} \qquad {}^{\alpha}\chi := {}^{B_+ + \delta\alpha}\rho\colon\ H_R \to H_R \quad \textit{such that} \quad {}^{\alpha}\chi \circ B_+ = [B_+ + \delta\alpha] \circ {}^{\alpha}\chi. $$

EXAMPLE A.8. *The action on the simplest trees yields*

$$ {}^{\alpha}\chi(\bullet) = {}^{\alpha}\chi \circ B_+(\mathbb{1}) = B_+(\mathbb{1}) + (\delta\alpha)(\mathbb{1}) = B_+(\mathbb{1}) = \bullet, $$
$$ {}^{\alpha}\chi(\mathbf{I}) = {}^{\alpha}\chi \circ B_+(\bullet) = (B_+ + \delta\alpha)\, {}^{\alpha}\chi(\bullet) = \mathbf{I} + \delta\alpha(\bullet) = \mathbf{I} + \alpha(\mathbb{1})\bullet, $$
$$ {}^{\alpha}\chi\left(\vdots\right) = \vdots + 2\alpha(\mathbb{1})\mathbf{I} + \left\{[\alpha(\mathbb{1})]^2 + \alpha(\bullet)\right\}\bullet \quad \textit{and} \quad {}^{\alpha}\chi(\Lambda) = \Lambda + 2\alpha(\bullet)\bullet + \alpha(\mathbb{1})\bullet\bullet. $$

These morphisms capture how ${}^{L}\rho$ reacts to variation of $L$ by a coboundary in

THEOREM A.9. *Let $H$ denote a bialgebra, $L \in HZ^1_\varepsilon(H)$ a 1-cocycle and further $\alpha \in H'$ a functional. Then for ${}^{L}\rho, {}^{L+\delta\alpha}\rho\colon\ H_R \to H$ given through theorem* A.3 *and ${}^{\alpha\circ{}^{L}\rho}\chi\colon\ H_R \to H_R$ from definition* A.7*, we have*

$$ \text{(A.6)} \qquad {}^{L+\delta\alpha}\rho = {}^{L}\rho \circ {}^{[\alpha \circ {}^{L}\rho]}\chi, \quad \textit{equivalently} \quad \begin{array}{ccc} H_R & \xrightarrow{{}^{L+\delta\alpha}\rho} & H \\ {\scriptstyle {}^{\alpha\circ{}^{L}\rho}\chi}\downarrow & \nearrow{\scriptstyle {}^{L}\rho} & \\ H_R & & \end{array} \quad \textit{commutes.} $$

PROOF. As both sides of (A.6) are algebra morphisms, it suffices to prove it inductively for trees: Let it be true for a forest $w \in \mathcal{F}$, then it holds as well for the tree $B_+(w)$ by

$$ {}^{L}\rho \circ {}^{[\alpha\circ{}^{L}\rho]}\chi \circ B_+(w) \underset{(A.2)}{=} {}^{L}\rho \circ \left[B_+ + \delta\left(\alpha \circ {}^{L}\rho\right)\right] \circ {}^{[\alpha\circ{}^{L}\rho]}\chi(w) $$
$$ = \left\{L \circ {}^{L}\rho + (\delta\alpha) \circ {}^{L}\rho\right\} \circ {}^{[\alpha\circ{}^{L}\rho]}\chi(w) = \{L + \delta\alpha\} \circ \underbrace{{}^{L}\rho \circ {}^{[\alpha\circ{}^{L}\rho]}\chi(w)}_{{}^{L+\delta\alpha}\rho(w)} \underset{(A.2)}{=} {}^{L+\delta\alpha}\rho \circ B_+(w). $$

We used $(\delta\alpha)\circ{}^{L}\rho = {}^{L}\rho\circ\delta\left(\alpha \circ {}^{L}\rho\right)$, following from ${}^{L}\rho$ being a morphism of bialgebras. □

[15] By $w_v$ we denote the subtree of $w$ rooted at the node $v \in V(w)$.

THEOREM A.10. *The map* ${}^{\cdot}\chi\colon H_R' \to \mathrm{End}_{Hopf}(H_R)$, *taking values in the space of Hopf algebra endomorphisms of* $H_R$, *fulfills the following properties:*

(1) *For any* $w \in \mathcal{F}$ *and* $\alpha \in H_R'$, ${}^{\alpha}\chi(w)$ *differs from* $w$ *only by lower order forests:*

$$ {}^{\alpha}\chi(w) \in w + H_R^{|w|-1} = w + \bigoplus_{n=0}^{|w|-1} H_{R,n}. \tag{A.7}$$

(2) ${}^{\cdot}\chi$ *maps* $H_R'$ *into the Hopf algebra automorphisms* $\mathrm{Aut}_{Hopf}(H_R)$. *Its image is closed under composition, as for any* $\alpha, \beta \in H_R'$ *we have*

$$ {}^{\alpha}\chi \circ {}^{\beta}\chi = {}^{\gamma}\chi \quad \text{upon setting} \quad \gamma = \alpha + \beta \circ {}^{\alpha}\chi^{-1}. \tag{A.8}$$

(3) *The maps* $\delta\colon H_R' \to HZ_{\varepsilon}^1(H_R)$ *and* ${}^{\cdot}\chi\colon H_R' \to \mathrm{Aut}_{Hopf}(H_R)$ *are injective, thus the subgroup* $\mathrm{im}\,{}^{\cdot}\chi = \{{}^{\alpha}\chi\colon\ \alpha \in H_R'\} \subset \mathrm{Aut}_{Hopf}(H_R)$ *induces a group structure on* $H_R'$ *with neutral element* $0$ *and group law* $\triangleright$ *given by*

$$ \alpha \triangleright \beta := {}^{\cdot}\chi^{-1}\left({}^{\alpha}\chi \circ {}^{\beta}\chi\right) \underset{(A.8)}{=} \alpha + \beta \circ {}^{\alpha}\chi^{-1} \quad \text{and} \quad \alpha^{\triangleright -1} = -\alpha \circ {}^{\alpha}\chi. \tag{A.9}$$

PROOF. Statement (A.7) is an immediate consequence of $\delta\alpha(H_R^n) \subseteq H_R^n$: Starting from ${}^{\alpha}\chi(\bullet) = \bullet$, suppose inductively (A.7) to hold for forests $w, w' \in \mathcal{F}$. Then it obviously also holds for $w \cdot w'$ as well and even so for $B_+(w)$ through

$$ {}^{\alpha}\chi \circ B_+(w) = [B_+ + \delta\alpha] \circ {}^{\alpha}\chi(w) \subseteq [B_+ + \delta\alpha]\left(w + H_R^{|w|-1}\right) \subseteq B_+(w) + H_R^{|w|}.$$

This already implies bijectivity of ${}^{\alpha}\chi$, but applying (A.6) to $L = B_+ + \delta\alpha$ and ${}^{\tilde{\alpha}}\chi$ for $\tilde{\alpha} := -\alpha \circ {}^{\alpha}\chi$ shows $\mathrm{id} = {}^{\alpha}\chi \circ {}^{\tilde{\alpha}}\chi$ directly. We deduce bijectivity of all ${}^{\alpha}\chi$ and thus ${}^{\alpha}\chi \in \mathrm{Aut}_{\mathrm{Hopf}}(H_R)$ with the inverse ${}^{\alpha}\chi^{-1} = {}^{\tilde{\alpha}}\chi$. Now (A.8) follows from

$$ {}^{\left[\alpha + \beta \circ {}^{\alpha}\chi^{-1}\right]}\chi = {}^{[B_+ + \delta\alpha] + \delta\left(\beta \circ {}^{\alpha}\chi^{-1}\right)}\rho \underset{(A.6)}{=} {}^{[B_+ + \delta\alpha]}\rho \circ {}^{\left[\beta \,\circ\, {}^{\alpha}\chi^{-1} \,\circ\, {}^{(B_+ + \delta\alpha)}\rho\right]}\chi = {}^{\alpha}\chi \circ {}^{\beta}\chi.$$

Finally consider $\alpha, \beta \in H_R'$ with ${}^{\alpha}\chi = {}^{\beta}\chi$, then $0 = ({}^{\alpha}\chi - {}^{\beta}\chi) \circ B_+ = \delta \circ (\alpha - \beta) \circ {}^{\alpha}\chi$ reduces the injectivity of ${}^{\cdot}\chi$ to that of $\delta$. But if $\delta\alpha = 0$, for all $n \in \mathbb{N}_0$

$$ 0 = \delta\alpha\left(\bullet^{n+1}\right) = \sum_{i=0}^{n} \binom{n+1}{i} \alpha\left(\bullet^{i}\right) \bullet^{n+1-i} \quad \text{implies} \quad \alpha\left(\bullet^{n}\right) = 0.$$

Given an arbitrary forest $w \in \mathcal{F}$ and $n \in \mathbb{N}$, the expression

$$ \begin{aligned} 0 = \delta\alpha\left(\bullet^{n} w\right) = w \underbrace{\alpha\left(\bullet^{n}\right)}_{0} &+ \sum_{w} \sum_{i=0}^{n} \binom{n}{i} \bullet^{i} w' \alpha\left(\bullet^{n-i} w''\right) \\ &+ \sum_{i=1}^{n} \binom{n}{i} \left[\bullet^{i} w \underbrace{\alpha\left(\bullet^{n-i}\right)}_{0} + \bullet^{i} \alpha\left(w \bullet^{n-i}\right)\right] \end{aligned}$$

simplifies upon projection onto $\mathbb{K}\bullet$ to $\alpha\left(w \bullet^{n-1}\right) = -\frac{1}{n} \sum_{w\colon\ w' = \bullet} \alpha\left(\bullet^{n} w''\right)$. Iterating this formula exhibits $\alpha(w)$ as a scalar multiple of $\alpha\left(\bullet^{|w|}\right) = 0$ and proves $\alpha = 0$. $\square$

**A.2. Decorated rooted trees.** Our observations generalize straightforwardly to the Hopf algebra $H_R(\mathcal{D})$ of rooted trees with decorations drawn from a set $\mathcal{D}$. In this case, the universal property assigns to each $\mathcal{D}$-indexed family $L_\cdot\colon \mathcal{D} \to \operatorname{End}(\mathcal{A})$ the unique algebra morphism

$$^{L_\cdot}\rho\colon H_R(\mathcal{D}) \to \mathcal{A} \quad \text{such that} \quad {}^{L_\cdot}\rho \circ B_+^d = L_d \circ {}^{L_\cdot}\rho \quad \text{for any } d \in \mathcal{D}.$$

For cocycles $\operatorname{im} L_\cdot \subseteq HZ_\varepsilon^1(\mathcal{A})$ this is a morphism of bialgebras and even of Hopf algebras (should $\mathcal{A}$ be Hopf). For a family $\alpha_\cdot\colon \mathcal{D} \to H_R'(\mathcal{D})$ of functionals, setting $L_d^{\alpha_\cdot} := B_+^d + \delta\alpha_d$ yields an automorphism $^{\alpha_\cdot}\chi := {}^{L^{\alpha_\cdot}}\rho$ of the Hopf algebra $H_R(\mathcal{D})$. Theorems A.9 and A.10 generalize in the obvious way.

## Appendix B. The Hopf algebra of polynomials

LEMMA B.1. *Requiring* $\Delta(x) = x \otimes \mathbb{1} + \mathbb{1} \otimes x$ *induces a unique Hopf algebra structure on the polynomials* $\mathbb{K}[x]$. *It is graded by degree, connected, commutative and cocommutative with* $\Delta(x^n) = \sum_{i=0}^n \binom{n}{i} x^i \otimes x^{n-i}$ *and the primitive elements are precisely* $\operatorname{Prim}(\mathbb{K}[x]) = \mathbb{K} \cdot x$.

The *integration operator* $\int_0\colon x^n \mapsto \frac{1}{n+1}x^{n+1}$ is a cocycle $\int_0 \in \mathrm{HZ}_\varepsilon^1(\mathbb{K}[x])$ as

$$\begin{aligned}\Delta \textstyle\int_0 \left(\frac{x^n}{n!}\right) &= \Delta\left(\frac{x^{n+1}}{(n+1)!}\right) = \sum_{k=0}^{n+1} \frac{x^k}{k!} \otimes \frac{x^{n+1-k}}{(n+1-k)!} \\ &= \frac{x^{n+1}}{(n+1)!} \otimes \mathbb{1} + \sum_{k=0}^{n} \frac{x^k}{k!} \otimes \textstyle\int_0 \left(\frac{x^{n-k}}{(n-k)!}\right) = \left[\int_0 \otimes \mathbb{1} + \left(\mathrm{id} \otimes \int_0\right) \circ \Delta\right] \left(\frac{x^n}{n!}\right),\end{aligned}$$

and is not a coboundary since $\int_0 1 = x \neq 0$. In fact it generates the cohomology by

THEOREM B.2. $HH_\varepsilon^1(\mathbb{K}[x]) = \mathbb{K} \cdot [\int_0]$ *is one-dimensional as the 1-cocycles are*

$$HZ_\varepsilon^1(\mathbb{K}[x]) = \mathbb{K} \cdot \textstyle\int_0 \oplus \delta\left(\mathbb{K}[x]'\right) = \mathbb{K} \cdot \int_0 \oplus HB_\varepsilon^1(\mathbb{K}[x]). \tag{B.1}$$

PROOF. For an arbitrary cocycle $L \in \mathrm{HZ}_\varepsilon^1(\mathbb{K}[x])$, lemma 1.1 ensures $L(1) = xa_{-1}$ where $a_{-1} := \partial_0 L(1)$. Hence $\tilde{L} := L - a_{-1} \int_0 \in \mathrm{HZ}_\varepsilon^1$ fulfills $\tilde{L}(1) = 0$, so $L_0 := \tilde{L} \circ \int_0 \in \mathrm{HZ}_\varepsilon^1$ by

$$\Delta \circ L_0 = (\mathrm{id} \otimes \tilde{L}) \circ \Delta \circ \textstyle\int_0 + (\tilde{L} \otimes 1) \circ \int_0 = (\mathrm{id} \otimes L_0) \circ \Delta + L_0 \otimes 1 + \tilde{L}(1) \cdot \int_0.$$

Repeating the argument inductively yields $a_n := \partial_0 L_n(1) = \partial_0 \circ L \circ \int_0^{n+1}(1) \in \mathbb{K}$ and $L_{n+1} := (L_n - a_n \int_0) \circ \int_0 \in \mathrm{HZ}_\varepsilon^1$, so for any $n \in \mathbb{N}_0$ we may read off from

$$\begin{aligned}L \circ \textstyle\int_0^n(1) = a_{-1} \int_0^{n+1}(1) + \ldots + a_{n-2} \int_0^2(1) + L_{n-1}(1) &= a_{-1} \textstyle\int_0 \left(\int_0^n 1\right) \\ &\quad + \sum_{j=0}^{n-1} a_j \textstyle\int_0^{n-j}(1)\end{aligned}$$

that indeed $L = a_{-1} \int_0 + \delta\alpha$ for the functional $\alpha := \partial_0 \circ L \circ \int_0$ with $\alpha(\frac{x^n}{n!}) = a_n$. □

LEMMA B.3. *Up to subtraction* $P = \delta\varepsilon = \mathrm{id} - \mathrm{ev}_0 : \mathbb{K}[x] \twoheadrightarrow \ker\varepsilon = x\mathbb{K}[x]$ *of the constant part, direct computation exhibits* $\delta\alpha$ *as the differential operator*

$$\delta\alpha = P \circ \sum_{n \in \mathbb{N}_0} \alpha\left(\tfrac{x^n}{n!}\right) \partial^n \in \operatorname{End}(\mathbb{K}[x]) \quad \textit{for any} \quad \alpha \in \mathbb{K}[x]'. \tag{B.2}$$

Lemma B.4. *As any character* $\phi \in G_{\mathbb{K}}^{\mathbb{K}[x]}$ *of* $\mathbb{K}[x]$ *is fixed by* $\lambda := \phi(x)$*, they are the group* $G_{\mathbb{K}}^{\mathbb{K}[x]} = \{\mathrm{ev}_\lambda \colon\ \lambda \in \mathbb{K}\}$ *of evaluations (the counit* $\varepsilon = \mathrm{ev}_0$ *equals the neutral element)*

$$\mathbb{K}[x] \ni p(x) \mapsto \mathrm{ev}_\lambda(p) := p(\lambda) \quad \textit{with the product} \quad \mathrm{ev}_a \star \mathrm{ev}_b = \mathrm{ev}_{a+b}. \tag{B.3}$$

Proof. Note $[\mathrm{ev}_a \star \mathrm{ev}_b](x^n) = [\mathrm{ev}_a(1) \cdot \mathrm{ev}_b(x) + \mathrm{ev}_a(x) \cdot \mathrm{ev}_b(1)]^n = (b+a)^n$. ☐

Lemma B.5. *The isomorphism* $(\mathbb{K}, +) \ni a \mapsto \mathrm{ev}_a \in G_{\mathbb{K}}^{\mathbb{K}[x]}$ *of groups is generated by the functional* $\partial_0 = \mathrm{ev}_0 \circ \partial \in \mathfrak{g}_{\mathbb{K}}^{\mathbb{K}[x]}$*, meaning* $\log_\star \mathrm{ev}_a = a\partial_0$ *and* $\mathrm{ev}_a = \exp_\star(a\partial_0)$.

Proof. Expanding the exponential series reveals $\exp_\star(a\partial_0)(x^n) = a^n$ as a direct consequence of $\partial_0^{\star k} = \varepsilon \circ \partial^{\star k} = \varepsilon \circ \partial^k$:

$$\partial_0^{\star k}\left(\frac{x^n}{n!}\right) = \sum_{i_1+\ldots+i_k=n} \left(\partial_0 \tfrac{x^{i_1}}{i_1!}\right) \cdots \left(\partial_0 \tfrac{x^{i_k}}{i_k!}\right) = \sum_{i_1+\ldots+i_k=n} \delta_{1,i_1} \cdots \delta_{1,i_k} = \delta_{k,n} = \partial^k\big|_0 \left(\frac{x^n}{n!}\right). \quad \square$$

## Appendix C. The Dynkin operator $D = S \star Y$

We briefly present the crucial properties of $D$ which are employed in this text and further recommend in particular section 4 of [**11**] as well as [**31**].

Definition C.1. *For some fixed connected graduation* $Y$ *of* $H$*, define operators* $D_Y := S \star Y$ *and* $\pi_Y := Y^{-1} \circ D_Y = D_Y \circ Y^{-1}$.

As $Y$ is invertible on $\ker \varepsilon$, $\pi_Y$ is well-defined. Note that each of $\{S, D_Y, \pi_Y\}$ commutes with $Y$ and $Y^{-1}$.

Proposition C.2. $D_Y, \pi_Y \in \mathfrak{g}_H^H$ *are infinitesimal characters with* $\mathbb{K} \cdot \mathbb{1} \oplus (\ker \varepsilon)^2 = \ker D_Y = \ker \pi_Y$ *and* $D_Y - Y, \pi_Y - P$ *map into* $(\ker \varepsilon)^2$.

Proof. Clearly, $\mathbb{K} \cdot \mathbb{1} \oplus (\ker \varepsilon)^2 \subseteq \ker D_Y$ is an immediate consequence of

$$\begin{aligned} D_Y \circ m &= m \circ (S \otimes Y) \circ \Delta \circ m = m \circ (S \otimes Y) \circ (m \otimes m) \circ \tau_{2,3} \circ (\Delta \otimes \Delta) \\ &= m^3 \circ (S \otimes S \otimes Y \otimes \mathrm{id} + S \otimes S \otimes \mathrm{id} \otimes Y) \circ \tau_{1,2} \circ \tau_{2,3} \circ (\Delta \otimes \Delta) \\ &= m \circ [(S \star Y) \otimes (S \star \mathrm{id}) + (S \star \mathrm{id}) \otimes (S \star Y)] = D_Y \otimes \varepsilon + \varepsilon \otimes D_Y. \end{aligned}$$

The reverse inclusion follows from $D_Y(x) = Yx + \sum_x (Sx')(Yx'') = Yx \mod (\ker \varepsilon)^2$ for $x \in \ker \varepsilon$. ☐

Corollary C.3. $V_Y := \operatorname{im} D_Y = \operatorname{im} \pi_Y$ *generates* $H$ *as an algebra and contains the primitive elements* $\mathrm{Prim}(H) \subseteq V_Y$.

Proof. $H_{n+1} \subseteq V_Y^{\cdot} := \sum_{n \geq 0} V_Y^n$ follows inductively from $H^n = \bigoplus_{i \leq n} H_i \subseteq V_Y^{\cdot}$ using $x \in \pi_Y(x) + m(H^n \otimes H^n)$ for any $x \in H_{n+1}$. A primitive $p$ yields $D_Y(p) = S(p) \cdot 0 + S(\mathbb{1}) \cdot Y(p) = Y(p)$. ☐

Proposition C.4. $\pi_Y^2 = \pi_Y$ *is a projection, hence its image complements the square of the augmentation ideal:* $H = \mathbb{K} \cdot \mathbb{1} \oplus V_Y \oplus (\ker \varepsilon)^2, \ker \varepsilon = V_Y \oplus (\ker \varepsilon)^2$.

Proof. Expand $D_Y^2 = m \circ (S \otimes Y) \circ (m \otimes m) \circ \tau_{2,3} \circ (\Delta \otimes \Delta) \circ (S \otimes Y) \circ \Delta$ to

$$\begin{aligned} D_Y^2 &= m^3 \circ [S \otimes S \otimes Y \otimes \mathrm{id} + S \otimes S \otimes \mathrm{id} \otimes Y] \circ \tau_{1,2} \circ \tau_{2,3} \circ (\Delta \otimes \Delta) \circ (S \otimes Y) \circ \Delta \\ &= m \circ [(S \star Y) \otimes e + e \otimes (S \star Y)] \circ (S \otimes Y) \circ \Delta = D_Y \circ Y. \end{aligned}$$

☐

PROPOSITION C.5. *From* $\Delta \circ D_Y = \mathbb{1} \otimes D_Y + [D_Y \otimes m \circ (S \otimes \mathrm{id})] \circ \tau_{1,2} \circ \Delta^2$ *we deduce that* $\mathbb{K} \cdot \mathbb{1} \oplus V_Y$ *is a right-coideal. Further,* $\pi_Y$ *and* $D_Y$ *map co-commutative elements to primitives as then* $\Delta \circ \pi_Y = \mathbb{1} \otimes \pi_Y + \pi_Y \otimes \mathbb{1}$.

PROOF. Apply $S \star \mathrm{id} = e = \mathbb{1} \cdot \varepsilon$ and $(\mathrm{id} \otimes \varepsilon) \circ \Delta = \mathrm{id}$ to

$$\Delta \circ D_Y = (m \otimes m) \circ \tau_{2,3} \circ \tau_{1,2} \circ [S \otimes S \otimes Y \otimes \mathrm{id} + S \otimes S \otimes \mathrm{id} \otimes Y] \circ \Delta^3$$

$$= \Big\{ (S \star Y) \otimes [m \circ (S \otimes \mathrm{id})] \Big\} \circ \tau_{1,2} \circ \Delta^2 + \Big\{ (S \star \mathrm{id}) \otimes [m \circ (S \otimes Y)] \Big\} \circ \tau_{1,2} \circ \Delta^2.$$

□

COROLLARY C.6. *For cocommutative* $H$, $\exp_\star(\pi_Y) \in G_H^H$ *is a character that coincides with* id *on the generating subspace* $\mathrm{im}\,(\pi_Y) = \mathrm{Prim}(H)$, *hence*

$$\exp_\star(\pi_Y) = \mathrm{id}, \qquad \textit{equivalently} \qquad \log_\star(\mathrm{id}) = \pi_Y. \tag{C.1}$$

In particular note that in this case $\pi_Y = \log_\star(\mathrm{id})$ does not depend on the choice of grading $Y$. Recalling that by the Milnor-Moore theorem for this case $H = S(\mathrm{Prim}(H))$ is just the symmetric algebra, $\pi_Y$ is nothing but the projection on $\mathrm{Prim}(H)$ corresponding to

$$H = \bigoplus_{n \geq 0} \mathrm{Prim}(H)^{\otimes n}. \tag{C.2}$$

But also in the non-cocommutative case we have

PROPOSITION C.7. $V_Y$ *generates* $H$ *as a free algebra:* $S(V_Y) = H$ *(as algebras).*

PROOF. The inclusion $V_Y \hookrightarrow H$ induces a unique morphism $\nu : S(V_Y) \twoheadrightarrow H$ of algebras which is surjective by C.3. For $n, m \in \mathbb{N}_0$ and $v_1, \ldots, v_m \in H$,

$$\pi_Y^{\star n}(v_1 \cdots v_m) = \sum_{i_1 + \ldots + i_m = n} \binom{n}{i_1 \cdots i_m} \pi_Y^{\star i_1}(v_1) \cdots \pi_Y^{\star i_m}(v_m) \tag{$*$}$$

results from iteration of $\pi_Y \circ m = m \circ (e \otimes \pi_Y + \pi_Y \otimes e)$ and proves

$$\pi_Y^{\star n}(V_Y^m) = 0 \quad \text{for any} \quad 0 \leq n < m$$

as in $(*)$ some $i_k$ must vanish and $\pi_Y^{\star 0} = e$ annihilates $V_Y \subset \ker \varepsilon$. In the case $n = m$ we find $\pi_Y^{\star n}|_{V_Y^n} = n! \cdot \mathrm{id}|_{V_Y^n}$ by $i_1 = \ldots = i_n = 1$. Therefore a finite sum $0 = \sum_{n \geq 0} x_n \in H$ with $x_n \in V_Y^n$ implies $x_0 = \frac{1}{0!}\pi_Y^{\star 0}(x) = 0$, then $x_1 = \frac{1}{1!}\pi_Y^{\star 1}(x) = 0$ and hence iteratively $x_n = 0$ for any $n$.

Thus $H = \bigoplus_{n \geq 0} V_Y^n$ is a direct sum with $V_Y^n = \nu(S_n(V_Y))$ upon the decomposition $S(V_Y) = \bigoplus_{n \geq 0} S_n(V_Y)$ into the homogeneous polynomials $S_n(V_Y)$ of degree $n$. Since $\pi_Y^{\otimes n} \circ \Delta^{n-1}(v_1 \cdots v_n) = \sum_\sigma v_{\sigma(1)} \otimes \ldots \otimes v_{\sigma(n)}$ for $v_1, \ldots, v_n \in V_Y$ delivers an inverse to $\nu|_{S_n(V_Y)}$, $\nu$ is injective. □

Hence all commutative connected graded Hopf algebras are free, and $V_Y \supseteq \mathrm{Prim}(H)$ delivers a generating coideal. Note that we defined the algebra of rooted trees as $S(\mathrm{lin}\,\mathcal{T})$, but $V_Y$ is a much more special generator as it contains all primitives and hence roughly speaking many elements with simple coproducts. Especially in the cocommutative case $V_Y$ delivers the simplest possible generator $V_Y = \mathrm{Prim}(H)$ and $S(V_Y) = H$ becomes an isomorphism of Hopf algebras.

COROLLARY C.8. *The map* $\tilde{R}: G^H_{\mathcal{A}} \to \mathfrak{g}^H_{\mathcal{A}}, \varphi \mapsto \varphi^{\star-1}\star(\varphi\circ Y) = \varphi \circ D_Y$ *defined in* [**30**] *is a bijection, since the character* $\varphi$ *is already determined by its restriction on* $V_Y = \operatorname{im} D_Y$ *through corollary* C.3.

THEOREM C.9 (*scattering formula* from [**10**] in the form of [**30**]). *The inverse of* $\tilde{R}$ *is given by* $\tilde{R}^{-1}(\beta) = \lim_{t\to\infty} e^{-tZ_0} e^{t(Z_0+\beta)}$ *for any* $\beta \in \mathfrak{g}^H_{\mathcal{A}}$. *This equation is to be understood in the Lie group associated to the semidirect sum* $\mathfrak{g}^H_{\mathcal{A}} \rtimes \mathbb{C}\cdot Z_0$, *adjoining the derivation* $Z_0(\beta) := \beta \circ Y$.

Another description of $\tilde{R}^{-1}$ can be given as follows (see also [**11**, **31**]): Denote by $\Psi \in \operatorname{End}(\operatorname{End}(H))$ the map $\operatorname{End}(H) \ni \alpha \mapsto \Psi(\alpha) := (\alpha \star D_Y)\circ Y^{-1}$ and exploit $\mathrm{id} = e + Y \circ Y^{-1}$ as well as $Y = \mathrm{id}\star S \star Y = \mathrm{id}\star D_Y$ repeatedly, then

$$\begin{aligned}\mathrm{id} &= e + (\mathrm{id}\star D_Y)\circ Y^{-1} = e + \left\{\left[e + (\mathrm{id}\otimes D_Y)\circ Y^{-1}\right]\star D_Y\right\}\circ Y^{-1}\\ &= \cdots = \sum_{n\geq 0}\Psi^n(e).\end{aligned} \tag{C.3}$$

Observe that $\Psi^n(e)$ vanishes on any element $x \in H$ of coradical degree less than $n$, wherefore this series is pointwise finite. Hence given $\beta := \tilde{R}(\varphi)$ we can reconstruct

$$\begin{aligned}\varphi = \varphi\circ\mathrm{id} &= e + \beta\circ Y^{-1} + \left[(\beta\circ Y^{-1})\star\beta\right]\circ Y^{-1}\\ &\quad + \left(\left\{\left[(\beta\circ Y^{-1})\star\beta\right]\circ Y^{-1}\right\}\star\beta\right)\circ Y^{-1} + \ldots.\end{aligned} \tag{C.4}$$

## References


[1] Christoph Bergbauer and Dirk Kreimer, *Hopf algebras in renormalization theory: locality and Dyson-Schwinger equations from Hochschild cohomology*, Physics and number theory, IRMA Lect. Math. Theor. Phys., vol. 10, Eur. Math. Soc., Zürich, 2006, pp. 133–164, DOI 10.4171/028-1/4. MR2277758 (2007k:81138)

[2] I. Bierenbaum, D. Kreimer, and S. Weinzierl, *The next-to-ladder approximation for Dyson-Schwinger equations*, Phys. Lett. B **646** (2007), no. 2–3, 129–133.

[3] Spencer Bloch and Dirk Kreimer, *Mixed Hodge structures and renormalization in physics*, Commun. Number Theory Phys. **2** (2008), no. 4, 637–718, DOI 10.4310/CNTP.2008.v2.n4.a1. MR2492196 (2010h:81101)

[4] D. J. Broadhurst and D. Kreimer, *Renormalization automated by Hopf algebra*, J. Symbolic Comput. **27** (1999), no. 6, 581–600, DOI 10.1006/jsco.1999.0283. MR1701096 (2000h:81167)

[5] D. J. Broadhurst and D. Kreimer, *Exact solutions of Dyson-Schwinger equations for iterated one-loop integrals and propagator-coupling duality*, Nucl. Phys. B **600** (2001), no. 2, 403–422.

[6] Francis Brown and Dirk Kreimer, *Angles, scales and parametric renormalization*, Lett. Math. Phys. **103** (2013), no. 9, 933–1007, DOI 10.1007/s11005-013-0625-6. MR3077961

[7] John C. Collins, *Renormalization*, Cambridge Monographs on Mathematical Physics, Cambridge University Press, Cambridge, 1984. An introduction to renormalization, the renormalization group, and the operator-product expansion. MR778558 (86k:81093)

[8] Alain Connes and Dirk Kreimer, *Hopf algebras, renormalization and noncommutative geometry*, Comm. Math. Phys. **199** (1998), no. 1, 203–242, DOI 10.1007/s002200050499. MR1660199 (99h:81137)

[9] Alain Connes and Dirk Kreimer, *Renormalization in quantum field theory and the Riemann-Hilbert problem. I. The Hopf algebra structure of graphs and the main theorem*, Comm. Math. Phys. **210** (2000), no. 1, 249–273, DOI 10.1007/s002200050779. MR1748177 (2002f:81070)

[10] Alain Connes and Dirk Kreimer, *Renormalization in quantum field theory and the Riemann-Hilbert problem. II. The β-function, diffeomorphisms and the renormalization group*, Comm. Math. Phys. **216** (2001), no. 1, 215–241, DOI 10.1007/PL00005547. MR1810779 (2002f:81071)

[11] Kurusch Ebrahimi-Fard, José M. Gracia-Bondía, and Frédéric Patras, *A Lie theoretic approach to renormalization*, Comm. Math. Phys. **276** (2007), no. 2, 519–549, DOI 10.1007/s00220-007-0346-8. MR2346399 (2009b:81135)

[12] Kurusch Ebrahimi-Fard and Frédéric Patras, *Exponential renormalization*, Ann. Henri Poincaré **11** (2010), no. 5, 943–971, DOI 10.1007/s00023-010-0050-7. MR2736528 (2012b:81164)
[13] Kurusch Ebrahimi-Fard and Frédéric Patras, *Exponential renormalisation. II. Bogoliubov's R-operation and momentum subtraction schemes*, J. Math. Phys. **53** (2012), no. 8, 083505, 15, DOI 10.1063/1.4742185. MR3012649
[14] L. Foissy, *Finite-dimensional comodules over the Hopf algebra of rooted trees*, J. Algebra **255** (2002), no. 1, 89–120, DOI 10.1016/S0021-8693(02)00110-2. MR1935036 (2003j:16047)
[15] L. Foissy, *Les algèbres de Hopf des arbres enracinés décorés. I* (French, with English and French summaries), Bull. Sci. Math. **126** (2002), no. 3, 193–239, DOI 10.1016/S0007-4497(02)01108-9. MR1905177 (2003d:16049)
[16] L. Foissy, *Les algèbres de Hopf des arbres enracinés décorés. II* (French, with English and French summaries), Bull. Sci. Math. **126** (2002), no. 4, 249–288, DOI 10.1016/S0007-4497(02)01113-2. MR1909461 (2003e:16047)
[17] Loïc Foissy, *Faà di Bruno subalgebras of the Hopf algebra of planar trees from combinatorial Dyson-Schwinger equations*, Adv. Math. **218** (2008), no. 1, 136–162, DOI 10.1016/j.aim.2007.12.003. MR2409411 (2009d:16057)
[18] Loïc Foissy, *Classification of systems of Dyson-Schwinger equations of the Hopf algebra of decorated rooted trees*, Adv. Math. **224** (2010), no. 5, 2094–2150, DOI 10.1016/j.aim.2010.01.024. MR2646118 (2011j:81220)
[19] Loïc Foissy, *General Dyson-Schwinger equations and systems*, Comm. Math. Phys. **327** (2014), no. 1, 151–179, DOI 10.1007/s00220-014-1941-0. MR3177935
[20] Maxim Kontsevich and Don Zagier, *Periods*, Mathematics unlimited—2001 and beyond, Springer, Berlin, 2001, pp. 771–808. MR1852188 (2002i:11002)
[21] Thomas Krajewski, Vincent Rivasseau, and Adrian Tanasa, *Combinatorial Hopf algebraic description of the multi-scale renormalization in quantum field theory*, Sém. Lothar. Combin. **70** (2013), Art. B70c, 23. MR3261807
[22] T. Krajewski and R. Wulkenhaar, *On Kreimer's Hopf algebra structure of Feynman graphs*, Eur. Phys. J. C Part. Fields **7** (1999), no. 4, 697–708, DOI 10.1007/s100520050439. MR1737771 (2001f:81115)
[23] Dirk Kreimer, *On the Hopf algebra structure of perturbative quantum field theories*, Adv. Theor. Math. Phys. **2** (1998), no. 2, 303–334. MR1633004 (99e:81156)
[24] D. Kreimer, *Chen's iterated integral represents the operator product expansion*, Adv. Theor. Math. Phys. **3** (1999), no. 3, 627–670. MR1797019 (2003b:81128)
[25] Dirk Kreimer, *On overlapping divergences*, Comm. Math. Phys. **204** (1999), no. 3, 669–689, DOI 10.1007/s002200050661. MR1707611 (2000f:81128)
[26] Dirk Kreimer, *Factorization in quantum field theory: an exercise in Hopf algebras and local singularities*, Frontiers in number theory, physics, and geometry. II, Springer, Berlin, 2007, pp. 715–736, DOI 10.1007/978-3-540-30308-4_14. MR2290771 (2008g:81167)
[27] *Traces in number theory, geometry and quantum fields*, Aspects of Mathematics, E38, Friedr. Vieweg & Sohn, Wiesbaden, 2008. A publication of the Max-Planck-Institute for Mathematics, Bonn; Edited by Sergio Albeverio, Matilde Marcolli, Sylvie Paycha and Jorge Plazas. MR2494193 (2009j:00008)
[28] D. Kreimer and E. Panzer, *Renormalization and Mellin transforms*, Computer Algebra in Quantum Field Theory (C. Schneider and J. Blümlein, eds.), Texts & Monographs in Symbolic Computation, vol. XII, Springer Wien, September 2013, pp. 195–223.
[29] Dirk Kreimer and Karen Yeats, *An étude in non-linear Dyson-Schwinger equations*, Nuclear Phys. B Proc. Suppl. **160** (2006), 116–121, DOI 10.1016/j.nuclphysbps.2006.09.036. MR2255485 (2007i:81162)
[30] Dominique Manchon, *Hopf algebras in renormalisation*, Handbook of algebra. Vol. 5, Handb. Algebr., vol. 5, Elsevier/North-Holland, Amsterdam, 2008, pp. 365–427, DOI 10.1016/S1570-7954(07)05007-3. MR2523455 (2011b:81166)
[31] Frédéric Menous and Frédéric Patras, *Logarithmic derivatives and generalized Dynkin operators*, J. Algebraic Combin. **38** (2013), no. 4, 901–913, DOI 10.1007/s10801-013-0431-3. MR3119364
[32] E. Panzer, *Hopf-algebraic Renormalization of Kreimer's toy model*, Master's thesis, Humboldt-Universität zu Berlin, July 2011.

[33] Moss E. Sweedler, *Hopf algebras*, Mathematics Lecture Note Series, W. A. Benjamin, Inc., New York, 1969. MR0252485 (40 #5705)

[34] Guillaume van Baalen, Dirk Kreimer, David Uminsky, and Karen Yeats, *The QED $\beta$-function from global solutions to Dyson-Schwinger equations*, Ann. Physics **324** (2009), no. 1, 205–219, DOI 10.1016/j.aop.2008.05.007. MR2493334 (2010c:81174)

[35] Steven Weinberg, *High-energy behavior in quantum field-theory*, Phys. Rev. (2) **118** (1960), 838–849. MR0116953 (22 #7736)

[36] Karen Yeats, *Rearranging Dyson-Schwinger equations*, Mem. Amer. Math. Soc. **211** (2011), no. 995, x+82, DOI 10.1090/S0065-9266-2010-00612-4. With a foreword by Dirk Kreimer. MR2791969 (2012d:81239)

[37] W. Zimmermann, *Convergence of Bogoliubov's method of renormalization in momentum space*, Comm. Math. Phys. **15** (1969), 208–234. MR0255162 (40 #8367)

Institutes of Physics and Mathematics, Humboldt-Universität zu Berlin, Unter den Linden 6, 10099 Berlin, Germany

*Current address*: Institute des Hautes Études Scientifiques, 35 Route de Chartres, 91440 Bures-sur-Yvette, France

*E-mail address*: erikpanzer@ihes.fr

*URL*: https://www.math.hu-berlin.de/~panzer/

# Multiple zeta value cycles in low weight

Ismaël Soudères

Abstract. In a recent work, the author has constructed two families of algebraic cycles in Bloch's cycle algebra over $\mathbb{P}^1 \backslash \{0, 1, \infty\}$ that are expected to correspond to multiple polylogarithms in one variable and have a good specialization at 1 related to multiple zeta values.

This is a short presentation, by the way of toy examples in low weight ($\leqslant 5$), of this construction and could serve as an introduction to the general setting. Working in low weight also makes it possible to push ("by hand") the construction further. In particular, we will not only detail the construction of the cycles but we will also associate to these cycles explicit elements in the bar construction over the cycle algebra and make as explicit as possible the "bottom-left" coefficient of the Hodge realization period matrix. That is, in a few relevant cases we will associated to each cycle an integral showing how the specialization at 1 is related to multiple zeta values. We will be particularly interested in a new weight 3 example corresponding to $-\zeta(2, 1)$.

Contents



This work has been partially supported by DFG grant SFB/TR45 and by Prof. Levine Humboldt Professorship.

I would like to thank P. Cartier for his attention to my work and H. Gangl for all his help, his comments and his stressing demand for me using tree related structures which allowed me to understand the Lie-like underlying combinatorics. Finally, none of this would have been possible without M. Levine patience, his explanations and deep inputs. This paper has been finalized during my stay at the MPIM and I am grateful to the MPIM for providing ideal working conditions and support.

## 1. Introduction

The multiple polylogarithm functions were defined in [**Gon95**] by the power series

$$Li_{k_1,\ldots,k_m}(z_1,\ldots,z_m) = \sum_{n_1>\cdots>n_m>0} \frac{z_1^{n_1}}{n_1^{k_1}}\frac{z_2^{n_2}}{n_2^{k_2}}\cdots\frac{z_m^{n_m}}{n_m^{k_m}} \qquad (z_i \in \mathbb{C},\ |z_i|<1).$$

They admit an analytic continuation to a Zariski open subset of $\mathbb{C}^m$. The case $m=1$ is nothing but the classical *polylogarithm* functions. The case $z_1 = z$ and $z_2 = \cdots = z_m = 1$ gives a one variable version of multiple polylogarithm functions

$$Li^{\mathbb{C}}_{k_1,\ldots,k_m}(z) = Li_{k_1,\ldots,k_m}(z,1,\ldots,1) = \sum_{n_1>\cdots>n_m>0} \frac{z^{n_1}}{n_1^{k_1}n_2^{k_2}\cdots n_m^{k_m}}.$$

When $k_1$ is greater or equal to 2, the series converge as $z$ goes to 1 and one recovers the multiple zeta value

$$\zeta(k_1,\ldots,k_m) = Li^{\mathbb{C}}_{k_1,\ldots,k_m}(1) = Li_{k_1,\ldots,k_m}(1,\ldots,1) = \sum_{n_1>\cdots>n_m>0} \frac{1}{n_1^{k_1}n_2^{k_2}\cdots n_m^{k_m}}.$$

To the tuple of integers $(k_1,\ldots,k_m)$ of weight $n=\sum k_i$, we can associate a tuple of 0's and 1's

$$(\varepsilon_n,\ldots,\varepsilon_1) := (\underbrace{0,\ldots,0}_{k_1-1 \text{ times}},1,\ldots,\underbrace{0,\ldots,0}_{k_m-1 \text{ times}},1)$$

which allows to write multiple polylogarithms as iterated integrals ($z_i \neq 0$ for all $i$):

$$Li^{\gamma}_{k_1,\ldots,k_m}(z_1,\ldots,z_m) = (-1)^m \int_{\Delta_\gamma} \frac{dt_1}{t_1-\varepsilon_1 x_1}\wedge\cdots\wedge\frac{dt_n}{t_n-\varepsilon_n x_n}$$

where $\gamma$ is a path from 0 to 1 in $\mathbb{C}\setminus\{x_1,\ldots,x_n\}$, the integration domain $\Delta_\gamma$ is the associated real simplex consisting of all $n$-tuples of points $(\gamma(t_1),\ldots,\gamma(t_n))$ with $t_i < t_j$ for $i<j$ and where we have set $x_n = z_1^{-1}$, $x_1 = (z_1\cdots z_m)^{-1}$ and where, for all $i$ such that $k_1+\cdots+k_{l-1}+1 \leqslant i < k_1+\cdots+k_l$, we have set $x_{n-i} = (z_1\cdots z_l)^{-1}$. Classically, $\gamma$ is the straight path from 0 to 1 : $\gamma(t)=t$ and in this case the superscript will be omitted..

Bloch and Kriz in [**BK94**] have constructed an algebraic cycle avatar of the classical polylogarithm function. More recently in [**GGL09**], Gangl, Goncharov and Levin, using a combinatorial approach, have built algebraic cycles corresponding to the multiple polylogarithm values $Li_{k_1,\ldots,k_m}(z_1,\ldots,z_m)$ with parameters $z_i$ satisfying in particular that all the $z_i$ but $z_1$ have to be different from 1 and their methods do not give algebraic cycles corresponding to multiple zeta values.

The goal of the article [**Sou12**] was to develop a geometric construction for multiple polylogarithm cycles removing the previous obstruction which will allow to have multiple zeta cycles.

A general idea underlying this project consists of looking for cycles fibered over a larger base and not just point-wise cycles for some fixed parameter $(z_1,\ldots,z_m)$. Levine in [**Lev11**] shows that there exists a short exact sequence relating the Bloch-Kriz Hopf algebra over $\mathrm{Spec}(\mathbb{Q})$, its relative version over $\mathbb{P}^1\setminus\{0,1,\infty\}$ and the Hopf algebra associated to Goncharov and Deligne's motivic fundamental group over $\mathbb{P}^1\setminus\{0,1,\infty\}$ which contains motivic avatars of iterated integrals associated to the multiple polylogarithms in one variable.

As this one variable version of multiple polylogarithms gives multiple zeta values for $z = 1$, it is natural to investigate first the case of the Bloch-Kriz construction over $\mathbb{P}^1 \setminus \{0, 1, \infty\}$ in order to obtain algebraic cycles corresponding to multiple polylogarithms in one variable with a "good specialization" at 1.

This paper presents the main geometric tools in order to construct such algebraic cycles and applies the general construction described in [**Sou12**] to concrete examples up to weight 5. In these particular cases, one can easily go further in the description, lifting the obtained cycles to the bar constructions over the Bloch's cycle algebra, describing the corresponding Bloch-Kriz motive and computing some associated integrals related to the Hodge realization. Those integrals give back multiple polylogarithms in one variable and their specialization at 1 give multiple zeta values.

The structure of the paper is organized as follows. In section 2 we review shortly the combinatorial context as it provides interesting relations for the bar elements associated to the cycles and an interesting relation with Goncharov's motivic coproduct for motivic iterated integrals. Section 3 is devoted to the geometric situation and to the construction of the cycles after a presentation of the Bloch's cycle algebra. Section 4, presents a combinatorial representation of the constructed cycles as parametrized cycles.

Section 5 recalls the definition of the bar construction over a commutative differential graded algebra and associates elements in the bar constructions (and a corresponding motive in the Bloch-Kriz construction) to the low weight examples of cycles. Finally in section 6, I follow Gangl, Goncharov and Levin's algorithm associating an integral to some of the low weight algebraic cycles previously described.

## 2. Combinatorial situation

In this paper a *tree* is a planar finite tree whose internal vertices have valency $\geqslant 3$ and where at each vertex a cyclic ordering of the incident edges is given. A *rooted* tree has a distinguished external vertex called the *root* and a *forest* is a disjoint union of trees.

Trees will be drawn with the convention that the cyclic ordering of the edges around an internal vertex is displayed in counterclockwise direction. The root vertex in the case of a rooted tree is displayed at the top.

**2.1. Trees, Lie algebras and Lyndon words.** Let $\mathcal{T}^{tri}$ be the $\mathbb{Q}$-vector space generated by rooted trivalent trees with leaves decorated by 0 and 1 modulo the relation

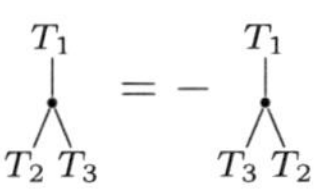


where the $T_i$'s are subtrees (and $T_1$ contains the root of the global tree). Note that in the above definition, the root is not decorated.

Define on $\mathcal{T}^{tri}$ the internal law ⅄ by

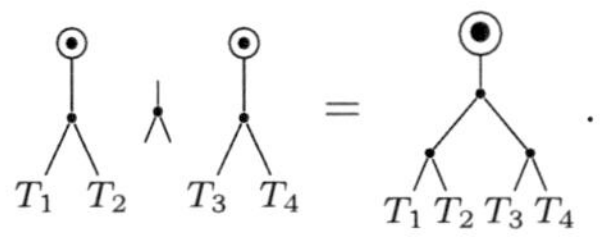

and extend it by bilinearity. One remarks that by definition $\curlywedge$ is antisymmetric. Identifying $\{0,1\}$ with $\{X_0, X_1\}$ by the obvious morphism and using the correspondence $\curlywedge \leftrightarrow [,]$, this internal law allows us to identify the free Lie algebra $\mathrm{Lie}(X_0, X_1)$ with $\mathcal{T}^{tri}$ modulo the Jacobi identity. Thus one can identify the (graded) dual of $\mathrm{Lie}(X_0, X_1)$ as a subspace of $\mathcal{T}^{tri}$

$$\mathrm{Lie}(X_0, X_1)^* \subset \mathcal{T}^{tri}.$$

A Lyndon word in 0 and 1 is a word in 0 and 1 strictly smaller than any of its nonempty proper right factors for the lexicographic order with $0 < 1$ (for more details, see [**Reu93**]). The standard factorization $[W]$ of a Lyndon word $W$ is defined inductively by $[0] = X_0$, $[1] = X_1$ and otherwise by $[W] = [[U],[V]]$ with $W = UV$, $U$ and $V$ nontrivial and such that $V$ is minimal. The sets of Lyndon brackets $\{[W]\}$, that is Lyndon words in standard factorization, form a basis of $\mathrm{Lie}(X_0, X_1)$ which can then be used to write the Lie bracket

$$[[U],[V]] = \sum_{\substack{W \text{ Lyndon} \\ \text{words}}} \alpha_{U,V}^{W} [W].$$

with $U < V$ Lyndon words.

EXAMPLE 2.1. Lyndon words in letters $0 < 1$ in lexicographic order are up to weight 5:

$$0 < 00001 < 0001 < 00011 < 001 < 00101 < 0011 < 00111 <$$
$$01 < 01011 < 011 < 0111 < 01111 < 1$$

The above identification of $\mathrm{Lie}(X_0, X_1)$ as a quotient of $\mathcal{T}^{tri}_{\cdot}$ and the basis of Lyndon brackets allows us to define a family of trees dual to the Lyndon bracket basis beginning with $T_{0^*} =$  and $T_{1^*} =$  and then setting

(1) $$T_{W^*} = \sum_{U<V} \alpha_{U,V}^{W} T_{U^*} \curlywedge T_{V^*}.$$

EXAMPLE 2.2. We give below the corresponding dual trees in weight 1, 2 and 3

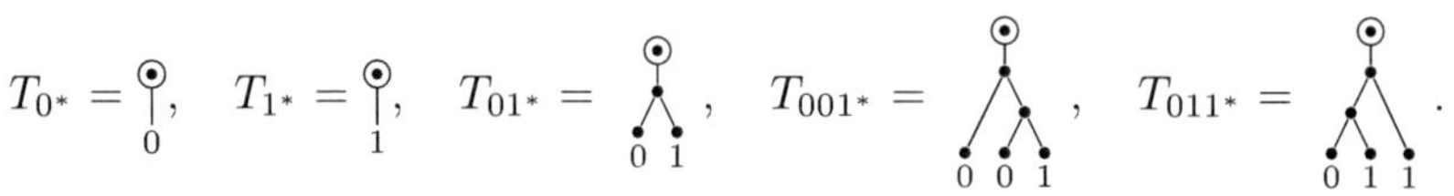

In weight 4 appears the first linear combination

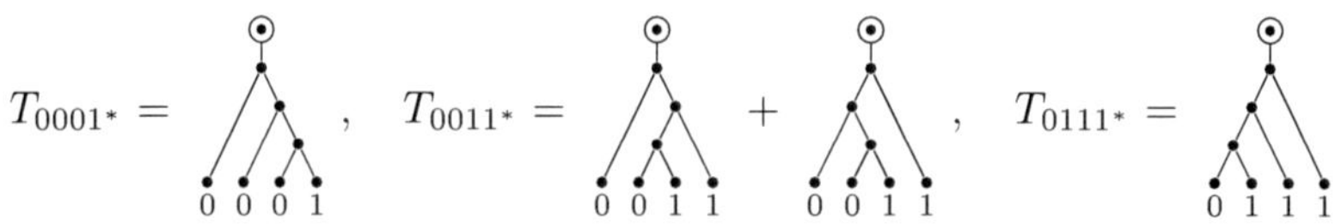

due to the fact that both $[0] \wedge [011]$ and $[001] \wedge [1]$ are mapped onto $[0011] = [X_0, [[X_0, X_1], X_1]]$ under the bracket map.

In weight 5, we will concentrate our attention to the two following examples

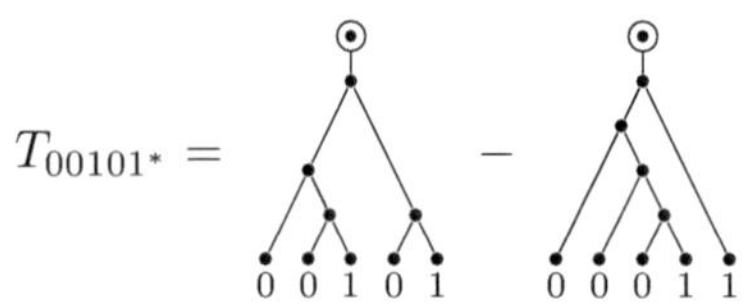

and

$T_{01011^*} =$ 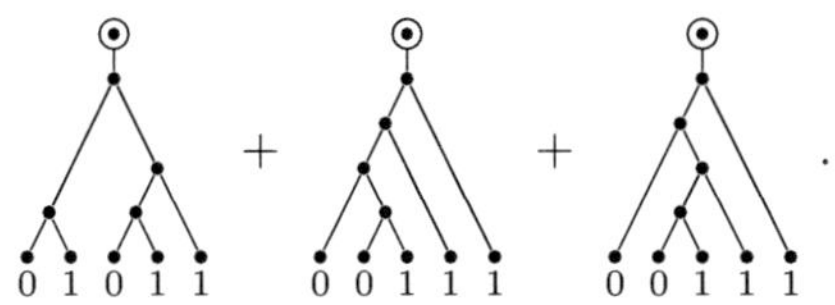

**2.2. Another differential on trees.** In **[GGL09]**, Gangl, Goncharov and Levin introduced a differential $d_{cy}$ on trees which reflects the differential in the Bloch's cycle algebra $\mathcal{N}_{\mathrm{Spec}(\mathbb{Q})}$ (see Section 3). In their work they have shown that some particular linear combinations of trivalent trees attached to decompositions of polygons have decomposable differential. More precisely, the differential of these particular linear combinations of trees is a linear combination of products of the same type of linear combinations of trees. The elements $T_{W^*}$ have a similar behavior under $d_{cy}$.

One begins by endowing trees with an extra structure.

DEFINITION 2.3.
- An *orientation* $\omega$ of a tree $T$ (or a forest) is a numbering of the edges. That is if $T$ has $n$ edges and if $E(T)$ denotes its set of edges , $\omega$ is a map $E(T) \longrightarrow \{1, \ldots, n\}$.
- Let $e(T)$ denote the cardinality of $E(T)$, that is the number of edges of $T$, and let $we(T)$, the weight of $T$, be the number of leaves of $T$. The degree of $T$ is defined by $deg(T) = 2we(T) - e(T)$. We extend these definitions to forests by linearity.

DEFINITION 2.4.
- Let $V^t$ be the $\mathbb{Q}$-vector space generated by a unit 1 and oriented forests of rooted trees $T$ with root vertex decorated by: $t$, 0 or 1 and leaves decorated by 0 or 1.
- Let $\cdot$ denote the product induced by the disjoint union of the trees and shift of the numbering for the orientation of the second factor. That is the product of $(F_1, \omega_1)$ and $(F_2, \omega_2)$ is the forest $F = F_1 \sqcup F_2$ together with the numbering $\omega$ satisfying $\omega|_{E(F_1)} = \omega_1$ and $\omega|_{E(F_2)} = \omega_2 + n_1$ where $n_1 = e(F_1)$ is the number of edges in $F_1$. Note that here, by convention, the empty tree is 0 and the unit for $\cdot$ is the extra generator 1.
- Define $\mathcal{F}_{\mathbb{Q}}^{\bullet}$ to be the algebra $V^t$ modulo the relations:

$(T, \sigma(\omega)) = \varepsilon(\sigma)(T, \omega),$ 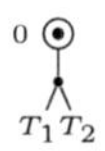 $= 0$ and 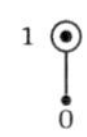 $= 0.$

for any permutation $\sigma$ and where $\varepsilon(\sigma)$ denotes the usual signature of the permutation $\sigma$.

The algebra $\mathcal{F}_{\mathbb{Q}}^{\bullet}$ endowed with the product $\cdot$ is graded commutative because the orientation introduces signs into the usual disjoint union. Note that for any forest $F$ one has $(-1)^{\deg(F)} = (-1)^{e(F)}$.

REMARK 2.5.
(1) There is an obvious direction on the edges of a rooted tree: away from the root.
(2) A rooted tree comes with a canonical numbering, starting from the root edge and induced by the cyclic ordering at each vertex.

EXAMPLE 2.6. With our convention, an example of this canonical ordering is shown at Figure 1; we recall that by convention we draw trees with the root at the top and the cyclic order at internal vertices counterclockwise.

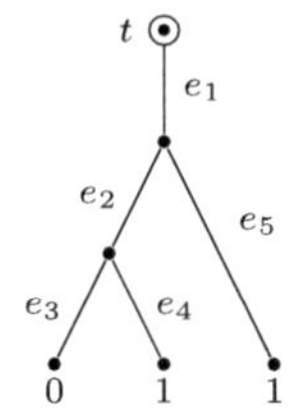


FIGURE 1. A tree with its canonical orientation, that is the canonical numbering of its edges.

Now, we define on $\mathcal{F}_{\mathbb{Q}}^{\bullet}$ a differential satisfying $d^2 = 0$ and the Leibniz rule

$$d((F_1,\omega_1)\cdot(F_2,\omega_2)) = d((F_1,\omega_1))\cdot(F_2,\omega_2) + (-1)^{e(F_1)}(F_1,\omega_1)\cdot d((F_2,\omega_2)).$$

The set of rooted planar trees decorated as above endowed with their canonical orientation forms a set of representatives for the permutation relation and it generates $\mathcal{F}_{\mathbb{Q}}^{\bullet}$ as an algebra. Hence, we will define this differential first on these trees and then extend the definition by the Leibniz rule.

The differential of an oriented tree $(T,\omega)$ is a linear combination of oriented forests where the trees appearing arise by contracting an edge of $T$ and fall into two types depending on whether the edge is internal or not. We will need the notion of splitting.

DEFINITION 2.7. A *splitting* of a tree $T$ at an internal vertex $v$ is the disjoint union of the trees which arise as $T_i \cup v$ where the $T_i$ are the connected components of $T \setminus v$. Moreover

- the planar structure of $T$ and its decorations of leaves induce on each $T_i \cup v$ a planar structure and decorations of leaves ;
- an ordering of the edges of $T$ induces an orientation of the forest $\sqcup_i (T_i \cup v)$;
- if $T$ has a root $r$ then $v$ becomes the root for all $T_i \cup v$ which do not contain $r$, and if $v$ has a decoration then it keeps its decoration in all the $T_i \cup v$.

DEFINITION 2.8. Let $e$ be an edge of a tree $T$. The contraction of $T$ along $e$ denoted $T/e$ is given as follows:

(1) If the tree consists of a single edge, its contraction is the empty tree.
(2) If $e$ is an internal edge, then $T/e$ is the tree obtained from $T$ by contracting $e$ and identifying the incident vertices to a single vertex.
(3) If $e$ is the edge containing the root vertex then $T/e$ is the forest obtained by first contracting $e$ to the internal incident vertex $w$ (which inherits the decoration of the root) and then by splitting at $w$; $w$ becoming the new root of all trees in the forest $T/e$.
(4) If $e$ is an external edge not containing the root vertex then $T/e$ is the forest obtained as follows: first one contracts $e$ to the internal incident vertex $w$ (which inherits the decoration of the leaf) and then one performs a splitting at $w$.

(5) If $T$ is endowed with its canonical orientation $\omega$ there is a natural orientation $i_e\omega$ on $T/e$ given as follows :

$$\forall f \in E(T/e) \quad \begin{array}{ll} i_e\omega(f) = \omega(f) & \text{if } \omega(f) < \omega(e) \\ i_e\omega(f) = \omega(f) - 1 & \text{if } \omega(f) > \omega(e). \end{array}$$

EXAMPLE 2.9. Two examples are given below. In Figure 2, one contracts the root vertex and in Figure 3, a leaf is contracted.

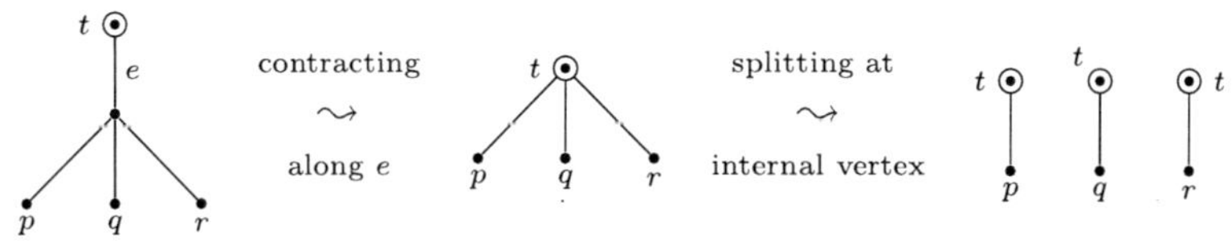


FIGURE 2. Contracting the root

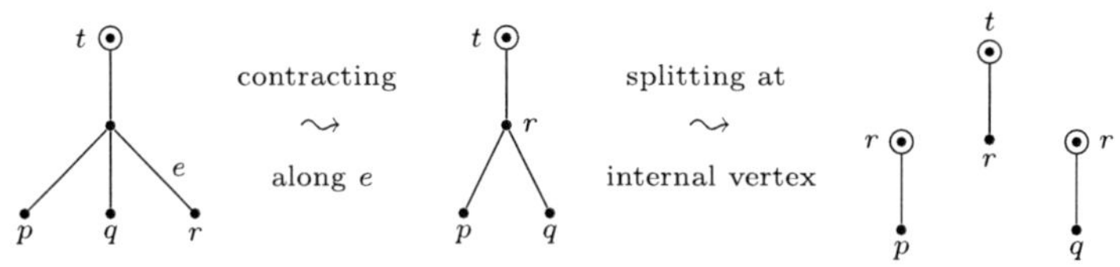


FIGURE 3. Contracting a leaf

DEFINITION 2.10. Let $(T,\omega)$ be a tree endowed with its canonical orientation, one defines $d_{cy}(T,w)$ as

$$d_{cy}(T,\omega) = \sum_{e \in E(T)} (-1)^{\omega(e)-1}(T/e, i_e\omega).$$

One extends $d_{cy}$ to all oriented trees by the relation $d_{cy}(T,\sigma\circ\omega) = \varepsilon(\sigma)d_{cy}(T,\omega)$ and to $\mathcal{F}_{\mathbb{Q}}^{\bullet}$ by linearity and the Leibniz rule.

In particular $d_{cy}$ maps a tree with at most one edge to 0 (which corresponds by convention to the empty tree).

As proved in [**GGL09**], $d_{cy}$, extended with the Leibniz rule, induces a differential on $\mathcal{F}_{\mathbb{Q}}^{\bullet}$.

PROPOSITION 2.11. *The map* $d_{cy} : \mathcal{F}_{\mathbb{Q}}^{\bullet} \longrightarrow \mathcal{F}_{\mathbb{Q}}^{\bullet}$ *makes* $\mathcal{F}_{\mathbb{Q}}^{\bullet}$ *into a commutative differential graded algebra. In particular* $d_{cy}^2 = 0$.

By an abuse of notation, for any Lyndon word $U$ the image of $T_{U^*}$ in $\mathcal{F}_{\mathbb{Q}}^{\bullet}$ with root vertex decorated by $t$ and canonical orientation is also denoted by $T_{U^*}$. The image of $T_{U^*}$ in $\mathcal{F}_{\mathbb{Q}}^{\bullet}$ with root vertex decorated by 1 and canonical orientation is denoted by $T_{U^*}(1)$.

The main result of the combinatorial aspects is the following.

THEOREM 2.12. *Let* $W$ *be a Lyndon word. Then the following equality holds in* $\mathcal{F}_{\mathbb{Q}}^{\bullet}$*:*

$$\text{(ED-T)} \qquad d_{cy}(T_{W^*}) = \sum_{U<V} \alpha_{U,V}^{W} T_{U^*} \cdot T_{V^*} + \sum_{U,V} \beta_{U,V}^{W} T_{U^*} \cdot T_{V^*}(1)$$

*where the* $\alpha_{U,V}^{W}$ *are the ones from Equation* (1). *In the above equation, coefficients* $\alpha$*'s and* $\beta$*'s are in* $\mathbb{Z}$.

For a detailed proof, we refer to [**Sou12**]. The theorem mainly follows by induction from the combinatorics of the free Lie algebra $\mathrm{Lie}(X_0, X_1)$:

- The terms in the first sum in the R.H.S of (ED-T) come from the contraction of the root edge which is nothing but the differential $d_{Lie}$ dual to the bracket of $\mathrm{Lie}(X_0, X_1)$.
- Using the inductive definition of $T_{W^*}$ (cf. Equation (1)), one shows by induction that, as $d_{Lie}^2 = 0$, internal edges do not contribute.
- Terms in $T_{V^*}(1)$ arise from leaves decorated by 1. The fact that terms arising from leaves decorated by 1 can be regrouped as a product $T_{U^*} \cdot T_{V^*}(1)$ is due to a particular decomposition of some specific brackets in terms of the Lyndon basis.

EXAMPLE 2.13. As mentioned before, the trees are endowed with their canonical numbering. First on remarks that trees with only one edge are mapped to 0 so

$d_{cy}(T_{0^*}) = d_{cy}($  $) = 0$ and $d_{cy}(T_{1^*}) = d_{cy}($  $) = 0.$

We recall that a tree with root decorated by 0 is 0 in $\mathcal{F}_{\mathbb{Q}}^{\bullet}$. As applying an odd permutation to the numbering changes the sign of the tree, the trivalency of the tree $T_{W^*}$ shows that some trees coming from the computation of $d_{cy}$ are 0 in $\mathcal{F}_{\mathbb{Q}}^{\bullet}$ because they contain a symmetric subtree; that is they contain a subtree of the form

where $T$ is a trivalent tree.

Using the fact that the tree  is 0 in $\mathcal{F}_{\mathbb{Q}}^{\bullet}$, one computes in weight 2

$d_{cy}(T_{01^*}) = d_{cy}($  $) =$  $\cdot$  $= T_{0^*}\, T_{1^*}.$

In weight 3, one has

$d_{cy}(T_{001^*}) = d_{cy}($ 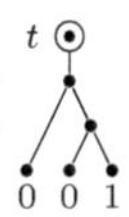 $) =$  $\cdot$  $= T_{0^*}\, T_{01^*}$

and

$d_{cy}(T_{011^*}) = d_{cy}\Bigg($ 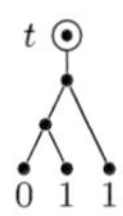 $\Bigg) =$  $\cdot$  $+$  

$$= T_{01^*}\, T_{1^*} + T_{1^*}\, T_{01^*}(1).$$

In weight 4, one can easily check that

$d_{cy}(T_{0001^*}) = T_{0^*}\, T_{001^*}$ and $d_{cy}(T_{0111^*}) = T_{011^*}\, T_{1^*} + T_{1^*} T_{011^*}(1).$

The example of $T_{0011^*}$ is more interesting.

(2) 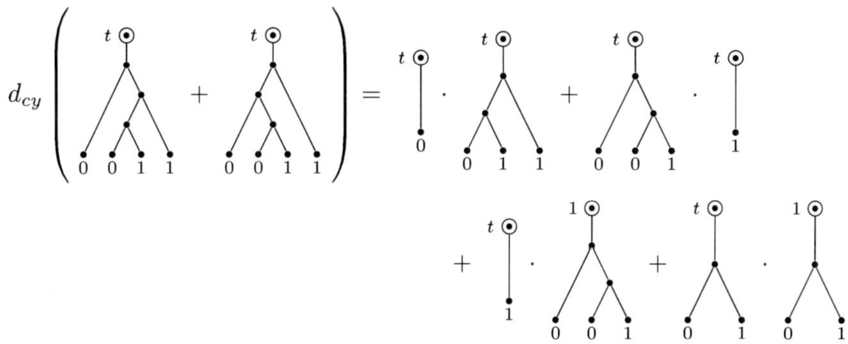


That is:

$$d_{cy}(T_{0011^*}) = T_{0^*}\, T_{011^*} + T_{001^*}\, T_{1^*} + T_{1^*}\, T_{001^*}(1) + T_{01^*} T_{01^*}(1) \tag{3}$$

In the above equations, the term in $T_{01^*}\, T_{01^*}(1)$ is coming form the last edge of the tree

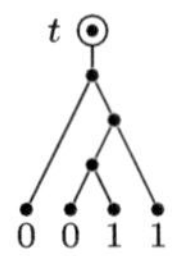


appearing in $T_{0011^*}$.

Computing $d_{cy}^2(T_{0011^*})$ (which is 0), the differential $d_{cy}(T_{01^*}\, T_{01^*}(1))$ cancels with the term in $T_{0^*}\, T_{1^*} T_{01^*}(1)$ arising from $d_{cy}(T_{0^*}\, T_{011^*})$. It can be thought of as the propagation of the weight 3 correction term $T_{1^*} T_{01^*}(1)$ appearing in $d_{cy}(T_{011^*})$.

We give below an example in weight 5, $d_{cy}(T_{01011^*})$ :

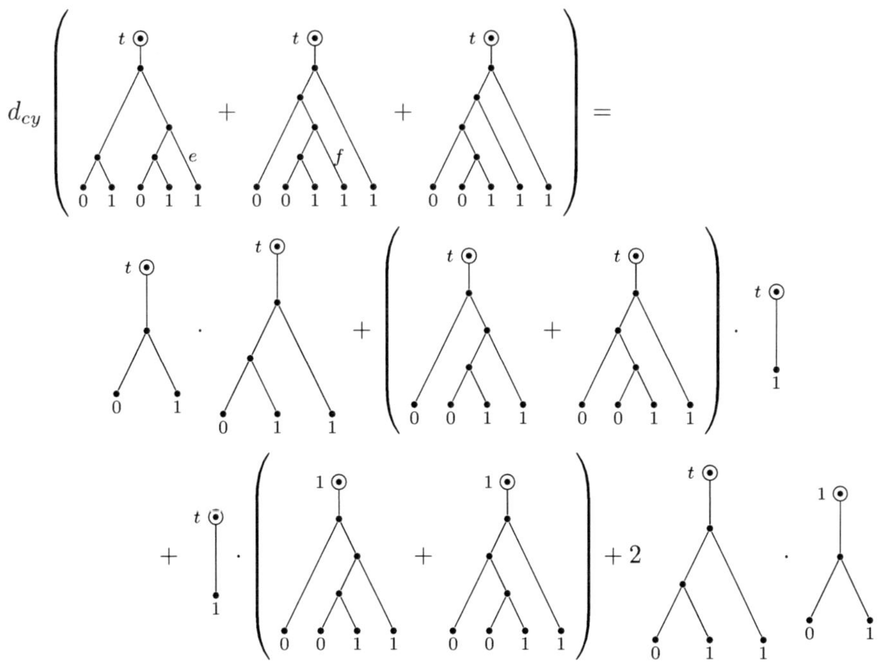

where the last term arises from the part of the differential associated to edges $e$ and $f$. The above equation can be written as

$$T_{01011^*} = T_{01^*} \cdot T_{011^*} + T_{0011^*} \cdot T_{1^*} + T_{1^*} \cdot T_{0011^*}(1) + 2T_{011^*} \cdot T_{01^*}(1) \tag{4}$$

## 3. Algebraic cycles

This section begins with the construction of the cycle complex (or cycle algebra) as presented in [**Blo86, Blo97, BK94, Lev94**]. Then, we give some properties of equidimensional cycles groups over $\mathbb{A}^1$ and build some algebraic cycles corresponding to multiple polylogarithms in one variable.

Here the base field is $\mathbb{Q}$ and the various structures have $\mathbb{Q}$ coefficients.

**3.1. Construction of the cycle algebra.** Let $\square^n$ be the algebraic $n$-cube

$$\square^n = (\mathbb{P}^1 \setminus \{1\})^n.$$

Insertion morphisms $s_i^\varepsilon : \square^{n-1} \longrightarrow \square^n$ are given by the identification

$$\square^{n-1} \simeq \square^{i-1} \times \{\varepsilon\} \times \square^{n-i}$$

for $\varepsilon = 0, \infty$. A *face* $F$ of codimension $p$ of $\square^n$ is given by the equation $x_{i_k} = \varepsilon_k$ for $k$ in $\{1, \dots, p\}$ and $\varepsilon_k$ in $\{0, \infty\}$ where $x_1, \dots, x_n$ are the usual affine coordinates on $\mathbb{P}^1$. In particular, codimension 1 faces are given by the images of insertion morphisms.

Now, let $X$ be a smooth irreducible quasi-projective variety over $\mathbb{Q}$.

DEFINITION 3.1. Let $p$ and $n$ be non-negative integers. Let $\mathcal{Z}^p(X, n)$ be the free group generated by closed irreducible sub-varieties of $X \times \square^n$ of codimension $p$ which intersect all faces $X \times F$ properly (where $F$ is a face of $\square^n$). That is:

$$\mathbb{Z}\left\langle Z \subset X \times \square^n \text{ such that} \begin{cases} Z \text{ is closed and irreducible} \\ \operatorname{codim}_{X\times F}(Z \cap (X \times F)) = p \\ \text{or } Z \cap (X \times F) = \emptyset \end{cases} \right\rangle$$

A sub-variety $Z$ of $X \times \square^n$ as above is *admissible*. The insertion morphisms $s_i^\varepsilon$ induce a well defined pull-back $s_i^{\varepsilon\,*} : \mathcal{Z}^p(X, n) \to \mathcal{Z}^p(X, n-1)$ and a differential:

$$\partial = \sum_{i=1}^{n} (-1)^{i-1} (s_i^{0\,*} - s_i^{\infty\,*}) : \mathcal{Z}^p(X, n) \longrightarrow \mathcal{Z}^p(X, n-1).$$

The permutation group $\mathfrak{S}_n$ acts on $\square^n$ by permutation of the factors. This action extends to an action of the semi-direct product $G_n = (\mathbb{Z}/2\mathbb{Z})^n \rtimes \mathfrak{S}_n$ where each $\mathbb{Z}/2\mathbb{Z}$ acts on $\square^1$ by sending the usual affine coordinates $x$ to $1/x$. The sign representation of $\mathfrak{S}_n$ extends to a sign representation $G_n \longrightarrow \{\pm 1\}$. Let $\mathcal{A}lt_n \in \mathbb{Q}[G_n]$ be the corresponding projector; when the context is clear enough, we may drop the subscript $n$.

DEFINITION 3.2. Let $p$ and $k$ be integers as above. One defines

$$\mathcal{N}_X^k(p) = \mathcal{A}lt_{2p-k}(\mathcal{Z}^p(X, 2p-k) \otimes \mathbb{Q}).$$

We will refer to $k$ as the *cohomological degree* and to $p$ as the *weight*.

For our purpose, we will not only need admissible cycles but cycles in $X \times \square^n$ whose fibers over $X$ are also admissible.

DEFINITION 3.3 (Equidimensionality). Let $X$ be an irreducible smooth variety.

- Let $\mathcal{Z}^p_{eq}(X,n)$ denote the free abelian group generated by irreducible closed subvarieties $Z \subset X \times \square^n$ such that for any face $F$ of $\square^n$, the intersection $Z \cap (X \times F)$ is empty or the restriction of $p_1 : X \times \square^n \longrightarrow X$ to

$$Z \cap (X \times F) \longrightarrow X$$

is equidimensional of relative dimension $\dim(F) - p$.
- We say that elements of $\mathcal{Z}^p_{eq}(X,n)$ are *equidimensional over $X$ with respect to any face* or simply *equidimensional.*
- Following the definition of $\mathcal{N}^k_X(p)$, let $\mathcal{N}^{eq,\,k}_X(p)$ denote

$$\mathcal{N}^{eq,\,k}_X(p) = \mathcal{A}lt_{2p-k}\left(\mathcal{Z}^p_{eq}(X, 2p-k) \otimes \mathbb{Q}\right).$$

- If $Z$ is an irreducible closed subvariety of $X \times \square^n$ satisfying the above condition, $Z|_{t=x}$ will denote the fiber over the point $x \in X$ of $p_1$ restricted to $Z$ that is $Z \cap (\{x\} \times \square^n)$.

  Let $C = \mathcal{A}lt(\sum q_i Z_i)$ be an element in $\mathcal{N}^{eq,\,\bullet}_X$ with the $Z_i$ as above and $q_i$'s in $\mathbb{Q}$. For a point $x \in X$, we will denote by $C|_{t=x}$ the element of $\mathcal{N}^\bullet_X$

$$C|_{t=x} = \mathcal{A}lt(\sum q_i Z_i|_{t=x})$$

  which is well defined in both $\mathcal{N}^\bullet_X$ and $\mathcal{N}^\bullet_{\{x\}}$ by definition of the $Z_i$.

EXAMPLE 3.4. Consider the graph of the identity $\mathbb{A}^1 \overset{t\mapsto t}{\longrightarrow} \mathbb{A}^1$ restricted to $\mathbb{A}^1 \times \mathbb{A}^1 \setminus \{1\}$. Let $\Gamma_0$ be its embedding in $\mathbb{A}^1 \times \square^1$. Then $\Gamma_0$ is of codimension 1 in $\mathbb{A}^1 \times \square^1$ and is admissible as the intersection with the face $x_1 = \infty$ is empty and the intersection with the face $x_1 = 0$ is $\{0\} \times \{0\}$ which is of codimension 1 in $\mathbb{A}^1 \times \{0\}$.

However, $\Gamma_0$ is not equidimensional as

$$\Gamma_0 \cap \left(\mathbb{A}^1 \times \{0\}\right) = \{0\} \times \{0\}$$

is neither equidimensional over $\mathbb{A}^1$ nor empty as the condition would require.

Applying the projector $\mathcal{A}lt$ gives an element $\overline{\mathcal{L}_0}$ in $\mathcal{N}^1_{\mathbb{A}^1}(1)$. Using the definition of $\Gamma_0$ as a graph, one obtains a parametric representation (where the projector $\mathcal{A}lt$ is omitted):

$$\overline{\mathcal{L}_0} = [t; t] \qquad \subset \mathbb{A}^1 \times \square^1.$$

In the above notation the semicolon separates the base space coordinates from the cubical coordinates.

The morphisms $s_i^{\varepsilon\,*}$ induce morphisms $\partial_i^\varepsilon : \mathcal{N}^k_X(p) \longrightarrow \mathcal{N}^{k+1}_X(p)$ and the above differential $\partial = \sum_i (-1)^{i-1}(\partial_i^0 - \partial_i^\infty)$ gives a complex

$$\mathcal{N}^\bullet_X(p): \qquad \cdots \longrightarrow \mathcal{N}^k_X(p) \overset{\partial}{\longrightarrow} \mathcal{N}^{k+1}_X(p) \longrightarrow \cdots$$

DEFINITION 3.5. One defines the cycle complex as

$$\mathcal{N}^\bullet_X = \bigoplus_{p\geqslant 0} \mathcal{N}^\bullet_X(p) = \mathbb{Q} \oplus \bigoplus_{p\geqslant 1} \mathcal{N}^\bullet_X(p)$$

and as the differential restricts to equidimensional cycles, one also defines

$$\mathcal{N}^{eq,\,\bullet}_X = \bigoplus_{p\geqslant 0} \mathcal{N}^{eq,\,\bullet}_X(p).$$

The author refers sometimes to $\mathcal{N}_X^\bullet$ as the cycle algebra because of another natural structure coming with this cubical cycle complex: the product structure.

Levine has shown in [**Lev94**][§5] or [**Lev11**][Example 4.3.2] the following proposition.

PROPOSITION 3.6. *Concatenation of the cube factors and pull-back by the diagonal*

$$X \times \square^n \times X \times \square^m \xrightarrow{\sim} X \times X \times \square^n \times \square^m \xrightarrow{\sim} X \times X \times \square^{n+m} \xleftarrow{\Delta_X} X \times \square^{n+m}$$

*induce, after applying the Alt projector, a well-defined product:*

$$\mathcal{N}_X^k(p) \otimes \mathcal{N}_X^l(q) \longrightarrow \mathcal{N}_X^{k+l}(p+q)$$

*denoted by* $\cdot$

*The complex* $\mathcal{N}_X^{eq,\bullet}$ *is stable under this product law.*

REMARK 3.7. The smoothness hypothesis on $X$ allows us to consider the pull-back by the diagonal $\Delta_X : X \longrightarrow X \times X$ which is, in this case, of local complete intersection.

One has the following theorem (stated in [**BK94, Blo97**] for $X = \operatorname{Spec}(\mathbb{Q})$).

THEOREM 3.8 ([**Lev94**]). *The cycle complex* $\mathcal{N}_X^\bullet$ *is an Adams graded, commutative differential graded algebra (Adams graded, c.d.g.a.). In weight* $p$*, its cohomology groups are the higher Chow groups of* $X$*:*

$$\mathrm{H}^k(\mathcal{N}_X(p)) = \mathrm{CH}^p(X, 2p-k)_{\mathbb{Q}},$$

*where* $\mathrm{CH}^p(X, 2p-k)_{\mathbb{Q}}$ *stands for* $\mathrm{CH}^p(X, 2p-k) \otimes \mathbb{Q}$.

Moreover $\mathcal{N}_X^{eq,\bullet}$ turns into a sub-Adams graded, c.d.g.a. Note that, in the graded algebra context, *commutative* always means *graded commutative*.

One has natural flat pull-backs and proper push-forwards on $\mathcal{N}_X^\bullet$ (and on $\mathcal{N}_X^{eq,\bullet}$). Comparison with higher Chow groups also gives on the cohomology groups both $\mathbb{A}^1$-homotopy invariance and the long exact sequence associated to an open and its closed complement. Writing $\mathbb{P}^1 \setminus \{0,1,\infty\}$ as $\mathbb{A}^1 \setminus \{0,1\}$, one obtains the following description of $\mathrm{H}^*(\mathcal{N}_{\mathbb{P}^1\setminus\{0,1,\infty\}}^\bullet(p))$ :

$$\mathrm{H}^k(\mathcal{N}_{\mathbb{P}^1\setminus\{0,1,\infty\}}^\bullet(p)) \simeq \mathrm{H}^k(\mathcal{N}_{\mathbb{Q}}^\bullet(p)) \oplus \\ \mathrm{H}^{k-1}(\mathcal{N}_{\mathbb{Q}}^\bullet(p-1)) \otimes \mathbb{Q}\mathcal{L}_0 \oplus \mathrm{H}^{k-1}(\mathcal{N}_{\mathbb{Q}}^\bullet(p-1)) \otimes \mathbb{Q}\mathcal{L}_1,$$

where $\mathcal{L}_0$ and $\mathcal{L}_1$ are in cohomological degree 1 and weight 1 (that is of codimension 1). Their explicit description will be given later on.

Comparing the situation over $\mathbb{P}^1 \setminus \{0,1,\infty\}$ and over $\mathbb{A}^1$ comes as an important idea in our project as the desired cycles over $\mathbb{P}^1 \setminus \{0,1,\infty\}$ need to admit a natural specialization at 1. In particular, we will need to work with equidimensional cycle and some of their properties are given in the next subsection.

**3.2. Equidimensional cycles.** The following result given in [**Sou12**] essentially follows from the definition and makes it easy to compare both situations.

PROPOSITION 3.9. *Let* $X_0$ *be an open dense subset of* $X$ *an irreducible smooth variety and let* $j : X_0 \longrightarrow X$ *be the inclusion. Then the restriction of cycles from*

*$X$ to $X_0$ induces a morphism of c.d.g.a. preserving the weight (that is the Adams grading)*

$$j^* : \mathcal{N}_X^{eq,\bullet} \longrightarrow \mathcal{N}_{X_0}^{eq,\bullet}.$$

*Moreover, let $C$ be in $\mathcal{N}_{X_0}^{\bullet}$ and write $C$ in terms of the generators of the group $\oplus \mathcal{Z}^{\star}(X_0, \bullet)$ as*

$$C = \sum_{i \in I} q_i Z_i, \qquad q_i \in \mathbb{Q}$$

*where $I$ is a finite set. Assume that, for any $i$, the Zariski closure $\overline{Z_i}$ of $Z_i$ in $X \times \square^{n_i}$ intersected with any face $X \times F$ of $X \times \square^{n_i}$ is equidimensional over $X$ of relative dimension $\dim(F) - p_i$. Define $C'$ as*

$$C' = \sum_{i \in I} q_i \overline{Z_i},$$

*then*

$$C' \in \mathcal{N}_X^{eq,\bullet} \qquad \text{and} \qquad C = j^*(C') \in \mathcal{N}_{X_0}^{eq,\bullet}.$$

Below, we describe the main geometric fact that allows the construction of our cycles: pulling back by the multiplication induces a homotopy between identity and the zero section on the cycle algebra over $\mathbb{A}^1$.

Let $m : \mathbb{A}^1 \times \mathbb{A}^1 \longrightarrow \mathbb{A}^1$ be the multiplication map sending $(x, y)$ to $xy$ and let $\tau : \square^1 = \mathbb{P}^1 \setminus \{1\} \longrightarrow \mathbb{A}^1$ be the isomorphism sending the affine coordinate $u$ to $\frac{1}{1-u}$. The map $\tau$ sends $\infty$ to $0$, $0$ to $1$ and extends as a map from $\mathbb{P}^1$ to $\mathbb{P}^1$ sending $1$ to $\infty$.

The maps $m$ and $\tau$ are in particular flat and equidimensional of relative dimension 1 and 0, respectively.

Consider the following commutative diagram for a positive integer $n$

$$\begin{array}{ccc}
\mathbb{A}^1 \times \square^1 \times \square^n & \xrightarrow{(m \circ (\mathrm{id}_{\mathbb{A}^1} \times \tau)) \times \mathrm{id}_{\square^n}} & \mathbb{A}^1 \times \square^n \\
\Big\downarrow {\scriptstyle p_{\mathbb{A}^1 \times \square^1}} & & \Big\downarrow {\scriptstyle p_{\mathbb{A}^1}} \\
\mathbb{A}^1 \times \square^1 & \xrightarrow{m \circ (\mathrm{id}_{\mathbb{A}^1} \times \tau)} & \mathbb{A}^1 \\
\Big\downarrow {\scriptstyle p_{\mathbb{A}^1}} & & \\
\mathbb{A}^1 & &
\end{array}$$

Proposition 3.10 (multiplication and equidimensionality). *In the following statement, $p$, $k$ and $n$ will denote positive integers subject to the relation $n = 2p - k$*

- *the composition $\widetilde{m} = (m \circ (\mathrm{id}_{\mathbb{A}^1} \times \tau)) \times \mathrm{id}_{\square^n}$ induces a group morphism*

$$\mathcal{Z}^p_{eq}(\mathbb{A}^1, n) \xrightarrow{\widetilde{m}^*} \mathcal{Z}^p_{eq}(\mathbb{A}^1 \times \square^1, n)$$

  *which extends to a morphism of complexes for any $p$*

$$\mathcal{N}_{\mathbb{A}^1}^{eq,\bullet}(p) \xrightarrow{\widetilde{m}^*} \mathcal{N}_{\mathbb{A}^1 \times \square^1}^{eq,\bullet}(p).$$

- *Moreover, one has a natural group morphism*

$$h^p_{\mathbb{A}^1, n} : \mathcal{Z}^p_{eq}(\mathbb{A}^1 \times \square^1, n) \longrightarrow \mathcal{Z}^p_{eq}(\mathbb{A}^1, n+1)$$

  *given by regrouping the $\square^1$ factors (as $\square^n = (\square^1)^n$).*

- *The composition* $\mu^* = h^p_{\mathbb{A}^1,n} \circ \widetilde{m}^*$ *gives a linear map*
$$\mu^* : \mathcal{N}^{eq,\,k}_{\mathbb{A}^1}(p) \longrightarrow \mathcal{N}^{eq,\,k-1}_{\mathbb{A}^1}(p)$$
*sending equidimensional cycles with empty fiber at* $0$ *to equidimensional cycles with empty fiber at* $0$.
- *Let* $\theta : \mathbb{A}^1 \longrightarrow \mathbb{A}^1$ *be the involution sending the natural affine coordinate* $t$ *to* $1-t$. *Twisting the map* $\widetilde{m}$ *by* $\theta$ *gives a map* $\widehat{m}$ *via*
$$\begin{array}{ccc} \mathbb{A}^1 \times \square^1 \times \square^n & \overset{\widehat{m}}{\dashrightarrow} & \mathbb{A}^1 \times \square^n \\ {\scriptstyle \theta\times \mathrm{id}_{\square^{n+1}}}\downarrow & & \uparrow{\scriptstyle \theta\times\mathrm{id}_{\square^n}} \\ \mathbb{A}^1 \times \square^1 \times \square^n & \xrightarrow{\widetilde{m}} & \mathbb{A}^1 \times \square^n \end{array}$$
*and induces a linear map*
$$\nu^* : \mathcal{N}^{eq,\,k}_{\mathbb{A}^1}(p) \longrightarrow \mathcal{N}^{eq,\,k-1}_{\mathbb{A}^1}(p)$$
*sending equidimensional cycles with empty fiber at* $1$ *to equidimensional cycles with empty fiber at* $1$.

PROOF. It is enough to work with generators of $\mathcal{Z}^p_{eq}(\mathbb{A}^1, n)$. Let $Z$ be an irreducible subvariety of $\mathbb{A}^1 \times \square^n$ such that for any face $F$ of $\square^n$, the first projection
$$p_{\mathbb{A}^1} : Z \cap (\mathbb{A}^1 \times F) \longrightarrow \mathbb{A}^1$$
is equidimensional of relative dimension $\dim(F) - p$ or empty. Let $F$ be a face of $\square^n$. We want first to show that under the projection $\mathbb{A}^1 \times \square^1 \times \square^n \longrightarrow \mathbb{A}^1 \times \square^1$,
$$\widetilde{m}^{-1}(Z) \cap (\mathbb{A}^1 \times \square^1 \times F) \longrightarrow \mathbb{A}^1 \times \square^1$$
is equidimensional of relative dimension $\dim(F) - p$ or empty. This follows from the fact that $Z \cap (\mathbb{A}^1 \times F)$ is equidimensional over $\mathbb{A}^1$ and $m$ is flat and equidimensional of relative dimension 1 (hence are $m \times \tau$ and $\widetilde{m}$). The map $\widetilde{m}$ is the identity on the $\square^n$ factor, thus for $Z \subset \mathbb{A}^1 \times \square^n$ as above and a codimension 1 face $F$ of $\square^n$, $\widetilde{m}^{-1}(Z)$ satisfies
$$\widetilde{m}^{-1}(Z) \cap (\mathbb{A}^1 \times \square^1 \times F) = \widetilde{m}^{-1}(Z \cap (\mathbb{A}^1 \times F))$$
which makes $\widetilde{m}^*$ into a morphism of complexes.

Moreover, assuming that the fiber of $Z$ at $0$ is empty, as $\widetilde{m}$ restricted to
$$\{0\} \times \square^1 \times \square^n$$
factors through the inclusion $\{0\} \times \square^n \longrightarrow \mathbb{A}^1 \times \square^n$, the intersection
$$\widetilde{m}^{-1}(Z) \cap \big(\{0\} \times \square^1 \times \square^n\big)$$
is empty. Hence the fiber of $\widetilde{m}^{-1}(Z)$ over $\{0\} \times \square^1$ by $p_{\mathbb{A}^1 \times \square^1}$ is empty and the same holds for the fiber over $\{0\}$ by $p_{\mathbb{A}^1} \circ p_{\mathbb{A}^1 \times \square^1}$.

Now, let $Z$ be an irreducible subvariety of $\mathbb{A}^1 \times \square^1 \times \square^n$ such that for any face $F$ of $\square^n$
$$Z \cap (\mathbb{A}^1 \times \square^1 \times F) \longrightarrow \mathbb{A}^1 \times \square^1$$
is equidimensional of relative dimension $\dim(F) - p$. Let $F'$ be a face of
$$\square^{n+1} = \square^1 \times \square^n.$$

The face $F'$ is either of the form $\square^1 \times F$ or of the form $\{\varepsilon\} \times F$ with $F$ a face of $\square^n$ and $\varepsilon \in \{0, \infty\}$. If $F'$ is of the first type, as

$$Z \cap (\mathbb{A}^1 \times \square^1 \times F) \longrightarrow \mathbb{A}^1 \times \square^1$$

is equidimensional and, as $\mathbb{A}^1 \times \square^1 \longrightarrow \mathbb{A}^1$ is equidimensional of relative dimension 1, the projection

$$Z \cap (\mathbb{A}^1 \times \square^1 \times F) \longrightarrow \mathbb{A}^1$$

is equidimensional of relative dimension

$$\dim(F) - p + 1 = \dim(F') - p.$$

If $F'$ is of the second type, by symmetry of the role of 0 and $\infty$, we can assume that $\varepsilon = 0$. Then the intersection

$$Z \cap (\mathbb{A}^1 \times \{0\} \times F)$$

is nothing but the fiber of $Z \cap (\mathbb{A}^1 \times \square^1 \times F)$ over $\mathbb{A}^1 \times \{0\}$. Hence, it has pure dimension $\dim(F) - p + 1$.

Moreover, denoting with a subscript the fiber, the composition

$$Z \cap (\mathbb{A}^1 \times \{0\} \times F) = \big(Z \cap (\mathbb{A}^1 \times \square^1 \times F)\big)_{\mathbb{A}^1 \times \{0\}} \longrightarrow \mathbb{A}^1 \times \{0\} \longrightarrow \mathbb{A}^1$$

is equidimensional of relative dimension

$$\dim(F) - p = \dim(F') - p.$$

This shows that $h^p_{\mathbb{A}^1,n}$ gives a well defined morphism and that it preserves the fiber at a point $x$ in $\mathbb{A}^1$; in particular, if $Z$ has an empty fiber at 0, so does $h^p_{\mathbb{A}^1,n}(Z)$.

Finally, the last part of the proposition is deduced from the fact that $\theta$ exchanges the role of 0 and 1. $\square$

REMARK 3.11. We have remarked that $\widetilde{m}$ sends cycles with empty fiber at 0 to cycles with empty fiber at any point in $\{0\} \times \square^1$. Similarly $\widetilde{m}$ sends cycles with empty fiber at 0 to cycles that also have an empty fiber at any point in $\mathbb{A}^1 \times \{\infty\}$.

From the proof of Levine's Proposition 4.2 in [**Lev94**], we deduce that $\mu^*$ gives a homotopy between $p_0^* \circ i_0^*$ and id where $i_0$ is the zero section $\{0\} \to \mathbb{A}^1$ and $p_0$ the projection onto the point $\{0\}$.

PROPOSITION 3.12. *Notations are the ones from Proposition* 3.10 *above. Let $i_0$ (resp. $i_1$) be the inclusion of* 0 *(resp.* 1*) in $\mathbb{A}^1$:*

$$i_0 : \{0\} \longrightarrow \mathbb{A}^1, \qquad i_1 : \{1\} \longrightarrow \mathbb{A}^1.$$

*Let $p_0$ and $p_1$ be the corresponding projections $p_\varepsilon : \mathbb{A}^1 \longrightarrow \{\varepsilon\}$ for $\varepsilon = 0, 1$.*

*Then $\mu^*$ provides a homotopy between*

$$p_0^* \circ i_0^* \text{ and } \mathrm{id} : \mathcal{N}^{eq,\bullet}_{\mathbb{A}^1} \longrightarrow \mathcal{N}^{eq,\bullet}_{\mathbb{A}^1}$$

*and similarly $\nu^*$ provides a homotopy between*

$$p_1^* \circ i_1^* \text{ and } \mathrm{id} : \mathcal{N}^{eq,\bullet}_{\mathbb{A}^1} \longrightarrow \mathcal{N}^{eq,\bullet}_{\mathbb{A}^1}.$$

*In other words, one has*

$$\partial_{\mathbb{A}^1} \circ \mu^* + \mu^* \circ \partial_{\mathbb{A}^1} = \mathrm{id} - p_0^* \circ i_0^* \quad \text{and} \quad \partial_{\mathbb{A}^1} \circ \nu^* + \nu^* \circ \partial_{\mathbb{A}^1} = \mathrm{id} - p_1^* \circ i_1^*.$$

The proposition follows from computing the different compositions involved and the relation between the differential on $\mathcal{N}^{eq,\bullet}_{\mathbb{A}^1 \times \square^1}$ and the one on $\mathcal{N}^{eq,\bullet}_{\mathbb{A}^1}$ via the map $h^p_{\mathbb{A}^1,n}$.

PROOF. We denote by $i_{0,\square}$ and $i_{\infty,\square}$ the zero section and the infinity section $\mathbb{A}^1 \longrightarrow \mathbb{A}^1 \times \square^1$. The action of $\theta$ only exchanges the role of 0 and 1 in $\mathbb{A}^1$, hence it is enough to prove the statement for $\mu^*$. As previously, in order to obtain the proposition for $\mathcal{N}_{\mathbb{A}^1}^{eq,k}(p)$, it is enough to work on the generators of $\mathcal{Z}_{eq}^p(\mathbb{A}^1, n)$ with $n = 2p - k$.

By the previous proposition 3.10, the morphism $\widetilde{m}^*$ commutes with the differential on $\mathcal{Z}_{eq}^p(\mathbb{A}^1, \bullet)$ and on $\mathcal{Z}_{eq}^p(\mathbb{A}^1 \times \square^1, \bullet)$. As the morphism $\mu^*$ is defined by $\mu^* = h_{\mathbb{A}^1,n}^p \circ \widetilde{m}^*$, the proof relies on computing $\partial_{\mathbb{A}^1} \circ h_{\mathbb{A}^1,n}^p$. Let $Z$ be a generator of $\mathcal{Z}_{eq}^p(\mathbb{A}^1 \times \square^1, n)$. In particular,

$$Z \subset \mathbb{A}^1 \times \square^1 \times \square^n$$

and $h_{\mathbb{A}^1,n}^p(Z)$ is also given by $Z$ but viewed in

$$\mathbb{A}^1 \times \square^{n+1}.$$

The differentials denoted by $\partial_{\mathbb{A}^1}^{n+1}$ on $\mathcal{Z}_{eq}^p(\mathbb{A}^1, n+1)$ and $\partial_{\mathbb{A}^1\times\square^1}^n$ on $\mathcal{Z}_{eq}^p(\mathbb{A}^1 \times \square^1, n)$ are both given by intersections with the codimension 1 faces but the first $\square^1$ factor in $\square^{n+1}$ gives two more faces and introduces a change of sign. Namely, using an extra subscript to indicate in which cycle groups the intersections take place, one has:

$$\begin{aligned}
\partial_{\mathbb{A}^1}^{n+1}(h_{\mathbb{A}^1,n}^p(Z)) =& \sum_{i=1}^{n+1}(-1)^{i-1}\left(\partial_{i,\mathbb{A}^1}^0(Z) - \partial_{i,\mathbb{A}^1}^\infty(Z)\right)\\
=&\partial_{1,\mathbb{A}^1}^0(Z) - \partial_{1,\mathbb{A}^1}^\infty(Z) - \sum_{i=2}^{n+1}(-1)^{i-2}\left(\partial_{i,\mathbb{A}^1}^0(Z) - \partial_{i,\mathbb{A}^1}^\infty(Z)\right)\\
=&i_{0,\square}^*(Z) - i_{\infty,\square}^*(Z) - \sum_{i=1}^{n}(-1)^{i-1}\left(\partial_{i+1,\mathbb{A}^1}^0(Z) - \partial_{i+1,\mathbb{A}^1}^\infty(Z)\right).
\end{aligned}$$

Hence one gets

$$\begin{aligned}
\partial_{\mathbb{A}^1}^{n+1}(h_{\mathbb{A}^1,n}^p(Z)) = & i_{0,\square}^*(Z) - i_{\infty,\square}^*(Z)\\
& - \sum_{i=1}^{n}(-1)^{i-1}\left(h_{\mathbb{A}^1,n-1}^p\left(\partial_{i,\mathbb{A}^1\times\square^1}^0(Z) - \partial_{i,\mathbb{A}^1\times\square^1}^\infty(Z)\right)\right)
\end{aligned}$$

which can be written has

$$\partial_{\mathbb{A}^1}^{n+1}(h_{\mathbb{A}^1,n}^p(Z)) = i_{0,\square}^*(Z) - i_{\infty,\square}^*(Z) - h_{\mathbb{A}^1,n-1}^p \circ \partial_{\mathbb{A}^1\times\square^1}^n(Z).$$

Thus one can compute $\partial_{\mathbb{A}^1} \circ \mu^* + \mu^* \circ \partial_{\mathbb{A}^1}$ on $\mathcal{Z}_{eq}^p(\mathbb{A}^1, n)$ as

$$\begin{aligned}
\partial_{\mathbb{A}^1} \circ \mu^* + \mu^* \circ \partial_{\mathbb{A}^1} =& \partial_{\mathbb{A}^1} \circ h_{\mathbb{A}^1,n} \circ \widetilde{m}^* + h_{\mathbb{A}^1,n-1} \circ \widetilde{m}^* \circ \partial_{\mathbb{A}^1}\\
=& i_{0,\square}^* \circ \widetilde{m}^* - i_{\infty,\square}^* \circ \widetilde{m}^* - h_{\mathbb{A}^1,n-1} \circ \partial_{\mathbb{A}^1} \circ \widetilde{m}^*\\
& + h_{\mathbb{A}^1,n-1} \circ \partial_{\mathbb{A}^1} \circ \widetilde{m}^*\\
=& i_{0,\square}^* \circ \widetilde{m}^* - i_{\infty,\square}^* \circ \widetilde{m}^*.
\end{aligned}$$

The morphism $i_{\infty,\square}^* \circ \widetilde{m}^*$ is induced by

$$\begin{array}{ccccccc}
\mathbb{A}^1 & \xrightarrow{i_{\infty,\square}} & \mathbb{A}^1 \times \square^1 & \xrightarrow{\tau} & \mathbb{A}^1 \times \mathbb{A}^1 & \xrightarrow{m} & \mathbb{A}^1\\
x & \longmapsto & (x, \infty) & \longmapsto & (x, 0) & \longmapsto & 0
\end{array}$$

which factors through

$$\begin{array}{ccccccc} \mathbb{A}^1 & \xrightarrow{i_{\infty,\square}} & \mathbb{A}^1\times\square^1 & \xrightarrow{\tau} & \mathbb{A}^1\times\mathbb{A}^1 & \xrightarrow{m} & \mathbb{A}^1 \\ \downarrow{\scriptstyle p_0} & & & & & & \downarrow{\scriptstyle \mathrm{id}_{\mathbb{A}^1}} \\ \{0\} & & & \xrightarrow{i_0} & & & \mathbb{A}^1 \end{array}$$

Thus,

$$i_{\infty,\square}^* \circ \widetilde{m}^* = (i_0 \circ p_0)^* = p_0^* \circ i_0^*.$$

Similarly $i_{0,\square}^* \circ \widetilde{m}^*$ is induced by

$$\begin{array}{ccccccc} \mathbb{A}^1 & \xrightarrow{i_{\infty,\square}} & \mathbb{A}^1\times\square^1 & \xrightarrow{\tau} & \mathbb{A}^1\times\mathbb{A}^1 & \xrightarrow{m} & \mathbb{A}^1 \\ x & \longmapsto & (x,0) & \longmapsto & (x,1) & \longmapsto & x \end{array}$$

which factors through $\mathrm{id}_{\mathbb{A}^1} : \mathbb{A}^1 \longrightarrow \mathbb{A}^1$ and one has

$$i_{0,\square}^* \circ \widetilde{m}^* = \mathrm{id}$$

which concludes the proof of the proposition. □

**3.3. Weight 1, weight 2 and polylogarithm cycles.** For now on, we set $X = \mathbb{P}^1 \setminus \{0,1,\infty\}$.

3.3.1. *Two weight 1 cycles generating the* $\mathrm{H}^1$. As mentioned before, there is a decomposition of $\mathrm{H}^1(\mathcal{N}_X^\bullet(p))$ as

$$\mathrm{H}^1(\mathcal{N}_X^\bullet(p)) \simeq \mathrm{H}^1(\mathcal{N}_{\mathbb{Q}}^\bullet(p)) \oplus \mathrm{H}^0(\mathcal{N}_{\mathbb{Q}}^\bullet(p-1)) \otimes \mathbb{Q}\mathcal{L}_0 \oplus \mathrm{H}^0(\mathcal{N}_{\mathbb{Q}}^\bullet(p-1)) \otimes \mathbb{Q}\mathcal{L}_1$$

and $\mathcal{L}_0$ and $\mathcal{L}_1$ (which are in weight 1 and degree 1) generates the $\mathrm{H}^*(\mathcal{N}_X^\bullet)$ relatively to $\mathrm{H}^*(\mathcal{N}_{\mathbb{Q}}^\bullet)$. Explicit expression for $\mathcal{L}_0$ and $\mathcal{L}_1$ are given below.

In Example 3.4, a cycle $\overline{\mathcal{L}_0}$ was constructed using the graph of $t \longmapsto t$ from $\mathbb{A}^1 \longrightarrow \mathbb{A}^1$. Taking its restriction to $X \times \square^1$, and using the same convention, one gets a cycle

$$\mathcal{L}_0 = [t;t] \subset X \times \square^1, \qquad \mathcal{L}_0 \in \mathcal{N}_X^1(1). \tag{5}$$

Similarly, using the graph of $t \longmapsto 1-t$, one gets

$$\mathcal{L}_1 = [t;1-t] \subset X \times \square^1, \qquad \mathcal{L}_1 \in \mathcal{N}_X^1(1). \tag{6}$$

One notices that the cycles $\mathcal{L}_0$ and $\mathcal{L}_1$ are both equidimensional over $X = \mathbb{P}^1 \setminus \{0,1,\infty\}$ but not equidimensional over $\mathbb{A}^1$.

Moreover, as

$$\mathcal{L}_0 \cap (X \times \{\varepsilon\}) = \mathcal{L}_0 \cap (\mathbb{P}^1 \setminus \{0,1,\infty\} \times \{\varepsilon\}) = \emptyset$$

for $\varepsilon = 0, \infty$, the above intersection tells us that $\partial(\mathcal{L}_0) = 0$. Similarly, one shows that $\partial(\mathcal{L}_1) = 0$. Thus $\mathcal{L}_0$ and $\mathcal{L}_1$ give two well defined classes in $\mathrm{H}^1(\mathcal{N}_X^\bullet(1))$.

In order to show that they are non-trivial and that they give the above decomposition of the $\mathrm{H}^1(\mathcal{N}_X^1)$, one shows that, in the localization sequence, their images under the boundary map

$$\mathrm{H}^1(\mathcal{N}_X^\bullet(1)) \xrightarrow{\delta} \mathrm{H}^0(\mathcal{N}_{\{0\}}^\bullet(0)) \oplus \mathrm{H}^0(\mathcal{N}_{\{1\}}^\bullet(0))$$

are non-zero. It is enough to treat the case of $\mathcal{L}_0$. Recall that $\overline{\mathcal{L}_0}$ is the closure of $\mathcal{L}_0$ in $\mathbb{A}^1 \times \square^1$ and is given by the parametrized cycle

$$\overline{\mathcal{L}_0} = [t; t] \subset \mathbb{A}^1 \times \square^1.$$

Its intersection with the face $u_1 = 0$ is of codimension 1 in $\mathbb{A}^1 \times \{0\}$ and the intersection with $u_1 = \infty$ is empty. Hence $\overline{\mathcal{L}_0}$ is admissible.

Thus, considering the definition of $\delta$, $\delta(\mathcal{L}_0)$ is given by the intersection of the differential of $\overline{\mathcal{L}_0}$ with $\{0\}$ and $\{1\}$ on the first and second factor, respectively. The above discussion on the admissibility of $\overline{\mathcal{L}_0}$ tells us that $\delta(\mathcal{L}_0)$ is non-zero on the factor $\mathrm{H}^0(\mathcal{N}^\bullet_{\{0\}}(0))$ and 0 on the other factor as the admissibility condition is trivial for $\mathcal{N}^\bullet_{\{0\}}(0)$ and the restriction of $\overline{\mathcal{L}_0}$ to 1 is empty. The situation is reverse for $\mathcal{L}_1$ using its closure $\overline{\mathcal{L}_1}$ in $\mathbb{A}^1 \times \square^1$.

Hence, even if the differentials of $\mathcal{L}_0$ and $\mathcal{L}_1$ are 0 in $\mathcal{N}^\bullet_X$, the differentials of their closure in $\mathbb{A}^1$ are non-zero in $\mathcal{N}^\bullet_{\mathbb{A}^1}$ and have a particular behavior when multiplied by an equidimensional cycle (see Lemma 3.13 below and Equation (11) for an example). We consider here only equidimensional cycles as it is needed to work with such cycles in order to pull-back by the multiplication. We use below notations of propositions 3.10 and 3.12.

LEMMA 3.13. *Let $C$ be an element in $\mathcal{N}^{eq,\bullet}_{\mathbb{A}^1}$, then*

$$\partial_{\mathbb{A}^1}(\overline{\mathcal{L}_0})\, C = C|_{t=0} \qquad \textit{and} \qquad \partial_{\mathbb{A}^1}(\overline{\mathcal{L}_1})\, C = C|_{t=1}$$

*where the notation $C|_{t=0}$ (resp. $C|_{t=1}$) denotes, as in Definition 3.3 the (image under the projector $\mathcal{A}lt$ of the) fiber at 0 (resp. 1) of the irreducible closed subvarieties composing the formal sum that defines $C$.*

PROOF. It is enough to assume that $C$ is given by $C = \mathcal{A}lt(Z)$ where $Z$ is an irreducible closed subvariety of $\mathbb{A}^1 \times \square^n$ such that for any face $F$ of $\square^n$, the intersection $Z \cap (X \times F)$ is empty or the restriction of $p_1 : \mathbb{A}^1 \times \square^n \longrightarrow \mathbb{A}^1$ to

$$Z \cap (\mathbb{A}^1 \times F) \longrightarrow \mathbb{A}^1$$

is equidimensional of relative dimension $\dim(F) - p$.

Remark that for $\varepsilon = 0, 1$ the cycle $\partial_{\mathbb{A}^1}(\overline{\mathcal{L}_\varepsilon})$ is given by the point

$$\{\varepsilon\} \in \mathbb{A}^1$$

which is of codimension 1 in $\mathbb{A}^1$. In order to compute the product $\partial_{\mathbb{A}^1}(\overline{\mathcal{L}_\varepsilon})\, C$, one considers first the product in $\mathbb{A}^1 \times \mathbb{A}^1 \times \square^n$:

$$\{\varepsilon\} \times Z \subset \mathbb{A}^1 \times \mathbb{A}^1 \times \square^n.$$

Let $\Delta$ denote the image of the diagonal $\mathbb{A}^1 \longrightarrow \mathbb{A}^1 \times \mathbb{A}^1$. The equidimensionality of $Z$ insures that for any face $F$ of $\square^n$

$$(\{\varepsilon\} \times Z) \bigcap (\Delta \times F) \simeq (Z \cap (\{\varepsilon\} \times \square^n)) \bigcap (\mathbb{A}^1 \times F)$$

is of codimension $p+1$. Thus the product $\partial_{\mathbb{A}^1}(\overline{\mathcal{L}_\varepsilon})\, C$ is simply the image under $\mathcal{A}lt$ of

$$Z \cap (\{\varepsilon\} \times \square^n) = Z|_{t=\varepsilon} \subset \mathbb{A}^1 \times \square^n.$$

$\square$

3.3.2. *A weight 2 example: the Totaro cycle.* One considers the linear combination

$$b = \mathcal{L}_0 \cdot \mathcal{L}_1 \in \mathcal{N}_X^2(2).$$

It is given as a parametrized cycle by

$$b = [t; t, 1-t] \subset X \times \square^2$$

or in terms of defining equations by

$$T_1V_1 - U_1T_2 = 0 \qquad \text{and} \qquad U_1V_2 + U_2V_1 = V_1V_2$$

where $T_1$ and $T_2$ denote the homogeneous coordinates on $X = \mathbb{P}^1 \setminus \{0,1,\infty\}$ and $U_i$, $V_i$ the homogeneous coordinates on each factor $\square^1 = \mathbb{P}^1 \setminus \{1\}$ of $\square^2$. One sees that the intersection of $b$ with faces $U_i$ or $V_i = 0$ for $i = 1, 2$ is empty because $T_1$ and $T_2$ are different from 0 in $X$ and because $U_i$ is different from $V_i$ in $\square^1$. Thus it tells us that

$$\partial(b) = 0.$$

Now, let $\overline{b}$ denote the algebraic closure of $b$ in $\mathbb{A}^1 \times \square^2$. As previously, its expression as parametrized cycle is

$$\overline{b} = \overline{\mathcal{L}_0}\,\overline{\mathcal{L}_1} = [t; t, 1-t] \subset \mathbb{A}^1 \times \square^2$$

and the intersection with $\mathbb{A}^1 \times F$ for any codimension 1 face $F$ of $\square^2$ is empty. Writing, as before, $\partial_{\mathbb{A}^1}$ for the differential in $\mathcal{N}_{\mathbb{A}^1}$, one has $\partial_{\mathbb{A}^1}(\overline{b}) = 0$.

As $\overline{\mathcal{L}_0}$ (resp. $\overline{\mathcal{L}_1}$) is equidimensional over $\mathbb{A}^1 \setminus \{0\}$ (resp. over $\mathbb{A}^1 \setminus \{1\}$), the cycle $\overline{b}$ is equidimensional over $\mathbb{A}^1 \setminus \{0,1\}$. Moreover, as $\overline{\mathcal{L}_0}$ (resp. $\overline{\mathcal{L}_1}$) has an empty fiber at 1 (resp. at 0), $\overline{b}$ has empty fiber at both 0 and 1. So $\overline{b}$ is equidimensional over $\mathbb{A}^1$ with empty fibers at 0 and 1. Following notations of Proposition 3.12, one defines two elements in $\mathcal{N}_{\mathbb{A}^1}^1(2)$ by pull back by the multiplication ( resp. twisted multiplication):

$$\overline{\mathcal{L}_{01}} = \mu^*(\overline{b}) \qquad \text{and} \qquad \overline{\mathcal{L}_{01}^1} = \nu^*(\overline{b}). \tag{7}$$

One also defines their restrictions to $X$

$$\mathcal{L}_{01} = j^*(\overline{\mathcal{L}_{01}}) \qquad \text{and} \qquad \mathcal{L}_{01}^1 = j^*(\overline{\mathcal{L}_{01}^1}). \tag{8}$$

Now, direct application of Proposition 3.12 shows that

$$\partial_{\mathbb{A}^1}(\overline{\mathcal{L}_{01}}) = -\mu^*(\partial_{\mathbb{A}^1}(\overline{b})) + \overline{b} - p_0^* \circ i_0^*(b) = -0 + \overline{\mathcal{L}_0}\,\overline{\mathcal{L}_1} - 0$$

because $\overline{b}$ has empty fiber at 0 and is 0 under $\partial_{\mathbb{A}^1}$. More generally, as $j^*$ is a morphism of c.d.g.a., Proposition 3.12 gives the following.

LEMMA 3.14. *Cycles $\mathcal{L}_{01}$, $\overline{\mathcal{L}_{01}}$, $\mathcal{L}_{01}^1$ and $\overline{\mathcal{L}_{01}^1}$ satisfy the following properties*

(1) *$\mathcal{L}_{01}$ and $\mathcal{L}_{01}^1$ (resp. $\overline{\mathcal{L}_{01}}$ and $\overline{\mathcal{L}_{01}^1}$) are equidimensional over $X$, that is elements in $\mathcal{N}_X^{eq,1}(2)$ (resp. equidimensional over $\mathbb{A}^1$).*

(2) *They satisfy the following differential equations*

$$\partial(\mathcal{L}_{01}) = \partial(\mathcal{L}_{01}^1) = b = \mathcal{L}_0\,\mathcal{L}_1$$

*and $\partial_{\mathbb{A}^1}(\overline{\mathcal{L}_{01}}) = \partial_{\mathbb{A}^1}(\overline{\mathcal{L}_{01}^1}) = \overline{b} = \overline{\mathcal{L}_0}\,\overline{\mathcal{L}_1}$.*

(3) *By the definition given in Equation (8), the cycle $\overline{\mathcal{L}_{01}}$ (resp. $\overline{\mathcal{L}_{01}^1}$) extends $\mathcal{L}_{01}$ (resp. $\mathcal{L}_{01}^1$) over $\mathbb{A}^1$ and has an empty fiber at 0 (resp. at 1).*

Moreover, one can explicitly compute the two pull-backs and obtain parametric representations

(9) $$\mathcal{L}_{01} = [t; 1 - \frac{t}{x}, x, 1-x], \qquad \mathcal{L}^1_{01} = [t; \frac{x-t}{x-1}, x, 1-x].$$

The multiplication map inducing $\mu^*$ is given by

$$\mathbb{A}^1 \times \square^1 \times \square^2 \longrightarrow A^1 \times \square^2 , \qquad [t; u_1, u_2, u_3] \longmapsto [\tfrac{t}{1-u_1}; u_2, u_3] .$$

In order to compute the pull-back, one should remark that if $u = 1 - t/x$ then

$$\frac{t}{1-u} = x.$$

Computing the pull-back by $\mu^*$ is then just rescaling the new $\square^1$ factor which arrives in first position. The case of $\nu^*$ is similar but using the fact that for $u = \frac{x-t}{x-1}$ one has

$$\frac{t-u}{1-u} = x.$$

REMARK 3.15. The cycle $\mathcal{L}_{01}$ is nothing but Totaro's cycle [**Tot92**], already described in [**BK94, Blo91**].

Moreover, $\mathcal{L}_{01}$ corresponds to the function $t \mapsto Li_2^{\mathbb{C}}(t)$ as shown in [**BK94**].

One recovers the value $\zeta(2)$ by specializing at $t = 1$ using the extension of $\mathcal{L}_{01}$ to $\mathbb{A}^1$.

3.3.3. *Polylogarithm cycles.* By induction one can build cycles $\mathrm{Li}_n^{cy} = \mathcal{L}_{0\cdots01}$ ($n-1$ zeros and one 1). We define $\mathrm{Li}_1^{cy}$ to be equal to $\mathcal{L}_1$.

LEMMA 3.16. *For any integer $n \geqslant 2$ there exists an equidimensional cycle over $X$, $\mathrm{Li}_n^{cy}$ in $\mathcal{N}_X^{eq,1}(n) \subset \mathcal{N}_X^1(n)$ satisfying*

1. *There is an equidimensional cycle over $\mathbb{A}^1$, $\overline{\mathrm{Li}_n^{cy}}$ in $\mathcal{N}_{\mathbb{A}^1}^{eq,1}(n)$, such that $\mathrm{Li}_n^{cy} = j^*(\overline{\mathrm{Li}_n^{cy}})$ (it has in particular a well defined fiber at 1).*
2. *The cycle $\overline{\mathrm{Li}_n^{cy}}$ has empty fiber at 0.*
3. *The cycles $\mathrm{Li}_n^{cy}$ and $\overline{\mathrm{Li}_n^{cy}}$ satisfy the differential equations*
$$\partial(\mathrm{Li}_n^{cy}) = \mathcal{L}_0 \cdot \mathrm{Li}_{n-1}^{cy} \qquad \text{and} \qquad \partial(\overline{\mathrm{Li}_n^{cy}}) = \overline{\mathcal{L}_0} \cdot \overline{\mathrm{Li}_{n-1}^{cy}}.$$
4. *$\mathrm{Li}_n^{cy}$ is explicitly given as a parametrized cycle by*
$$[t; 1 - \frac{t}{x_{n-1}}, x_{n-1}, 1 - \frac{x_{n-1}}{x_{n-2}}, x_{n-2}, \dots, 1 - \frac{x_2}{x_1}, x_1, 1-x_1] \subset X \times \square^{2n-1}.$$

PROOF. For $n = 2$, we have already defined $\mathrm{Li}_2^{cy} = \mathcal{L}_{01}$ satisfying the expected properties.

Assume that one has built the cycles $\mathrm{Li}_k^{cy}$ for $2 \leqslant k < n$. One considers in $\mathcal{N}^2_{\mathbb{A}^1}(n)$ the product

$$\overline{b} = \overline{\mathcal{L}_0} \cdot \overline{\mathrm{Li}_{n-1}^{cy}} = [t; t, 1 - \frac{t}{x_{n-2}}, x_{n-2}, 1 - \frac{x_{n-2}}{x_{n-3}}, x_{n-3}, \dots, 1 - \frac{x_2}{x_1}, x_1, 1 - x_1].$$

As $\overline{\mathcal{L}_0}$ is equidimensional over $\mathbb{A}^1 \setminus \{0\}$ and as $\overline{\mathrm{Li}_{n-1}^{cy}}$ is equidimensional over $\mathbb{A}^1$, $\overline{b}$ is equidimensional over $\mathbb{A}^1 \setminus \{0\}$. Moreover, as $\overline{\mathrm{Li}_{n-1}^{cy}}$ has empty fiber at 0, $\overline{b}$ is equidimensional over $\mathbb{A}^1$ with empty fiber at 0.

Computing the differential with the Leibniz rule and Lemma 3.13, one gets

$$\partial_{\mathbb{A}^1}(\overline{b}) = \overline{\mathrm{Li}^{cy}_{n-1}}|_{t=0} - \overline{\mathcal{L}_0} \cdot \overline{\mathcal{L}_0} \cdot \overline{\mathrm{Li}^{cy}_{n-2}} = 0.$$

One concludes using Proposition 3.12. The same argument used to obtain the parametrized representation for $\mathcal{L}_{01}$ at Equation (9) shows that

$$\mathrm{Li}^{cy}_n = [t; 1 - \frac{t}{x_{n-1}}, x_{n-1}, 1 - \frac{x_{n-1}}{x_{n-2}}, x_{n-2}, \ldots, 1 - \frac{x_2}{x_1}, x_1, 1 - x_1] \subset \mathbb{A}^1 \times \square^{2n-1}.$$

□

REMARK 3.17.

- One retrieves the expression given in [**BK94**].
- Moreover, $\mathrm{Li}^{cy}_n$ corresponds to the function $t \mapsto Li^{\mathbb{C}}_n(t)$ as shown in [**BK94**] (or in [**GGL09**]).
- $\overline{\mathcal{L}_0}$ having an empty fiber at 1, one can also pull-back by the twisted multiplication and obtain similarly cycles $\mathcal{L}^1_{0\cdots 01}$ satisfying $\partial(\mathcal{L}^1_{0\cdots 01}) = \partial(\mathcal{L}_{0\cdots 01})$. In some sense, they correspond to $\mathcal{L}_{0\cdots 01} - p^* \circ i_1^*(\mathcal{L}_{0\cdots 01})$ which in terms of integrals corresponds to $Li^{\mathbb{C}}_n(t) - \zeta(n)$.

### 3.4. Some higher weight examples for multiple polylogarithm cycles.

3.4.1. *Weight* 3. The cycle $\mathcal{L}_{01}$ was defined previously, so was the cycle $\mathcal{L}_{001} = \mathrm{Li}^{cy}_3$ by considering the product

$$b = \mathcal{L}_0 \cdot \mathcal{L}_{01}.$$

Now, in weight 3, one could also consider the product

$$\mathcal{L}_{01} \cdot \mathcal{L}_1 \quad \in \mathcal{N}^2_X(3). \tag{10}$$

However the above product does not lead by similar arguments to a new cycle. Before explaining how to follow the strategy used in weight 2 and for the polylogarithms in order to obtain another weight 3 cycle, the author would like to spend a little time on the obstruction occurring with the product in Equation (10) as it enlightens in particular the need of the cycle $\mathcal{L}^1_{01}$ previously built.

Thus let $b = \mathcal{L}_{01} \cdot \mathcal{L}_1$ be the above product in $\mathcal{N}^2_X(3)$, given as a parametrized cycle by

$$b = [t; 1 - \frac{t}{x_1}, x_1, 1 - x_1, 1 - t] \quad \subset X \times \square^4.$$

From this expression, one sees that $b$ is admissible and that $\partial(b) = 0$ because $t \in X$ can not be equal to 1.

Let $\overline{b}$ be the closure of the defining cycle of $b$ in $\mathbb{A}^1 \times \square^4$, that is the image under the projector $\mathcal{A}lt$ of

$$\left\{(t, 1 - \frac{t}{x_1}, x_1, 1 - x_1, 1 - t) \text{ such that } t \in \mathbb{A}^1,\, x_1 \in \mathbb{P}^1\right\} \cap \mathbb{A}^1 \times \square^4.$$

Let $u_i$ denote the coordinate on the $i$-th factor $\square^1$. As most of the intersections of $\overline{b}$ with face $\mathbb{A}^1 \times F$ are empty, in order to prove that $\overline{b}$ is admissible and gives an element in $\mathcal{N}^2_{\mathbb{A}^1}(3)$, it is enough to check the (co)dimension condition on the three faces : $u_1 = 0$, $u_4 = 0$ and $u_1 = u_4 = 0$. The intersection of $\overline{b}$ with the face $u_1 = u_4 = 0$ is empty as $u_2 \neq 1$. The intersection $\overline{b}$ with the face defined by $u_1 = 0$ or $u_4 = 0$ is 1 dimensional and so of codimension 3 in $\mathbb{A}^1 \times F$.

Computing the differential in $\mathcal{N}^\bullet_{\mathbb{A}^1}$, using Lemma 3.13 or the fact that the intersection with $u_1 = 0$ is killed by the projector $\mathcal{A}lt$, gives

$$\partial_{\mathbb{A}^1}(\bar{b}) = \partial_{\mathbb{A}^1}(\overline{\mathcal{L}_{01}\mathcal{L}_1}) = -\overline{\mathcal{L}_{01}}|_{t=1} \neq 0 \tag{11}$$

and the homotopy trick used previously will not work as it relies (partly) on beginning with a cycle $\bar{b}$ satisfying $\partial_{\mathbb{A}^1}(\bar{b}) = 0$.

In order to bypass this, one could introduce the constant cycle $\mathcal{L}_{01}(1) = p^* \circ i_1^*(\overline{\mathcal{L}_{01}})$ and consider the linear combination

$$\bar{b} = (\overline{\mathcal{L}_{01}} - \overline{\mathcal{L}_{01}}(1)) \cdot \overline{\mathcal{L}_1} \in \mathcal{N}^2_{\mathbb{A}^1}(3). \tag{12}$$

and its equivalent in $\mathcal{N}^2_X(3)$. Now, the correction by $-\overline{\mathcal{L}_{01}}(1) \cdot \overline{\mathcal{L}_1}$ insures that $\partial_{\mathbb{A}^1}(\bar{b}) = 0$.

However, it is still not good enough as the use of the homotopy property for the pull-back by the multiplication requires to work with equidimensional cycles which is not the case for $\bar{b}$ (the problem comes from the fiber at 1).

The fact that $\overline{\mathcal{L}_1}$ is not equidimensional over $\mathbb{A}^1$ but equidimensional on $\mathbb{A}^1 \setminus \{1\}$ requires to multiply it by a cycle with an empty fiber at 1 which insures that the fiber of the product at 1 is empty. Thus one considers the product in $\mathcal{N}^{eq,\,2}_{\mathbb{A}^1}(3)$

$$\bar{b} = \overline{\mathcal{L}^1_{01}}\,\overline{\mathcal{L}_1} = -\overline{\mathcal{L}_1}\,\overline{\mathcal{L}^1_{01}}$$

which has an empty fiber at 0 and 1. Moreover the Leibniz rule and Lemma 3.13 imply that

$$\partial_{\mathbb{A}^1}(\bar{b}) = \partial_{\mathbb{A}^1}(\overline{\mathcal{L}^1_{01}})\,\overline{\mathcal{L}_1} - \overline{\mathcal{L}^1_{01}}\,\partial(\overline{\mathcal{L}_1}) = \overline{\mathcal{L}_1}\,\overline{\mathcal{L}_0\mathcal{L}_1} - \overline{\mathcal{L}^1_{01}}|_{t=1} = 0.$$

Thus one defines

$$\overline{\mathcal{L}_{011}} = \mu^*(\overline{\mathcal{L}^1_{01}}\,\overline{\mathcal{L}_1}) \qquad \text{and} \qquad \overline{\mathcal{L}^1_{011}} = \nu^*(\overline{\mathcal{L}^1_{01}}\,\overline{\mathcal{L}_1}) \tag{13}$$

and their restrictions to $X = \mathbb{P}^1 \setminus \{0, 1, \infty\}$

$$\mathcal{L}_{011} = j^*(\overline{\mathcal{L}_{011}}) \qquad \text{and} \qquad \mathcal{L}^1_{011} = j^*(\overline{\mathcal{L}^1_{011}}). \tag{14}$$

As previously, propositions 3.12 and 3.10 insure the following.

LEMMA 3.18. *The cycles $\mathcal{L}_{011}$, $\overline{\mathcal{L}_{011}}$, $\mathcal{L}^1_{011}$ and $\overline{\mathcal{L}^1_{011}}$ satisfy the following properties*

1. *$\mathcal{L}_{011}$ and $\mathcal{L}^1_{011}$ (resp. $\overline{\mathcal{L}_{011}}$ and $\overline{\mathcal{L}^1_{011}}$) are in $\mathcal{N}^{eq,\,1}_X(2)$ (resp. in $\mathcal{N}^{eq,\,1}_{\mathbb{A}^1}(2)$).*
2. *They satisfy the following differential equations*
$$\partial(\mathcal{L}_{011}) = \partial(\mathcal{L}^1_{011}) = \mathcal{L}^1_{01}\,\mathcal{L}_1 = -\mathcal{L}_1\,\mathcal{L}^1_{01}$$
*and $\partial_{\mathbb{A}^1}(\overline{\mathcal{L}_{011}}) = \partial_{\mathbb{A}^1}(\overline{\mathcal{L}^1_{011}}) = \overline{\mathcal{L}^1_{01}}\,\overline{\mathcal{L}_1}$.*
3. *The cycle $\overline{\mathcal{L}_{011}}$ (resp. $\overline{\mathcal{L}^1_{011}}$) has an empty fiber at 0 (resp. at 1).*

3.4.2. *Weight* 4. In weight 4 the first linear combination appears. The situation in weight 4 is given by the following Lemma

LEMMA 3.19. *Let $W$ be one of the Lyndon words* 0001, 0011 *or* 0111. *There exist cycles $\mathcal{L}_W$, $\mathcal{L}^1_W$ in $\mathcal{N}^{eq,\,1}_X(4)$ and cycles $\overline{\mathcal{L}_W}$, $\overline{\mathcal{L}^1_W}$ in $\mathcal{N}^{eq,\,1}_{\mathbb{A}^1}(4)$ which satisfy the following properties*

1. *$\mathcal{L}_W = j^*(\overline{\mathcal{L}_W})$ and $\mathcal{L}^1_W = j^*(\overline{\mathcal{L}^1_W})$*
2. *$\overline{\mathcal{L}_W}$ (resp. $\overline{\mathcal{L}^1_W}$) has an empty fiber at 0 (resp. at 1)*

(3) *Cycles $\mathcal{L}_W$ and $\mathcal{L}^1_W$ for $W = 0001$, $0011$ and $0111$ satisfy the following differential equations derived from the differential equations satisfied by $\overline{\mathcal{L}_W}$ and $\overline{\mathcal{L}^1_W}$*

$$\partial(\mathcal{L}_{0001}) = \partial(\mathcal{L}^1_{0001}) = \mathcal{L}_0\,\mathcal{L}_{001}, \tag{15}$$

$$\partial(\mathcal{L}_{0011}) = \partial(\mathcal{L}^1_{0011}) = \mathcal{L}_0\,\mathcal{L}_{011} + \mathcal{L}^1_{001}\,\mathcal{L}_1 - \mathcal{L}_{01}\,\mathcal{L}^1_{01} \tag{16}$$

*and*

$$\partial(\mathcal{L}_{0111}) = \partial(\mathcal{L}^1_{0111}) = \mathcal{L}^1_{011}\,\mathcal{L}_1. \tag{17}$$

PROOF. The proof goes as before as the main difficulty is to "guess" the differential equations. The case of $\mathcal{L}_{0001} = \mathrm{Li}_4^{cy}$ and $\mathcal{L}^1_{0001}$ has already been treated in Lemma 3.16 and the remark afterward. The case of $\mathcal{L}_{0111}$ and $\mathcal{L}^1_{0111}$ is extremely similar to the case of $\mathcal{L}_{011}$. We will only describe the case of $\mathcal{L}_{0011}$. Let $\bar{b}$ be the element in $\mathcal{N}^2_{\mathbb{A}^1}(4)$ defined by:

$$\bar{b} = \overline{\mathcal{L}_0}\,\overline{\mathcal{L}_{011}} + \overline{\mathcal{L}^1_{001}}\,\overline{\mathcal{L}_1} - \overline{\mathcal{L}_{01}}\,\overline{\mathcal{L}^1_{01}}.$$

All the cycles involved are equidimensional over $\mathbb{A}^1 \setminus \{0, 1\}$. As the products in the above equation always involve a cycle with empty fiber at 0 and one with empty fiber at 1, the product has empty fiber at 0 and 1 and is equidimensional over $\mathbb{A}^1$.

This shows that $\bar{b}$ is equidimensional over $\mathbb{A}^1$ with empty fiber at 0 and 1. One computes $\partial_{\mathbb{A}^1}(\bar{b})$ using the Leibniz rule, Lemma 3.13 and the previously obtained differential equations:

$$\partial_{\mathbb{A}^1}(\bar{b}) = -\overline{\mathcal{L}_0}\,\overline{\mathcal{L}^1_{01}}\,\overline{\mathcal{L}_1} + \overline{\mathcal{L}_0}\,\overline{\mathcal{L}_{01}}\,\overline{\mathcal{L}_1} - \overline{\mathcal{L}_0}\,\overline{\mathcal{L}_1}\,\overline{\mathcal{L}^1_{01}} + \overline{\mathcal{L}_{01}}\,\overline{\mathcal{L}_0}\,\overline{\mathcal{L}_1} = 0$$

One can thus define

$$\overline{\mathcal{L}_{0011}} = \mu^*(\bar{b}) \qquad \text{and} \qquad \overline{\mathcal{L}^1_{0011}} = \nu^*(\bar{b})$$

and conclude with propositions 3.12 and 3.10. □

3.4.3. *General statement and a weight 5 example.* In weight 5 there are six Lyndon words and the combinatorics of equation (ED-T) leads to six cycles with empty fiber at 0 and six cycles with empty fiber at 1. The general statement proved in [**Sou12**] is given below.

THEOREM 3.20. *For any Lyndon word $W$ of length $p$ greater or equal to 2, there exist two cycles $\mathcal{L}_W$ and $\mathcal{L}^1_W$ in $\mathcal{N}^1_X(p)$ such that :*

- *$\mathcal{L}_W$, $\mathcal{L}^1_W$ are elements in $\mathcal{N}^{eq,\,1}_X(p)$.*
- *There exist cycles $\overline{\mathcal{L}_W}$, $\overline{\mathcal{L}^1_W}$ in $\mathcal{N}^{eq,\,1}_{\mathbb{A}^1}(p)$ such that*

$$\mathcal{L}_W = j^*(\overline{\mathcal{L}_W}) \qquad \text{and} \qquad \mathcal{L}^1_W = j^*(\overline{\mathcal{L}^1_W}).$$

- *The restriction of $\overline{\mathcal{L}_W}$ (resp. $\overline{\mathcal{L}^1_W}$) to the fiber $t = 0$ (resp. $t = 1$) is empty.*
- *The cycle $\mathcal{L}_W$ satisfies the equation*

$$\partial(\mathcal{L}_W) = \sum_{U<V} a^W_{U,V}\mathcal{L}_U\mathcal{L}_V + \sum_{U,V} b^W_{U,V}\mathcal{L}_U\mathcal{L}^1_V \tag{18}$$

*and resp. $\mathcal{L}^1_W$ satisfies*

$$\partial(\mathcal{L}^1_W) = \sum_{0<U<V} {a'}^W_{U,V}\mathcal{L}^1_U\mathcal{L}^1_V + \sum_{U,V} {b'}^W_{U,V}\mathcal{L}_U\mathcal{L}^1_V + \sum_V a'_{0,V}\mathcal{L}_0\mathcal{L}_V \tag{19}$$

*and the same holds for their extensions* $\overline{\mathcal{L}_W}$ *and* $\overline{\mathcal{L}^1_W}$ *to* $\mathcal{N}^{eq,1}_{\mathbb{A}^1}$. *In the above equations* $U$ *and* $V$ *are Lyndon words of smaller length than* $W$ *and the coefficients* $a^W_{U,V}$, $b^W_{U,V}$, $a'^w_{U,V}$ *and* $b'^W_{U,V}$ *are integers derived from equation* (ED-T).

REMARK 3.21. Without giving a proof which works by induction on the length of $W$, the author would like to stress that the construction of the cycles $\overline{\mathcal{L}_W}$ (resp. $\overline{\mathcal{L}^1_W}$) relies on a geometric argument that has already been described and used here: the pull-back by the (twisted) multiplication $\mu^*$ (resp. $\nu^*$) gives a homotopy between the identity and $p^* \circ i_0^*$ (resp. $p^* \circ i_1^*$). Thus, defining

$$\overline{A_W} = \sum_{U<V} a^W_{U,V} \overline{\mathcal{L}_U}\,\overline{\mathcal{L}_V} + \sum_{U,V} b^W_{U,V} \overline{\mathcal{L}_U}\,\overline{\mathcal{L}^1_V}$$

and

$$\overline{A^1_W} = \sum_{0<U<V} a'^W_{U,V} \overline{\mathcal{L}^1_U}\,\overline{\mathcal{L}^1_V} + \sum_{U,V} b'^W_{U,V} \overline{\mathcal{L}_U}\,\overline{\mathcal{L}^1_V} + \sum_{V} a'_{0,V} \overline{\mathcal{L}_0}\,\overline{\mathcal{L}_V},$$

the cycle $\overline{\mathcal{L}_W}$ and $\overline{\mathcal{L}^1_W}$ are defined by

$$\overline{\mathcal{L}_W} = \mu^*(\overline{A_W}) \qquad \text{and} \qquad \overline{\mathcal{L}^1_W} = \nu^*(\overline{A^1_W}). \tag{20}$$

The fact that $\overline{A_W}$ (resp. $\overline{A^1_W}$) is equidimensional over $\mathbb{A}^1$ with empty fiber at 0 (resp. 1) is essentially a consequence of the induction. The main problem is to show that $\partial_{\mathbb{A}^1}(\overline{A_W}) = \partial_{\mathbb{A}^1}(\overline{A^1_W}) = 0$ which in [**Sou12**] is deduced after a long preliminary work from the combinatorial situation given by the trees $T_{W^*}$.

In weight 5 appears the need of two distinct differential equations and the first example with coefficient different from $\pm 1$.

EXAMPLE 3.22. The two cycles associated to the Lyndon word 01011 satisfy

$$\partial(\mathcal{L}_{01011}) = -\,\mathcal{L}_{01}\,\mathcal{L}_{011} - \mathcal{L}_1\,\mathcal{L}^1_{0011} - 2\mathcal{L}_{011}\,\mathcal{L}^1_{01} \tag{21}$$

$$\partial(\mathcal{L}^1_{01011}) = \mathcal{L}^1_{01}\,\mathcal{L}^1_{011} - \mathcal{L}_{011}\,\mathcal{L}^1_{01} - \mathcal{L}_{01}\,\mathcal{L}^1_{011} - \mathcal{L}_1\,\mathcal{L}^1_{0011}. \tag{22}$$

The factor 2 in the last term of $\partial(\mathcal{L}_{01011})$ is related to the factor 2 appearing in $d_{cy}(T_{01011^*})$ presented in Equation (4). The term

$$2\mathcal{L}_{011}\mathcal{L}_0\mathcal{L}_1$$

which is equal to $\partial(-2\mathcal{L}_{011}\mathcal{L}^1_{01})$ cancels with one term in $-\mathcal{L}_0\mathcal{L}_1\mathcal{L}_{011}$ coming from $\partial(-\mathcal{L}_{01} \cdot \mathcal{L}_{011})$ and one term in $\mathcal{L}_1\mathcal{L}_0\mathcal{L}_{011}$ coming from $\partial(\mathcal{L}_1 \cdot \mathcal{L}_{0011})$. The whole computation can in fact be done over $\mathbb{A}^1$ and $\overline{\mathcal{L}_{01011}}$ is defined as previously as the pull-back by $\mu^*$ of

$$\bar{b} = -\overline{\mathcal{L}_{01}}\,\overline{\mathcal{L}_{011}} - \overline{\mathcal{L}_1}\,\overline{\mathcal{L}^1_{0011}} - 2\overline{\mathcal{L}_{011}}\,\overline{\mathcal{L}^1_{01}}.$$

The cycle $\mathcal{L}_{01011}$ is then its restriction to $X$. The above linear combination has an empty fiber at 0 (which allows the use of $\mu^*$). However its fiber at 1 is nonempty and given by

$$-\overline{\mathcal{L}_{01}}|_{t=1}\,\overline{\mathcal{L}_{011}}|_{t=1}$$

and its pull-back by the twisted multiplication $\nu^*$ satisfies

$$\partial_{\mathbb{A}^1}(\nu^*(\bar{b})) = \bar{b} + p^* \circ i_1^*(\bar{b}) \neq \bar{b}.$$

That is why we have introduced the linear combination

$$\mathcal{L}^1_{01}\,\mathcal{L}^1_{011} - \mathcal{L}_{011}\,\mathcal{L}^1_{01} - \mathcal{L}_{01}\,\mathcal{L}^1_{011} - \mathcal{L}_1\,\mathcal{L}^1_{0011}$$

whose extension to $\mathbb{A}^1$ has empty fiber at 1 (but not at 0). This allows us to define

$$\overline{\mathcal{L}^1_{01011}} = \nu^* \left(\mathcal{L}^1_{01}\,\mathcal{L}^1_{011} - \mathcal{L}_{011}\,\mathcal{L}^1_{01} - \mathcal{L}_{01}\,\mathcal{L}^1_{011} - \mathcal{L}_1\,\mathcal{L}^1_{0011}\right).$$

## 4. Parametric and combinatorial representation for the cycles: trees with colored edges

One can give a combinatorial approach to describe cycles $\mathcal{L}_W$ and $\mathcal{L}^1_W$ as parametrized cycles using trivalent trees with two types of edge.

DEFINITION 4.1. Let $\mathcal{T}^{||}$ be the $\mathbb{Q}$ vector space spanned by rooted trivalent trees such that

- the edges can be of two types: $|$ or $\|$;
- the root vertex is decorated by $t$ ;
- other external vertices are decorated by 0 or 1.

We say that such a tree is a rooted colored tree or simply a colored tree.

We define two bilinear maps $\mathcal{T}^{||} \otimes \mathcal{T}^{||} \longrightarrow \mathcal{T}^{||}$ as follows on the colored trees:

- Let $T_1 \overset{|}{\wedge} T_2$ be the colored tree given by joining the two roots of $T_1$ and $T_2$ and adding a new root and a new edge of type $|$ :

$T_1 \overset{|}{\wedge} T_2 =$ 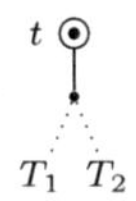


where the dotted edges denote either type of edges.

- Let $T_1 \overset{\|}{\wedge} T_2$ be the colored tree given by joining the two roots of $T_1$ and $T_2$ and adding a new root and a new edge of type $\|$ :

$T_1 \overset{\|}{\wedge} T_2 =$ 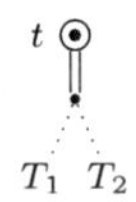


where the dotted edges denote either type of edges.

DEFINITION 4.2. Let $\mathfrak{T}_0$ and $\mathfrak{T}_1$ be the colored trees defined by

$\mathfrak{T}_0 =$ 

and $\mathfrak{T}_1 =$ 

.

For any Lyndon word $W$ of length greater or equal to 2, let $\mathfrak{T}_W$ (resp. $\mathfrak{T}^1_W$) be the linear combination of colored trees given by

$$\mathfrak{T}_W = \sum_{U<V} a^W_{U,V}\,\mathfrak{T}_U \overset{|}{\wedge} \mathfrak{T}_V + \sum_{U,V} b^W_{U,V}\,\mathfrak{T}_U \overset{|}{\wedge} \mathfrak{T}^1_V,$$

and respectively by

$$\mathfrak{T}^1_W = \sum_{0<U<V} {a'}^W_{U,V}\,\mathfrak{T}^1_U \overset{\|}{\wedge} \mathfrak{T}^1_V + \sum_{U,V} {b'}^W_{U,V}\,\mathfrak{T}_U \overset{\|}{\wedge} \mathfrak{T}^1_V + \sum_{V} {a'}_{0,V}\,\mathfrak{T}_0 \overset{\|}{\wedge} \mathfrak{T}_V.$$

where the coefficients appearing are the ones from Theorem 3.20.

To a colored tree $T$ with $p$ external leaves and a root, one associates a function $f_T : X \times (\mathbb{P}^1)^{p-1} \longrightarrow X \times (\mathbb{P}^1)^{2p-1}$ as follows :

- Endow $T$ with its natural order as trivalent tree.
- This induces a numbering of the edges of $T$ : $(e_1, e_2, \ldots, e_{2p-1})$.
- The edges being oriented away from the root, the numbering of the edges induces a numbering of the vertices $(v_1, v_2, \ldots, v_{2p})$ such that the root is $v_1$.
- Associate variables $x_1, \ldots, x_{p-1}$ to each internal vertices such that the numbering of the variables is opposite to the order induced by the numbering of the vertices (first internal vertex has variable $x_{p-1}$, second internal vertex has variable $x_{p-2}$ and so on).
- For each edge $e_i = \begin{smallmatrix} a \bullet \\ \vdots \\ \bullet b \end{smallmatrix}$ oriented from $a$ to $b$, define a function

$$f_i(a,b) = \begin{cases} 1 - \dfrac{a}{b} & \text{if } e_i \text{ is of type } | \,, \\[2ex] \dfrac{b-a}{b-1} & \text{if } e_i \text{ is of type } \| \,. \end{cases}$$

- Finally $f_T : X \times (\mathbb{P}^1)^{p-1} \longrightarrow X \times (\mathbb{P}^1)^{2p-1}$ is defined by

$$f_T(t, x_1, \ldots x_{p-1}) = (t, f_1, \ldots, f_{2p-1}).$$

Let $\Gamma(T)$ be the intersection of the image of $f_T$ with $X \times \square^{2p-1}$. One extends the definition of $\Gamma$ to $\mathcal{T}^{||}$ by linearity and thus obtains a *twisted forest cycling map* similar to the one defined by Gangl, Goncharov and Levin in [**GGL09**].

The map $\Gamma$ satisfies:

- $\mathcal{A}lt(\Gamma(\mathfrak{T}_0)) = \mathcal{L}_0$ and $\mathcal{A}lt(\Gamma(\mathfrak{T}_1)) = \mathcal{L}_1$.
- For any Lyndon word of length $p \geqslant 2$,

$$\mathcal{A}lt(\Gamma(\mathfrak{T}_W)) = \mathcal{L}_W \qquad \text{and} \qquad \mathcal{A}lt(\Gamma(\mathfrak{T}^1_W)) = \mathcal{L}^1_W.$$

The fact that $\Gamma(\mathfrak{T}_0)$ (resp. $\Gamma(\mathfrak{T}_1)$) is the graph of $t \mapsto t$ (resp. $t \mapsto 1-t$) follows from the definition. Thus one already has $\Gamma(\mathfrak{T}_0)$ (resp. $\Gamma(\mathfrak{T}_1)$) in $\mathcal{Z}^1_{eq}(X,1)$ and

$$\mathcal{A}lt(\Gamma(\mathfrak{T}_0)) = \mathcal{L}_0 \qquad \text{and} \qquad \mathcal{A}lt(\Gamma(\mathfrak{T}_1)) = \mathcal{L}_1.$$

Then the above property is deduced by induction. Recall that the defining equation (20) for the cycle $\overline{\mathcal{L}_W}$ is

$$\overline{\mathcal{L}_W} = \mu^* \left( \sum_{U<V} a^W_{U,V} \overline{\mathcal{L}_U} \, \overline{\mathcal{L}_V} + \sum_{U,V} b^W_{U,V} \overline{\mathcal{L}_U} \, \overline{\mathcal{L}^1_V} \right).$$

As already remarked in Example 9, in order to compute the pull-back by $\mu^*$ one sets the former parameter $t$ to a new variable $x_n$ and parametrizes the new $\square^1$ factor arriving in first position by $1 - \frac{t}{x_n}$; $t$ is again the parameter over $X$ or $\mathbb{A}^1$ depending if one considers cycles over $\mathbb{A}^1$ or their restriction to $X = \mathbb{P}^1 \setminus \{0, 1, \infty\}$. Thus the expression of $\mathcal{L}_W$, restriction of $\overline{\mathcal{L}_W}$ to $X$ is exactly given by

$$\mathcal{L}_W = \mathcal{A}lt(\Gamma(\mathfrak{T}_W)).$$

The case of $\nu^*$ is similar but parametrizing the new $\square^1$ factor by $\frac{x_n - t}{x_n - 1}$.

For the previously built examples, we give below the corresponding colored trees and expressions as parametrized cycles (omitting the projector $\mathcal{A}lt$). We also recall the corresponding differential equations as given by Theorem 3.20.

EXAMPLE 4.3 (Weight 1).

$\mathfrak{T}_0 =$ 

, $\mathfrak{T}_1 =$ 

and $\partial(\mathcal{L}_0) = \partial(\mathcal{L}_1) = 0.$

We recall below how cycles $\mathcal{L}_0$ and $\mathcal{L}_1$ are expressed in terms of parametrized cycles:

$$\mathcal{L}_0 = [t; t] \quad \subset X \times \square^1 \qquad \text{and} \qquad \mathcal{L}_1 = [t; 1-t] \quad \subset X \times \square^1.$$

EXAMPLE 4.4 (Weight 2).

$\mathfrak{T}_{01} =$ 

, $\mathfrak{T}_{01}^1 =$ 

and $\partial(\mathcal{L}_{01}) = \partial(\mathcal{L}_{01}^1) = \mathcal{L}_0\mathcal{L}_1$

We have seen in Equation (9) that cycles $\mathcal{L}_{01}$ and $\mathcal{L}_{01}^1$ are given (in $X \times \square^3$) by

$$\mathcal{L}_{01} = [t; 1-\frac{t}{x_1}, x_1, 1-x_1] \qquad \text{and} \qquad \mathcal{L}_{01}^1 = [t; \frac{x_1-t}{x_1-1}, x_1, 1-x_1].$$

EXAMPLE 4.5 (Weight 3).

$$\partial(\mathcal{L}_{001}) = \partial(\mathcal{L}_{001}^1) = \mathcal{L}_0\mathcal{L}_{01}, \quad \partial(\mathcal{L}_{011}) = \partial(\mathcal{L}_{011}^1) = -\mathcal{L}_1\mathcal{L}_{01}^1.$$

$\mathfrak{T}_{001} =$ 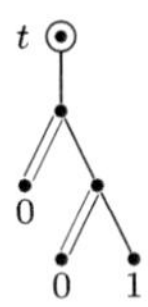

, $\mathfrak{T}_{001}^1 =$ 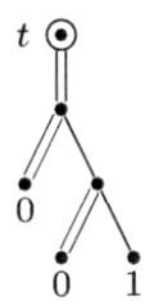

, $\mathfrak{T}_{011} = -$ 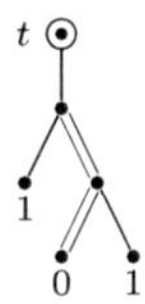

, $\mathfrak{T}_{011}^1 = -$ 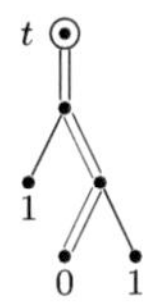


The corresponding expression as parametrized cycles are given below (following our "twisted forest cycling map"):

$$\mathcal{L}_{001} = [t; 1-\frac{t}{x_2}, x_2, 1-\frac{x_2}{x_1}, x_1, 1-x_1] \quad \subset X \times \square^5,$$
$$\mathcal{L}_{001}^1 = [t; \frac{x_2-t}{x_2-1}, x_2, 1-\frac{x_2}{x_1}, x_1, 1-x_1] \quad \subset X \times \square^5$$

and

$$\mathcal{L}_{011} = -[t; 1-\frac{t}{x_2}, 1-x_2, \frac{x_1-x_2}{x_1-1}, x_1, 1-x_1] \quad \subset X \times \square^5,$$
$$\mathcal{L}_{011}^1 = -[t; \frac{x_2-t}{x_2-1}, 1-x_2, \frac{x_1-x_2}{x_1-1}, x_1, 1-x_1] \quad \subset X \times \square^5$$

EXAMPLE 4.6 (Weight 4). The differential equations satisfied by the weight 4 cycles are:

$$\begin{aligned}
\partial(\mathcal{L}_{0001}) = \partial(\mathcal{L}_{0001}^1) &= \mathcal{L}_0\mathcal{L}_{001} \\
\partial(\mathcal{L}_{0011}) = \partial(\mathcal{L}_{0011}^1) &= \mathcal{L}_0\mathcal{L}_{011} - \mathcal{L}_1\mathcal{L}_{001}^1 - \mathcal{L}_{01}\mathcal{L}_{01}^1 \\
\partial(\mathcal{L}_{0111}) = \partial(\mathcal{L}_{0111}^1) &= -\mathcal{L}_1\mathcal{L}_{011}^1
\end{aligned}$$

The corresponding colored trees are given by:

$\mathfrak{T}_{0001} =$ 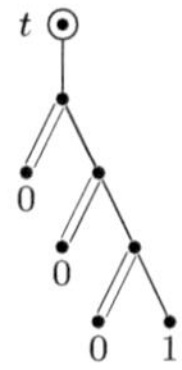

, $\mathfrak{T}^1_{0001} =$ 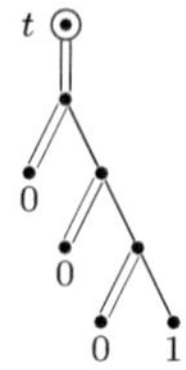

, $\mathfrak{T}_{0111} =$ 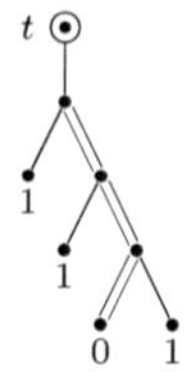

, $\mathfrak{T}^1_{0111} =$ 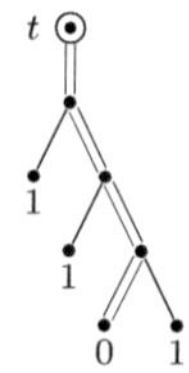

,

$\mathfrak{T}_{0011} = -$ 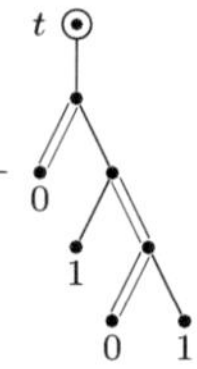

$-$ 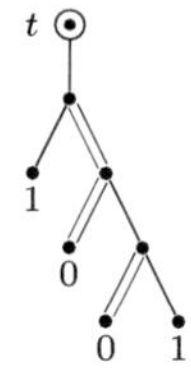

$-$ 

and

$\mathfrak{T}^1_{0011} = -$ 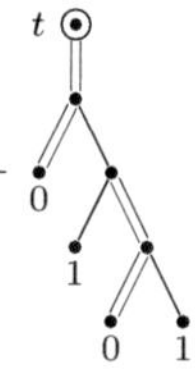

$-$ 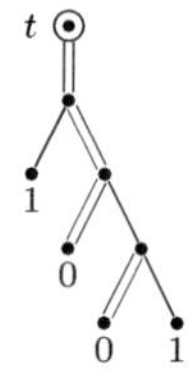

$-$ 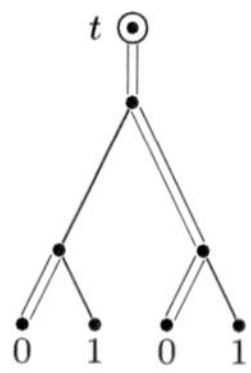

.

The expressions as parametrized cycles of $\mathcal{L}_{0001}$, $\mathcal{L}^1_{0001}$, $\mathcal{L}_{0111}$ and $\mathcal{L}^1_{0111}$ are given below (in $X \times \square^7$):

$$\mathcal{L}_{0001} = [t; 1 - \frac{t}{x_3}, x_3, 1 - \frac{x_3}{x_2}, x_2, 1 - \frac{x_2}{x_1}, x_1, 1 - x_1],$$
$$\mathcal{L}^1_{0001} = [t; \frac{x_3 - t}{x_3 - 1}, x_3, 1 - \frac{x_3}{x_2}, x_2, 1 - \frac{x_2}{x_1}, x_1, 1 - x_1],$$

$$\mathcal{L}_{0111} = [t; 1 - \frac{t}{x_3}, 1 - x_3, \frac{x_2 - x_3}{x_2 - 1}, 1 - x_2, \frac{x_1 - x_2}{x_1 - 1}, x_1, 1 - x_1],$$
$$\mathcal{L}^1_{0111} = [t; \frac{x_3 - t}{x_3 - 1}, 1 - x_3, \frac{x_2 - x_3}{x_2 - 1}, 1 - x_2, \frac{x_1 - x_2}{x_1 - 1}, x_1, 1 - x_1],$$

while the expressions for $\mathcal{L}_{0011}$ and $\mathcal{L}^1_{0011}$ involved linear combinations:

$$\begin{aligned}\mathcal{L}_{0011} = &-[t; 1 - \frac{t}{x_3}, x_3, 1 - \frac{x_3}{x_2}, 1 - x_2, \frac{x_1 - x_2}{x_1 - 1}, x_1, 1 - x_1] \\ &- [t; 1 - \frac{t}{x_3}, 1 - x_3, \frac{x_2 - x_3}{x_2 - 1}, x_2, 1 - \frac{x_2}{x_1}, x_1, 1 - x_1] \\ &- [t; 1 - \frac{t}{x_3}, 1 - \frac{x_3}{x_2}, x_2, 1 - x_2, \frac{x_1 - x_3}{x_1 - 1}, x_1, 1 - x_1]\end{aligned}$$

and

$$\begin{aligned}\mathcal{L}^1_{0011} = &-[t; \frac{x_3 - t}{x_3 - 1}, x_3, 1 - \frac{x_3}{x_2}, 1 - x_2, \frac{x_1 - x_2}{x_1 - 1}, x_1, 1 - x_1] \\ &- [t; \frac{x_3 - t}{x_3 - 1}, 1 - x_3, \frac{x_2 - x_3}{x_2 - 1}, x_2, 1 - \frac{x_2}{x_1}, x_1, 1 - x_1] \\ &- [t; \frac{x_3 - t}{x_3 - 1}, 1 - \frac{x_3}{x_2}, x_2, 1 - x_2, \frac{x_1 - x_3}{x_1 - 1}, x_1, 1 - x_1].\end{aligned}$$

EXAMPLE 4.7 (Weight 5). The differential equations satisfied by $\mathcal{L}_{01011}$ and $\mathcal{L}^1_{01011}$ are:

$$\begin{aligned}\partial(\mathcal{L}_{01011}) &= -\mathcal{L}_{01}\cdot\mathcal{L}_{011} - \mathcal{L}_1\mathcal{L}^1_{0011} - 2\mathcal{L}_{011}\mathcal{L}^1_{01}\\ \partial(\mathcal{L}^1_{01011}) &= \mathcal{L}^1_{01}\cdot\mathcal{L}^1_{011} - \mathcal{L}_{011}\cdot\mathcal{L}^1_{01} - \mathcal{L}_{01}\cdot\mathcal{L}^1_{011} - \mathcal{L}_1\cdot\mathcal{L}^1_{0011}.\end{aligned}$$

The corresponding colored trees are given by:

$\mathfrak{T}_{01011} =$

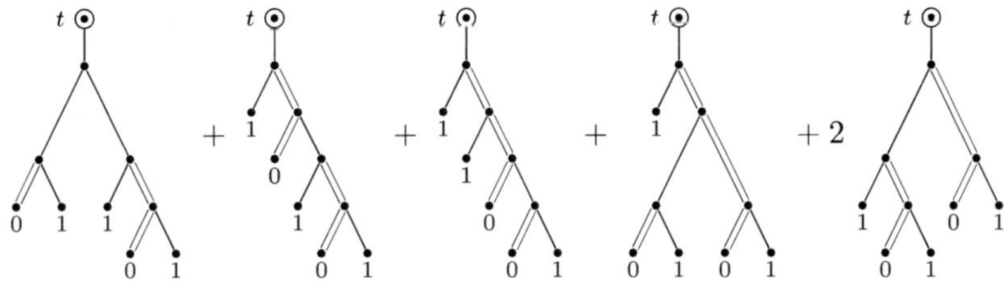

and

$\mathfrak{T}^1_{01011} = -$

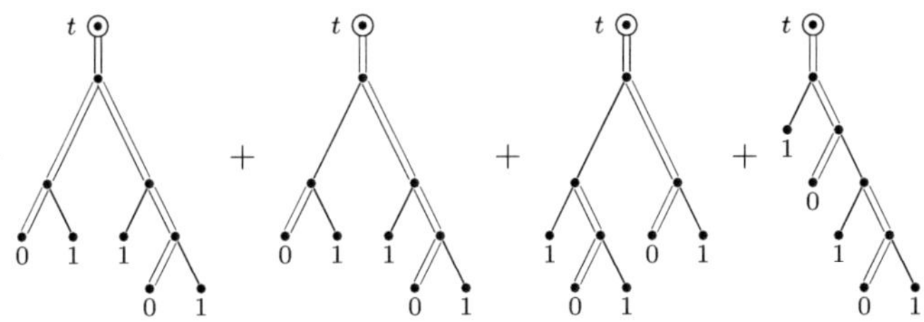

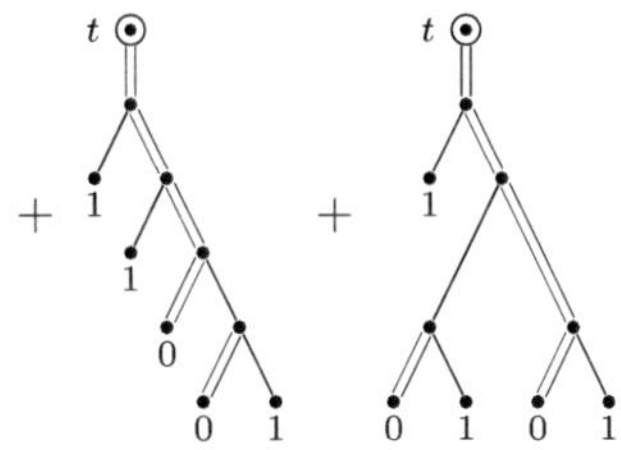

The corresponding expression as parametrized cycles are given below (in $X \times \square^9$):

$$\begin{aligned}\mathcal{L}_{01011} = {} & [t; 1-\frac{t}{x_4}, 1-\frac{x_4}{x_3}, x_3, 1-x_3, 1-\frac{x_4}{x_2}, 1-x_2, \frac{x_1-x_2}{x_2-1}, x_1, 1-x_1]\\ & + [t; 1-\frac{t}{x_4}, 1-x_4, \frac{x_3-x_4}{x_3-1}, x_3, 1-\frac{x_3}{x_2}, 1-x_2, \frac{x_1-x_2}{x_1-1}, x_1, 1-x_1]\\ & + [t; 1-\frac{t}{x_4}, 1-x_4, \frac{x_3-x_4}{x_3-1}, 1-x_3, \frac{x_2-x_3}{x_2-1}, x_2, 1-\frac{x_2}{x_1}, x_1, 1-x_1]\\ & + [t; 1-\frac{t}{x_4}, 1-x_4, \frac{x_3-x_4}{x_3-1}, 1-\frac{x_3}{x_2}, x_2, 1-x_2, \frac{x_1-x_3}{x_1-1}, x_1, 1-x_1]\\ & + 2[t; 1-\frac{t}{x_4}, 1-\frac{x_4}{x_3}, 1-x_3, \frac{x_2-x_3}{x_2-1}, x_2, 1-x_2, \frac{x_1-x_4}{x_1-1}, x_1, 1-x_1],\end{aligned}$$

and

$$\begin{aligned}\mathcal{L}^1_{01011} = &-[t; \frac{x_4-t}{x_4-1}, \frac{x_3-x_4}{x_3-1}, x_3, 1-x_3, \frac{x_2-x_4}{x_2-1}, 1-x_2, \frac{x_1-x_2}{x_2-1}, x_1, 1-x_1]\\ &+[t; \frac{x_4-t}{x_4-1}, 1-\frac{x_4}{x_3}, x_3, 1-x_3, \frac{x_2-x_4}{x_2-1}, 1-x_2, \frac{x_1-x_2}{x_1-1}, x_1, 1-x_1]\\ &+[t; \frac{x_4-t}{x_4-1}, 1-\frac{x_4}{x_3}, 1-x_3, \frac{x_2-x_3}{x_2-1}, x_2, 1-x_2, \frac{x_1-x_4}{x_1-1}, x_1, 1-x_1]\\ &+[t; \frac{x_4-t}{x_4-1}, 1-x_4, \frac{x_3-x_4}{x_3-1}, x_3, 1-\frac{x_3}{x_2}, 1-x_2, \frac{x_1-x_2}{x_1-1}, x_1, 1-x_1]\\ &+[t; \frac{x_4-t}{x_4-1}, 1-x_4, \frac{x_3-x_4}{x_3-1}, 1-x_3, \frac{x_2-x_3}{x_2-1}, x_2, 1-\frac{x_2}{x_1}, x_1, 1-x_1]\\ &+[t; \frac{x_4-t}{x_4-1}, 1-x_4, \frac{x_3-x_4}{x_3-1}, 1-\frac{x_3}{x_2}, x_2, 1-x_2, \frac{x_1-x_3}{x_1-1}, x_1, 1-x_1].\end{aligned}$$

## 5. Bar construction settings

In the cycle motives setting, a motive over $X$ is a comodule on the $\mathrm{H}^0$ of the bar construction over $\mathcal{N}^\bullet_X$ modulo shuffle products. For more details, one can look at the works of Bloch and Kriz [**BK94**], Spitzweck [**Spi**, **Spi01**] (i.e. as presented in [**Lev05**]) and Levine [**Lev11**].

In this context, the cycles constructed above, which are expected to correspond to multiple polylogarithms (as outlined in Section 6), induce elements in this $\mathrm{H}^0$ and naturally gives rise to an associated comodule, thus to mixed Tate motives corresponding to multiple polylogarithms.

Before, giving explicit expressions for the induced elements in the bar construction, the beginning of the section is devoted to a short review of the bar construction.

**5.1. Bar construction.** As there does not seem to exist a global sign convention for the various operations on the bar construction, the main definitions in the cohomological setting are recalled below following the (homological) description given in [**LV12**].

Let $A$ be a commutative differential graded algebra (c.d.g.a.) with augmentation $\varepsilon : A \longrightarrow \mathbb{Q}$, with product $\mu_A$ and let $A^+$ be the augmentation ideal $A^+ = \ker(\varepsilon)$. Note again that *commutative* in this context stand for *graded commutative.*

In order to understand the sign convention below and the "bar grading", one should think of the bar construction as built on the tensor coalgebra over the shifted (suspended) graded vector space $A^+[1]$.

DEFINITION 5.1. The bar construction $B(A)$ over $A$ is the tensor coalgebra over the suspension of $A^+$.

- In particular, as vector space $B(A)$ is given by :

$$B(A) = T(A^+) = \bigoplus_{n\geqslant 0} (A^+)^{\otimes n}.$$

- A homogeneous element $\mathbf{a}$ of tensor degree $n$ is denoted using the *bar* notation, that is

$$\mathbf{a} = [a_1|\dots|a_n]$$

and its degree is

$$\deg_B(\mathbf{a}) = \sum_{i=1}^{n} (\deg_A(a_i) - 1)\,.$$

- The coalgebra structure comes from the natural deconcatenation coproduct, that is

$$\Delta([a_1|\dots|a_n]) = \sum_{i=0}^{n} [a_1|\dots|a_i] \otimes [a_{i+1}|\dots|a_n].$$

REMARK 5.2. This construction can be seen as a simplicial total complex associated to the complex $A$ (Cf. [**BK94**]). The augmentation makes it possible to use directly $A^+$ without referring to the tensor coalgebra over $A$ and without the need of killing the degeneracies.

However this simplicial presentation usually masks the need of working with the shifted complex.

We associate to any bar element $[a_1|\dots|a_n]$ the function $\eta(i)$ giving its "partial" degree

$$\eta(i) = \sum_{k=1}^{i} (\deg_A(a_k) - 1).$$

The original differential $d_A$ induces a differential $D_1$ on $B(A)$ given by

$$D_1([a_1|\dots|a_n]) = -\sum_{i=1}^{n} (-1)^{\eta(i-1)} [a_1|\dots|d_A(a_i)|\dots|a_n]$$

where the initial minus sign comes from the fact that the differential on the shifted complex $A[1]$ is $-d_A$. Moreover, the multiplication on $A$ induces another differential $D_2$ on $B(A)$ given by

$$D_2([a_1|\dots|a_n]) = -\sum_{i=1}^{n-1} (-1)^{\eta(i)} [a_1|\dots|\mu_A(a_i, a_{i+1})|\dots|a_n]$$

where the signs are coming from Koszul commutation rules (due to the shifting). One checks that the two differentials anticommute providing $B(A)$ with a total differential.

DEFINITION 5.3. The total differential on $B(A)$ is defined by

$$d_{B(A)} = D_1 + D_2.$$

The last structure arising with the bar construction is the graded shuffle product

$$[a_1|\dots|a_n] ш [a_{n+1}|\dots|a_{n+m}] = \sum_{\sigma \in sh(n,m)} (-1)^{\varepsilon_{gr}(\sigma)} [a_{\sigma(1)}, \dots, a_{\sigma(n+m)}]$$

where $sh(n,m)$ denotes the permutation of $\{1,\dots,n+m\}$ such that if $1 \leqslant i < j \leqslant n$ or $n+1 \leqslant i < j \leqslant n+m$ then $\sigma(i) < \sigma(j)$. The sign is the graded signature of the permutation (for the degree in $A^+[1]$) given by

$$\varepsilon_{gr}(\sigma) = \sum_{\substack{i<j \\ \sigma(i)>\sigma(j)}} (\deg_A(a_i) - 1)(\deg_A(a_j) - 1).$$

With these definitions, one can explicitly check the following

PROPOSITION 5.4. *Let $A$ be a (Adams/weight graded) c.d.g.a. The operations $\Delta$, $d_{B(A)}$ and ш together with the obvious unit and counit give $B(A)$ a structure of (Adams graded) commutative graded differential Hopf algebra.*

*In particular, these operations induce on $\mathrm{H}^0(B(A))$, and more generally on $\mathrm{H}^*(B(A))$, a (Adams graded) commutative Hopf algebra structure. This (Adams graded) algebra is cohomologically graded in the case of $\mathrm{H}^*(B(A))$ and cohomologically graded concentrated in degree 0 in the case of $\mathrm{H}^0(B(A))$.*

We recall that the set of indecomposable elements of an augmented c.d.g.a. is defined as the augmentation ideal $I$ modulo products, that is $I/I^2$. Applying a general fact about Hopf algebras, the coproduct structure on $\mathrm{H}^0(B(A))$ induces a coLie algebra structure on its set of indecomposable elements.

**5.2. Bar elements.** Considering the bar construction over $\mathcal{N}_X^\bullet$, part of the issue is to associate to any cycle $\mathcal{L}_W$ and $\mathcal{L}_W^1$ a corresponding element in $\mathrm{H}^0(B(\mathcal{N}_X^\bullet))$.

As the weight 1 cycles $\mathcal{L}_0$ and $\mathcal{L}_1$ have 0 differential in $\mathcal{N}_X^\bullet$, there are obvious corresponding bar elements:

$$\mathcal{L}_0^B = [\mathcal{L}_0] \qquad \text{and} \qquad \mathcal{L}_1^B = [\mathcal{L}_1]. \tag{23}$$

Let $\mathcal{M}_X$ denote the indecomposable elements of $\mathrm{H}^0(B(\mathcal{N}_X^\bullet))$ and let $\tau$ be the morphism exchanging the two factors of $\mathrm{H}^0(B(\mathcal{N}_X^\bullet)) \otimes \mathrm{H}^0(B(\mathcal{N}_X^\bullet))$. We denote by

$$d_\Delta = \frac{1}{2}(\Delta - \tau\Delta)$$

the differential on the coLie algebra $\mathcal{M}_X$ induced by the coproduct on $\mathrm{H}^0(B(\mathcal{N}_X^\bullet))$. In general, one should have the following.

CLAIM. *For any Lyndon word $W$ (of length greater or equal to 2), there exist elements $\mathcal{L}_W^B$ and $\mathcal{L}_W^{1,B}$ in $B(\mathcal{N}_X^\bullet)$ of bar degree 0 satisfying:*

- *Let $d_B$ denotes the total bar differential $d_B = d_{B(\mathcal{N}_X^\bullet)}$. Then one has:*
$$d_B(\mathcal{L}_W^B) = d_B(\mathcal{L}_W^{1,B}) = 0.$$
- *The tensor degree 1 part of $\mathcal{L}_W^B$ (resp. $\mathcal{L}_W^{1,B}$) is given by $[\mathcal{L}_W]$ (resp. $[\mathcal{L}_W^1]$).*
- *The elements $\mathcal{L}_W^B$ (resp. $\mathcal{L}_W^{1,B}$) satisfy the differential equation* (18) *(resp.* (19)*) in $\mathcal{M}_X$. That is*

$$d_\Delta(\mathcal{L}_W^B) = -\left(\sum_{U<V} a_{U,V}^W \mathcal{L}_U^B \mathcal{L}_V^B + \sum_{U,V} b_{U,V}^W \mathcal{L}_U^B \mathcal{L}_V^{1,B}\right) \qquad \in \mathcal{M}_X \wedge \mathcal{M}_X$$

*and*

$$d_\Delta(\mathcal{L}_W^{1,B}) = -\left(\sum_{0<U<V} a'^W_{U,V} \mathcal{L}_U^{1,B} \mathcal{L}_V^{1,B} + \sum_{U,V} b'^W_{U,V} \mathcal{L}_U^B \mathcal{L}_V^{1,B} + \sum_V a'_{0,V} \mathcal{L}_0^B \mathcal{L}_V^B\right) \in \mathcal{M}_X \wedge \mathcal{M}_X$$

*where the overall minus sign is due to shifting reasons.*

The obstruction for proving the general statement lies in the control of the global combinatorics relating $D_1$, $D_2$ and the two systems of differential equations.

Below, one finds some elements $\mathcal{L}_W^B$ and $\mathcal{L}_W^{1,B}$ corresponding to the previously described examples together with some relations among those elements. Once the element $\mathcal{L}_W^B$ are explicitly described, it is a straightforward computation to check that it lies in the kernel of $d_B$ and this verification will be omitted.

Note that all cycles $\mathcal{L}_W$ and $\mathcal{L}_W^1$ are in cohomological degree 1, that is in $\mathcal{N}_X^1$. Thus, signs appearing in the operations on the bar construction are much simpler as all terms in $\deg_A(a_i) - 1$ are 0.

EXAMPLE 5.5 (Weight 2). Cycles $\mathcal{L}_{01}$ and $\mathcal{L}_{01}^1$ satisfy $\partial(\mathcal{L}_{01}) = \partial(\mathcal{L}_{01}^1) = \mathcal{L}_0\mathcal{L}_1$. Thus one can define

(24)

$$\mathcal{L}_{01}^B = [\mathcal{L}_{01}] - \frac{1}{2}\left([\mathcal{L}_0|\mathcal{L}_1] - [\mathcal{L}_1|\mathcal{L}_0]\right) \quad \text{and} \quad \mathcal{L}_{01}^{1,B} = [\mathcal{L}_{01}^1] - \frac{1}{2}\left([\mathcal{L}_0|\mathcal{L}_1] - [\mathcal{L}_1|\mathcal{L}_0]\right).$$

Remark that, looking at things modulo products, that is in $\mathcal{M}_X$, the tensor degree 2 involves some choices. Instead of

$$-\frac{1}{2}\left([\mathcal{L}_0|\mathcal{L}_1] - [\mathcal{L}_1|\mathcal{L}_0]\right),$$

we could have used

$$-[\mathcal{L}_0|\mathcal{L}_1] \qquad \text{or} \qquad [\mathcal{L}_1|\mathcal{L}_0]$$

and obtained the same elements in $\mathcal{M}_X$ as

$$-\frac{1}{2}\left([\mathcal{L}_0|\mathcal{L}_1] - [\mathcal{L}_1|\mathcal{L}_0]\right) = -[\mathcal{L}_0|\mathcal{L}_1] + \frac{1}{2}\mathcal{L}_0^B \sqcup\!\sqcup \mathcal{L}_1^B = [\mathcal{L}_1|\mathcal{L}_0] - \frac{1}{2}\mathcal{L}_0^B \sqcup\!\sqcup \mathcal{L}_1^B.$$

The above choice reflects in some sense that there is no preferred choice for either

$$\partial(\mathcal{L}_{01}) = \partial(\mathcal{L}_{01}^1) = \mathcal{L}_0\mathcal{L}_1 \quad \text{or} \quad \partial(\mathcal{L}_{01}) = \partial(\mathcal{L}_{01}^1) = -\mathcal{L}_1\mathcal{L}_0.$$

Recall that we have defined a cycle $\mathcal{L}_{01}(1)$ in $\mathcal{N}_X^1$ by

$$\mathcal{L}_{01}(1) = j^*(p^* \circ i_1^*(\overline{\mathcal{L}_{01}}))$$

Building the cycle $\mathcal{L}_{011}$, we have introduced the cycle $\mathcal{L}_{01}^1$ instead of using the difference $\mathcal{L}_{01} - \mathcal{L}_{01}(1)$ in order to keep working with equidimensional cycles. The "correspondence"

$$\mathcal{L}_{01}^1 \leftrightarrow \mathcal{L}_{01} - \mathcal{L}_{01}(1)$$

becomes an equality in $\mathrm{H}^0(B(\mathcal{N}_X^\bullet))$.

More precisely, using either the commutation of the above morphisms with the differential or the expression of $\mathcal{L}_{01}(1)$ as parametrized cycle, one sees that $\partial(\mathcal{L}_{01}(1)) = 0$ and one defines

$$\mathcal{L}_{01}^B(1) = [\mathcal{L}_{01}(1)].$$

A direct computation shows that

$$\mathcal{L}_{01}(1) = [t; 1 - \frac{1}{x_1}, x_1, 1 - x_1].$$

Now, from the expressions of $\mathcal{L}_{01}$, $\mathcal{L}_{01}(1)$ and $\mathcal{L}_{01}^1$ as parametrized cycles, one checks that in $\mathcal{N}_X^1$

$$\mathcal{L}_{01} - \mathcal{L}_{01}(1) = \mathcal{L}_{01}^1 + \partial(C_{01})$$

where $C_{01}$ is the element of $\mathcal{N}_X^0$ defined by

$$C_{01} = -[t; y, \frac{y - \frac{x_1 - t}{x_1}}{y - \frac{x_1 - 1}{x_1}}, x_1, 1 - x_1] \subset X \times \square^4.$$

The bar element $C_{01}^B = [C_{01}]$ is of bar degree $-1$ and gives in $B(\mathcal{N}_X^\bullet)$

$$\mathcal{L}_{01}^B - \mathcal{L}_{01}^B(1) = \mathcal{L}_{01}^{1,B} - d_B(C_{01}^B)$$

and thus, the equality $\mathcal{L}_{01}^B - \mathcal{L}_{01}^B(1) = \mathcal{L}_{01}^{1,B}$ in the $\mathrm{H}^0$.

For these weight 2 examples , computing the deconcatenation coproduct is trivial and gives the expected relation

$$d_\Delta(\mathcal{L}_{01}^B) = d_\Delta(\mathcal{L}_{01}^{1,B}) = -\mathcal{L}_0^B \wedge \mathcal{L}_1^B.$$

Finally the motive corresponding to $\mathcal{L}_{01}$ is the comodule generated by $\mathcal{L}_{01}^B$, that is the subvector space of $\mathcal{M}_X$ spanned by $\mathcal{L}_{01}^B$, $\mathcal{L}_0^B$ and $\mathcal{L}_1^B$.

EXAMPLE 5.6 (Weight 3). The differentials of $\mathcal{L}_{001}$, $\mathcal{L}_{001}^1$, $\mathcal{L}_{011}$, $\mathcal{L}_{011}^1$ allow us to easily write down the corresponding tensor degree 1 and 2. The expressions below try to keep a symmetric presentation for the part in tensor degree 3.

In the equations below, cycles $\mathcal{L}_W$ are simply denoted by $W$ and cycles $\mathcal{L}_W^1$ simply by $\overline{W}$. We will also use this abuse of notation later on in weight 4. One defines

$$\mathcal{L}_{001}^B = [001] - \frac{1}{2}([0|01] - [01|0]) + \frac{1}{4}([0|0|1] - [0|1|0] + [1|0|0]),$$
$$\mathcal{L}_{001}^{1,B} = [\overline{001}] - \frac{1}{2}([0|01] - [01|0]) + \frac{1}{4}([0|0|1] - [0|1|0] + [1|0|0])$$

and

$$\mathcal{L}_{011}^B = [011] - \frac{1}{2}([\overline{01}|1] - [1|\overline{01}]) + \frac{1}{4}([0|1|1] - [1|0|1] + [1|1|0]),$$
$$\mathcal{L}_{011}^{1,B} = [\overline{011}] - \frac{1}{2}([\overline{01}|1] - [1|\overline{01}]) + \frac{1}{4}([0|1|1] - [1|0|1] + [1|1|0]).$$

As the cycles $\mathcal{L}_{001}$ and $\mathcal{L}_{001}^1$ (resp. $\mathcal{L}_{011}$ and $\mathcal{L}_{011}^1$) differ only by their first $\square^1$ factors, the arguments used to compare $\mathcal{L}_{01}^B$ and $\mathcal{L}_{01}^{1,B}$ apply here and give:

$$\mathcal{L}_{001}^B - \mathcal{L}_{001}^B(1) = \mathcal{L}_{001}^{1,B} \quad \text{and} \quad \mathcal{L}_{001}^B - \mathcal{L}_{001}^B(1) = \mathcal{L}_{001}^{1,B} \qquad \in \mathcal{M}_X.$$

The "correction" cycles giving the explicit relations between the cycles are

$$C_{001} = -[t; s, \frac{s - \frac{x_2 - t}{x_2}}{s - \frac{x_2 - 1}{x_2}}, x_2, 1 - \frac{x_2}{x_1}, x_1, 1 - x_1]$$

and

$$C_{011} = [t; s, \frac{s - \frac{x_2 - t}{x_2}}{s - \frac{x_2 - 1}{x_2}}, 1 - x_2, \frac{x_1 - x_2}{x_1 - 1}, x_1, 1 - x_1].$$

Now, computing the reduced coproduct $\Delta' = \Delta - 1 \otimes \mathrm{id} - \mathrm{id} \otimes 1$ of $\mathcal{L}_{001}^B$ gives:

$$\begin{aligned}\Delta'(\mathcal{L}_{001}^B) = -\frac{1}{2}&([0] \otimes [01] - [01] \otimes [0]) \\ &+ \frac{1}{4}([0] \otimes [0|1] - [0] \otimes [1|0] + [1] \otimes [0|0] \\ &\qquad + [0|0] \otimes [1] - [0|1] \otimes [0] + [1|0] \otimes [0]).\end{aligned}$$

As $[0|0] = 1/2\mathcal{L}_0^B \sqcup\!\sqcup \mathcal{L}_0^B$, one has modulo products

$$\Delta'(\mathcal{L}_{001}^B) = -\frac{1}{2}\left([0]\otimes\left([01]-\frac{1}{2}([0|1]-[1|0])\right)-\left([01]-\frac{1}{2}([0|1]-[1|0])\right)\otimes[0]\right).$$

Similar computations apply to $\mathcal{L}_{001}^{1,B}$, $\mathcal{L}_{011}^B$ and $\mathcal{L}_{011}^{1,B}$ and give in $\mathcal{M}_X\wedge\mathcal{M}_X$:

$$d_\Delta(\mathcal{L}_{001}^B) = d_\Delta(\mathcal{L}_{001}^{1,B}) = -\mathcal{L}_0^B\wedge\mathcal{L}_{01}^B \quad\text{and}\quad d_\Delta(\mathcal{L}_{011}^B) = d_\Delta(\mathcal{L}_{011}^{1,B}) = -\mathcal{L}_{01}^{1,B}\wedge\mathcal{L}_1^B.$$

One should remark that the equality $\mathcal{L}_{01}^{1,B} = \mathcal{L}_{01}^B - \mathcal{L}_{01}^B(1)$ in the $\mathrm{H}^0$ implies

$$d_\Delta(\mathcal{L}_{011}^B) = -\left(\mathcal{L}_{01}^B - \mathcal{L}_{01}^B(1)\right)\wedge\mathcal{L}_1^B \tag{25}$$

which is (up to a global minus sign) the equation satisfied by $T_{011^*}$ as shown at Example 2.13.

Finally, the corresponding comodules giving motives associated to the cycle $\mathcal{L}_{001}$ and $\mathcal{L}_{011}$ are the subvector spaces of $\mathcal{M}_X$ generated respectively by

$$\left\langle\mathcal{L}_{001}^B, \mathcal{L}_{01}^B, \mathcal{L}_0^B, \mathcal{L}_1^B\right\rangle$$

and

$$\left\langle\mathcal{L}_{011}^B, \mathcal{L}_{01}^B, \mathcal{L}_{01}^B(1), \mathcal{L}_0^B, \mathcal{L}_1^B\right\rangle.$$

The above arguments apply similarly in weight 4. Hence, we will describe below the case of $\mathcal{L}_{0011}$ as it gives a "preview" of the combinatorial difficulties related to the bar construction context.

EXAMPLE 5.7 (Weight 4: $\mathcal{L}_{0011}^B$). We give below an element $\mathcal{L}_{0011}^B$ in the bar construction with zero differential and with tensor degree 1 part equal to $\mathcal{L}_{0011}$:

$$\begin{aligned}\mathcal{L}_{0011}^B =&[0011] - \frac{1}{2}\left([0|011]-[0|0\overline{11}]+[\overline{001}|1]-[1|\overline{001}]-[01|\overline{01}]+[\overline{01}|01]\right)\\ &+\frac{1}{4}\left(-[0|1|\overline{01}]+[0|1|01]-[01|0|1]+[\overline{01}|0|1]-[\overline{01}|1|0]+[01|1|0]\right.\\ &\left.-[1|0|01]+[1|0|\overline{01}]+[1|\overline{01}|0]+[1|01|0]+[0|\overline{01}|1]+[0|01|1]\right)\\ &-\frac{1}{2}\left([0|0|1|1]-[1|1|0|0]\right).\end{aligned}$$

Identifying the reduced coproduct of $\mathcal{L}_{0011}^B$ with

$$\begin{aligned}-\frac{1}{2}\Big(&\mathcal{L}_0^B\otimes\mathcal{L}_{011}^B-\mathcal{L}_{011}^B\otimes\mathcal{L}_0^B+\mathcal{L}_{001}^{1,B}\otimes\mathcal{L}_1^B-\mathcal{L}_1^B\otimes\mathcal{L}_{001}^{1,B}\\ &-\mathcal{L}_{01}^B\otimes\mathcal{L}_{01}^{1,B}+\mathcal{L}_{01}^{1,B}\otimes\mathcal{L}_{01}^B\Big)\end{aligned} \tag{26}$$

is more difficult than in the previous cases. First of all, one remarks that in the above expression the terms in $(\mathcal{N}_X^1)^{\otimes 2}\otimes(\mathcal{N}_X^1)^{\otimes 2}$ are coming only from $-\mathcal{L}_{01}^B\otimes\mathcal{L}_{01}^{1,B}$ and $\mathcal{L}_{01}^{1,B}\otimes\mathcal{L}_{01}^B$ and cancel each other. Thus the expression (26) has no term in $(\mathcal{N}_X^1)^{\otimes 2}\otimes(\mathcal{N}_X^1)^{\otimes 2}$. On the other hand, the terms in $(\mathcal{N}_X^1)^{\otimes 2}\otimes(\mathcal{N}_X^1)^{\otimes 2}$ coming from $\Delta'(\mathcal{L}_{0011}^B)$ are given by

$$\begin{aligned}-\frac{1}{2}&\left([0|0]\otimes[1|1]-[1|1]\otimes[0|0]\right)=\\ &-\frac{1}{8}\left(\mathcal{L}_0^B\sqcup\!\sqcup\mathcal{L}_0^B\otimes\mathcal{L}_1^B\sqcup\!\sqcup\mathcal{L}_1^B-\mathcal{L}_1^B\sqcup\!\sqcup\mathcal{L}_1^B\otimes\mathcal{L}_0^B\sqcup\!\sqcup\mathcal{L}_0^B\right)\end{aligned}$$

and thus are zero in $\mathcal{M}_X\wedge\mathcal{M}_X$.

The terms in $\mathcal{N}_X^1 \otimes \mathcal{N}_X^1$ in the above expression (26) obviously agree with the corresponding terms of $\Delta'(\mathcal{L}^B_{0011})$ as the term of tensor degree 2 of $\mathcal{L}^B_{0011}$ is written down that way.

Computations below are done in $B(\mathcal{N}_X^\bullet) \otimes B(\mathcal{N}_X^\bullet)$. They will induce the expected relation in $\mathcal{M}_X \wedge \mathcal{M}_X$ after going to the $\mathrm{H}^0$ and taking the quotient modulo shuffle product. Let $\pi_n : B(\mathcal{N}_X^\bullet) \longrightarrow (\mathcal{N}_X^\bullet)^{\otimes n}$ be the projection to the $n$-th tensor factor. From the above discussion it is enough to compute $\Delta'(\pi_3(\mathcal{L}^B_{0011}))$ and part of $\Delta'(\pi_4(\mathcal{L}^B_{0011}))$.

First the definition of the coproduct gives

$$\begin{aligned}
4\Delta'(\pi_3(\mathcal{L}^B_{0011})) = {} & \left([0] \otimes [\overline{01}|1] + [0|01] \otimes [1] - [01] \otimes [0|1] - [0|1] \otimes [\overline{01}]\right) \\
& + \left(-[0] \otimes [1|\overline{01}] - [01|0] \otimes [1] + [01] \otimes [1|0] + [1|0] \otimes [\overline{01}]\right) \\
& + (-1)\left([\overline{01}|1] \otimes [0] + [1] \otimes [0|01] - [0|1] \otimes [01] - [\overline{01}] \otimes [0|1]\right) \\
& + (-1)\left(-[1|\overline{01}] \otimes [0] - [1] \otimes [01|0] + [1|0] \otimes [01] + [\overline{01}] \otimes [1|0]\right) \\
& + \left([0] \otimes ([01|1] + [1|01]) + ([01|1] + [1|01]) \otimes [0]\right) \\
& + \left([1] \otimes \left([\overline{01}|0] + [0|\overline{01}]\right) + \left([\overline{01}|0] + [0|01]\right) \otimes [0]\right).
\end{aligned}$$

The four first lines of the above equality correspond to 4 times the terms of Equation (26) in $\mathcal{N}_X^\bullet \otimes \mathcal{N}_X^{\bullet\,\otimes 2} \oplus \mathcal{N}_X^{\bullet\,\otimes 2} \otimes \mathcal{N}_X^\bullet$. The two last lines can be written as

$$[0] \otimes ([01] \sqcup\!\sqcup [1]) + ([01] \sqcup\!\sqcup [1]) \otimes [0] + [0] \otimes \left([\overline{01}] \sqcup\!\sqcup [0]\right) + \left([\overline{01}] \sqcup\!\sqcup [0]\right) \otimes [1]$$

which covers many terms of a shuffle between terms in the $\mathrm{H}^0$ of the bar construction. As an example,

$$[0] \otimes ([01] \sqcup\!\sqcup [1])$$

covers many terms of

$$[\mathcal{L}^B_0] \otimes \left(\mathcal{L}^B_{01} \sqcup\!\sqcup \mathcal{L}^B_1\right) = [0] \otimes \left([01] \sqcup\!\sqcup [1] - \frac{1}{2}([0|1] - [1|0]) \sqcup\!\sqcup [1]\right).$$

Computing the reduced coproduct of $\pi_4(\mathcal{L}^B_{0011})$ gives

$$\begin{aligned}
\Delta'(\pi_4(\mathcal{L}^B_{0011})) = \left(-\frac{1}{2}\right)\left(\frac{1}{4}\right) & (4[0] \otimes [0|1|1] + 4[0|0] \otimes [1|1] + 4[0|0|1] \otimes [1] \\
& -4[1] \otimes [1|0|0] + 4[1|1] \otimes [0|0] - 4[1|1|0] \otimes [0]).
\end{aligned}$$

where the factors $1/4$ and 4 make it easier to relate $\Delta'(\pi_4(\mathcal{L}^B_{0011}))$ with shuffle products and the corresponding terms in Equation (26).

We have already remarked that the terms $[0|0] \otimes [1|1]$ and $[1|1] \otimes [0|0]$ in the equation above can be expressed as shuffles. The four other terms are similar. Hence we will only discuss the case of $[0] \otimes [0|1|1]$. One can write

$$\begin{aligned}
4[0|1|1] &= 2[0|1|1] + 2[1|1|0] + 2[0|1|1] - 2[1|1|0] \\
&= [0|1|1] + [1|1|0] - [1|0|1] + [0|1|1] + [1|0|1] + [1|1|0] + 2[0|1|1] - 2[1|1|0].
\end{aligned}$$

Now, one remarks that

$$[0|1|1] + [1|0|1] + [1|1|0] = \frac{1}{2}[0] \sqcup\!\sqcup [1] \sqcup\!\sqcup [1] = \frac{1}{2}\mathcal{L}^B_0 \sqcup\!\sqcup \mathcal{L}^B_1 \sqcup\!\sqcup \mathcal{L}^B_1$$

and that the tensor degree 3 part of $-\frac{1}{2}\mathcal{L}^B_{01} \sqcup\!\sqcup \mathcal{L}^B_1$ is equal to

$$\frac{-1}{2}\,\frac{-1}{2}\left([0|1] \sqcup\!\sqcup [1] - [1|0] \sqcup\!\sqcup [1]\right) = \frac{1}{4}\left(2[0|1|1] - 2[1|1|0]\right).$$

Then one can conclude that

$$\Delta'(\mathcal{L}^B_{0011}) = -\frac{1}{2}\Big(\mathcal{L}^B_0 \otimes \mathcal{L}^B_{011} - \mathcal{L}^B_{011} \otimes \mathcal{L}^B_0 + \mathcal{L}^{1,B}_{001} \otimes \mathcal{L}^B_1 - \mathcal{L}^B_1 \otimes \mathcal{L}^{1,B}_{001}$$
$$- \mathcal{L}^B_{01} \otimes \mathcal{L}^{1,B}_{01} + \mathcal{L}^{1,B}_{01} \otimes \mathcal{L}^B_{01}\Big)$$
$$+ \frac{1}{4}\Big(\mathcal{L}^B_{01} ш \mathcal{L}^B_1 \otimes \mathcal{L}^B_0 + \mathcal{L}^{1,B}_{01} ш \mathcal{L}^B_0 \otimes \mathcal{L}^B_1 + \mathcal{L}^B_0 \otimes \mathcal{L}^B_{01} ш \mathcal{L}^B_1$$
$$+ \mathcal{L}^B_1 \otimes \mathcal{L}^{1,B}_{01} ш \mathcal{L}^B_0\Big)$$
$$\frac{-1}{16}\big(\ \mathcal{L}^B_0 ш \mathcal{L}^B_1 ш \mathcal{L}^B_1 \otimes \mathcal{L}^B_0 - \mathcal{L}^B_0 ш \mathcal{L}^B_0 ш \mathcal{L}^B_1 \otimes \mathcal{L}^B_1$$
$$+ \mathcal{L}^B_0 \otimes \mathcal{L}^B_0 ш \mathcal{L}^B_1 ш \mathcal{L}^B_1 + \mathcal{L}^B_1 \otimes \mathcal{L}^B_0 ш \mathcal{L}^B_0 ш \mathcal{L}^B_1\big).$$

In $\mathcal{M}_X \wedge \mathcal{M}_X$, one simply gets

$$d_\Delta(\mathcal{L}^B_{0011}) = -\Big(\mathcal{L}^B_0 \wedge \mathcal{L}^B_{011} + \mathcal{L}^{1,B}_{001} \wedge \mathcal{L}^B_1 - \mathcal{L}^B_{01} \wedge \mathcal{L}^{1,B}_{01}\Big) \tag{27}$$

and using the relations between $\mathcal{L}^B_W$ and $\mathcal{L}^{1,B}_W$, one recovers (up to a global minus sign) the differential equation associated to $T_{0011^*}$

$$d_\Delta(\mathcal{L}^B_{0011}) = -\left(\mathcal{L}^B_0 \wedge \mathcal{L}^B_{011} + (\mathcal{L}^B_{001} - \mathcal{L}^B_{001}(1)) \wedge \mathcal{L}^B_1 + \mathcal{L}^B_{01} \wedge \mathcal{L}^B_{01}(1)\right). \tag{28}$$

The associated motive is as above the sub-vector space of $\mathcal{M}_X$ generated by

$$\mathcal{L}^B_0,\ \mathcal{L}^{1,B}_{001},\ \mathcal{L}^B_{0011},\ \mathcal{L}^B_{01},\ \mathcal{L}^B_{011},\ \mathcal{L}^{1,B}_{01},\ \mathcal{L}^B_1.$$

**5.3. Goncharov's motivic coproduct.** In this subsection, we would like to illustrate how the differential equation satisfied by the elements $\mathcal{L}^B_W$, written using its "tree differential form" (that is using the elements $\mathcal{L}^B_U(1)$ instead of the elements $\mathcal{L}^{1,B}_U$), gives another expression for Goncharov's motivic coproduct.

Work of Levine [**Lev11**] insures that the above differential coincides with Goncharov's motivic coproduct for motivic iterated integrals (modulo products). We will not review this theory here but only recall some of the needed properties satisfied by Goncharov's motivic iterated integrals [**Gon05**]. A short exposition of the combinatorics involved is also recalled in [**GGL09**][Section 8].

For our purpose, it is enough to consider motivic iterated integrals as degree $n$ generating elements $I(a_0; a_1, \ldots, a_n; a_{n+1})$ of a Hopf algebra with $a_i$ in $\mathbb{A}^1(\mathbb{Q})$. They are subject to the following relations.

**Path composition:** for $x$ in $\mathbb{A}^1(\mathbb{Q})$, one has

$$I(a_0; a_1, \ldots, a_n; a_{n+1}) = \sum_{k=0}^{n} I(a_0; a_1, \ldots, a_k; x) I(x; a_{k+1}, \ldots, a_n; a_{n+1}).$$

**Inversion:** which relates $I(a_0; a_1, \ldots, a_n; a_{n+1})$ and $I(a_{n+1}; a_n, \ldots, a_1; a_0)$

$$I(a_{n+1}; a_n, \ldots, a_1; a_0) = (-1)^n I(a_0; a_1, \ldots, a_n; a_{n+1}).$$

**Unit and neutral identities:**

$$\text{for } a \neq b \quad I(a; b) = 1 \qquad \text{and} \qquad I(a_0; a_1, \ldots, a_n; a_0) = 0.$$

**Rescaling:** If $a_{n+1}$ and at least one of the $a_i$ is not zero then

$$I(0; a_1, \ldots, a_n, a_{n+1}) = I(0; a_1/a_{n+1}, \ldots, a_n/a_{n+1}, 1).$$

**Regularization:**

$$I(0;1;1) = I(0;0;1) = 0.$$

The product is given by the shuffle relations

$$I(a;a_1,\dots,a_n;b)I(a;a_{n+1},\dots,a_{n+m};b) = \sum_{\sigma\in sh(n,m)} I(a;a_{\sigma(1)},\dots,a_{\sigma(n+m)};b)$$

where $sh(n,m)$ denotes the set of permutations preserving the order of the ordered subset $\{1,\dots,n\}$ and $\{n+1,\dots,n+m\}$. Such a motivic iterated integral corresponds formally to the iterated integral

$$\int_{\Delta_{a_0,a_{n+1}}} \frac{dt}{t-a_1}\wedge\cdots\wedge\frac{dt}{t-a_n}$$

with $\Delta_{a_0,a_{n+1}}$ the image of the standard simplex induced by a path from $a_0$ to $a_{n+1}$. The above relations reflect the relations satisfied by the integrals.

The coproduct is given by the formula

$$\Delta^M(I(a_0;a_1\dots,a_n;a_{n+1})) =$$

$$\sum_{\substack{(a_{k_1},\dots,a_{k_r})\\ \{k_1,\dots,k_r\}\subset\{1,\dots,n\}}} I(a_0;a_{k_1},\dots,a_{k_r};a_{n+1})\otimes\prod_{l=0}^{r} I(a_{k_l};a_{k_l+1},\dots,a_{k_{l+1}-1};a_{k_{l+1}})$$

with the convention that $r$ runs from 0 to $n$ and that $k_0 = 0$ and $k_{r+1} = n+1$.

Now, considering the reduced coproduct $\Delta'^M = \Delta^M - (1\otimes\mathrm{id} + \mathrm{id}\otimes 1)$ on the space of indecomposable elements (that is modulo products), the above formula reduces to

$$\Delta'^M(I(a_0;a_1\dots,a_n;a_{n+1})) =$$

$$\sum_{\substack{k<l\\ k\neq l-1}} I(a_0;a_1,\dots,a_k,a_l,a_{l+1},\dots a_n;a_{n+1})\otimes I(a_k;a_{k+1},\dots,a_{l-1},a_l).$$

This formula can be pictured placing the $a_i$ on a semicircle in the order dictated by their indices. Then a term in the above sums corresponds to a non-trivial chord between to vertices:

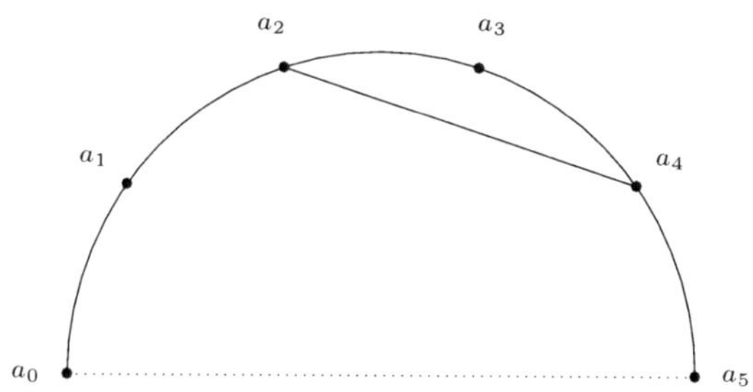


Considering the relation between multiple polylogarithms and iterated integrals, we want to related our expression of the differential of $\mathcal{L}^B_{011}$ at $t$ to the reduced coproduct for the motivic iterated integral $I(0;0,x,x;1)$ for $x = t^{-1}$. From the semicircle representation, one sees that there are five terms to consider:

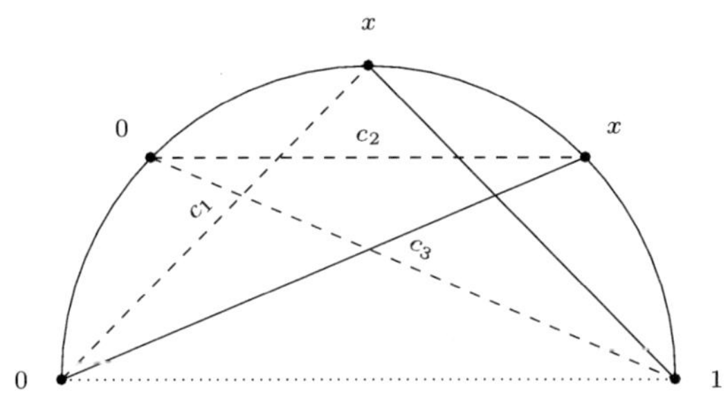


However, the chord $c_1$ gives a zero term modulo products as $I(0; x, x; 1) = \frac{1}{2}I(0; x; 1)I(0; x; 1)$ and chord $c_2$ and $c_3$ give terms equal to 0 using the regularization relations. There are thus only two terms to consider

$$I(0; 0, x; 1) \otimes I(x; x; 1) \qquad \text{and} \qquad I(0; x; 1) \otimes I(0; 0, x; x).$$

Using path composition, inversion and regularization relations, in the set of indecomposable elements, one has

$$I(x; x; 1) = I(0; x; 1) + I(x; x; 0) = I(0; x; 1) - I(0; x; x) = I(0; x; 1).$$

Thus the first term equals

$$I(0; 0, x; 1) \otimes I(x; x; 1) = I(0; 0, x; 1) \otimes I(0; x; 1).$$

From the rescaling relation, the second term equals

$$I(0; x; 1) \otimes I(0; 0, 1; 1)$$

and one can write modulo products

$$\Delta'^{M}(I(0; 0, x, x; 1)) = I(0; 0, x; 1) \otimes I(0; x; 1) + I(0; x; 1) \otimes I(0; 0, 1; 1). \tag{29}$$

Keeping in mind that, for $x = t^{-1}$, $I(0; x; 1)$ corresponds to the fiber at $t$ of $\mathcal{L}_1^B$ ($t \neq 1$) and that $I(0; 0, x; 1)$ corresponds to the fiber at $t$ of $\mathcal{L}_{01}^B$ (any $t$), the above formula (29) corresponds exactly to Equation (25):

$$d_\Delta(\mathcal{L}_{011}^B) = -\left(\mathcal{L}_{01}^B - \mathcal{L}_{01}^B(1)\right) \wedge \mathcal{L}_1^B = -\left(\mathcal{L}_{01}^B \wedge \mathcal{L}_1^B + \mathcal{L}_1^B \wedge \mathcal{L}_{01}^B(1)\right).$$

The fact that, as in the above example, Goncharov's motivic iterated integrals corresponding to multiple polylogarithms in one variable, satisfy the tree differential equations (ED-T) is easily checked for Lyndon words with one 1 (that is for the classical polylogarithms) but can also be checked for some Lyndon words with two 1's or more. It seems to be a general behavior. However, the above example and the example of $I(0; 0, 0, x, x; 1)$ corresponding to $\mathcal{L}_{0011}^B$ below show that it involves using all the relations in order to pass from Goncharov's formula to the shape of Equation (ED-T).

The case of $\mathcal{L}_{0011}^B$ involves more computations but works essentially as the case of $\mathcal{L}_{011}^B$. The reduced coproduct for $I(0; 0, 0, x, x; 1)$ modulo products gives nine terms corresponding to the nine chords below:

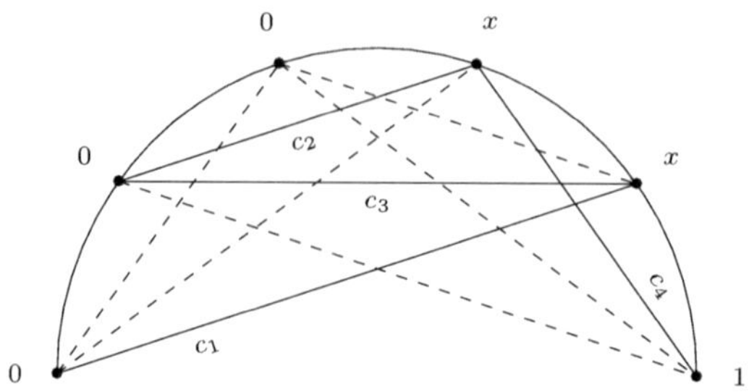


The five dashed chords give terms equal to 0 for one of the following reasons: $I(a;\ldots;a)=0$, regularization relations or shuffle relations. Hence we are left with four terms. The chord $c_1$ gives, using the rescaling relation,

$$I(0;x;1)\otimes I(0;0,0,1;1)$$

corresponding to $\mathcal{L}_1^B\wedge\mathcal{L}_{001}^B(1)$. The chord $c_2$ gives a term in

$$I(0;0,x,x;1)\otimes I(0;0;x)$$

corresponding to $\mathcal{L}_0^B\wedge\mathcal{L}_{011}^B$. The chord $c_3$ gives, using the rescaling relation, a term in

$$I(0;0,x;1)\otimes I(0;0,x;x)=I(0;0,x;1)\otimes I(0;0,1;1)$$

corresponding to $\mathcal{L}_{01}^B\wedge\mathcal{L}_{01}^B(1)$. Finally the chord $c_4$ gives, using the path composition and regularization relations

$$\begin{aligned}I(0;0,0,x;1)\otimes I(x;x;1)&=I(0;0,0,x;1)\otimes I(0;x;1)+I(0;0,0,x;1)\otimes I(x;x;0)\\&=I(0;0,0,x;1)\otimes I(0;x;1).\end{aligned}$$

Finally, $\Delta'^{M}(I(0;0,0,x,x;1))$ can be written as

(30)
$$\begin{aligned}\Delta'^{M}(I(0;0,0,x,x;1))=I(0;0,x,x;1)\otimes I(0;0;x)+I(0;0,0,x;1)\otimes I(0;x;1)+\\I(0;x;1)\otimes I(0;0,0,1;1)+I(0;0,x;1)\otimes I(0;0,1;1).\end{aligned}$$

This expression corresponds to Equation (28):

$$d_\Delta(\mathcal{L}_{0011}^B)=-\left(\mathcal{L}_0^B\wedge\mathcal{L}_{011}^B+(\mathcal{L}_{001}^B-\mathcal{L}_{001}^B(1))\wedge\mathcal{L}_1^B+\mathcal{L}_{01}^B\wedge\mathcal{L}_{01}^B(1)\right).$$

## 6. Integrals and multiple zeta values

We present here a sketch of how to associate an integral to cycles $\mathcal{L}_{01}$, $\mathcal{L}_{01}^1$ and $\mathcal{L}_{011}$. The author will directly follow the algorithm described in [**GGL09**][Section 9] and put in detailed practice in [**GGL07**]. There will be no general review of the direct Hodge realization from Bloch-Kriz motives [**BK94**][Section 8 and 9]. Gangl, Goncharov and Levin's construction seems to consist in setting particular choices of representatives in the intermediate Jacobians for their algebraic cycles.

However, the goal of this paper is not to formalize such an idea. That is why computations below are only outlined. In particular, the lack of precise knowledge of the "algebraico-topological cycle algebra" described in [**GGL09**] makes it difficult to control how "negligible" cycles are killed looking at the $\mathrm{H}^0$ of its bar construction.

**6.1. An integral associated to $\mathcal{L}_{01}$ and $\mathcal{L}_{01}^1$.** We recall the parametrized cycle expression for $\mathcal{L}_{01}$:

$$\mathcal{L}_{01} = [t; 1 - \frac{t}{x_1}, x_1, 1 - x_1] \qquad \subset X \times \square^3.$$

One wants to bound $\mathcal{L}_{01}$ by an algebraic-topological cycle in a larger bar construction (not described here) introducing topological variables $s_i$ in real simplices

$$\Delta_s^n = \{0 \leqslant s_1 \leqslant \cdots \leqslant s_n \leqslant 1\}.$$

Let $d^s : \Delta_s^n \to \Delta_s^{n-1}$ denotes the simplicial differential

$$d^s = \sum_{k=0}^{n} (-1)^k i_k^*$$

where $i_k : \Delta_s^{n-1} \to \Delta_s^n$ is given by the face $s_k = s_{k+1}$ in $\Delta_s^n$ with the usual conventions for $k = 0, n$.

Defining

$$C_{01}^{s,1} = -[t; 1 - \frac{s_2 t}{x_1}, x_1, 1 - x_1]$$

for $s_2$ going from 0 to 1, one sees that $d^s(C_{01}^{s,1}) = \mathcal{L}_{01}$ as $s_2 = 0$ implies that the first cubical coordinate is 1. The algebraic boundary $\partial$ of $C_{01}^{s,1}$ is given by the intersection with the faces of $\square^3$:

$$\partial(C_{01}^{s,1}) = -[t; s_2 t, 1 - s_2 t] \qquad \subset X \times \square^2.$$

This cycle is part of the boundary of a larger "simplicial" algebraic cycle

$$C_{01}^{s,2} = -[t; s_2 t, 1 - s_1 t].$$

Computing the simplicial differential of $C_{01}^{s,2}$ gives

$$d^s(C_{01}^{s,2}) = [t; s_2 t, 1 - s_2 t] - [t; t, 1 - s_1 t] \qquad \subset X \times \square^2$$

with $0 \leqslant s_2 \leqslant 1$ in the first term and $0 \leqslant s_1 \leqslant 1$ in the second term.

Note that the cycle $[t; t, 1 - s_1 t]$ is negligible as it is a product

$$[t; t, 1 - s_1 t] = \mathcal{L}_0\, [t; 1 - s_1 t]$$

and thus can be canceled in the bar construction setting as the multiplicative boundary of

$$-[\mathcal{L}_0 | [t, 1 - s_1 t]].$$

Thus, up to negligible terms,

$$(d^s + \partial)(C_{01}^{s,1} + C_{01}^{s,2}) = \mathcal{L}_{01}.$$

Now, we fix the situation at the fiber $t_0$ and following Gangl, Goncharov and Levin, we associate to the algebraic cycle $\mathcal{L}_{01}|_{t=t_0}$ the integral $I_{01}(t_0)$ of the standard volume form

$$\frac{1}{(2i\pi)^2} \frac{dz_1}{z_1} \wedge \frac{dz_2}{z_2}$$

over the simplex given by $C_{01}^{s,2}$. That is :

$$\begin{aligned} I_{01}(t_0) &= -\frac{1}{(2i\pi)^2} \int_{0 \leqslant s_1 \leqslant s_2 \leqslant 1} \frac{ds_2}{s_2} \wedge \frac{-t_0\, ds_1}{1 - t_0 s_1} \\ &= \frac{-1}{(2i\pi)^2} \int_{0 \leqslant s_1 \leqslant s_2 \leqslant 1} \frac{ds_1}{t_0^{-1} - s_1} \wedge \frac{ds_2}{s_2} = \frac{-1}{(2i\pi)^2} Li_2^{\mathbb{C}}(t_0). \end{aligned}$$

In particular, this expression is valid for $t_0 = 1$, as is the cycle $\mathcal{L}_{01}|_{t=1}$, and gives $-1/(2\pi^2)\zeta(2)$.

Before presenting the weight 3 example of $\mathcal{L}_{011}$, we describe shortly below the situation for $\mathcal{L}^1_{01}$. In the bar construction the element $\mathcal{L}^{1,B}_{01}$ is equal to the difference $\mathcal{L}^B_{01} - \mathcal{L}^B_{01}(1)$. Associating an integral to $\mathcal{L}^1_{01}$ works in the same way as the cycle $\mathcal{L}_{01}$ but it also reflects the correspondence with $\mathcal{L}^B_{01} - \mathcal{L}^B_{01}(1)$.

The expression of $\mathcal{L}^1_{01}$ in terms of parametrized cycle is given by

$$\mathcal{L}^1_{01} = [t; \frac{x_1 - t}{x_1 - 1}, x_1, 1 - x_1]$$

and can be bounded using the "simplicial" algebraic cycle

$$C^{s,1}_{\overline{01}} = -[t; \frac{x_1 - s_2 t}{x_1 - s_2}, x_1, 1 - x_1].$$

Now, the algebraic boundary of $C^{s,1}_{\overline{01}}$ gives two terms

$$\partial(C^{s,1}_{\overline{01}}) = -[t; s_2 t, 1 - s_2 t] + [t; s_2, 1 - s_2].$$

Then one defines $C^{s,2}_{\overline{01}}$ for simplicial variables $0 \leqslant s_1 \leqslant s_2 \leqslant 1$ as

$$C^{s,2}_{\overline{01}} = -[t; s_2 t, 1 - s_1 t] + [t; s_2, 1 - s_1]$$

whose simplicial boundary cancels $\partial(C^{s,1}_{\overline{01}})$ up to negligible cycles. Again, fixing a fiber $t_0$ the integral associated to $\mathcal{L}^1_{01}|_{t=t_0}$ is the integral of the standard volume form over the $C^{s,2}_{\overline{01}}$:

$$I_{\overline{01}}(t_0) = -\frac{1}{(2i\pi)^2}\left(\int_{0\leqslant s_1\leqslant s_2\leqslant 1} \frac{ds_2}{s_2} \wedge \frac{-t_0\, ds_1}{1 - t_0 s_1} + \int_{0\leqslant s_1\leqslant s_2\leqslant 1} \frac{ds_2}{s_2} \wedge \frac{-ds_1}{1 - s_1}\right).$$

This expression is exactly the difference

$$I_{\overline{01}}(t_0) = \frac{-1}{(2i\pi)^2}\left(Li_2^{\mathbb{C}}(t_0) - Li_2^{\mathbb{C}}(1)\right).$$

**6.2. An integral associated to $\mathcal{L}_{011}$.** Let's recall the expression of $\mathcal{L}_{011}$ as parametrized cycle:

$$\mathcal{L}_{011} = -[t; 1 - \frac{t}{x_2}, 1 - x_2, \frac{x_1 - x_2}{x_1 - 1}, x_1, 1 - x_1].$$

As previously, one wants to bound $\mathcal{L}_{011}$ by an algebraic-topological cycle. Hence we define

$$C^{s,1}_{011} = [t; 1 - \frac{s_3 t}{x_2}, 1 - x_2, \frac{x_1 - x_2}{x_1 - 1}, x_1, 1 - x_1]$$

for $s_3$ going from 0 to 1. Then $d^s(C^{s,1}_{011}) = \mathcal{L}_{011}$ as $s_3 = 0$ implies that the first cubical coordinate is 1.

Now the algebraic boundary $\partial$ of $C^{s,1}_{011}$ is given by the intersection with the codimension 1 faces of $\square^5$:

$$\partial(C^{s,1}_{011}) = [t; 1 - s_3 t, \frac{x_1 - s_3 t}{x_1 - 1}, x_1, 1 - x_1].$$

We can again bound this cycle by introducing a new simplicial variable $0 \leqslant s_2 \leqslant s_3$ and the cycle

$$C^{s,2}_{011} = [t; 1 - s_3 t, \frac{x_1 - s_2 t}{x_1 - s_2/s_3}, x_1, 1 - x_1].$$

The intersection with the face of the simplex $\{0 \leqslant s_2 \leqslant s_3 \leqslant 1\}$ given by $s_2 = 0$ leads to an empty cycle (as one cubical coordinate equals 1) while the intersection with face $s_3 = 1$ leads to a "negligible" cycle. Thus, the simplicial boundary of $C_{011}^{s,2}$ satisfies (up to a negligible term)

$$d^s(C_{011}^{s,2}) = -\partial(C_{011}^{s,1}) = -[t; 1 - s_3t, \frac{x_1 - s_3t}{x_1 - 1}, x_1, 1 - x_1].$$

Its algebraic boundary is given by

$$\partial(C_{011}^{s,2}) = -[t; 1 - s_3t, s_2t, 1 - s_2t] + [t; 1 - s_3t, \frac{s_2}{s_3}, 1 - \frac{s_2}{s_3}].$$

Finally, we introduce a last simplicial variable $0 \leqslant s_1 \leqslant s_2$ and a purely topological cycle

$$\widetilde{C_{011}}^{s,3} = -[t; 1 - s_3t, s_2t, 1 - s_1t] + [t; 1 - s_3t, \frac{s_2}{s_3}, 1 - \frac{s_1}{s_3}]$$

whose simplicial differential is (up to negligible terms) given in one hand by the face $s_1 = s_2$:

$$[t; 1 - s_3t, s_2t, 1 - s_2t] - [t; 1 - s_3t, \frac{s_2}{s_3}, 1 - \frac{s_2}{s_3}]$$

which is equal to $-\partial(C_{011}^{s,2})$; and in the other hand by the face $s_2 = s_3$:

$$-[t; 1 - s_3t, s_3t, 1 - s_1t].$$

In order to cancel this extra term, we defined $C_{011}^{s,3}$ by

$$C_{011}^{s,3} = \widetilde{C_{011}}^{s,3} + [t; 1 - s_2t, s_3t, 1 - s_1t]$$

whose algebraic boundary is 0 (up to negligible terms).

Finally one has

$$(d^s + \partial)(C_{011}^{s,1} + C_{011}^{s,2} + C_{011}^{s,3}) = \mathcal{L}_{011}$$

up to negligible terms.

Now, we fix the situation at the fiber $t_0$ and following Gangl, Goncharov and Levin, we associate to the algebraic cycle $\mathcal{L}_{011}|_{t=t_0}$ the integral $I_{011}(t_0)$ of the standard volume form

$$\frac{1}{(2i\pi)^3}\frac{dz_1}{z_1} \wedge \frac{dz_2}{z_2} \wedge \frac{dz_3}{z_3}$$

over the simplex given by $C_{011}^{s,3}$. That is :

$$\begin{aligned} I_{011}(t_0) = &-\frac{1}{(2i\pi)^3}\int_{0\leqslant s_1\leqslant s_2\leqslant s_3\leqslant 1} \frac{t_0\, ds_3}{1 - t_0s_3} \wedge \frac{ds_2}{s_2} \wedge \frac{t_0\, ds_1}{1 - t_0s_1} \\ &+ \frac{1}{(2i\pi)^3}\int_{0\leqslant s_3\leqslant 1} \frac{t_0\, ds_3}{1 - t_0s_3} \int_{0\leqslant s_1\leqslant s_2\leqslant 1} \frac{ds_2}{s_2} \wedge \frac{ds_1}{1 - s_1} \\ &+ \frac{1}{(2i\pi)^3}\int_{0\leqslant s_1\leqslant s_2\leqslant s_3\leqslant 1} \frac{t_0 ds_2}{1 - t_0s_2} \wedge \frac{ds_3}{s_3} \wedge \frac{t_0 ds_1}{1 - t_0s_1}. \end{aligned}$$

Taking care of the change of sign due to the numbering, the first term in the above sum is (for $t_0 \neq 0$ and up to the factor $(2i\pi)^{-3}$) equal to

$$Li_{1,2}^{\mathbb{C}}(t_0) = \int_{0\leqslant s_1\leqslant s_2\leqslant s_3\leqslant 1} \frac{ds_1}{t_0^{-1} - s_1} \wedge \frac{ds_2}{s_2} \wedge \frac{ds_3}{t_0^{-1} - s_3}$$

while (up to the same multiplicative factor) the second term is equal to

$$-Li_1^{\mathbb{C}}(t_0)Li_2^{\mathbb{C}}(1)$$

and the third term is equal to

$$Li_{2,1}^{\mathbb{C}}(t_0).$$

Globally the integral is well defined for $t_0 = 0$ and, which is the interesting part, also for $t_0 = 1$ as the divergencies as $t_0$ goes to 1 cancel each other in the above sum. A simple computation and the shuffle relation for $Li_1^{\mathbb{C}}(t_0)Li_2^{\mathbb{C}}(t_0)$ show that the integral associated to the fiber of $\mathcal{L}_{011}$ at $t_0 = 1$ is given by

$$(2i\pi)^3 I_{011}(1) = -Li_{2,1}^{\mathbb{C}}(1) = -\zeta(2,1).$$

## References

[BK94] Spencer Bloch and Igor Kříž, *Mixed Tate motives*, Ann. of Math. (2) **140** (1994), no. 3, 557–605, DOI 10.2307/2118618. MR1307897 (96c:11062)

[Blo86] Spencer Bloch, *Algebraic cycles and higher K-theory*, Adv. in Math. **61** (1986), no. 3, 267–304, DOI 10.1016/0001-8708(86)90081-2. MR852815 (88f:18010)

[Blo91] Spencer Bloch, *Algebraic cycles and the Lie algebra of mixed Tate motives*, J. Amer. Math. Soc. **4** (1991), no. 4, 771–791, DOI 10.2307/2939288. MR1102577 (92i:14004)

[Blo97] Spencer Bloch, *Lectures on mixed motives*, Algebraic geometry—Santa Cruz 1995, Proc. Sympos. Pure Math., vol. 62, Amer. Math. Soc., Providence, RI, 1997, pp. 329–359. MR1492527 (99g:11073)

[GGL07] H. Gangl, A. B. Goncharov, and A. Levin, *Multiple logarithms, algebraic cycles and trees*, Frontiers in number theory, physics, and geometry. II, Springer, Berlin, 2007, pp. 759–774, DOI 10.1007/978-3-540-30308-4_16. MR2290773 (2007m:14011)

[GGL09] H. Gangl, A. B. Goncharov, and A. Levin, *Multiple polylogarithms, polygons, trees and algebraic cycles*, Algebraic geometry—Seattle 2005. Part 2, Proc. Sympos. Pure Math., vol. 80, Amer. Math. Soc., Providence, RI, 2009, pp. 547–593, DOI 10.1090/pspum/080.2/2483947. MR2483947 (2009m:11100)

[Gon95] Alexander B. Goncharov, *Polylogarithms in arithmetic and geometry*, Proceedings of the International Congress of Mathematicians, Vol. 1, 2 (Zürich, 1994), Birkhäuser, Basel, 1995, pp. 374–387. MR1403938 (97h:19010)

[Gon05] A. B. Goncharov, *Galois symmetries of fundamental groupoids and noncommutative geometry*, Duke Math. J. **128** (2005), no. 2, 209–284, DOI 10.1215/S0012-7094-04-12822-2. MR2140264 (2007b:11094)

[Lev94] Marc Levine, *Bloch's higher Chow groups revisited*, Astérisque **226** (1994), 10, 235–320. K-theory (Strasbourg, 1992). MR1317122 (96c:14008)

[Lev05] Marc Levine, *Mixed motives*, Handbook of K-theory. Vol. 1, 2, Springer, Berlin, 2005, pp. 429–521, DOI 10.1007/3-540-27855-9_10. MR2181828 (2007b:14040)

[Lev11] *Cycles, motives and Shimura varieties*, Tata Institute of Fundamental Research Studies in Mathematics, Published for the Tata Institute of Fundamental Research, Mumbai; Narosa Publishing House, New Delhi, 2010. Edited by V. Srinivas; International distribution by American Mathematical Society. MR2906018 (2012i:14002)

[LV12] Jean-Louis Loday and Bruno Vallette, *Algebraic operads*, Grundlehren der Mathematischen Wissenschaften [Fundamental Principles of Mathematical Sciences], vol. 346, Springer, Heidelberg, 2012. MR2954392

[Reu93] Christophe Reutenauer, *Free Lie algebras*, London Mathematical Society Monographs. New Series, vol. 7, The Clarendon Press, Oxford University Press, New York, 1993. Oxford Science Publications. MR1231799 (94j:17002)

[Sou12] Ismael Soudères, *Cycle complex over $\mathbb{P}^1$ minus 3 points : toward multiple zeta values cycles*, http://arxiv.org/abs/1210.4653, 2012.

[Spi] Markus Spitzweck, *Some constructions for Voevodsky's triangulated categories of motives*, preprint.

[Spi01] Markus Spitzweck, *Operads, algebras and modules in model categories and motives*, Ph.D. thesis, Bonn Universität, http://hss.ulb.uni-bonn.de/2001/0241/0241.htm, 2001.

[Tot92] Burt Totaro, *Milnor K-theory is the simplest part of algebraic K-theory*, K-Theory **6** (1992), no. 2, 177–189, DOI 10.1007/BF01771011. MR1187705 (94d:19009)

Fachbereich Mathematik, Universität Duisburg-Essen, Campus Essen, Universitätsstrasse 2, 45117 Essen, Germany
*E-mail address*: ismael.souderes@uni-due.de

Max-Planck-Institut für Mathematik, Bonn, Vivatsgasse 7, 53111 Bonn, Germany
*E-mail address*: souderes@mpim-bonn.mpg.de

# Periods and Hodge structures in perturbative quantum field theory

Stefan Weinzierl

ABSTRACT. There is a fruitful interplay between algebraic geometry on the one side and perturbative quantum field theory on the other side. I review the main relevant mathematical concepts of periods, Hodge structures and Picard-Fuchs equations and discuss the connection with Feynman integrals.

## 1. Periods

**1.1. Periodic functions.** Let us consider a meromorphic function $f$ of a complex variable $z$. A period $\omega$ of the function $f$ is a constant such that

$$f(z+\omega) = f(z) \tag{1.1}$$

for all $z$. Let us exclude the trivial case that $f$ is constant. The set of all periods of $f$ forms a lattice, which is either trivial, a simple lattice or a double lattice. The trivial lattice consists of $\omega = 0$ only. A simple lattice is generated by one element. An example for a singly periodic function is given by

$$\exp(z). \tag{1.2}$$

In this case the simple lattice is generated by $\omega = 2\pi i$. A double lattice is generated by two periods $\omega_1$ and $\omega_2$ with $\text{Im}(\omega_2/\omega_1) \neq 0$. It is common practice to order these two periods such that $\text{Im}(\omega_2/\omega_1) > 0$. An example for a doubly periodic function is given by Weierstrass's $\wp$-function. Let $\Lambda$ be the lattice generated by $\omega_1$ and $\omega_2$

$$\Lambda = \{n_1\omega_1 + n_2\omega_2 | n_1, n_2 \in \mathbb{Z}\}. \tag{1.3}$$

Then

$$\wp(z) = \frac{1}{z^2} + \sum_{\omega \in \Lambda \backslash \{0\}} \left( \frac{1}{(z+\omega)^2} - \frac{1}{\omega^2} \right). \tag{1.4}$$

$\wp(z)$ is periodic with periods $\omega_1$ and $\omega_2$. Of particular interest are also the corresponding inverse functions. These are in general multivalued functions. In the case of the exponential function $x = \exp(z)$, the inverse function is given by

$$z = \ln(x). \tag{1.5}$$

The inverse function to Weierstrass's elliptic function $x = \wp(z)$ is an elliptic integral given by

$$z = \int\limits_x^\infty \frac{dt}{\sqrt{4t^3 - g_2 t - g_3}} \tag{1.6}$$

with

$$g_2 = 60 \sum_{\omega \in \Lambda \backslash \{0\}} \frac{1}{\omega^4}, \qquad g_3 = 140 \sum_{\omega \in \Lambda \backslash \{0\}} \frac{1}{\omega^6}. \tag{1.7}$$

In both examples the periods can be expressed as integrals involving only algebraic functions. For the first example of the exponential function we have

$$2\pi i = 4i \int\limits_0^1 \frac{dt}{\sqrt{1-t^2}}. \tag{1.8}$$

For the second example of Weierstrass's $\wp$-function let us assume that $g_2$ and $g_3$ are two given algebraic numbers. Then

$$\omega_1 = 2 \int\limits_{t_1}^{t_2} \frac{dt}{\sqrt{4t^3 - g_2 t - g_3}}, \qquad \omega_2 = 2 \int\limits_{t_3}^{t_2} \frac{dt}{\sqrt{4t^3 - g_2 t - g_3}}, \tag{1.9}$$

where $t_1$, $t_2$ and $t_3$ are the roots of the cubic equation $4t^3 - g_2 t - g_3 = 0$.

**1.2. Numerical periods.** In the following we will denote by $\mathbb{Q}$ the set of rational numbers and by $\bar{\mathbb{Q}}$ the set of algebraic numbers. An algebraic number is a solution of a polynomial equation with rational coefficients. $\bar{\mathbb{Q}}$ is a countable subset of $\mathbb{C}$. Numbers which are not algebraic are called transcendental. The sets $\mathbb{R}$, $\mathbb{C}$ and the set of transcendental numbers are uncountable.

The representation of the periods of $\exp(z)$ and $\wp(z)$ in the form of eq. (1.8) and eq. (1.9) is the motivation for the following generalisation, due to Kontsevich and Zagier [**22**]:

A numerical period is a complex number whose real and imaginary parts are values of absolutely convergent integrals of rational functions with rational coefficients, over domains in $\mathbb{R}^n$ given by polynomial inequalities with rational coefficients. Domains defined by polynomial inequalities with rational coefficients are called semi-algebraic sets.

We denote the set of numerical periods by $\mathbb{P}$. We may replace in the above definition every occurrence of "rational function" with "algebraic function" and every occurrence of "rational number" with "algebraic number" without changing the set of numbers $\mathbb{P}$. Then it is clear, that the integrals in eq. (1.8) and eq. (1.9) are numerical periods in the sense of the above definition.

The algebraic numbers are contained in the set of periods, $\bar{\mathbb{Q}} \subset \mathbb{P}$. The numerical periods $\mathbb{P}$ are a countable set of numbers, lying between $\bar{\mathbb{Q}}$ and $\mathbb{C}$. The set of numerical periods $\mathbb{P}$ is a $\bar{\mathbb{Q}}$-algebra. In particular the sum and the product of two numerical periods are again numerical periods. The transcendental number $\pi$ is contained in $\mathbb{P}$, as can be seen from eq. (1.8). However it is not known, whether $1/\pi$ belongs to $\mathbb{P}$ or not. As multiplication with (positive or negative) powers of $(2\pi i)$ is a common operation in number theory, one often considers the algebra

$\mathbb{P}[1/(2\pi i)]$ with $1/(2\pi i)$ adjoined. On the other hand, it is conjectured that the basis of the natural logarithm $e$ and Euler's constant $\gamma_E$ are not periods.

Let us now turn to dimensionally regulated Feynman integrals. We denote the dimension of space-time by $D$. We consider a Feynman graph $G$ with $m$ external edges, $n$ internal edges and $l$ loops. For each internal edge $e_j$ we choose an orientation and associate to the edge a momentum $q_j$, a mass $m_j$ and a positive integer number $\nu_j$, giving the power to which the propagator occurs. At each vertex we impose momentum conservation: The sum of all momenta flowing into the vertex equals the sum of all momenta flowing out of the vertex. Then we can express the momenta flowing through the internal lines as a linear combination of the independent loop momenta $k_1$, ..., $k_l$ and the external momenta $p_1$, ..., $p_m$ as

$$(1.10) \qquad q_i = \sum_{j=1}^{l} \rho_{ij} k_j + \sum_{j=1}^{m} \sigma_{ij} p_j, \qquad \rho_{ij}, \sigma_{ij} \in \{-1, 0, 1\}.$$

We define the Feynman integral by

$$(1.11) \qquad I_G = \frac{\prod\limits_{j=1}^{n} \Gamma(\nu_j)}{\Gamma(\nu - lD/2)} \left(\mu^2\right)^{\nu - lD/2} \int \prod_{r=1}^{l} \frac{d^D k_r}{i\pi^{\frac{D}{2}}} \prod_{j=1}^{n} \frac{1}{(-q_j^2 + m_j^2)^{\nu_j}},$$

with $\nu = \nu_1 + ... + \nu_n$. $\mu$ is an arbitrary scale introduced to make the integral dimensionless. After Feynman parametrisation we obtain

$$(1.12) \qquad I_G = \int\limits_{\Delta} \omega_0 \left( \prod_{j=1}^{n} x_j^{\nu_j - 1} \right) \frac{\mathcal{U}^{\nu - (l+1)D/2}}{\mathcal{F}^{\nu - lD/2}}.$$

The prefactors in the defintion of the Feynman integral in eq. (1.11) are chosen such that after Feynman parametrisation we obtain an expression without prefactors, as can be seen from eq. (1.12). In eq. (1.12) the integration is over

$$(1.13) \qquad \Delta = \left\{ [x_1 : x_2 : ... : x_n] \in \mathbb{P}^{n-1} | x_i \geq 0, 1 \leq i \leq n \right\}.$$

Here, $\mathbb{P}^{n-1}$ denotes the real projective space with $n-1$ dimensions. $\omega_0$ is a differential $(n-1)$-form given by

$$(1.14) \qquad \omega_0 = \sum_{j=1}^{n} (-1)^{j-1} \, x_j \, dx_1 \wedge ... \wedge \widehat{dx_j} \wedge ... \wedge dx_n,$$

where the hat indicates that the corresponding term is omitted. The functions $\mathcal{U}$ and $\mathcal{F}$ are obtained from first writing

$$(1.15) \qquad \sum_{j=1}^{n} x_j(-q_j^2 + m_j^2) = -\sum_{r=1}^{l} \sum_{s=1}^{l} k_r M_{rs} k_s + \sum_{r=1}^{l} 2k_r \cdot Q_r - J,$$

where $M$ is a $l \times l$ matrix with scalar entries and $Q$ is a $l$-vector with $D$-vectors as entries. We then have

$$(1.16) \qquad \mathcal{U} = \det(M), \qquad \mathcal{F} = \det(M) \left( -J + Q M^{-1} Q \right) / \mu^2.$$

$\mathcal{U}$ and $\mathcal{F}$ are the first and second graph polynomial of the Feynman graph $G$. Both polynomials are homogeneous in the Feynman parameters, $\mathcal{U}$ is of degree $l$, $\mathcal{F}$ is of degree $l+1$. The polynomial $\mathcal{U}$ is linear in each Feynman parameter. If all internal masses are zero, then also $\mathcal{F}$ is linear in each Feynman parameter. In the polynomial $\mathcal{F}$ Lorentz invariant quantities like

$$s = \left( \sum_j p_j \right)^2 \tag{1.17}$$

appear, where the sum runs over a subset of the external momenta. We denote the set of all Lorentz invariants appearing in $\mathcal{F}$ by $S_G$. There is an alternative definition of the two graph polynomials in terms of spanning forests of the graph $G$ [10].

We would like to discuss one special case of eq. (1.12). Suppose that (i) the graph has no external lines or all invariants from $S_G$ are zero, (ii) all internal masses $m_j$ are equal to $\mu$ and (iii) all propagators occur with power 1, i.e. $\nu_j = 1$ for all $j$. Then the Feynman parameter integral reduces to

$$I_G = \int_{\Delta} \omega_0 \left( \sum_{j=1}^{n} x_j \right)^{lD/2-n} \mathcal{U}^{-D/2}. \tag{1.18}$$

This integral is a Igusa local zeta function (when viewed as a function of $D/2$) and has been studied by Belkale and Brosnan in [4].

The Feynman integral defined in eq. (1.12) has an expansion as a Laurent series in the parameter $\varepsilon = (4 - D)/2$ of dimensional regularisation:

$$I_G = \sum_{j=-2l}^{\infty} c_j \varepsilon^j. \tag{1.19}$$

The Laurent series of an $l$-loop integral can have poles in $\varepsilon$ up to the order $(2l)$. The poles in $\varepsilon$ correspond to ultraviolet or infrared divergences. The coefficients $c_j$ are functions of the Lorentz invariants $s \in S_G$, the masses $m_i$ and (in a trivial way) of the arbitrary scale $\mu$. Suppose for the moment that (i) all kinematical invariants $s \in S_G$ are negative or zero, (ii) all masses $m_i$ and $\mu$ are positive or zero ($\mu \neq 0$) and (iii) all ratios of invariants and masses are rational, then it can be shown that all coefficients $c_j$ of the Laurent expansion are numerical periods. For the special case of Igusa local zeta functions in eq. (1.18) this has been proven by Belkale and Brosnan [4], the more general case of eq. (1.12) in [9]. The non-trivial part of this statement comes from the fact that in four space-time dimensions Feynman integrals can diverge. A value of $D \neq 4$ acts as a regulator. In general the expansion in $\varepsilon$ does not commute with the integration. In order to be able to write the coefficients $c_j$ as manifestly absolute convergent integrals one uses a resolution of the singularities of the two graph polynomials [21, 27]. In the physics community a constructive algorithm to do this is known as sector decomposition [5, 8].

**1.3. Abstract periods.** There is a more formal definition of periods as follows [22]: Let $X$ be a smooth algebraic variety of dimension $n$ defined over $\mathbb{Q}$ and $D \subset X$ a divisor with normal crossings. (A normal crossing divisor is a subvariety of dimension $n-1$, which looks locally like a union of coordinate hyperplanes.) Further let $\omega$ be an algebraic differential form on $X$ of degree $n$ and $\Delta$ a singular $n$-chain on the complex manifold $X(\mathbb{C})$ with boundary on the divisor $D(\mathbb{C})$. We thus have a quadruple $(X, D, \omega, \Delta)$. To each quadruple we can associate a complex number $P(X, D, \omega, \Delta)$ called the period of the quadruple and given by the integral

$$(1.20) \qquad P(X, D, \omega, \Delta) = \int\limits_{\Delta} \omega.$$

It is clear that the period of the quadruple is an element of $\mathbb{P}$, and that to any element $p \in \mathbb{P}$ one can find a quadruple, such that $P(X, D, \omega, \Delta) = p$. The period map is therefore surjective. The interesting question is whether the period map is also injective. As it stands above, the period map is certainly not injective for trivial reasons. For example, a simple change of variables can lead to a different quadruple, but does not change the period. One therefore considers equivalence classes of quadruples modulo relations induced by linearity in $\omega$ and $\Delta$, changes of variables, Stokes' formula and Fubini's formula. The vector space over $\mathbb{Q}$ of these equivalence classes is called the space of effective periods and denoted by $\mathcal{P}$. $\mathcal{P}$ is an algebra. It is conjectured that the period map from $\mathcal{P}$ to $\mathbb{P}$ is injective and therefore an isomorphism [**3, 20, 22**]. This would imply that all relations between numerical periods are due to linearity, change of variables, Stokes and Fubini.

In order to make the definition more concrete, we consider as an example the quadruple given by $X = \mathbb{A}^1_{\mathbb{Q}} \backslash \{0\}$, $D = \emptyset$, $\omega = dz/z$ and $\Delta$ the path along the unit circle in the counter-clockwise direction. The corresponding period is $(2\pi i)$.

As in the case of numerical periods it is not known whether there is a quadruple in $\mathcal{P}$, whose period is $1/(2\pi i)$. One therefore adjoins to $\mathcal{P}$ formally the inverse of the element whose period is $(2\pi i)$ and writes $\mathcal{P}[1/(2\pi i)]$ for the so obtained algebra. Elements of $\mathcal{P}[1/(2\pi i)]$ are called abstract periods.

## 2. Hodge structures

Hodge structures have their origin in the study of compact Kähler manifolds. Let $M$ be a complex manifold with complex structure $J$. A Riemannian metric $g$ on $M$ is called Hermitian, if it is compatible with the complex structure $J$, in other words

$$(2.1) \qquad g(JX, JY) = g(X, Y).$$

$X$ and $Y$ denote vector fields on $M$. For a Hermitian manifold one defines an associated differential two-form by

$$(2.2) \qquad K(X, Y) = g(JX, Y).$$

$K$ is called the Kähler form. (In the literature also the letter $\omega$ is frequently used to denote the Kähler form.) A Hermitian manifold is called a Kähler manifold, if the two-form $K$ is closed:

$$(2.3) \qquad dK = 0.$$

An example of compact Kähler manifolds is provided by compact Riemann surfaces. Riemann surfaces are complex manifolds of complex dimension one and the Kähler form of any Hermitian metric is necessarily closed. A second example is given by the complex projective space $\mathbb{P}^n(\mathbb{C})$. As a third example we mention complex submanifolds of Kähler manifolds. These submanifolds are again Kähler.

On a compact Kähler manifold we have the following decomposition of the cohomology groups

$$H^k(X) \otimes \mathbb{C} = \bigoplus_{p+q=k} H^{p,q}(X), \qquad \overline{H^{p,q}(X)} = H^{q,p}(X). \tag{2.4}$$

For a fixed $k$ this provides an example of a pure Hodge structure of weight $k$. A pure Hodge structure of weight $k$ on a $\mathbb{Z}$-module $V_\mathbb{Z}$ of finite rank is a direct sum decomposition

$$V_\mathbb{C} = V_\mathbb{Z} \otimes_\mathbb{Z} \mathbb{C} = \bigoplus_{p+q=n} V^{p,q} \quad \text{with} \quad \overline{V^{p,q}} = V^{q,p}. \tag{2.5}$$

If one replaces $\mathbb{Z}$ by $\mathbb{Q}$ or $\mathbb{R}$, one speaks about a rational or real Hodge structure, respectively. The numbers

$$h^{p,q}(V) = \dim V^{p,q} \tag{2.6}$$

are called the Hodge numbers. There is a second definition of a pure Hodge structure, which is more adapted for generalisations. The second definition is based on a Hodge filtration: Let $F^\bullet V_\mathbb{C}$ be a finite decreasing filtration:

$$V_\mathbb{C} \supseteq ... \supseteq F^{p-1}V_\mathbb{C} \supseteq F^p V_\mathbb{C} \supseteq F^{p+1}V_\mathbb{C} \supseteq ... \supseteq (0) \tag{2.7}$$

such that

$$V_\mathbb{C} = F^p V_\mathbb{C} \oplus \overline{F^{k-p+1}V_\mathbb{C}}. \tag{2.8}$$

Then $V$ carries a pure Hodge structure of weight $k$. These two definitions are equivalent: Given the Hodge decomposition, we can define the corresponding Hodge filtration by

$$F^p V_\mathbb{C} = \bigoplus_{j \geq p} V^{j,k-j}. \tag{2.9}$$

Conversely, given a Hodge filtration we obtain the Hodge decomposition by

$$V^{p,q} = F^p V_\mathbb{C} \cap \overline{F^q V_\mathbb{C}}. \tag{2.10}$$

Pure Hodge structures are relevant for smooth projective algebraic varieties, these are necessarily compact. If one gives up the requirement of smoothness or compactness one is lead to a generalisation called mixed Hodge structure [**14–16**]. A mixed Hodge structure is given by a $\mathbb{Z}$-module $V_\mathbb{Z}$ of finite rank, a finite increasing filtration on $V_\mathbb{Q} = V_\mathbb{Z} \otimes_\mathbb{Z} \mathbb{Q}$, called the weight filtration:

$$(0) \subseteq ... \subseteq W_{k-1}V_\mathbb{Q} \subseteq W_k V_\mathbb{Q} \subseteq W_{k+1}V_\mathbb{Q} \subseteq ... \subseteq V_\mathbb{Q}, \tag{2.11}$$

and a finite decreasing filtration on $V_\mathbb{C} = V_\mathbb{Z} \otimes_\mathbb{Z} \mathbb{C}$, called the Hodge filtration:

$$V_\mathbb{C} \supseteq ... \supseteq F^{p-1}V_\mathbb{C} \supseteq F^p V_\mathbb{C} \supseteq F^{p+1}V_\mathbb{C} \supseteq ... \supseteq (0), \tag{2.12}$$

such that $F^\bullet$ induces a pure Hodge structure of weight $k$ on

$$\mathrm{Gr}_k^W V_\mathbb{Q} = W_k V_\mathbb{Q} / W_{k-1} V_\mathbb{Q}. \tag{2.13}$$

In addition, we can consider a family of Hodge structures, parametrised by a manifold. If in addition a few technical conditions are met, this leads to a variation of a Hodge structure. The precise definition of a variation of a Hodge structure can be found in the literature [**18,19**]. Suppose first that the variation with the parameters is smooth. Then the Hodge numbers $h^{p,q}$, being integers, have to remain constant. If the variation with the parameters has singularities, one can study what happens

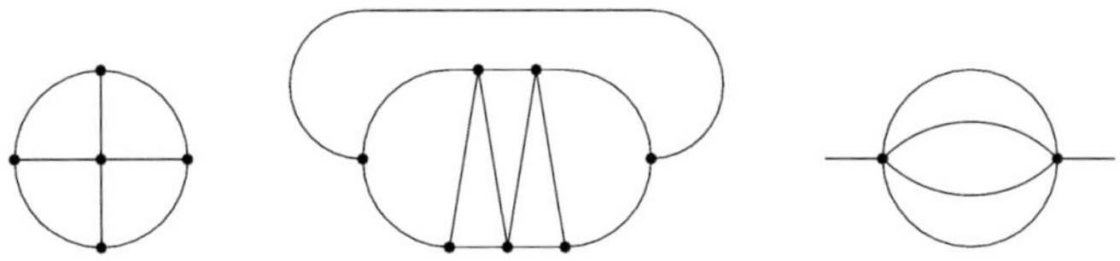

FIGURE 1. The wheel with four spokes (left), the 6-loop zigzag graph (middle) and the 3-loop banana graph (right).

as one approaches the singularity. This gives a limiting mixed Hodge structure. For an exact definition the reader is again referred to the literature [**25**, **28**].

In order to discuss the connection of Hodge structures with periods we focus on a smooth algebraic variety over the rational numbers $\mathbb{Q}$. To this variety we can on the one hand associate the de Rham cohomology $H^k_{\mathrm{DR}}(X)$, as well as the Betti cohomology $H^k_{\mathrm{B}}(X)$. Let us denote by $\omega_j$ a basis for $H^k_{\mathrm{DR}}(X)$ and by $\gamma_i$ a basis for the Betti homology $H^{\mathrm{B}}_k(X)$. A basis for $H^k_{\mathrm{B}}(X)$ is then given by the duals $\gamma_j^*$ and satisfies $\gamma_i \gamma_j^* = \delta_{ij}$. There is an isomorphism between the de Rham and Betti cohomology,

$$H^k_{\mathrm{B}}(X) \otimes \mathbb{C} \to H^k_{\mathrm{DR}}(X) \otimes \mathbb{C} \tag{2.14}$$

given by

$$\omega_i = \sum_j p_{ij} \gamma_j^*, \qquad p_{ij} = \int_{\gamma_j} \omega_i. \tag{2.15}$$

The coefficients $p_{ij}$ are called the periods of $X$. The abstract periods are then the equivalence classes of the quadruples $(X, \emptyset, \omega_i, \gamma_j)$.

Let us now come back to the Feynman integral in eq. (1.12) and focus on finite Feynman integrals, which do not need to be regulated. We consider two special cases: The first case is given by $D = 4$, $\nu_j = 1$ and $n = 2l$. Then

$$I_G \quad = \quad \int_\Delta \frac{\omega_0}{\mathcal{U}^2}. \tag{2.16}$$

Examples for this case are the wheel with $l$ spokes and the family of zigzag graphs [**6**, **12**, **17**]. Examples of corresponding Feynman graphs are shown in fig. (1) on the left and in the middle, respectively. In the second case we consider $D = 2$, $\nu_j = 1$ and $n = l + 1$. Then

$$I_G \quad = \quad \int_\Delta \frac{\omega_0}{\mathcal{F}}. \tag{2.17}$$

Examples for this case are the family of banana graphs [**1**, **23**]. An example of a corresponding Feynman graph is shown in fig. (1) on the right.

In both cases the integrand is a rational differential form. We would like to think of these integrals as periods of mixed Hodge structures. There are two complications to this. First of all, the integrand becomes singular when the polynomials in the denominator vanish. Secondly, the region of integration $\Delta$ has a boundary $\partial\Delta$. It is therefore clear, that there are two geometric objects of interest in eq. (2.16) or eq. (2.17): On the one hand the domain of integration $\Delta$ and on the other hand the algebraic variety $X$ defined by the zero set of the polynomial in the denominator.

In the first example, $X$ is defined by $\mathcal{U}^2 = 0$, in the second example $X$ is defined by $\mathcal{F} = 0$. The two objects $X$ and $\Delta$ may intersect. Since $\mathcal{U}$ is a sum of monomials with all coefficients equal to 1, intersections of $X$ and $\Delta$ may happen in the case of the first example only on the boundary $\partial\Delta$. The same holds true in the second example if we restrict ourselves for the kinematics to the Euclidean region, already encountered after eq. (1.19): If (i) all kinematical invariants $s \in S_G$ are negative or zero and (ii) all masses $m_i$ and $\mu$ are positive or zero ($\mu \neq 0$), then the zero set of $\mathcal{F}$ intersects $\Delta$ only on the boundary. If $X \cap \partial\Delta$ is non-empty, we blow-up $\mathbb{P}^{n-1}$ in this region. Let us denote the blow-up by $P$. We further denote the strict transform of $X$ by $Y$ and the total transform of the set $\{x_1 x_2 ... x_n = 0\}$ by $B$. With these notations we can now consider the mixed Hodge structure (or the motive) given by the relative cohomology group [**2**, **6**, **7**, **11**, **13**, **26**]

$$H^{n-1}\left(P \backslash Y, B \backslash B \cap Y\right). \tag{2.18}$$

The Feynman integral is then a period of this cohomology class. Let us denote by

$$S = S_G \cup \{m_1^2, ..., m_n^2\} \tag{2.19}$$

the set of all Lorentz invariants $s$ and masses squared appearing in $\mathcal{F}$. The polynomial $\mathcal{F}$ depends on these variables and we actually have in the example of eq. (2.17) a variation of a mixed Hodge structure.

## 3. Picard-Fuchs equations

Do the formal considerations of the previous section have practical applications ? Yes, they do. We will give an example. Periods of variations of mixed Hodge structures are expected to satisfy differential equations of Picard-Fuchs type. In the case of example (2.17) the Feynman integral is a period of a variation of a mixed Hodge structure and should satisfy for any given choice $t \in S$ an ordinary linear differential equation of Picard-Fuchs type. Such a differential equation is a useful tool for computing the Feynman integral, once the boundary values and the inhomogeneous terms in the differential equation are known. Below we will sketch an algorithm for finding for a given variable $t \in S$ an ordinary linear differential equation of Picard-Fuchs type of minimal order [**24**]. The algorithm has the additional benefit that we can work with dimensionally regulated integrals as in eq. (1.12) and are no longer forced to restrict ourselves to rational integrands as in eq. (2.16) or in eq. (2.17).

Starting from eq. (1.12) we pick one $t \in S$ and set

$$\omega_t = f\omega_0, \qquad f = \left(\prod_{j=1}^{n} x_j^{\nu_j - 1}\right) \frac{\mathcal{U}^{\nu-(l+1)D/2}}{\mathcal{F}^{\nu-lD/2}}. \tag{3.1}$$

We look for a differential equation of the form

$$L^{(r)}\omega_t = d\beta, \tag{3.2}$$

where

$$L^{(r)} = \sum_{j=0}^{r} p_j \left(\mu^2 \frac{d}{dt}\right)^j \tag{3.3}$$

is a Picard-Fuchs operator of order $r$. The coefficients $p_j$ may depend on the kinematical invariants from the set $S$, the scale $\mu$, the space-time dimension $D$ and the exponents $\nu_i$, but not on the Feynman parameters $x_i$. We normalise the Picard-Fuchs operator such that $p_r = 1$. We recall that a differential operator of the form as in eq. (3.3) is said to be of Fuchsian type, if all coefficients $p_j$ are meromorphic functions of $t$ and if $p_j$ has at most poles of order $(r - j)$. $\beta$ is a $(n - 2)$-form, depending on the Feynman parameters $x_i$. The differential $d$ is with respect to the Feynman parameters $x_i$. For $\beta$ we make the following ansatz

$$\beta = \frac{f}{\mathcal{F}^{r-1}} \sum_{j_1<j_2} (-1)^{j_1+j_2} \left[-x_{j_1} a_{j_2} + x_{j_2} a_{j_1}\right] dx_1 \wedge ... \wedge \widehat{dx_{j_1}} \wedge ... \wedge \widehat{dx_{j_2}} \wedge ... \wedge dx_n, \tag{3.4}$$

where the $a_i$ are homogeneous polynomials of degree $(rl + r - l)$ in the variables $x_i$. The ansatz is based on the fact, that the singularities of the integrand $\omega_t$ are given by powers of the graph polynomials $\mathcal{U}$ and $\mathcal{F}$. Acting with $L^{(r)}$ on the integrand $\omega_t$ will only increase the power of $\mathcal{F}$ in the denominator by $r$, but will not introduce singularities on new algebraic varieties. Each polynomial $a_i$ contains only a finite number of monomials in the Feynman parameters with a priori unknown coefficients. We therefore take these coefficients and the variables $p_0$, ..., $p_{r-1}$ as the set of our unknown variables. Plugging the ansatz (3.4) in eq. (3.2) gives a linear system of equations for the unknown variables. This system may or may not have a solution. In order to find the differential equation of minimal order we start at $r = 1$ and try to solve the linear system of equations. If no solution is found, we increase $r$ by one and repeat the exercise, until a solution is found. This is then the solution of minimal order $r$.

Integration of eq. (3.2) yields then

$$L^{(r)} I_G = \int\limits_{\Delta} d\beta = \int\limits_{\partial\Delta} \beta, \tag{3.5}$$

where we used Stokes. Eq. (3.5) is the sought-after ordinary linear differential equation for the Feynman integral $I_G$. The right-hand side is given as a sum of integrals with $(n - 1)$ Feynman parameters. These integrals correspond to graphs with one propagator less and can be considered as simpler.

To give an example we consider the one-loop two-point function with equal internal masses in two space-time dimensions. This integral is given by

$$I_G = \int\limits_{\Delta} \frac{x_1 dx_2 - x_2 dx_1}{\mathcal{F}}, \qquad \mathcal{F} = -\frac{t}{\mu^2} x_1 x_2 + (x_1 + x_2)^2 \frac{m^2}{\mu^2}. \tag{3.6}$$

This integral has a first-order Picard-Fuchs operator

$$L^{(1)} = \mu^2 \frac{d}{dt} + \mu^2 \frac{t - 2m^2}{t(t - 4m^2)} \tag{3.7}$$

and a possible solution for $\beta$ is

$$\beta = \frac{1}{\mathcal{F}} \frac{\mu^2}{t(t - 4m^2)} \left[(t - 2m^2) x_1 x_2 - 2m^2 x_2^2\right]. \tag{3.8}$$

The boundary of $\Delta$ is given by the two points $[1:0]$ and $[0:1]$. The integration of the inhomogeneous term yields

$$\int\limits_{\partial\Delta} \beta = -\frac{2\mu^4}{t(t-4m^2)}. \tag{3.9}$$

Putting everything together, we obtain the differential equation

$$\left( t(t-4m^2)\frac{d}{dt} + t - 2m^2 \right) I_G = -2\mu^2. \tag{3.10}$$

## References


[1] Paolo Aluffi and Matilde Marcolli, *Feynman motives of banana graphs*, Commun. Number Theory Phys. **3** (2009), no. 1, 1–57, DOI 10.4310/CNTP.2009.v3.n1.a1. MR2504753 (2010d:81109)

[2] Paolo Aluffi and Matilde Marcolli, *Parametric Feynman integrals and determinant hypersurfaces*, Adv. Theor. Math. Phys. **14** (2010), no. 3, 911–963. MR2801414 (2012e:81075)

[3] Joseph Ayoub, *Une version relative de la conjecture des périodes de Kontsevich-Zagier*, preprint K-theory/1010, (2011).

[4] Prakash Belkale and Patrick Brosnan, *Periods and Igusa local zeta functions*, Int. Math. Res. Not. **49** (2003), 2655–2670, DOI 10.1155/S107379280313142X. MR2012522 (2004h:11097)

[5] T. Binoth and G. Heinrich, *An automatized algorithm to compute infrared divergent multi-loop integrals*, Nuclear Phys. B **585** (2000), no. 3, 741–759, DOI 10.1016/S0550-3213(00)00429-6. MR1784111 (2001i:81180)

[6] Spencer Bloch, Hélène Esnault, and Dirk Kreimer, *On motives associated to graph polynomials*, Comm. Math. Phys. **267** (2006), no. 1, 181–225, DOI 10.1007/s00220-006-0040-2. MR2238909 (2008a:81074)

[7] Spencer Bloch and Dirk Kreimer, *Mixed Hodge structures and renormalization in physics*, Commun. Number Theory Phys. **2** (2008), no. 4, 637–718, DOI 10.4310/CNTP.2008.v2.n4.a1. MR2492196 (2010h:81101)

[8] Christian Bogner and Stefan Weinzierl, *Resolution of singularities for multi-loop integrals*, Comput. Phys. Comm. **178** (2008), no. 8, 596–610, DOI 10.1016/j.cpc.2007.11.012. Associated computer program available online. MR2672132

[9] Christian Bogner and Stefan Weinzierl, *Periods and Feynman integrals*, J. Math. Phys. **50** (2009), no. 4, 042302, 16, DOI 10.1063/1.3106041. MR2513973 (2010g:81110)

[10] Christian Bogner and Stefan Weinzierl, *Feynman graph polynomials*, Internat. J. Modern Phys. A **25** (2010), no. 13, 2585–2618, DOI 10.1142/S0217751X10049438. MR2651267 (2012b:81071)

[11] Francis C. S. Brown, *Multiple zeta values and periods of moduli spaces* $\overline{\mathfrak{M}}_{0,n}$ (English, with English and French summaries), Ann. Sci. Éc. Norm. Supér. (4) **42** (2009), no. 3, 371–489. MR2543329 (2010f:32013)

[12] Francis Brown and Oliver Schnetz, *A K3 in* $\phi^4$, Duke Math. J. **161** (2012), no. 10, 1817–1862, DOI 10.1215/00127094-1644201. MR2954618

[13] Alain Connes and Matilde Marcolli, *Quantum fields and motives*, J. Geom. Phys. **56** (2006), no. 1, 55–85, DOI 10.1016/j.geomphys.2005.04.004. MR2170600 (2006h:81175)

[14] Pierre Deligne, *Théorie de Hodge. I* (French), Actes du Congrès International des Mathématiciens (Nice, 1970), Gauthier-Villars, Paris, 1971, pp. 425–430. MR0441965 (56 #356)

[15] Pierre Deligne, *Théorie de Hodge II*, Publ. Math. Inst. Hautes Études Sci. **40** (1971), 5–58.

[16] Pierre Deligne, *Théorie de Hodge III*, Publ. Math. Inst. Hautes Études Sci. **44** (1974), 5–77.

[17] Dzmitry Doryn, *Cohomology of graph hypersurfaces associated to certain Feynman graphs*, Commun. Number Theory Phys. **4** (2010), no. 2, 365–415, DOI 10.4310/CNTP.2010.v4.n2.a3. MR2725055 (2012b:81072)

[18] Phillip A. Griffiths, *Periods of integrals on algebraic manifolds. I. Construction and properties of the modular varieties*, Amer. J. Math. **90** (1968), 568–626. MR0229641 (37 #5215)

[19] Phillip A. Griffiths, *Periods of integrals on algebraic manifolds II*, Amer. J. Math. **90** (1968), 808–865.

[20] A. Grothendieck, *On the de Rham cohomology of algebraic varieties*, Inst. Hautes Études Sci. Publ. Math. **29** (1966), 95–103. MR0199194 (33 #7343)

[21] Heisuke Hironaka, *Resolution of singularities of an algebraic variety over a field of characteristic zero. I, II*, Ann. of Math. (2) 79 (1964), 109–203; ibid. (2) **79** (1964), 205–326. MR0199184 (33 #7333)

[22] Maxim Kontsevich and Don Zagier, *Periods*, Mathematics unlimited—2001 and beyond, Springer, Berlin, 2001, pp. 771–808. MR1852188 (2002i:11002)

[23] Stefan Müller-Stach, Stefan Weinzierl, and Raphael Zayadeh, *A second-order differential equation for the two-loop sunrise graph with arbitrary masses*, Commun. Number Theory Phys. **6** (2012), no. 1, 203–222, DOI 10.4310/CNTP.2012.v6.n1.a5. MR2955935

[24] Stefan Müller-Stach, Stefan Weinzierl, and Raphael Zayadeh, *Picard-Fuchs equations for Feynman integrals*, arXiv:1212.4389, (2012).

[25] Wilfried Schmid, *Variation of Hodge structure: the singularities of the period mapping*, Invent. Math. **22** (1973), 211–319. MR0382272 (52 #3157)

[26] Oliver Schnetz, *Quantum periods: a census of $\phi^4$-transcendentals*, Commun. Number Theory Phys. **4** (2010), no. 1, 1–47, DOI 10.4310/CNTP.2010.v4.n1.a1. MR2679376 (2012d:81120)

[27] Mark Spivakovsky, *A solution to Hironaka's polyhedra game*, Arithmetic and geometry, Vol. II, Progr. Math., vol. 36, Birkhäuser Boston, Boston, MA, 1983, pp. 419–432. MR717618 (85m:14021)

[28] Joseph Steenbrink, *Limits of Hodge structures*, Invent. Math. **31** (1975/76), no. 3, 229–257. MR0429885 (55 #2894)

PRISMA Cluster of Excellence, Institut für Physik, Johannes Gutenberg-Universität Mainz, D - 55099 Mainz, Germany

*E-mail address*: stefanw@thep.physik.uni-mainz.de

# Some combinatorial interpretations in perturbative quantum field theory

Karen Yeats

ABSTRACT. This paper will describe how combinatorial interpretations can help us understand the algebraic structure of two aspects of perturbative quantum field theory, namely analytic Dyson-Schwinger equations and periods of scalar Feynman graphs. The particular examples which will be looked at are, a better reduction to geometric series for Dyson-Schwinger equations, a subgraph which yields extra denominator reductions in scalar Feynman integrals, and an explanation of a trick of Brown and Schnetz to get one extra step in the denominator reduction of an important particular graph.

## 1. Introduction

In the subtle worlds of periods and quantum field theory it can often be hard to clearly see one's way. Combinatorial interpretations can help by giving us explicit objects to get our hands dirty with. This paper will discuss two situations with such interpretations in perturbative quantum field theory.

The first situation, which is discussed in Section 2, concerns combinatorial understanding of analytic aspects of Dyson-Schwinger equations. One example of this is the recent chord diagram expansion for a class of Dyson-Schwinger equations given by Nicolas Marie and the author in [**9**]. For those readers who came to this proceedings volume looking for information related to the conference talk, the results of [**9**] are summarized. Another example is the reduction process of [**18, 19**]. In this paper I will prove an observation which provides a conceptual clean-up to the reduction to geometric series of [**18, 19**].

The second situation, which is discussed in Section 3, concerns understanding when Brown's denominator reduction approach to calculating periods of scalar Feynman graphs [**2**] yields results which are nicer than expected. In particular we will look at when we get a "free" factorization, and one circumstance where we can proceed an extra step after denominator reduction fails, using a change of variables, as was done by Brown and Schnetz in section 6.2 of [**3**] in order to find a K3 surface. Note that tricks such as the Brown-Schnetz change of variables were found without such combinatorial interpretations, however the interpretation can tell us something about *why* the change of variables does its job.

The new material is Theorem 2.18, Proposition 3.25, and Theorem 3.32, but also interesting is the overall story of nice interpretations involving these results

2010 *Mathematics Subject Classification.* Primary 81T16, 81T18.

woven together with results of Brown [**2**], Brown and Schnetz [**3**], the author with Nicolas Marie [**9**], and the author [**18, 19**].

## 2. Dyson-Schwinger equations and the chord diagram expansion

**2.1. The Hopf algebra of rooted trees.** Let $\mathcal{H}$ be a renormalization Hopf algebra. It will suffice for our purposes here to take $\mathcal{H}$ to be the Connes-Kreimer Hopf algebra of rooted trees [**6**], but this framework functions in the same way for Hopf algebras of Feynman graphs [**18, 19**].

As an algebra, the Connes-Kreimer Hopf algebra is the polynomial algebra over the set of rooted trees. Monomials can also be viewed as disjoint unions of trees and hence are forests. Rooted trees are taken without regards to planar embedding, and with no restrictions on valence. The size of a tree is its number of vertices. This yields a grading on the algebra.

The coalgebra structure is not essential for the presentation of the main result of this section, however it underlies this entire theory and thus will be described briefly. To define the coproduct we need a few auxiliary definitions. An *admissible cut* on a rooted tree $T$ is a set of vertices (possibly empty) of $T$ with the property that no vertex is an ancestor of another. Given an admissible cut $c$ of a rooted tree $T$ the *pruned part*, $P_c(T)$ is the forest of trees rooted at the vertices of $c$; the *root part*, $R_c(T)$, is the tree resulting from removing $P_c(T)$ from $T$. Note that admissible cuts are usually defined based on edge cuts but the definitions are equivalent provided the trivial and full cuts are included as admissible cuts in the edge-based definitions. The coproduct for the Connes-Kreimer Hopf algebra is

$$\Delta(T) = \sum_{\substack{c \text{ admissible} \\ \text{cut of } T}} P_c(T) \otimes R_c(T)$$

on trees $T$ and extended as an algebra homomorphism. Note that the empty cut yields the term $1 \otimes T$ and the cut which consists of the root alone yields $T \otimes 1$. The counit is given by $\eta(1) = 1$ and $\eta(T) = 0$ for any tree $T$ and extended as an algebra homomorphism. This defines a graded bialgebra. The grading gives the antipode recursively from its defining property.

**2.2. Combinatorial Dyson-Schwinger equations.** Combinatorial Dyson-Schwinger equations are recurrences, sometimes systems of recurrences, with solutions which are formal power series in $\mathcal{H}$. Some examples in rooted trees follow the definitions. For Feynman graphs, the combinatorial Dyson-Schwinger equations are defined so that the solutions are sums of all graphs with a given external structure. Thus once we apply Feynman rules we obtain the Green functions of the theory; these are physically important quantities.

DEFINITION 2.1. In the Connes-Kreimer Hopf algebra let $B_+$ be the endomorphism which takes a forest $T_1 T_2 \cdots T_k$ to the rooted tree where the children of the root are $T_1, T_2, \ldots, T_k$.

For example

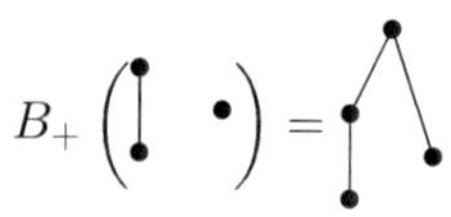

In other renormalization Hopf algebras there may be many $B_+$ operations indexed by primitive elements of the Hopf algebra. The $B_+$ must be Hochschild 1-cocycles, that is they satisfy

$$\Delta B_+ = (\mathrm{id} \otimes B_+)\Delta + B_+ \otimes 1$$

For the $B_+$ of rooted trees this equation can be seen to hold by considering the decomposition of a tree into the root and its subtrees. In other renormalization Hopf algebras it may be necessary to work in a quotient Hopf algebra in order to have the 1-cocycle property; this is the situation in gauge theories [**15**, **16**].

DEFINITION 2.2. *Combinatorial Dyson-Schwinger equations* are systems of equations of the form

$$X_1(x) = 1 \pm \sum_{k\geq 1} x^k B_+^{k,1}(P_{k,1}(X_1(x), \ldots, X_t(x)))$$
$$\vdots$$
$$X_t(x) = 1 \pm \sum_{k\geq 1} x^k B_+^{k,t}(P_{k,t}(X_1(x), \ldots, X_t(x)))$$

where the $B_+^{k,t}$ are 1-cocycles and the $P_{k,t}$ are rational functions.

There are two important restrictions to make, the first of which brings us to the key cases in physics, and the second of which brings us to the problems of this section.

First, let us restrict our interest to

$$P_{k,t}(X_1, \ldots, X_t) = X_t Q^k$$

where

$$Q = \prod_{r=1}^{t} X_r^{s_r}$$

with the $s_r$ integers. The signs in the Dyson-Schwinger equation are then taken to be the signs of the corresponding $s_r$; note that this is the opposite sign convention from [**18**, **19**]. The series $X_r$ all begin with 1 so taking formal inverses makes sense. Here $Q$ is the combinatorial avatar of the physicists' *invariant charge*. The fact that there is such a $Q$ in physically important cases, such as quantum chromodynamics, comes down to the same quotient which made the $B_+$s into 1-cocycles [**8**, section 2].

Second, let us further restrict to single equation Dyson-Schwinger equations

$$X(x) = 1 + \mathrm{sgn}(s) \sum_{k\geq 1} x^k B_+^k(X^{1+ks}(x)) \tag{2.1}$$

where $\mathrm{sgn}(s)$ is the sign of $s$.

Consider a few examples of such equations to get a feel for how they work.

EXAMPLE 2.3. Consider

$$X(x) = 1 + xB_+(X(x))$$

in the Connes-Kreimer Hopf algebra of rooted trees. This is $s = 0$ in (2.1). Expanding we get

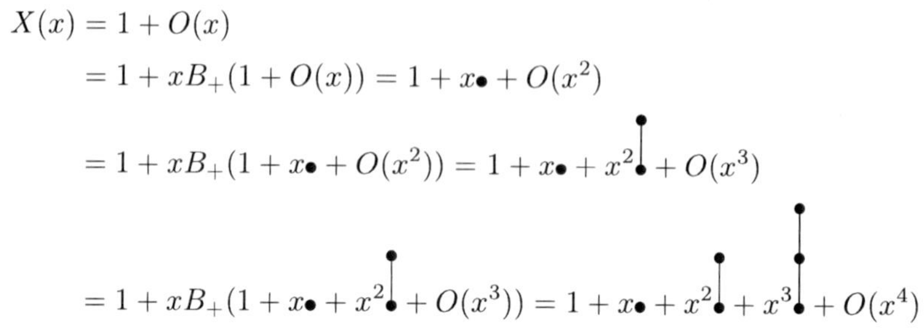

One can quickly see that the coefficient of $x^n$ in $X(x)$ will be the tree with $n$ vertices and no branching.

EXAMPLE 2.4. Consider

$$X(x) = 1 + xB_+(X^2(x))$$

in the Connes-Kreimer Hopf algebra. This is $s = 1$ in (2.1). Expanding we get

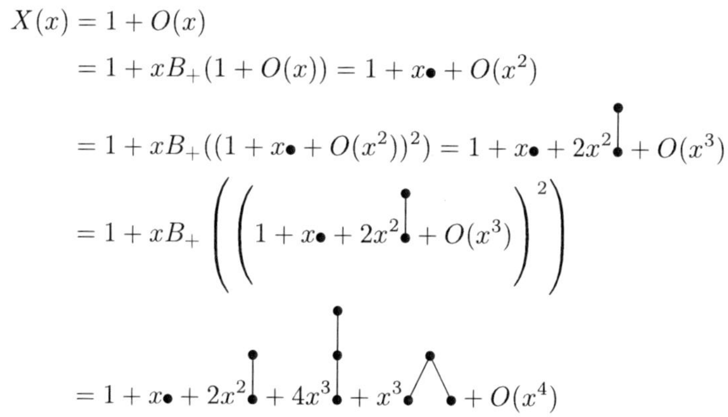

Here we are getting all binary trees weighted by the number of distinct ways that the children of every vertex can be assigned to be left or right children, with at most one of each for each vertex. This can be proved with a quick induction.

EXAMPLE 2.5. Consider

$$X(x) = 1 - xB_+(X^{-1}(x))$$

in the Connes-Kreimer Hopf algebra. This is $s = -2$ in (2.1). Expanding we get

$$
\begin{aligned}
X(x) &= 1 + O(x)\\
&= 1 - xB_+(1 + O(x)) = 1 - x\bullet + O(x^2)
\end{aligned}
$$

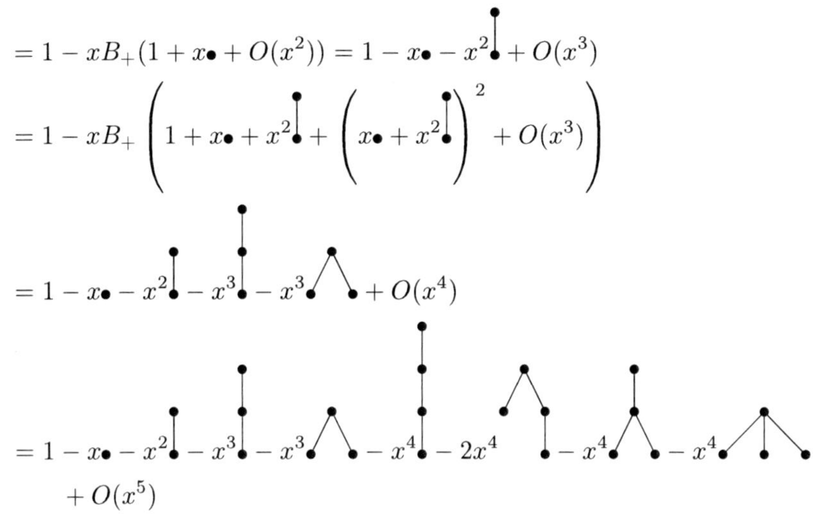

Here we are getting all rooted trees weighted by the number of distinct plane embeddings.

This is a very important example in what follows as it is the case for which we can solve the corresponding analytic Dyson-Schwinger equation using a chord diagram expansion.

**2.3. Analytic Dyson-Schwinger equations.** For a workshop on periods we cannot be satisfied with combinatorial Dyson-Schwinger equations as we need to have analytic information, thence periods.

We can no longer completely ignore the Feynman graphs as we will need the analytic structure of the primitives. For the purposes of this paper it suffices to consider situations where we insert in only one edge $e$ of each primitive $\gamma$. Take the Feynman integral of $\gamma$ and regularize it by raising the propagator associated to $e$ to the power $1 - \rho$. This converges for small nonzero values of $\rho$. Fix the external momenta and expand the result as a Laurent series in $\rho$. Call this series

$$F_\gamma(\rho)$$

Assume that $F_\gamma(\rho)$ has a first order pole at 0. This is the situation in the physical cases of interest since $\gamma$ is primitive.

DEFINITION 2.6. Given a combinatorial Dyson-Schwinger of the form

$$X(x) = 1 + \operatorname{sgn}(s)\sum_{k\geq 1} x^k B_+^k(X^{1+ks}(x))$$

the associated *analytic Dyson-Schwinger equation* is

$$G(x, L) = 1 + \operatorname{sgn}(s)\sum_{k\geq 1} x^k G\left(x, \frac{d}{d(-\rho)}\right)^{1+sk}(e^{-L\rho} - 1)F_k(\rho)\Bigg|_{\rho=0}$$

where

$$F_k(\rho) = \sum_{i=-1}^{\infty} f_{k,i+1}\rho^i$$

with the $f_{k,j}$ viewed as given (by physics).

Systems can be treated similarly [**18**, subsection 3.3.2], but it is messier and not important for the present purposes.

This definition doesn't look much like the usual form of Dyson-Schwinger equations that a physicist would be familiar with. The best way to see the connection between this definition and the more usual form is an example

EXAMPLE 2.7. This example is from [**1**]. Consider graphs built from

inserted into itself in all possible ways. This yields graphs such as

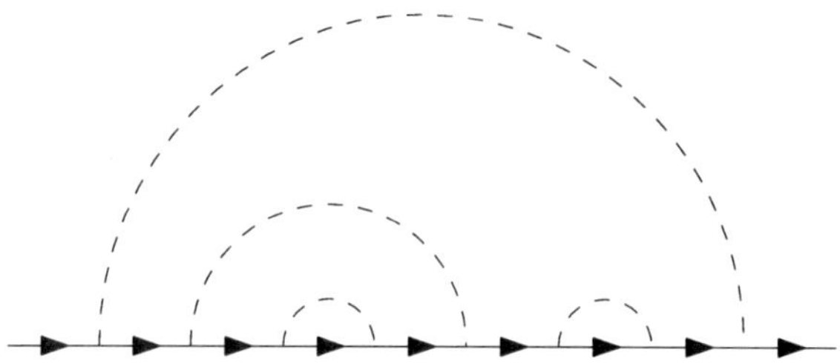

The tree structure of the insertions gives the combinatorial Dyson-Schwinger equation

$$X(x) = \mathbb{I} - xB_+\left(\frac{1}{X(x)}\right)$$

Applying Feynman rules we obtain an analytic Dyson-Schwinger equation in a form closer to what one would find in physics sources [**1**], specifically

$$G(x, L) = 1 - \frac{x}{q^2}\int d^4k \frac{k \cdot q}{k^2 G(x, \log k^2)(k+q)^2} - \cdots \bigg|_{q^2=\mu^2} \tag{2.2}$$

where $L = \log(q^2/\mu^2)$, $q$ is the external momentum, and $\mu$ is a reference momentum. See [**18**] Example 3.5 for further details. Note that the structure of this equation is quite similar to the combinatorial Dyson-Schwinger equation: $B_+$ has become an integral operator and those bits of the Feynman integral given by the primitive; the argument to $B_+$ has become the recursive appearance of $G(x, \log k^2)$. Note also that renormalization is taken care of by a single subtraction at a fixed value of the momentum because the renormalization of the subgraphs is done recursively by the equation itself.

This does not, however, look much like the analytic Dyson-Schwinger equations defined in Definition 2.6. Example 3.7 of [**18**] begins with the above example, proceeds to expand $G(x, L)$ in $L$, convert logarithms to powers using $\frac{d^k y^\rho}{d\rho^k}|_{\rho=0} = \log^k(y)$, swap the order of the operators, and thus obtains

$$G(x, L) = 1 - xG\left(x, \frac{d}{d(-\rho)}\right)^{-1}(e^{-L\rho} - 1)F(\rho)\Big|_{\rho=0} \tag{2.3}$$

where $F(\rho)$ is the Feynman integral of the primitive with the propagator we are inserting on regularized, and the integral evaluated at $q^2 = 1$. The calculations take about a page but do not contain any notable subtleties.

This example is the motivation for Definition 2.6. The basic steps which converted (2.2) to (2.3) could apply to any Dyson-Schwinger equation in its usual analytic form provided everything is sufficiently well-behaved analytically to swap the order of the integrals and derivatives as described. To avoid such analytic nuisance I simply *define* the analytic Dyson-Schwinger equations to be the formal outcome of this process, which is what is given by Definition 2.6. In this context the expansion of $F(\rho)$ is viewed as given by physics because, as in the example, $F(\rho)$ is the integral for the primitive regularized at the insertion place and with the external momenta fixed.

It is tempting, when we are considering the number theory or algebraic geometry of Feynman graphs, to focus on graphs individually. There is certainly a lot of interesting mathematics in each graph, and this will be the approach in Section 3. However, it is important that the mathematical investigation of Feynman graphs does not begin and end at the individual graph level. The solutions to Dyson-Schwinger equations are physically meaningful quantities while individual graphs are largely not. Thus the sums of graphs from Dyson-Schwinger equations yield the periods of greatest interest.

**2.4. Chord diagrams.** The main result of [**9**], which was the topic of the author's talk at the conference for which these are the proceedings, is a solution to the particular Dyson-Schwinger equation

$$G(x, L) = 1 - xG\left(x, \frac{d}{d(-\rho)}\right)^{-1} (e^{-L\rho} - 1)F(\rho)\Big|_{\rho=0} \tag{2.4}$$

as an expansion in $x$ and $L$ indexed by chord diagrams and with coefficients monomials in the $f_i$ where

$$F(\rho) = \sum_{i=-1}^{\infty} f_{i+1}\rho^i$$

[**9**] is joint work with Nicolas Marie.

The objects we need are rooted connected chord diagrams.

Definition 2.8. A *perfect matching* of a finite set $S$ is a set of pairs of elements of $S$ such that every element of $S$ is in exactly one pair.

Definition 2.9. A *rooted chord diagram* with $n$ chords is a perfect matching of $\{1, 2, \ldots, 2n\}$. The pairs of the perfect matching are called *chords* and are ordered by the order of their smaller elements. The *root chord* is the pair including 1.

To visualize a rooted chord diagram put the points $1, \ldots, 2n$ in counterclockwise order around a circle, let 1 be the root vertex, and draw chords through the circle joining the pairs of points of the matching. For example

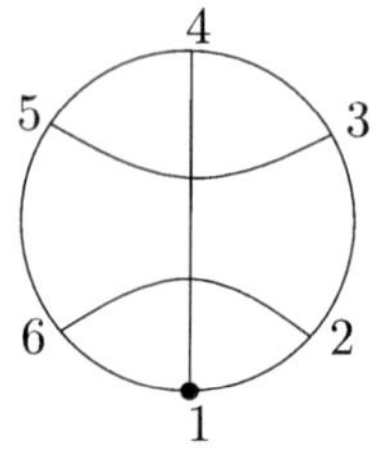


In the illustrations the root vertex will be marked with a dark dot.

The order of the chords is counterclockwise by their first appearance starting from the root.

DEFINITION 2.10. The *intersection graph* of a rooted chord diagram $C$ is the graph with one vertex for each chord of $C$ and an edge between vertices iff the chords cross, that is, iff the chords are pairs $\{a, b\}$, $\{c, d\}$ with $a < c < b < d$.

The *oriented intersection graph* of a rooted chord diagram $C$ is the intersection graph of $C$ with all edges oriented from the smaller chord to the larger chord.

For example the oriented intersection graph of the previous chord diagram is

DEFINITION 2.11. A rooted chord diagram $C$ is *connected* if the intersection graph is connected.

For example the previous chord diagram is connected while

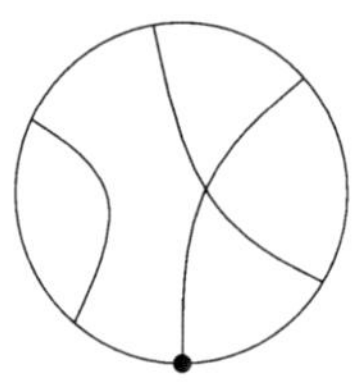

is not connected.

Next we need some more specialized definitions for our situation.

DEFINITION 2.12. A chord of a rooted connected chord diagram $C$ is called *terminal* if the corresponding vertex of the oriented intersection graph of $C$ has no outgoing edges.

There is another order which is important in the following. It is called the intersection order, and is defined recursively.

DEFINITION 2.13. Let $C$ be a rooted connected chord diagram. The *intersection order* of the chords of $C$ is defined by the following procedure.

- The root chord of $C$ is the first chord in the intersection order.
- Remove the root chord of $C$ and let $C_1, \ldots C_k$ be the connected components of the remaining chord diagram, ordered by the counterclockwise order of their first vertex.

- Order the chords of each of $C_1, \ldots C_k$ inductively in the intersection order. Order the chords of $C$ with the root chord first followed by all chords of $C_1$ in intersection order, then all chords of $C_2$ in intersection order, and so on.

For example the chords of the following chord diagrams are labelled in intersection order

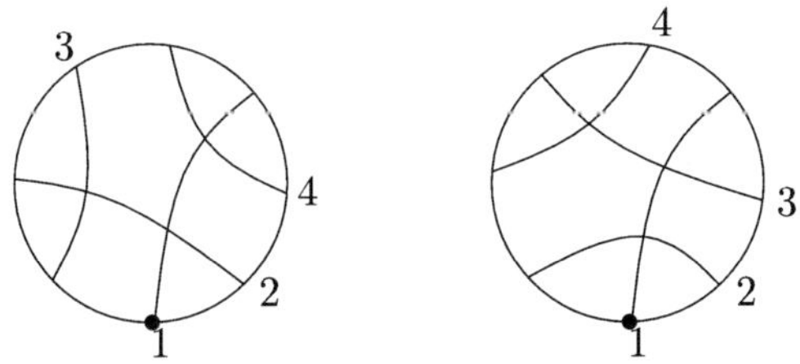


Note that the intersection order may or may not correspond to the counterclockwise order.

DEFINITION 2.14. Let $C$ be a rooted connected chord diagram with $n$ chords and let $f_0, f_1, \ldots$ be a countable set of indeterminates

(1) Let $T(C) = (i_1 < i_2 < \cdots < i_k)$ be the list in increasing order of indices of terminal chords of $C$ in the intersection order.
(2) Let $b(C)$ be the first element of $T(C)$
(3) Let $\delta(C) = (\underbrace{0, \ldots, 0}_{n-k \text{ times}}, i_2 - i_1, i_3 - i_2, \ldots, i_k - i_{k-1})$ be the list of differences of successive elements of $T(C)$ padded with 0s so that $|\delta(C)| = n - 1$.
(4) Let $f_C = \prod_{i \in \delta(C)} f_i$

The $f_C$ are what is needed to build up the monomials of the chord diagram expansion of the solution of (2.4).

The main result of [9] with Nicolas Marie is

THEOREM 2.15.

$$G(x, L) = 1 - \sum_{i \geq 1} \frac{(-L)^i}{i!} \sum_{\substack{C \\ b(C) \geq i}} x^{|C|} f_C f_{b(C)-i}$$

*where the sum is over rooted connected chord diagrams with the indicated restriction, solves the Dyson-Schwinger equation*

$$G(x, L) = 1 - xG\left(x, \frac{d}{d(-\rho)}\right)^{-1} (e^{-L\rho} - 1)F(\rho)\Big|_{\rho=0}$$

*where*

$$F(\rho) = \frac{f_0}{\rho} + f_1 + f_2\rho + f_3\rho^2 + \cdots$$

The proof of the result is the body of [9]. It involves two further recurrences, one of which generalizes a classical chord diagram recurrence of Stein [10] and one of which involves going through a rooted tree construction. This result is interesting because it gives the Green function $G(x, L)$ as a sort of multivariate generating function for chord diagrams. Expand

$$G(x, L) = 1 - \sum_{n \geq 1} \gamma_n(x) L^n$$

In this situation the analogue of the beta function of the theory is simply

$$\beta(x) = -2x\gamma_1(x)$$

and hence is also given by an expansion over chord diagrams. More physically realistic cases will be given by systems of Dyson-Schwinger equations and then the beta function will be a linear combination of the $\gamma_1$s for each Green function. There are some hints that similar combinatorial expansions will hold for other single equation Dyson-Schwinger equations and then ultimately for systems, but as of yet there are no further results along those lines.

**2.5. Reduction to geometric series.** Let us now return to the more general form of the Dyson-Schwinger equation defined in Definition 2.6

$$G(x, L) = 1 + \mathrm{sgn}(s) \sum_{k \geq 1} x^k G\left(x, \frac{d}{d(-\rho)}\right)^{1+sk} (e^{-L\rho} - 1) F_k(\rho) \Big|_{\rho=0} \tag{2.5}$$

where

$$F_k(\rho) = \sum_{i=-1}^{\infty} f_{k,i+1} \rho^i$$

This form of Dyson-Schwinger equation was chosen for its amenability to algebraic and combinatorial analysis.

We will need an important result from [**18, 19**]. Writing the renormalization group equation in this language gives (see [**18**, Section 4.1])

PROPOSITION 2.16. *With $G(x, L)$ as above, and writing*

$$G(x, L) = 1 + sgn(s) \sum_{n \geq 1} \gamma_n(x) L^n,$$

*we have*

$$k\gamma_k(x) = \gamma_1(x)\left(sgn(s) + |s|x\frac{d}{dx}\right)\gamma_{k-1}(x)$$

The next step of [**18, 19**] was to observe that there exists unique $r_k, r_{k,i} \in \mathbb{R}$ for $k \geq 1$, $1 \leq i < k$ such that

$$\begin{aligned} &\sum_{k \geq 1} x^k G\left(x, \frac{d}{d(-\rho)}\right)^{1+sk} (e^{-L\rho} - 1) F_k(\rho) \Big|_{\rho=0} \\ &= \sum_{k \geq 1} x^k G\left(x, \frac{d}{d(-\rho)}\right)^{1+sk} (e^{-L\rho} - 1)\left(\frac{r_k}{\rho(1-\rho)} + \sum_{1 \leq i < k} \frac{r_{k,i} L^i}{\rho}\right) \Bigg|_{\rho=0} \end{aligned}$$

and then use the $r_k$ and $r_{k,1}$ to build a series $P(x)$, which is the mysterious input to the differential equations studied in [**13, 14**]. The most unsatisfying thing about this entire framework for Dyson-Schwinger equations is the lack of understanding of $P(x)$. The chord diagram expansion is meant, among other things, as a contribution to the understanding of $P$ since in the case it applies to, it also gives $P$ as an explicit expansion over chord diagrams.

For a more general understanding we could first try to simply expand out and calculate the first few $r_k$ and $r_{k,i}$ in some examples. Here problems begin for the most prosaic of reasons: an error. Example 6.2 of [**18**] gives $r_k$ for $k \leq 5$ and $r_{k,i}$ for $k \leq 4$. Unfortunately it is wrong. The following example corrects this error.

EXAMPLE 2.17. This example corrects Example 6.2 of [**18**]. Note that the conventions of this paper differ from those of [**18**] in two ways, first the indexing of the $f_i$ has been shifted by 1 to make the chord diagram construction simpler, and second the sign of $s$ has been swapped.

Consider the case with $s = -2$ and a single $B_+$ at order $k = 1$. This is the case we now fully understand in terms of chord diagrams. Write

$$F(\rho) = \sum_{j=-1}^{\infty} f_{j+1}\rho^j$$

Then

$$\begin{aligned}
r_1 &= f_0 \\
r_2 &= f_0 f_1 - f_0^2 \\
r_{2,1} &= 0 \\
r_3 &= -4f_0^2 f_1 + 3f_2 f_0^2 + f_0 f_1^2 \\
r_{3,1} &= 0 \\
r_{3,2} &= 0 \\
r_4 &= 11 f_2 f_0^2 f_1 - 9f_0^2 f_1^2 - 18 f_2 f_0^3 + f_0 f_1^3 + 15 f_3 f_0^3 \\
r_{4,1} &= 0 \\
r_{4,2} &= 0 \\
r_{4,3} &= 0 \\
r_5 &= 86 f_3 f_0^3 f_1 - 120 f_3 f_0^4 - 16 f_0^2 f_1^3 + f_0 f_1^4 + 30 f_2^2 f_0^3 + 105 f_0^4 f_4 \\
&\quad - 112 f_2 f_0^3 f_1 + 26 f_2 f_0^2 f_1^2
\end{aligned}$$

The problem with the original computation was a sign error in the program. This new computation has been verified in two ways; first the example was independently computed by Erik Panzer, and second the $\gamma_i$ calculated by the program are now verified to satisfy Proposition 2.16.

The calculation brings up a very interesting point. Only the $r_i$ are needed. This is a very welcome development. The intuition of the rewriting with the $r_i$ and $r_{i,k}$ is that we are replacing the Mellin transforms of the primitives, $F_k(\rho)$, with geometric series; the sum term was simply a higher order hack to make things work out. What this author was not able to see at the time is that the hack is simply not needed. In fact any function of the form

$$g_k(\rho) = \frac{1}{\rho} + O(\rho^0)$$

can take the place of $1/(\rho(1-\rho))$ with no need for any $r_{k,i}$[1]. For the purposes of the program of [**18, 19**] the most useful $g_k(\rho)$ are $g_k(\rho) = 1/(\rho(1-\rho))$ or $g_k(\rho) = 1/(\rho(1+\rho))$ since the next steps of [**18, 19**] use $1 \pm \rho^2 g_k(\rho) = \rho g_k(\rho)$ to simultaneously keep the order of the ultimate differential equation low and the expression for $P$ relatively simple.

[1]Thanks to a referee for pointing out that this is the correct level of generality for this result and for pointing out the tidier proof included here.

THEOREM 2.18. *Let $s$ be an integer. Let $G(x,L) = 1 + sgn(s)\sum_{n\geq 1}\gamma_n(x)L^n$ solve the Dyson-Schwinger equation (2.5). Let $g_k(\rho) = \frac{1}{\rho} + O(\rho^0)$ Then there exists unique $r_k \in \mathbb{R}$ for $k \geq 1$ such that*

$$\sum_{k\geq 1} x^k G\left(x, \frac{d}{d(-\rho)}\right)^{1+sk} (e^{-L\rho} - 1)F_k(\rho)\Big|_{\rho=0}$$

$$= \sum_{k\geq 1} x^k G\left(x, \frac{d}{d(-\rho)}\right)^{1+sk} (e^{-L\rho} - 1)r_k g_k(\rho)\Big|_{\rho=0}$$

*In particular, the reduction to geometric series in* [**18, 19**] *can be a pure reduction to geometric series, without fudge factors at higher powers in $L$.*

PROOF. First note that it suffices to prove the result for $g_k = 1/(\rho(1-\rho))$ since the full result then follows by applying this smaller result twice, once with the $F_k$ and then again with the general $g_k$.

Next note that Proposition 2.16 holds independently of the values of the $f_{k,i}$, and hence in particular it holds when $F_k(\rho) = r_k/(\rho(1-\rho))$ for any $r_k \in \mathbb{R}$. Write $\gamma'_k$ for the $\gamma_k$ with these particular $F_k(\rho)$.

Write

$$P(x) = \sum_{k\geq 1} x^k G\left(x, \frac{d}{d(-\rho)}\right)^{1+sk} (\rho - \rho^2)F_k(\rho)\Big|_{\rho=0}$$

$$P'(x) = \sum_{k\geq 1} x^k G\left(x, \frac{d}{d(-\rho)}\right)^{1+sk} (\rho - \rho^2)\frac{r_k}{\rho(1-\rho)}\Big|_{\rho=0}$$

Then $P'(x) = \sum_{k\geq 1} r_k x^k$ and from (2.5)

$$P(x) = -\gamma_1(x) - 2\gamma_2(x) \quad \text{and} \quad P'(x) = -\gamma'_1(x) - 2\gamma'_2(x).$$

Choose the $r_k$ so that $P(x) = P'(x)$ as series expansions. Then by Proposition 2.16

$$-\gamma_1(x) - \gamma_1(x)\left(\text{sgn}(s) + |s|x\frac{d}{dx}\right)\gamma_1(x)$$

$$= P(x) = -\gamma'_1(x) - \gamma'_1(x)\left(\text{sgn}(s) + |s|x\frac{d}{dx}\right)\gamma'_1(x)$$

Taking this coefficient by coefficient we get

$$-\gamma_{1,i} - \sum_{j=1}^{i-1}\gamma_{1,j}(\text{sgn}(s) + |s|(i-j))\gamma_{1,i-j}$$

$$= r_i = -\gamma'_{1,i} - \sum_{j=1}^{i-1}\gamma'_{1,j}(\text{sgn}(s) + |s|(i-j))\gamma'_{1,i-j}$$

Therefore $\gamma_{1,1} = r_1 = \gamma'_{1,1}$ and inductively $\gamma_{1,i} = \gamma'_{1,i}$ for all $i$. The result follows. □

This theorem is conceptually important because it says that rather than performing an ad-hoc transformation what we are doing is converting our problem from the original theory with its analytically complicated primitives, to a modified theory with new primitives each of which is analytically simply a geometric series. In this way the analytic complexity of the Dyson-Schwinger equation has been unwound into the combinatorics.

## 3. Denominator reduction and special changes of variables

This section will investigate a different use of combinatorial interpretations in the study of periods in quantum field theory. Here we will consider the values of individual graphs, specifically primitive massless scalar $\phi^4$ graphs.

**3.1. Polynomials and denominator reduction.** For us a graph may have multiple edges and may have loops in the sense of graph theory. For any graph $G$, $E(G)$ represents the set of edges of $G$ and $V(G)$ represents the set of vertices of $G$. For $G$ connected $\ell(G)$ is the first Betti number of $G$, equivalently the number of independent cycles in $G$. $G/e$ denotes $G$ with the edge $e$ contracted, and $G \setminus e$ denotes $G$ with the edge $e$ deleted.

DEFINITION 3.1. Let $K$ be a connected 4-regular graph. Let $G$ be $K$ with one vertex removed and suppose $G$ is also connected. Then we say $G$ is a *4-point graph in* $\phi^4$.

DEFINITION 3.2. Let $G$ be a 4-point graph in $\phi^4$. $G$ is *primitive* if every proper subgraph $\gamma$ of $G$ with at least one edge satisfies $|E(\gamma)| > 2\ell(\gamma)$

Note that this is primitivity in the renormalization Hopf algebra because the condition in $\phi^4$ for $\gamma$ to be a subdivergence is $|E(\gamma)| \leq 2\ell(\gamma)$.

DEFINITION 3.3. A graph $H$ is *internally k-edge connected* if for every set $S$ of $k-1$ edges of $H$ either

- $H \setminus S$ is connected (so $S$ is not a cut set), or
- $H \setminus S$ has two components, one of which is an isolated vertex.

PROPOSITION 3.4. *Let $G$ be a 4-point graph in $\phi^4$. $G$ is primitive iff $K$ is internally 6-edge connected, where $K$ is as in Definition* 3.1.

The idea here is simply that an internal 4-edge cut is exactly a subdivergence.

PROOF. Let $v$ be the vertex of $K$ removed to create $G$. Since all vertices of $K$ have even degree $K$ has no cut sets of odd size. Thus $K$ not internally 6-edge connected is equivalent to there being an internal 4-edge cut.

Suppose $K$ is not internally 6-edge connected. Since $K$ is 4-regular, for any set $S$ of four edges of $K$, $K \setminus S$ can have at most one isolated vertex as a component. Furthermore, each connected component of $K \setminus S$ viewed as a subgraph of $K$ must be incident to an even number of edges of $S$, as if not then the set of vertices of $K \setminus S$ would be incident to an odd number of edges in $S$ contradicting 4-regularity.
Thus, by hypothesis, there is a set $S$ of four edges of $K$ such that $K \setminus S$ has at least two components neither of which are isolated vertices. Call these two components $K_1$ and $K_2$.
By Euler's formula $E(K_i) \leq 2\ell(K_i)$. At least one of $K_1$ or $K_2$ does not include $v$, but then this is a divergent subgraph of $G$ and so $G$ is not primitive.

Suppose $G$ is not primitive. Then there is a proper subgraph $\gamma$ of $G$ with at least one edge and with $|E(\gamma)| \leq 2\ell(\gamma)$. Consider $\gamma$ as a subgraph of $K$. By 4-regularity and Euler's formula there are at most 4 edges connecting $\gamma$ to the rest of $K$. These 4 edges form an edge cut of $K$. It is an internal cut since $\gamma$ has at least one edge and is a proper subgraph of $G$. Thus $K$ is not internally 6-edge connected. □

With those observations out of the way, let's move to some graph polynomials.

DEFINITION 3.5. Let $G$ be a graph. Take a variable $a_e$ for each edge $e$ of $G$. Then

$$\Psi_G = \prod_T \sum_{e \notin T} a_e$$

where the sum is over all spanning trees of $G$. Call $\Psi_G$ the *Kirchhoff polynomial* of $G$.

Note this is the dual definition to what is sometimes called the Kirchhoff polynomial.

For example if

$G =$ 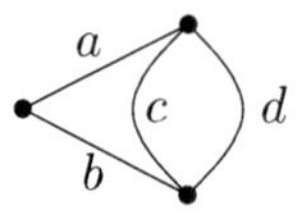


then $\Psi_G = cd + (a+b)(c+d)$. This example is not primitive, but was chosen for its smallness.

The period of $G$, its residue as a Feynman integral, is as follows.

DEFINITION 3.6. Let $G$ be a primitive 4-point graph in $\phi^4$. The *period* of $G$ is

$$\int_{a_i \geq 0} \frac{\sum_{i=1}^{|E(G)|} (-1)^i da_1 \wedge \cdots \widehat{da_i} \cdots \wedge da_{|E(G)|}}{\Psi_G^2}$$

This integral converges since $G$ is primitive. Really this integral should be considered as a projective integral over $a_i \geq 0$ in $\mathbb{PR}^{|E(G)|-1}$, but the above definition keeps things elementary and suffices for our needs.

By the matrix tree theorem $\Psi_G$ can also be expressed as a determinant.

PROPOSITION 3.7. *Orient $G$ and let $E$ be the $|V(G)| \times |E(G)|$ incidence matrix of $G$. Let $\widetilde{E}$ be $E$ with any one row removed and let $\Lambda$ be the diagonal matrix of the $a_e$ in the same order as the columns of $E$. Then*

$$\Psi_G = \det \begin{bmatrix} \Lambda & \widetilde{E}^T \\ -\widetilde{E} & 0 \end{bmatrix}$$

*and in particular does not depend on the choices used to construct the matrix.*

PROOF. There are a number of ways to approach this proof (see for example Proposition 21 of [**2**]) but in the end it always comes down to the matrix-tree theorem. The shortest proof the author knows is as follows (see [**17**]).

$\Lambda$ is invertible, so we can calculate the determinant using the Schur complement:

$$\det \begin{bmatrix} \Lambda & \widetilde{E}^T \\ -\widetilde{E} & 0 \end{bmatrix} = a_1 \cdots a_{|E(G)|} \det(0 - (-\widetilde{E}\Lambda^{-1}\widetilde{E}^T)) = a_1 \cdots a_{|E(G)|} \det(\widetilde{E}\Lambda^{-1}\widetilde{E}^T)$$

For an $m \times n$ matrix $A$ and $S \subseteq \{1, \cdots, n\}$ let $A_S$ be the submatrix of $A$ given by columns indexed by $S$. By the Cauchy-Binet formula

$$\det(\widetilde{E}\Lambda^{-1}\widetilde{E}^T) = \sum_{\substack{S \subseteq E(G) \\ |S|=|V(S)|-1}} \det(\widetilde{E}_S)\det((\Lambda^{-1}\widetilde{E}^T)_S^T)$$
$$= \sum_{\substack{S \subseteq E(G) \\ |S|=|V(S)|-1}} \prod_{i \in S} \frac{1}{a_i} \det(\widetilde{E}_S)^2$$

The matrix tree theorem says that $\det \widetilde{E}_S = \pm 1$ if $S$ is a spanning tree of $G$ and is 0 otherwise. The result follows. □

A key part of Brown's approach [**2**] to calculating the periods of these graphs is denominator reduction. First we need some polynomials built from the matrix of the previous proposition.

DEFINITION 3.8. Let $G$, $\Lambda$ and $\widetilde{E}$ be as in Proposition 3.7. Let

$$M = \begin{bmatrix} \Lambda & \widetilde{E}^T \\ -\widetilde{E} & 0 \end{bmatrix}.$$

Let $I$ and $J$ and $K$ be sets of edge indices. Let $M(I, J)$ be the matrix obtained from $M$ by removing the rows indexed by $I$ and the columns indexed by $J$. Suppose $|I| = |J|$. Then the polynomial

$$\Psi_{G,K}^{I,J} = \det M(I, J)|_{\substack{a_i=0 \\ i \in K}}$$

is called a *Dodgson polynomial.*

Dodgson polynomials satisfy a contraction-deletion relation.

PROPOSITION 3.9. *For* $\ell \notin I \cup J \cup K$

$$\Psi_{G,K}^{I,J} = \Psi_{G,K}^{I\ell,K\ell} a_\ell + \Psi_{G,K\ell}^{I,J}$$

*and*

$$\Psi_{G,K}^{I\ell,J\ell} = \Psi_{G\backslash\ell,K}^{I,J} \quad \Psi_{G,K\ell}^{I,J} = \Psi_{G/\ell,K}^{I,J}$$

PROOF. These follow from the form of the matrix defining the Dodgson polynomials. (See for example [**3**] subsection 2.2.) □

PROPOSITION 3.10. *Given 5 distinct edge indices* $i, j, k, l, m$

$$^5\Psi_G(i, j, k, l, m) = \pm(\Psi_{G,m}^{ij,kl}\Psi_G^{ikm,jlm} - \Psi_{G,m}^{ik,jl}\Psi_G^{ijm,klm})$$

*is independent (up to overall sign) on the order of* $i, j, k, l, m$.

For a proof see Lemma 87 in [**2**].

DEFINITION 3.11. $^5\Psi_G(i, j, k, l, m)$ is called the *5-invariant* of $G$ depending on edges $i, j, k, l, m$.

DEFINITION 3.12. Given $G$ with at least 5 edges, *denominator reduction* is a sequence of polynomials $D^5$, $D^6$, $\ldots D^k$, defined by

- $D_G^5(i_1, i_2, i_3, i_4, i_5) = {}^5\Psi_G(i_1, i_2, i_3, i_4, i_5)$

- If $D_G^j(i_1, \ldots, i_j)$ can be factored as
$$D_G^j(i_1, \ldots, i_j) = (Aa_\ell + B)(Ca_\ell + D)$$
where $A, B, C, D$ are polynomials (not necessarily nonzero) in the edge variables not involving $a_\ell$ then $D_G^{j+1}(i_1, \ldots, i_j, \ell) = \pm(AD - BC)$
- If $D_G^{j+1} = 0$ or $D_G^j$ cannot be factored then denominator reduction ends.

There are a number of things to observe about this definition. First, $D_G^j(i_1, \ldots, i_j)$ is defined up to overall sign.

Second, for a given graph, different edge orders will give a different sequence of polynomials, and such sequences may not all be the same length. The goal is to find long sequences because of the next observation.

Third, the sequences of polynomials given by denominator reduction calculate at least two useful things. Namely they give the denominators of the result of taking the Feynman integral of $G$ and integrating the indicated variables [**2**], and furthermore they carry a significant part of the information of the integral which can be seen from the fact that the $c_2$ invariant of $G$ [**3**], an arithmetic invariant, can be calculated from the denominators.

Fourth, we can begin the sequence with $D_G^4$ rather than $D_G^5$ at the expense of the fact that for any choice of four edges there are three generically distinct possible choices for $D_G^4$,
$$\Psi^{ij,kl}\Psi^{ik,jl}, \ \Psi^{ij,kl}\Psi^{il,jk}, \ \text{or} \ \Psi^{ik,jl}\Psi^{il,jk}$$
each of which yields the same $D_G^5$ and onwards following the denominator reduction algorithm. The denominators of the Feynman integral with fewer than 4 edges integrated also have nice explicit forms, see [**2**], but they don't relate by the same identity.

Denominator reduction is appealing in the context of this paper because the resulting polynomials usually have combinatorial interpretations as sums of products of spanning forest polynomials [**5**]. The true strength of denominator reduction is in the interplay of these many viewpoints.

To advance with these observations we need to define the $c_2$ invariant and spanning forest polynomials.

DEFINITION 3.13. Let $f_1, \ldots, f_k$ be polynomials in the variables $x_1, \ldots, x_n$. Let $q$ be a prime power. Let $V(f_1, \ldots, f_k)$ be the affine variety defined by the vanishing of the $f_i$. Define
$$[f_1, \ldots, f_k]_q$$
to be the number of points in $V(f_1, \ldots, f_k)$ over $\mathbb{F}_q$, the finite field with $q$ elements.

Brown and Schnetz showed [**3**, Theorems 2, 3, and 29]

PROPOSITION 3.14. *Let $G$ be any connected graph with at least 3 vertices, then*
$$[\Psi_G]_q \equiv c_2(G)_q q^2 \mod q^3$$
*for some $c_2(G)_q \in \mathbb{Z}/q\mathbb{Z}$.*

*If further $G$ has at least 5 edges, then*
$$c_2(G)_q \equiv (-1)^n [D_G^n(i_1, \ldots, i_n)]_q \mod q$$
*for any edges $i_1, \ldots, i_n$ of $G$ for which denominator reduction is defined.*

DEFINITION 3.15. Given a connected graph $G$ with at least 3 vertices, the sequence indexed by prime numbers

$$c_2(G) = (c_2(G)_2, c_2(G)_3, c_2(G)_5, \dots)$$

is the $c_2$ invariant of $G$.

In fact this is but one manifestation of the $c_2$ invariant, and not the deepest one [**4**], but it suffices for the purposes of this paper. The $c_2$ invariant is relatively easy to calculate but contains a lot of information about the period. For example if $c_2(G) = 0$ then the period of the graph has a drop in transcendental weight.

Spanning forest polynomials give a combinatorial interpretation for Dodgson polynomials and are very handy for reasoning about them.

DEFINITION 3.16. Let $G$ be a graph. Let $P$ be a partition of a subset of the vertices of $G$. Let $\mathcal{F}_{P,G}$ be the set of all subgraphs $F$ of $G$ with the properties that

- $V(F) = V(G)$
- $F$ is acyclic
- The number of connected components of $F$ is the number of parts of $P$
- For each part $p$ of $P$, there is a connected component of $F$ which contains all the vertices of $p$ and none of the other vertices of $P$.

Call the elements of $\mathcal{F}_{P,G}$ the *spanning forests of $F$ consistent* with $P$.

Note that the spanning forests may have isolated vertices, but only if those vertices are alone in a part of $P$.

DEFINITION 3.17. Let $G$ be a graph. Take a variable $a_e$ for each edge $e$ of $G$. Let $P$ be a partition of a subset of the vertices of $G$. Then

$$\Phi_G^P = \sum_{F \in \mathcal{F}_{P,G}} \prod_{e \notin F} a_e$$

Such polynomials are called *spanning forest polynomials.*

The number of parts of $P$ will be called the number of *colours* of $\Phi_G^P$; the idea is that we could draw the graph and colour those vertices of the partition according to which part they belonged to.

If $G$ has an isolated vertex $v$, then for $\Phi_G^P$ to be nonzero, $v$ must be alone in a part of $P$. Then as polynomials

$$\Phi_G^P = \Phi_{G-v}^{P-\{v\}}$$

but the former has one more colour than the latter.

The relationship between Dodgson polynomials and spanning forest polynomials is given in Proposition 12 from [**5**].

PROPOSITION 3.18. *Let $I, J, K$ be sets of edge indices of a connected graph $G$ with $|I| = |J|$, then*

$$\Psi_{G,K}^{I,J} = \sum_P \pm \Phi_{G\setminus(I\cup J\cup K)}^P \tag{3.1}$$

*where the sum runs over all set partitions $P$ of the end points of edges of $(I \cup J \cup K) \setminus (I \cap J)$ with the property that all the forests of $P$ become trees in both*

$$G\setminus I/(J \cup K) \quad \textit{and} \quad G\setminus J/(I \cup K)$$

Thus Dodgson polynomials are always signed sums over spanning forest polynomials. As we denominator reduce, the factorizations are not required to be expressible in terms of Dodgson polynomials, however in practice this is what occurs. Then the $D^n_G$ are typically signed sums of products of pairs of spanning forest polynomials.

**3.2. Free factorizations.** Nothing that has been described so far explains how many steps of denominator reduction will be possible in a given graph. Lets think of this in terms of free factorizations. If we make good choices of edges to reduce then the denominator may automatically be factored for combinatorial reasons. Call this a *free* factorization.

One important source of free factorizations, which would be known to anyone working with such Feynman integrals, and was first described in this language in [**2**, Section 7.4], is the following proposition.

PROPOSITION 3.19. *Let $I$, $J$, and $K$ be sets of edge indices of a graph $G$.*

(1) *Let $\{i,j,k\}$ be a triangle in $G$. Then if $\{i,j,k\} \subseteq (K \cup I) \setminus J$*

$$\Psi^{I,J}_{G,K} = 0$$

*While if $\{i,j\} \subseteq (K \cup I) \setminus J$ with $k \notin I \cup J \cup K$, then $\Psi^{I,J}_{G,K}$ is divisible by $a_k$.*

(2) *Let $\{i,j,k\}$ be a cut set in $G$ (that is $G \setminus \{i,j,k\}$ is disconnected). Then if $\{i,j,k\} \subseteq I$*

$$\Psi^{I,J}_{G,K} = 0$$

*While if $\{i,j\} \subseteq I$ with $k \notin I \cup J \cup K$, then $\Psi^{I,J}_{G,K}$ is independent of $a_k$.*

Since we are interested in $G$ primitive, the only 3-edge cuts will be 3-valent vertices, so the second point is only interesting in the case where $i, j, k$ meet at a common 3-valent vertex.

PROOF. By Proposition 3.9

$$\Psi^{I,J}_{G,K} = \Psi^{I\ell,J\ell}_{G,K} a_\ell + \Psi^{I,J}_{G,K\ell}.$$

for any edge index $\ell \notin I \cup J \cup J$. Hence in both (1) and (2) the first statement implies the second.

If $\{i,j,k\}$ is a cut set and $\{i,j,k\} \subseteq I$, then $G \backslash I/(J \cup K)$ is disconnected and hence $\Psi^{I,J}_{G,K} = 0$ by Proposition 3.18. (See [**2**] for a proof independent of spanning forest polynomials.)

Dually, if $\{i,j,k\}$ is a triangle and $\{i,j,k\} \subseteq (K \cup I) \setminus J$ then no spanning tree of $G \backslash J$ contains all three of the edges, Hence setting all three variables to 0, equivalently contracting them, gives $\Psi_{G\backslash J/(I \cup K)} = 0$ and hence $\Psi^{I,J}_{G,K} = 0$. □

The previous proposition gives us free factorizations because if we reduce two edges of the triangle or the 3-valent vertex, say in the initial 5-invariant, then by choosing an appropriate order we can make sure that in each term one of the factors either has no constant term (triangle case) or no linear term (cut set case) in the other edge. Thus the entire 5-invariant either has no constant term (triangle case) or has no quadratic term (cut set case). In either case we have a trivial sort of factorization and can continue denominator reducing.

Another important source of free factorizations is graphs which are very narrow in the following sense [**2**, section 1.4].

DEFINITION 3.20. Given a total order $<$ on the edges of a graph $G$, for $0 \leq i \leq |E(G)|$, let $L_i(G)$ be the first $i$ edges in the order and $R_i(G)$ be the remaining edges in the order. Let

$$V_i(G, <) = V(L_i(G)) \cap V(R_i(G))$$

Then the *vertex width* of $G$ is

$$\min_{<} \left( \max_{0 \leq i \leq |E(G)|} |V_i(G, <)| \right)$$

where the minimum runs over all total orders of edges of the graph.

This notion of vertex width is related to, but not equivalent to, the notion of path width from graph theory.

DEFINITION 3.21. Let $G$ be a graph. A *path decomposition* of $G$ is a sequence of sets of vertices of $G$,

$$V_0, V_1, \ldots, V_k$$

such that

- for every edge $e$ of $G$ there is a $V_i$ such that both ends of $e$ are in $V_i$ and
- for every vertex $v$ of $G$, the $V_i$ in which $v$ appear form a contiguous subsequence of the original sequence.

The *width* of the path decomposition is $\max_{0 \leq i \leq k} |V_i| - 1$.

DEFINITION 3.22. The *path width* of $G$ is the minimum width of a path decomposition of $G$.

Path width was first defined by Robertson and Seymour [**11**] and is an important notion in graph theory. It is a special case of the tree width of a graph.

PROPOSITION 3.23. *Let $G$ be a graph, let $vw(G)$ be the vertex width of $G$ and $pw(G)$ be the path width of $G$. Then*

$$vw(G) \geq pw(G)$$

PROOF. First note that both the path width and the vertex width of $G$ is the maximal path width, respectively vertex width, of a connected component of $G$, and so we may assume $G$ is connected. Next note that loops of $G$ affect neither the path width nor the vertex width, so we may assume $G$ has no loops in the sense of graph theory.

If $W$ is a set of vertices of $G$ use the notation $e \in W$ to say that both ends of $e$ are in $W$.

Suppose $G$ has vertex width $w$ and that $e_1 < e_2 < \ldots < e_{|E(G)|}$ is a total order on the edges of $G$ which gives width $w$. If the graph has only one edge then the result holds trivially. Otherwise let $E_i$ be the set of vertices which are ends of $e_i$.

Define

$$V_{2i-1} = V_{i-1}(G, <) \cup E_i \quad 1 \leq i \leq |E(G)|$$
$$V_{2i} = V_i(G, <) \quad 1 \leq i \leq |E(G)| - 1$$

We now want to check that the $V_j$ give a path decomposition of $G$. By construction $e_i \in V_{2i-1}$. By hypothesis the $V_j(G, <)$ all have size at most $w$.

If $e_i \in V_{i-1}(G,<)$ then $V_{2i-1} = V_{i-1}(G,<)$ which has size at most $w$. If one end of $e_i$ is in $V_{i-1}(G,<)$ then $V_{2i-1}$ has size at most $w+1$.

Suppose neither end of $e_i$ is in $V_{i-1}(G,<)$. Then each vertex of $V_{i-1}(G,<)$ must meet at least one edge $e_j$ with $j > i$. Thus $V_{i-1}(G,<) \subseteq V_i(G,<)$. Also any edges which meet an end of $e_i$ must have index greater than $i$. Finally by connectivity $e_i$ has at most one end of degree 1. Thus $|E_i \cap V_i(G,<)| \geq 1$, and so $V_{2i-1}$ has size at most $w+1$.

Consider a vertex $v$ of $G$. $v$ first appears in the $V_i(G,<)$ when the first edge incident to $v$ passes into $L_i(G)$ and last appears when the last edge incident to $v$ passes into $L_i(G)$. Thus $v$ appears in a contiguous subset of the $V_i(G,<)$. If $v \in E_i$ then $v \in V_{i-1}(G,<)$ or $v \in V_i(G,<)$. Therefore $v$ appears in a contiguous subset of the $V_j$ and so the $V_j$ give a path decomposition of $G$ with parts of size at most $w+1$. Hence $G$ has path width at most $w$. □

We can see that the path width and vertex width are not equal in general by considering $K_{3,3}$

EXAMPLE 3.24. Consider $K_{3,3}$ with the vertices labelled as follows

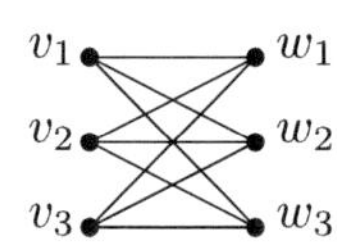


$K_{3,3}$ has path width at most 3, from the following path decomposition

$$\{v_1, w_1, w_2, w_3\}, \{v_2, w_1, w_2, w_3\}, \{v_3, w_1, w_2, w_3\}$$

Suppose $K_{3,3}$ had vertex width 3. Due to the automorphisms of $K_{3,3}$ it doesn't matter which edge we take first in the order, say we first take $\{v_1, w_1\}$. Then

$$V_1(G,<) = \{v_1, w_1\}$$

To force $V_2(G,<)$ to have size at most 3 we must pick as our next edge an edge incident to $v_1$ or to $w_1$. Without loss of generality take $\{v_1, w_2\}$ as the next edge. Then

$$V_2(G,<) = \{v_1, w_1, w_2\}$$

$w_1$ and $w_2$ both have two further edges incident to them. But we must lose a vertex from $V_2(G,<)$ for every vertex added, so as our next edge we must take $\{v_1, w_3\}$. Then

$$V_3(G,<) = \{w_1, w_2, w_3\}$$

Now no matter which edge we take next we will still have $w_1$,$w_2$, and $w_3$ in $V_4(G,<)$ along with either $v_1$ or $v_2$, and hence $K_{3,3}$ is not vertex width 3.

Brown in **[2]** proved that for every primitive 4-point graph in $\phi^4$ with vertex width at most 3 there is an edge order so that denominator reduction either terminates with a 0 or continues until no edges remain; the graph is sufficiently narrow that every step gives a free factorization.

Iain Crump in **[7]** showed that every 3-connected graph for which all 5-invariants factor due to one of the defining polynomials being zero has a particular structure which implies vertex width at most 3.

Triangles and 3-valent vertices are shapes in the graph which give free factorizations. There is one more shape which the author is aware of which yields free factorizations.

PROPOSITION 3.25. *Let $G$ be a graph with two 3-valent vertices $v_1$ and $v_2$ joined by an edge. Then after reducing the five edges incident to $v_1$ and $v_2$ and any other two edges of $G$, the denominator $D^7_G$ factors into two factors each at most linear in each edge variable, and expressible in terms of Dodgson polynomials.*

If $G$ is a 4-point graph in $\phi^4$ and $K$ is as in Definition 3.1, then $G$ satisfying the hypotheses of the proposition implies that $K$ contains a triangle. Thus, in view of Schnetz' completion results [**12**], is it not surprising that this shape yields a free factorization, however it is not a consequence of this since the missing edges from $K$ are not integrated.

Note also that a special case of Proposition 3.25 is Lemma 55 of [**3**].

PROOF. Label the edges of $G$ incident to $v_1$ and $v_2$ as illustrated

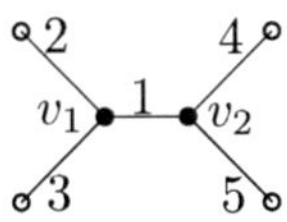


where the rest of the graph attaches at the white vertices. Let $i$ and $j$ be any other two edges of $G$. Then

$$^5\Psi_G(2,3,5,i,j) = \Psi^{23,ij}_{G,5}\Psi^{25i,35j}_{G} - \Psi^{235,ij5}_{G}\Psi^{2i,3j}_{G,5}$$

By Proposition 3.19, edge 1 cannot be cut in the first factor of either term, so

$$D^6_G(1,2,3,5,i,j) = \Psi^{23,ij}_{G,15}\Psi^{125i,135j}_{G} - \Psi^{235,ij5}_{G,1}\Psi^{12i,13j}_{G,5}$$

Again by Proposition 3.19, edge 4 cannot be cut in $\Psi^{125i,135j}_{G}$. Furthermore, edge 4 cannot be cut in $\Psi^{235,ij5}_{G,1}$ because when 2, 3 and 5 are cut, then cutting 4 would disconnect the graph giving 0 in the same way as in the proof of Proposition 3.19. Therefore

$$D^7_G(1,2,3,4,5,i,j) = \Psi^{234,4ij}_{G,15}\Psi^{125i,135j}_{G,4} - \Psi^{235,ij5}_{G,14}\Psi^{124i,134j}_{G,5}$$

Next observe that

$$\Psi^{125i,135j}_{G,4} = \Psi^{124i,134j}_{G,5}$$

since after cutting 1, $v_2$ is a two valent vertex, and hence it makes no difference whether we cut 4 and contract 5 or cut 5 and contract 4. In fact both these Dodgson polynomials are equal to

$$\Psi^{i,j}_{H}$$

where $H$ is $G$ with vertices $v_1$ and $v_2$ and their incident edges removed. Thus

$$D^7_G(1,2,3,4,5,i,j) = (\Psi^{234,4ij}_{G,15} - \Psi^{235,ij5}_{G,14})\Psi^{i,j}_{H}$$

□

It is the experience of the author that the sources of free factorizations from this subsection explain all the free factorizations which occur before the endgame of denominator reduction. As one example, lets look at a very important graph from [**3**].

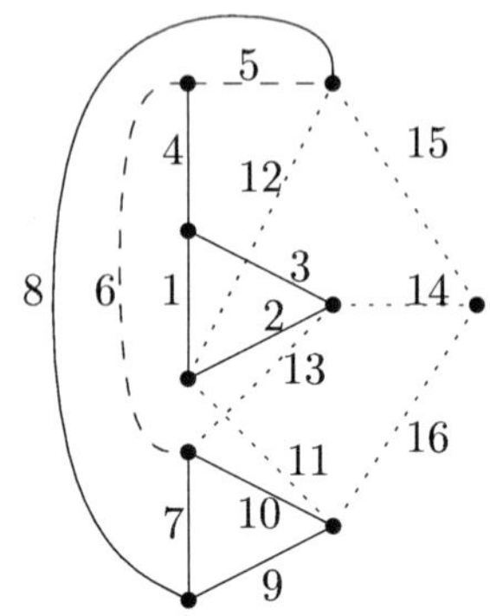


FIGURE 1. The original Brown Schnetz K3 graph marked to illustrate the initial reductions

EXAMPLE 3.26. In [3] Brown and Schnetz consider the graph

$G_{BS} =$ 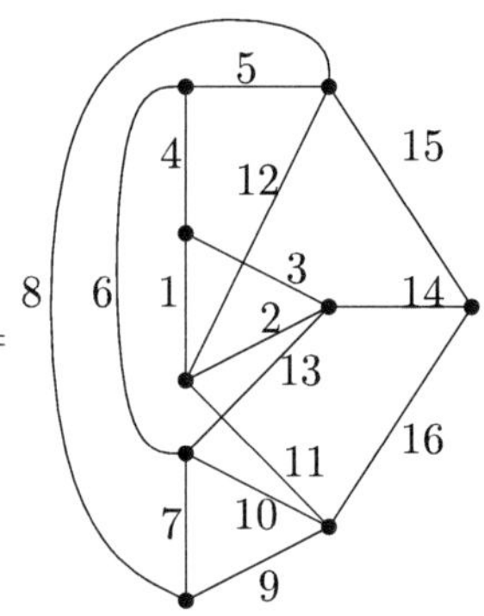


They denominator reduce edges $1, \ldots, 10$ to get

$$D^{10}_{G_{BS}}(1, \ldots, 10) = (Aa_{11} + B)(Ca_{11} + D) \tag{3.2}$$

where

$$\begin{aligned} A &= Q + a_{12}a_{13} + a_{16}a_{12} + a_{14}a_{12} + a_{15}a_{13} + a_{14}a_{13} \\ B &= a_{13}(Q + a_{16}a_{12} + a_{14}a_{12}) \\ C &= -a_{13}a_{15} \\ D &= a_{12}(Q + a_{13}a_{16}) \\ Q &= a_{14}a_{15} + a_{15}a_{16} + a_{14}a_{16} \end{aligned}$$

Then $D^{11}_{G_{BS}}(1, \ldots, 11)$ does not factor, so they proceed one step further by a clever trick which is the subject of the next subsection.

Now let's see why these first 11 denominator reduction steps work out in view of the results of this subsection. For each of the triangles with a 3-valent vertex as marked in full edges in Figure 1 we can take any two of the edges among our initial 5 edges and get the other 2 from triangles and 3-valent vertices. This takes care of 8 of the initial integrations. Taking either 5 or 6 as the fifth of the initial 5 edges we also get the other from the 3-valent vertex. We get one free factorization because of the shape made by edges $1, 3, 4, 5, 6$, so $D_{10}(1, \ldots, 10)$ factors and we may choose any of the remaining edges to proceed with.

EXAMPLE 3.27. Another interesting example is

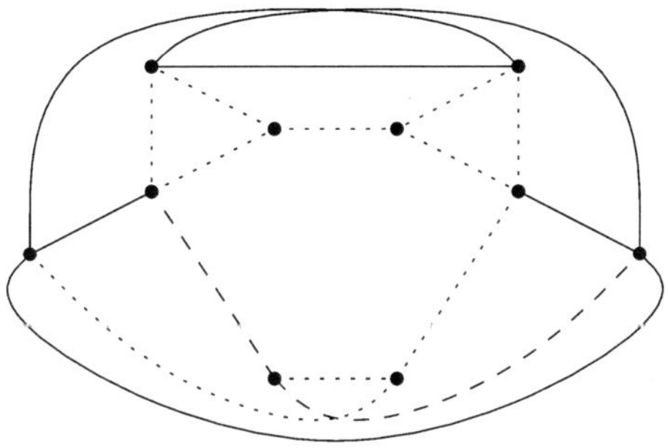

This graph is a decompletion of $P_{9,172}$ which appears in version 1 of [**12**]. Using triangles and 3-valent vertices we can choose an initial 5-invariant so as to denominator reduce all of the dotted edges. Note that we have, in the process, reduced two 3-valent vertices which are joined by an edge.

As our free reduction, reduce one of the dashed edges. This also gives the other dashed edge for free on account of the 3-valent vertex. Once again we have reduced two 3-valent vertices which are joined by an edge. Reducing as described we obtain the two factors

$$\begin{aligned} &zvu + wxu + yzx + ywv + wzu + yzu + yxu + zvx \\ &+ ywz + wvx + yvu + wvu + zxu + yvx + ywx + wzv \end{aligned}$$

and

$$zvu + wxu + wzu + yzu + ywz + wvu + zxu + wzv$$

The monomials of the second are contained in the first, but if we reduce $(Aa + B)((A + C)a + (B + D))$ we get $A(B + D) - B(A + C) = AD - BC$ which is the same as reducing $(Aa + B)(Ca + D)$ so it suffices to consider only the two factors

$$yzx + ywv + yxu + zvx + wvx + yvu + yvx + ywx$$
$$zvu + wxu + wzu + yzu + ywz + wvu + zxu + wzv$$

One more reduction leaves us with 5 variables.

**3.3. The Brown-Schnetz change of variables.** Continuing Example 3.26 following the analysis of Brown and Schnetz of $G_{BS}$ from [**3**], next they make the change of variables

$$a_{12} \mapsto a_{12}Q \quad a_{13} \mapsto a_{13}Q$$

Let $\widetilde{D}^{11}$ be $D^{11}_{G_{BS}}$ after this change of variables. Then

$$\begin{aligned} \widetilde{D}^{11} = Q^3((1 &+ Qa_{12}a_{13} + a_{16}a_{12} + a_{14}a_{12} + a_{15}a_{13} + a_{14}a_{13})a_{12}(1 + a_{13}a_{16}) \\ &+ a_{13}^2 a_{15}(1 + a_{16}a_{12} + a_{14}a_{12})) \end{aligned}$$

Let $R$ be the polynomial $\widetilde{D}^{11}/Q^3$. Viewing this back in the partially integrated Feynman integral, we have $\widetilde{D}^{11}$ in the denominator and some polylogarithms in the numerator along with a factor of $Q^2$ from the change of variables. Cancelling the $Q^2$ the denominator becomes $QR$.

$Q$ is linear in all its variables and $R$ is linear in $a_{14}$ and $a_{15}$ so we can hope to continue denominator reducing. At this stage in the argument we can't be certain whether or not we can continue – it will depend on the numerator. Rather than look into the form of the numerator, let's return to the approach of [**3**] by considering

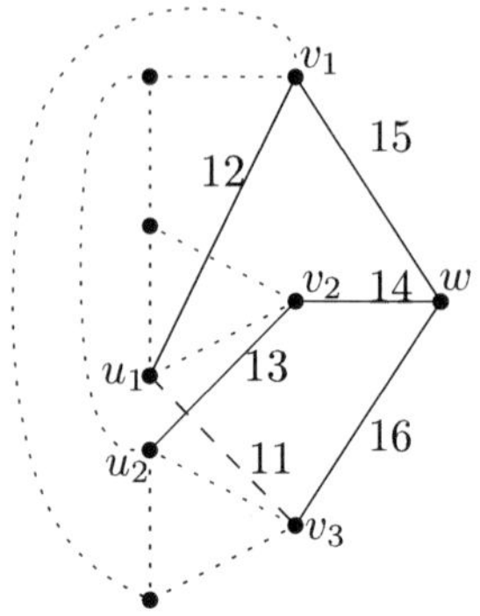


FIGURE 2. The original Brown Schnetz K3 graph marked to illustrate the change of variables

the $c_2$ invariant. The question, then, is whether or not $QR$ has the same point counts modulo $p$ as $D^{11}_{G_{BS}}$. The answer is that it does not, but that they differ only by a constant modulo $p$. This requires the following two facts

First before the change of variables Brown and Schnetz take the opportunity to dehomogenize by setting $a_{16} = 1$, so they need (see [**3**] Lemma 58 for slightly less)

LEMMA 3.28. $\left[D^{11}_{G_{BS}}|_{a_{16}=0}\right]_q$ *modulo* $q$ *is constant as a function of* $q$.

Then as the change of variables is an isomorphism off of $V(Q)$ we have that

$$[D^{11}_{G_{BS}}]_q - [Q, D^{11}_{G_{BS}}]_q = [R]_q - [Q, R]_q$$

Two of these are relatively easy to control

LEMMA 3.29 ([**3**] Lemma 59). $[Q, D^{11}_{G_{BS}}]_q$ *and* $[R]_q$ *modulo* $q$ *are constant as functions of* $q$.

Brown and Schnetz then proceed to use this to show that this particular graph is a counterexample to Kontesevich's conjecture that all such graphs should have $c_2$ invariants which are polynomial in $q$. Furthermore, taking the result of denominator reducing $QR|_{a_{16}=1}$ and changing variables one more time they are able to recognize this as defining a particular K3 surface.

Now let's see if we can get some insight into this change of variables and why it worked. The key is that there is a 3-valent vertex of the graph which has not been touched by the reductions so far – a spare 3-valent vertex.

Figure 2 illustrates what remains of the graph. Using the vertex labels from Figure 2, note that $Q$ is the Kirchhoff polynomial of the right hand side of the graph, edges 14, 15, and 16, with vertices $v_1$, $v_2$, and $v_3$ identified. Equivalently $Q$ is the spanning forest polynomial of the same subgraph defined by the partition $\{v_1\}, \{v_2\}, \{v_3\}$.

The change of variables consists of scaling each of the other edges (those remaining on the left hand side of Figure 2) by $Q$.

Each denominator is a sum of products of pairs of Dodgson polynomials and so is a sum of products of pairs of spanning forest polynomials.

DEFINITION 3.30. Given a product $\Phi^{P_1}\Phi^{P_2}\cdots\Phi^{P_k}$ of spanning forest polynomials, say that the *number of colours* of the product is $|P_1|+|P_2|+\cdots+|P_k|$ where $|P_i|$ is the number of parts of the partition $P_i$.

By homogeneity all terms of the denominator have the same number of colours, so we can speak of the number of colours of the denominator. By this definition $\Psi^2$ has two colours.

Observe that a spanning forest polynomial $\Phi^P$ of a connected graph $G$ has degree $\ell(G)+|P|-1$ since each cycle must be cut to get a tree and then the component trees must be detached to obtain a forest.

PROPOSITION 3.31. *Let $G$ be a primitive $\phi^4$ graph.*

*If $D_G^j$ has $c$ colours then $D_G^{j+1}$ has $c+1$ colours when viewed on the same underlying graph.*

*If $G-\{e_{i_1},\dots,e_{i_k}\}$ has an isolated vertex $v$ and $D_G^k(i_1,\dots,i_k)\neq 0$ has $c$ colours, then viewing the spanning forest polynomials defined with respect to the graph with $v$ removed, $D_{G\setminus v}^k(i_1,\dots,i_k)$ has $c-2$ colours.*

PROOF. Each step of denominator reduction decreases the degree by 1, so the corresponding spanning forests each have one fewer edge and hence one more component tree. This proves the first statement.

Since $v$ is isolated in $G-\{e_{i_1},\dots,e_{i_k}\}$, $v$ must have its own colour in each spanning forest polynomial contributing to $D_G^k$. Thus removing $v$ and this colour in each spanning spanning forest polynomial gives the same polynomial $D_{G\setminus v}^k$ but with two fewer colours. □

Returning to the Brown Schnetz calculation, ${}^5\Psi$ has 7 colours when viewed on all the vertices of the graph. In integrating the first 11 edges we isolate 3 vertices and so, by the previous proposition, the number of colours of $D_{G_{BS}}^{11}$, with isolated vertices removed, is $7+(11-5)-2\cdot 3=7$.

Recall that $Q$ is the spanning forest polynomial defined by $\{v_1\},\{v_2\},\{v_3\}$ on the subgraph consisting of edges 14, 15, and 16 and the change of variables consists of scaling each of the other remaining edges by $Q$.

Let $\Phi$ be any spanning forest polynomial involved in $D_{G_{BS}}^{11}$. Let $c$ be the number of colours of $\Phi$ when viewed on the graph with isolated vertices removed. $w$ has not been involved in the denominator reduction so $w$ is not in the partition of $\Phi$. Thus $c\leq 5$.

If $c=5$ then $e_{12}$ and $e_{13}$ both must be cut which contributes $Q^2$ after the change of variables, and $v_1$, $v_2$, and $v_3$ are different colours which contributes another $Q$, so there is a factor of $Q^3$ after the change of variables. If $c=4$ then either $e_{12}$ and $e_{13}$ are again cut, or exactly one of them is not cut and $v_1$, $v_2$, and $v_3$ are different colours; in both cases there is a factor of $Q^2$ after the change of variables. If $c=3$ then either at least one of $e_{12}$ and $e_{13}$ is cut or $v_1$, $v_2$, and $v_3$ are different colours; in both cases there is a factor of $Q$ after the change of variables.

Thus all pairs with 7 colours total lead to a factor of $Q^3$ after the change of variables, which explains the factorization of $\widetilde{D}^{11}$.

The general result to which the above calculation is a special case would be

THEOREM 3.32. *Let $G$ be a primitive 4-point graph in $\phi^4$. Suppose the edges of $G$ can be partitioned into 3 parts, $G_1$, $G_2$, and $G_3$. Suppose*

- *$G_2\cup G_3$ is connected.*

- *$G_1 \cup G_2$ and $G_3$ share $n$ vertices; call them $v_1, \dots, v_n$.*
- *There are $v$ vertices involved in $G_1$ but not $G_2 \cup G_3$.*
- *The edges of $G_1$ can be denominator reduced in the graph $G$, and the resulting denominator $D$ can be written as a sum of products of pairs of spanning forest polynomials of $G$ where the parts do not involve any vertices of $G_3$ other than possibly $v_1, \dots, v_n$.*
- $2\ell(G_2 \cup G_3) - 2\ell(G_3) + |G_1| - |G_2| - 2v - 2n + 3 \geq 0$

*Let $Q$ be the spanning forest polynomial for $G_3$ for the partition $\{v_1\}, \dots, \{v_n\}$. Let $\widetilde{D}$ be $D$ with the variables of $G_2$ scaled by $Q$. Then $Q^{|G_2|+1}$ divides $\widetilde{D}$.*

Note that in the example of $G_{BS}$, $G_1$ consists of edges 1 through 11, $G_2$ consists of edges 12 and 13, and $G_3$ consists of edges 14, 15, and 16.

Note also that the assumption that the spanning forest polynomials contributing to $D$ only involve $v_1, \dots, v_n$ among the vertices of $G_3$ is natural. This is because the other vertices of $G_3$ have not yet had any incident edges integrated, and so should not appear in any Dodgson polynomials or any spanning forest polynomials which have arisen in the usual way. However, the assumption is necessary since we cannot rule out an unexpected factorization which might yield such anomalous partitions.

PROOF. Count colours as in the example. Any 5-invariant of $G$ will have 7 colours by the form of a 5-invariant. Viewed on $G$, denominator reducing the edges of $G_1$ will give a denominator with $7 + (|G_1| - 5) = |G_1| + 2$ colours. Discarding each of the $v$ isolated vertices leaves $|G_1| + 2 - 2v$ colours in $D$.

Consider a spanning forest polynomial $\Phi$ which appears in $D$. Suppose $\Phi$ has $c$ colours viewed on the graph $G_2 \cup G_3$. Then $\Phi$ has degree $\ell(G_2 \cup G_3) + c - 1$. Let $\widetilde{\Phi}$ be $\Phi$ after scaling each variable of $G_2$ by $Q$.

Every spanning forest of $\Phi$ induces a partition on $\{v_1, \dots, v_n\}$ in the following way: restrict the forest to $G_3$ and partition $\{v_1, \dots, v_n\}$ by putting vertices in the same part iff they are in the same tree in the restriction. Gather together those terms of $\Phi$ corresponding to the same partition.

Let $P$ be such a partition with $k \leq n$ parts, $\Phi'$ the terms of $\Phi$ corresponding to $P$, and $\widetilde{\Phi}'$ those terms after the substitution. Since $P$ has $k$ parts we must cut each cycle of $G_3$ and then $k-1$ more edges of $G_3$ and no more. So $\Phi'$ is homogeneous of degree $\ell(G_3) + k - 1$ in the $G_3$ variables. Thus $\Phi'$ is homogeneous of degree

$$\ell(G_2 \cup G_3) + c - 1 - \ell(G_3) - k + 1 = \ell(G_2 \cup G_3) - \ell(G_3) + c - k$$

in the $G_2$ variables. If $k = n$ then $Q|\Phi'$ by definition of $Q$. So if $k = n$, $Q^{\ell(G_2 \cup G_3) - \ell(G_3) + c - n + 1}$ divides $\widetilde{\Phi}'$ and if $k < n$, $Q^{\ell(G_2 \cup G_3) - \ell(G_3) + c - k}$ divides $\widetilde{\Phi}'$, so in all cases at least

$$\ell(G_2 \cup G_3) - \ell(G_3) + c - n + 1$$

powers of $Q$ divide $\widetilde{\Phi}'$ and hence divide $\widetilde{\Phi}$. Recalling that $D$ has $|G_1| + 2 - 2v$ colours we get that at least

$$\begin{aligned} &2\ell(G_2 \cup G_3) - 2\ell(G_3) + |G_1| + 2 - 2v - 2n + 2 \\ &= 2\ell(G_2 \cup G_3) - 2\ell(G_3) + |G_1| - 2v - 2n + 4 \\ &\geq |G_2| + 1 \end{aligned}$$

powers of $Q$ divide $\widetilde{D}$, where the inequality is by hypothesis. □

Finally consider the constant mod $q$ conditions in the Brown-Schnetz example. Here there is not so much to say combinatorially.

Setting a variable to 0 corresponds to contracting it in the original graph (Proposition 3.9). Then $D^{11}_{G_{BS}}\big|_{a_{16}=0}$ is the same as $D^{11}_{G_{BS}/a_{16}}$. $G_{BS}/a_{16}$ is a strictly simpler graph which can be denominator reduced all the way and so the $c_2$ invariant is constant giving Lemma 3.28.

A key fact for $[R]_q$ is that $R$ is linear in $a_{14}$ and $a_{15}$. This fact is also necessary for denominator reducing $QR$ one more step. To see that the fact is true return to the decomposition of $D^{10}_{G_{BS}}$ in (3.2). Of the variables $a_{14}$, $a_{15}$ and $a_{16}$, $C$ depends only on $a_{15}$, $B$ depends only on $a_{14}$ and $a_{16}$ apart from the $Q$ which gets pulled out, and $D$ depends only on $a_{16}$ apart from the $Q$ which gets pulled out. Therefore $[R]_q$ depends only on the coefficient of $a_{14}a_{15}$ in $R$. The remainder of Lemma 3.29 is a computation for which this author has no particular insight.

Looking back at the whole calculation the key is that there is a decomposition of $G_{BS}$ into $G_1$, $G_2$, $G_3$ where $G_3$ is small and nice while still allowing $G_1$ to be large and denominator reducible. In other words a 3-valent vertex gives a nice $G_3$, as long as losing this vertex to $G_1$ still lets us reduce maximally far, so the key is to have an *extra* 3-valent vertex.

Let's consider a few other examples of using Theorem 3.32.

EXAMPLE 3.33. Begin with

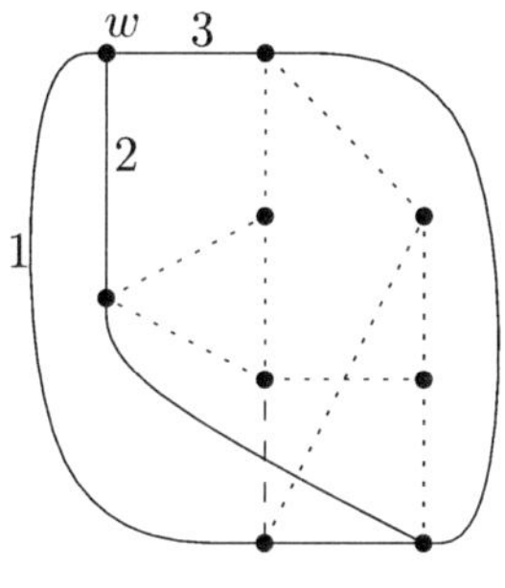


which is a decompletion of $P_{8,38}$ from [**12**]. Then, as in the previous subsection, we can denominator reduce the dotted edges and the result will factor. Additionally denominator reduce the dashed edge.

Now we are in a position to apply Theorem 3.32 with $G_1$ the dotted and dashed edges, $G_3$ the edges incident to $w$ and $G_2$ the remaining 3 edges. This gives that $n = 3$ and $v = 4$ in the statement of the theorem. Thus

$$\begin{aligned}&2\ell(G_2 \cup G_3) - 2\ell(G_3) + |G_1| - |G_2| - 2v - 2n + 3\\&= 4 - 0 + 10 - 3 - 8 - 6 + 3\\&= 0\end{aligned}$$

Then by the theorem we can make the change of variable $a_e \leftarrow a_e(a_1a_2+a_1a_3+a_2a_3)$ for $a_e \in G_2$ and the result will be divisible by $(a_1a_2 + a_1a_3 + a_2a_3)^4$. Let $R$ be the result of this division. Then it turns out $R$ is linear in $a_1$ and so we can perform one further reduction.

Note that we were able to do one fewer initial reduction compared to $G_{BS}$, so ultimately we have one more variable in play.

EXAMPLE 3.34. Now consider the graph

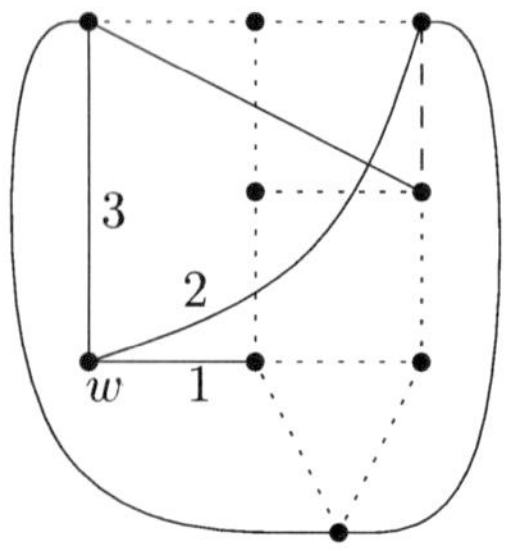


which is a decompletion of $P_{8,39}$ from [**12**]. Using the previous subsection we can denominator reduce the dotted edges and the result will factor. Additionally denominator reduce the dashed edge.

Apply Theorem 3.32 with $G_1$ the dotted and dashed edges, $G_3$ the edges incident to $w$ and $G_2$ the remaining 3 edges. This gives that $n = 3$ and $v = 3$ in the statement of the theorem. Thus

$$\begin{aligned}&2\ell(G_2 \cup G_3) - 2\ell(G_3) + |G_1| - |G_2| - 2v - 2n + 3\\&= 2 - 0 + 10 - 3 - 6 - 6 + 3\\&= 0\end{aligned}$$

Then once again by the theorem we can make the change of variable $a_e \leftarrow a_e(a_1a_2 + a_1a_3 + a_2a_3)$ for $a_e \in G_2$ and the result will be divisible by $(a_1a_2 + a_1a_3 + a_2a_3)^4$. Let $R$ be the result of this division. It turns out again that $R$ is linear in $a_1$ and so we can perform one further reduction.

This is not really a satisfactory answer for this graph since yet again we have one more variable remaining compared to $G_{BS}$ despite the fact that by the point counts this should not be necessary for this graph. Such changes of variables were known to Brown and Schnetz (personal communication, 2010); the difficulty is to get one further step. However, it is nice to know that at least this far the calculations aren't ad-hoc, but rather are directly analogous to the change of variables of [**3**].

## References


[1] D.J. Broadhurst and D. Kreimer. Exact solutions of Dyson-Schwinger equations for iterated one-loop integrals and propagator-coupling duality. *Nucl. Phys. B*, 600:403–422, 2001. arXiv:hep-th/0012146.

[2] Francis Brown. On the periods of some Feynman integrals. arXiv:0910.0114.

[3] Francis Brown and Oliver Schnetz, *A K3 in $\phi^4$*, Duke Math. J. **161** (2012), no. 10, 1817–1862, DOI 10.1215/00127094-1644201. MR2954618

[4] Francis Brown, Oliver Schnetz, and Karen Yeats, *Properties of $c_2$ invariants of Feynman graphs*, Adv. Theor. Math. Phys. **18** (2014), no. 2, 323–362. MR3273316

[5] Francis Brown and Karen Yeats, *Spanning forest polynomials and the transcendental weight of Feynman graphs*, Comm. Math. Phys. **301** (2011), no. 2, 357–382, DOI 10.1007/s00220-010-1145-1. MR2764991 (2012a:81102)

[6] Alain Connes and Dirk Kreimer, *Hopf algebras, renormalization and noncommutative geometry*, Comm. Math. Phys. **199** (1998), no. 1, 203–242, DOI 10.1007/s002200050499. MR1660199 (99h:81137)

[7] Iain Crump. Forbidden minors for 3-connected graphs with no non-splitting 5-configurations. Master's thesis, Simon Fraser University, 2012.

[8] Dirk Kreimer, *Anatomy of a gauge theory*, Ann. Physics **321** (2006), no. 12, 2757–2781, DOI 10.1016/j.aop.2006.01.004. MR2274653 (2008b:81187)

[9] Nicolas Marie and Karen Yeats, *A chord diagram expansion coming from some Dyson-Schwinger equations*, Commun. Number Theory Phys. **7** (2013), no. 2, 251–291, DOI 10.4310/CNTP.2013.v7.n2.a2. MR3164865

[10] Albert Nijenhuis and Herbert S. Wilf, *The enumeration of connected graphs and linked diagrams*, J. Combin. Theory Ser. A **27** (1979), no. 3, 356–359, DOI 10.1016/0097-3165(79)90023-2. MR555804 (82b:05074)

[11] Neil Robertson and P. D. Seymour, *Graph minors. I. Excluding a forest*, J. Combin. Theory Ser. B **35** (1983), no. 1, 39–61, DOI 10.1016/0095-8956(83)90079-5. MR723569 (85d:05148)

[12] Oliver Schnetz, *Quantum periods: a census of $\phi^4$-transcendentals*, Commun. Number Theory Phys. **4** (2010), no. 1, 1–47, DOI 10.4310/CNTP.2010.v4.n1.a1. MR2679376 (2012d:81120)

[13] Guillaume van Baalen, Dirk Kreimer, David Uminsky, and Karen Yeats, *The QED $\beta$-function from global solutions to Dyson-Schwinger equations*, Ann. Physics **324** (2009), no. 1, 205–219, DOI 10.1016/j.aop.2008.05.007. MR2493334 (2010c:81174)

[14] Guillaume van Baalen, Dirk Kreimer, David Uminsky, and Karen Yeats, *The QCD $\beta$-function from global solutions to Dyson-Schwinger equations*, Ann. Physics **325** (2010), no. 2, 300–324, DOI 10.1016/j.aop.2009.10.011. MR2585003 (2011i:81139)

[15] Walter D. van Suijlekom, *Renormalization of gauge fields: a Hopf algebra approach*, Comm. Math. Phys. **276** (2007), no. 3, 773–798, DOI 10.1007/s00220-007-0353-9. MR2350437 (2009d:81236)

[16] Walter D. van Suijlekom. Renormalization of gauge fields using Hopf algebras. In B. Fauser, J. Tolksdorf, and E. Zeidler, editors, *Recent Developments in Quantum Field Theory*. Birkhauser Verlag, 2008. arXiv:0801.3170v1.

[17] Aleksandar Vlasev and Karen Yeats, *A four-vertex, quadratic, spanning forest polynomial identity*, Electron. J. Linear Algebra **23** (2012), 923–941. MR3007197

[18] Karen Yeats, *Rearranging Dyson-Schwinger equations*, Mem. Amer. Math. Soc. **211** (2011), no. 995, x+82, DOI 10.1090/S0065-9266-2010-00612-4. With a foreword by Dirk Kreimer. MR2791969 (2012d:81239)

[19] Karen Amanda Yeats. *Growth estimates for Dyson-Schwinger equations*. PhD thesis, Boston University, 2008.

Department of Mathematics, Simon Fraser University, 8888 University Dr, Burnaby BC, Canada

*E-mail address*: `karen.yeats@sfu.ca`